Sustainable Industrial Processes

Edited by
Fabrizio Cavani, Gabriele Centi,
Siglinda Perathoner, and
Ferruccio Trifiró

Related Titles

E. Drioli, L. Giorni (Eds.)

Membrane Operations

Innovative Separations and Transformations

2009

ISBN: 978-3-527-32038-7

V. Hessel, A. Renken, J.C. Schouten, J.-i. Yoshida (Eds.)

Micro Process Engineering

A Comprehensive Handbook

2009

ISBN: 978-3-527-31550-5

P. Barbaro, C. Bianchini (Eds.)

Catalysis for Sustainable Energy Production

2009

ISBN: 978-3-527-32095-0

K.-V. Peinemann, S. Pereira-Nunes (Eds.)

Membranes Technology Series

Volume 2: Membranes for Energy Conversion

2008

ISBN: 978-3-527-31481-2

G. Centi, R.A. van Santen (Eds.)

Catalysis for Renewables

From Feedstock to Energy Production

2007

ISBN: 978-3-527-31788-2

R.A. Sheldon, I. Arends, U. Hanefeld

Green Chemistry and Catalysis

2007

ISBN: 978-3-527-30715-9

F.J. Keil (Ed.)

Modeling of Process Intensification

2007

ISBN: 978-3-527-31143-9

Sustainable Industrial Processes

Edited by
Fabrizio Cavani, Gabriele Centi,
Siglinda Perathoner, and Ferruccio Trifiró

WILEY-VCH Verlag GmbH & Co. KGaA

The Editors

Prof. Dr. Fabrizio Cavani
Alma Mater Studiorum –
University of Bologna
Department of Industrial Chemistry
and Materials Engineering
Viale Risorgimento 4
40136 Bologna
Italy

Prof. Dr. Gabriele Centi
University of Messina
Department of Industrial Chemistry
and Materials Engineering
Salita Sperone 31
98166 Messina
Italy

Prof. Dr. Siglinda Perathoner
University of Messina
Department of Industrial Chemistry
and Materials Engineering
Salita Sperone 31
98166 Messina
Italy

Prof. Dr. Ferruccio Trifirò
Alma Mater Studiorum –
University of Bologna
Department of Industrial Chemistry
and Materials Engineering
Viale Risorgimento 4
40136 Bologna
Italy

All books published by Wiley-VCH are carefully produced. Nevertheless, authors, editors, and publisher do not warrant the information contained in these books, including this book, to be free of errors. Readers are advised to keep in mind that statements, data, illustrations, procedural details or other items may inadvertently be inaccurate.

Library of Congress Card No.: applied for

British Library Cataloguing-in-Publication Data
A catalogue record for this book is available from the British Library.

Bibliographic information published by the Deutsche Nationalbibliothek
The Deutsche Nationalbibliothek lists this publication in the Deutsche Nationalbibliografie; detailed bibliographic data are available on the Internet at http://dnb.d-nb.de.

© 2009 WILEY-VCH Verlag GmbH & Co. KGaA, Weinheim

All rights reserved (including those of translation into other languages). No part of this book may be reproduced in any form – by photoprinting, microfilm, or any other means – nor transmitted or translated into a machine language without written permission from the publishers. Registered names, trademarks, etc. used in this book, even when not specifically marked as such, are not to be considered unprotected by law.

Printed in the Federal Republic of Germany
Printed on acid-free paper

Cover Design   Adam Design, Weinheim
Typesetting   Thomson Digital, Noida, India
Printing   Strauss GmbH, Mörlenbach
Bookbinding   Litges & Dopf Buchbinderei GmbH, Heppenheim

ISBN: 978-3-527-31552-9

Contents

Sustainable Industrial Processes. Edited by F. Cavani, G. Centi, S. Perathoner, and F. Trifiró
Copyright © 2009 WILEY-VCH Verlag GmbH & Co. KGaA, Weinheim
ISBN: 978-3-527-31552-9

Preface

The Russian economist Kondratieff, analyzing industrial development during the last two centuries, evidenced the existence of business cycles of around 55 years in which the interaction between technological development and economic determined a cyclical sequence of stages (renewal, prosperity, recession, depression), where the various industries come into harmony and mutually reinforce in determining the dynamics [1]. If we apply this theory to the chemical industrial sector, we see that the model applies well to describing the evolution of chemical production [2]. The last cycle, starting around 1950–1960, coincided with the development of the refining and petrochemicals industry, and it is characterized by two main elements:

1. An economy of scale that has lead to the construction of plants of increasing size.
2. Integration between the processes that led to the location of the chemical industry (and energy) in relatively few macro sites, whose impact on society and the environment is one of the causes contributing to the present crisis.

According to the theory of Kondratieff cycles we are beginning a new cycle of development. Together with the issue of how to tackle the environmental consequences of the development model of the chemical industry of this and previous cycles, we should ask ourselves which new development model of the chemical industry is necessary for a sustainable development.

Indeed, there is no doubt that the chemical industry has made a considerable effort in recent decades to reduce its impact on the environment, and consumption of raw materials, including energy. However, if we look in more detail at the breakdown of these costs, we see that they are mainly associated with a necessary adaptation to environmental regulations and security, and only a small part is dedicated to innovation process for the introduction of new technologies that have low environmental impact and/or are intrinsically safe.

Although scientific and technological progress has made available several new technologies/processes that allow a significant reduction in the environmental risk and/or in the consumption of resources (raw materials, energy) the rate of introduction of these technologies/processes in industrial production is still quite low. If we use as a benchmark the number of new processes introduced, we also observe a decrease in recent decades, due to various causes, including the massive reorganization of the entire chemical industry and uncertainties in the market that resulted in a

Sustainable Industrial Processes. Edited by F. Cavani, G. Centi, S. Perathoner, and F. Trifirò
Copyright © 2009 WILEY-VCH Verlag GmbH & Co. KGaA, Weinheim
ISBN: 978-3-527-31552-9

low propensity to new investments, at least in areas of high technology development such as Europe. The prospects for the future are no better.

The problem is connected to the high cost of investment for new processes in a situation of market instability. Most of the investments in new chemical plants are in Asia, although even in this area the number of novel sustainable technologies/ processes introduced is very limited, because of the need of urgent investment, the limited relevance to environmental issues, and so on.

It is therefore necessary, particularly in Europe, to adopt a new approach to industrial chemistry, leading to the development of technologies/processes that require lower development costs and investment, and may thus be introduced more quickly to the market, allowing at the same time a better sustainability of the entire chemical sector. Rethinking the development model of the chemical industry is not only a necessary element to a sustainable chemical production but also the key factor to ensuring the role of chemistry as a driver of innovation in developing countries, a statement often cited, but not always transformed into action.

This book originates from these considerations, which are shared by the great effort made in the last 3–4 years by the chemical industry in Europe to define a new vision and roadmap for industrial chemistry through the discussion made within the European Technology Platform for Sustainable Chemistry (www.suschem.org) sponsored by CEFIC (European Federation of Chemical Industry) and EuropaBio (European Association of Biotechnology).

Education is a key factor to achieving this new vision, and therefore, together with the definition of roadmaps, it is necessary to spread the concepts of sustainable industrial chemistry to students, scientists and managers in the sector. Even though many books have been published on "green chemistry," process intensification and other key elements to implement a sustainable industrial chemistry, we believe that a more global vision of this topic is necessary.

This book is thus organized in three parts. The first five chapters discuss the principles and tools needed to realize a sustainable industrial chemistry. Chapter 1 discusses the general principles and emphasizes the differences between "green" and sustainable industrial chemistry approaches. It is also an introductory chapter to the topic. Chapter 2 discusses the role of catalysis as a main enabling factor to achieve sustainability through chemistry. Several examples of homogeneous, heterogeneous and biocatalysis are discussed, with emphasis on industrial aspects, to provide a comprehensive view of the possibilities offered by this tool.

Chapter 3 gives an overview of the possibilities offered by new equipment for process intensification, including microdevices. Membranes – also a key tool to achieving sustainable development through process intensification – are covered in Chapter 4. Chapter 5 is dedicated to techno-economic aspects, and more specifically to the accounting for chemical sustainability, because sustainability requires the capability to assess different alternatives and to account for the effective global impact and benefits of the adoption of new solutions.

The second part (Chapters 6 and 7) discuss in depth two relevant successful examples of sustainable industrial chemistry: the syntheses of propene oxide and adipic acid, respectively. These two chapters were written together by academic and industrial scientists to better highlight the intertwining of the aspects necessary to achieve novel sustainable industrial processes. They also offer the possibility of a general analysis of the different possible process alternatives and of the factors driving the choice. Consequently, they represent excellent methodological examples to discuss sustainable chemistry from the perspective of industrial applications.

The third and final part (Chapters 8 to 15) is more specialist and industrially oriented, dealing with a series of case studies written from scientists working in companies that have developed interesting examples of sustainable chemical processes. Different areas are analyzed:

- *Greener energy and refinery:* Chapter 8, new green diesel production; Chapter 9, biodiesel using heterogeneous catalysis; Chapter 10, new gas treatment technologies; Chapter 11, BioETBE; Chapter 12, green technology for olefin/paraffin alkylation.

- *Petrochemistry:* Chapter 13, direct oxidation of benzene to phenol.

- *Speciality and fine chemicals:* Chapter 14, Friedel–Crafts acylation of aromatic ethers; Chapter 15, the production of nicotinates; Chapter 16, production of intermediate for resmethrins.

- *Polymerization:* Chapter 17, spherizone technology.

These case studies provide a good view of the issue of introducing new sustainable chemical processes in the various sectors of chemical production.

This book thus covers the new trends in designing industrial chemical processes that can reduce the impact on the environment and at the same time be sustainable from the economic point of view. Significant attention is also given to the problem of process intrinsic safety, and to the opportunities given by new technologies, catalysts and materials, and process control to develop sustainable industrial processes.

Aspects covered are the use of alternative or renewable raw materials (and biomasses), the reducing of the risks and number of steps in chemical processes, the reduction of energy consumption (and the use of alternative energy sources in chemical processes), the design of intrinsically safe processes, the downsizing of the chemical production and the opportunities offered by microreactor and integrated reaction/separation technologies (e.g., in changing from discontinuous to continuous fine chemical production), the possibilities deriving from process intensification, the reduction of waste by introducing new catalytic routes and/or solvents, process optimization deriving from *in silico* techniques and process control, and so on. Emphasis is generally given to the role of catalysis to develop new sustainable industrial processes, with also an overview of the new opportunities given by advanced catalytic materials and technologies.

The target audience of the book is: (i) R&D chemists or engineers who would like to apply sustainable chemical processes principles to their work; (ii) process development and production engineers in chemical and related industries, who would like to improve their environmental performances and apply new greener approaches; (iii) teachers and students in advanced university courses on chemical production; (iv) practitioners of associated industries (parachemicals, pharmaceuticals, paints and pigments, cosmetics, materials, etc.) and equipment suppliers; (v) managers who wish to contribute to sustainability and environmental goals; and (vi) decision-making managers in funding institutions that will have an overview of new trends in chemical processes. Additionally, this book will serve both as a guide for improving current industrial practices and for designing future industrial systems such as eco-industrial parks.

This book originates from the activities within the frame of the EU Network of Excellence (NoE) IDECAT (Integrated Design of Catalytic Nanomaterials for a Sustainable Production). The objective of this NoE is to strengthen research in catalysis by the creation of a coherent framework of research, know-how and training between the various disciplinary catalysis communities (heterogeneous, homogeneous and biocatalysis) with the objective of achieving a lasting integration between the main European Institutions in this area. IDECAT created in November 2008 the "European Research Institute on Catalysis" (ERIC), which is intended to be the main reference point for catalysis in Europe.

IDECAT focuses its research on (i) the synthesis and mastering of nano-objects, the materials of the future for catalysis, also integrating the concepts common to other nanotechnologies; (ii) bridging the gap between theory and modeling, surface science and kinetic/applied catalysis as well as between heterogeneous, homogeneous and biocatalytic approaches; and (iii) developing an integrated design of catalytic nanomaterials.

The book also originates from the intense discussions made within the frame of the Italian Technology Platform for Sustainable Chemistry (IT-SusChem), which was launched in Bologna (Italy) on November 2006 and which involves over 250 stakeholders. IT-SusChem reflects on a national level the objectives and structure of the European Technology Platform of Sustainable Chemistry that have inspired this book and which are discussed in several chapters.

Bologna and Messina
September 2009

Fabrizio Cavani
Gabriele Centi
Siglinda Perathoner
Ferruccio Trifirò

References

1 Solomou, S. (1990) *Phases of Economy Growth 1850–1973: Kondratieff Waves and Kuznets Swings*, Cambridge University Press, Cambridge.
2 Gent, C. (2002) *Chem. Commun.*, 2926.

List of Contributors

Stefano Alini
Radici Chimica S.p.A.
Via Fauser 50
28100 Novara
Italy

Philip J. Angevine
Technology Development Center
Lummus Technology
1515 Broad Street
Bloomfield, NJ 07003-3096
USA

Franco Baldiraghi
Eni S.p.A.
Refining & Marketing Division
Centro Ricerche
Via Maritano 26
20097 S. Donato Milanese
Italy

Marie-Pierre Belleville
Institut Européen des Membranes
UM II
Place Eugène Bataillon
Case Courrier 047
34095 Montpellier Cedex 5
France

Daniele Bianchi
Eni S.p.A.
Centro Ricerche per le Energie
Non Convenzionali
Istituto Eni Donegani
Via Fauser 4
28100 Novara
Italy

Rossella Bortolo
Eni S.p.A.
Centro Ricerche per le Energie
Non Convenzionali
Istituto Eni Donegani
Via Fauser 4
28100 Novara
Italy

Valerio Borzatta
ENDURA S.p.A.
Viale Pietramellara 5
40121 Bologna
Italy

Fabrizio Cavani
Alma Mater Studiorum –
University of Bologna
Department of Industrial Chemistry
and Materials Engineering
Viale Risorgimento 4
40136 Bologna
Italy

Sustainable Industrial Processes. Edited by F. Cavani, G. Centi, S. Perathoner, and F. Trifiró
Copyright © 2009 WILEY-VCH Verlag GmbH & Co. KGaA, Weinheim
ISBN: 978-3-527-31552-9

Gabriele Centi
University of Messina
Department of Industrial Chemistry
and Materials Engineering
Salita Sperone 31
98166 Messina
Italy

Roderick Chuck
Jesuitenweg 163
3902 Brig-Glis
Switzerland

Marco Di Girolamo
Saipem S.p.A.
Viale De Gasperi 16
20097 S. Donato Milanese
Italy

Marco Di Stanislao
Eni S.p.A.
Refining & Marketing Division
Centro Ricerche
Via Maritano 26
20097 S. Donato Milanese
Italy

Maurizio Dorini
Basell Poliolefine Italia srl
Centro Ricerche "Giulio Natta"
Piazzale Donegani 12
44100 Ferrara
Italy

Giovanni Faraci
Eni S.p.A.
Refining & Marketing Division
Centro Ricerche
Via Maritano 26
20097 S. Donato Milanese
Italy

Edouard Freund
Institut Francais du Petrole
1 & 4 Avenue de Bois Préau
92500 Rueil-Malmaison
France

Anne M. Gaffney
Technology Development Center
Lummus Technology
1515 Broad Street
Bloomfield, NJ 07003-3096
USA

Chris Gosling
UOP LLC
25 East Algonquin Road
Des Plaines, IL 60017-5017
USA

Roland Jacquot
Rhodia Opérations
Centre de Recherches et Technologies
de Lyon
85, rue des Frères Perret
69192 Saint Fons Cédex
France

Tom Kalnes
UOP LLC
25 East Algonquin Road
Des Plaines, IL 60017-5017
USA

Peter Kokayeff
UOP LLC
25 East Algonquin Road
Des Plaines, IL 60017-5017
USA

François Lallemand
Total E & P
Place de la Coupole, la Défense 6
92078 Paris Cedex
France

Rich Marinangeli
UOP LLC
25 East Algonquin Road
Des Plaines, IL 60017-5017
USA

Terry Marker
UOP LLC
25 East Algonquin Road
Des Plaines, IL 60017-5017
USA

Philippe Marion
Rhodia Opérations
Centre de Recherches et Technologies
de Lyon
85, rue des Frères Perret
69192 Saint Fons Cédex
France

Gabriele Mei
Basell Poliolefine Italia srl
Centro Ricerche "Giulio Natta"
Piazzale Donegani 12
44100 Ferrara
Italy

Ari Minkkinen
Institut Francais du Petrole
1 & 4 Avenue de Bois-Préau
92852 Rueil-Malmaison
France

Delphine Paolucci-Jeanjean
Institut Européen des Membranes
UM II
Place Eugène Bataillon
Case Courrier 047
34095 Montpellier Cedex 5
France

Siglinda Perathoner
University of Messina
Department of Industrial Chemistry
and Materials Engineering
Salita Sperone 31
98166 Messina
Italy

Carlo Perego
Eni S.p.A.
Centro Ricerche per le Energie
Non Convenzionali
Istituto Eni Donegani
Via Fauser 4
28100 Novara
Italy

Marco Ricci
Eni S.p.A.
Centro Ricerche per le Energie
Non Convenzionali
Istituto Eni Donegani
Via Fauser 4
28100 Novara
Italy

Paolo Righi
Alma Mater Studiorum –
Università di Bologna
Dipartimento di Chimica Organica
"A. Mangini"
Viale Risorgimento 4
40136 Bologna
Italy

Gilbert M. Rios
Institut Européen des Membranes
UM II
Place Eugène Bataillon
Case Courrier 047
34095 Montpellier Cedex 5
France

Goffredo Rosini
Alma Mater Studiorum –
Università di Bologna
Dipartimento di Chimica Organica
"A. Mangini"
Viale Risorgimento 4
40136 Bologna
Italy

José Sanchez
Institut Européen des Membranes
UM II
Place Eugène Bataillon
Case Courrier 047
34095 Montpellier Cedex 5
France

Domenico Sanfilippo
Saipem S.p.A.
Viale De Gasperi 16
20097 S. Donato Milanese
Italy

1
From Green to Sustainable Industrial Chemistry

Gabriele Centi and Siglinda Perathoner

1.1
Introduction

The concept of "Green Chemistry" was introduced in the early 1990s in the USA. After the introduction of the US Pollution Prevention Act, the Office of Pollution Prevention and Toxics (OPPT) explored the idea of developing new or improved chemical processes to decrease hazardous to human health and the environment. In 1991, OPPT launched a research grants program called "Alternative Synthetic Pathways for Pollution Prevention." This program was focused on pollution prevention in the design and synthesis of chemicals.

In 1993, the program was expanded to include other topics, such as greener solvents and safer chemicals, and was renamed "Green Chemistry." Since then, the Green Chemistry Program, led by the US Environmental Protection Agency (EPA), has built many collaborations with academia, industry, other government agencies and non-government organizations to promote the use of chemistry for pollution prevention through completely voluntary, non-regulatory partnerships. Further information can be found in the web site: http://www.epa.gov/gcc/index.html.

Notable among the initiatives is the "Presidential Green Chemistry Challenge" (launched in 1995), which promotes pollution prevention through partnerships with the chemistry community. The current focus of the challenge is the Presidential Green Chemistry Challenge Awards. Table 1.1 gives the 2008 award recipients to exemplify the initiative. Further details can be found in the web site cited above.

In addition, in the USA the Green Chemistry Institute (GCI) was created in 1997 – a not-for-profit corporation devoted to promoting and advancing green chemistry. In January 2001, GCI joined the American Chemical Society (ACS) in an increased effort to address global issues at the intersection of chemistry and the environment. The institute developed the Green Chemistry Resource Exchange as a place for users to exchange green chemistry information resources. Development was part of a

Sustainable Industrial Processes. Edited by F. Cavani, G. Centi, S. Perathoner, and F. Trifiró
Copyright © 2009 WILEY-VCH Verlag GmbH & Co. KGaA, Weinheim
ISBN: 978-3-527-31552-9

Table 1.1 Recipients of the 2008 US Presidential Green Chemistry Challenge Awards.

Topic	Award	Motivations
Greener Synthetic Pathways	Battelle for the development and commercialization of Biobased Toners	A soy-based toner that performs as well as traditional ones, but is much easier to remove
Greener Reaction Conditions	Nalco Company for 3D TRASAR Technology	3D TRASAR technology monitors the condition of cooling water continuously and adds appropriate chemicals only when needed, rather than on a fixed schedule
Designing Greener Chemicals	Dow AgroSciences LLC for Spinetoram: enhancing a natural product for insect control	The new insecticide replaces organo-phosphate pesticides for tree fruits, tree nuts, small fruits and vegetables. It reduces the risk of exposure throughout the supply chain
Small Business	SiGNa Chemistry, Inc. for new stabilized alkali metals for safer, sustainable syntheses	A new way to stabilize alkali metals by encapsulating them within porous, sand-like powders, while maintaining their usefulness in synthetic reactions
Academic	Robert E. Maleczka, Jr., Milton R. Smith, III Michigan State University for green chemistry for preparing boronic esters	New catalytic method to make in a greener way precursors with a carbon–boron bond for Suzuki "coupling" reactions

cooperative agreement with the US EPA Design for the Environment program. One of the notable results is the development of the "Green Chemistry Expert System" (GCES), which allows users to build a green chemical process or product, or survey the field of green chemistry. The system is equally useful for new and existing chemicals and their synthetic processes.

Education is of critical relevance to effectively incorporate green chemistry into chemical product and process designs. For green chemistry to enter widespread practice, chemists must be educated about green chemistry during their academic and professional training. To accomplish this goal the EPA, in collaboration with ACS, supports various educational efforts that include the development of materials and courses to assist in the training of professional chemists in industry and education of students. Consequently, the chemical industry has discovered that when their professional chemists are knowledgeable about pollution prevention concepts they are able to identify and implement effective pollution prevention technologies.

Over the years, the EPAs Green Chemistry Program has collaborated with several organizations internationally. The most important being the Organization for Economic Co-operation and Development (OECD), the International Union of

Pure and Applied Chemistry (IUPAC) and the G8 Ministers for Research (Carnegie Group). In 1999 the OECD adopted the following priority recommendations:

- supporting and promoting the research and development;
- recognizing sustainable chemistry accomplishments;
- disseminating related technical and event information, for example, on the Internet;
- developing guidance on implementing sustainable chemistry programs for OECD member countries and outreach to non-member international interests;
- incorporating sustainable chemistry principles into chemical education.

The OECD Sustainable Chemistry Initiative Steering Group includes over 40 representatives from ten countries.

In 2005, the G8 Ministers for Research founded a research and training network on green sustainable chemistry called the International Green Network (IGN). The Interuniversity Consortium "Chemistry for the Environment" (INCA, in Venice, Italy) was selected as the hub of the IGN. The goals of IGN are to sponsor, coordinate and provide information for scientific collaborations; provide training for young chemists; and support applications of green chemistry in developing nations.

In 1976, IUPAC set up a standing committee called CHEMRAWN (Chemistry Research Applied to World Needs). Over the years, there have been several CHEMRAWN conferences and projects to advance chemical technologies that help to achieve a sustainable society. IUPAC has also established in 2001 a working party on "Synthetic Pathways and Processes in Green Chemistry" and the Interdivisional sub-Committee on Green Chemistry.

Other multi-national organizations, including the United Nations, are now beginning to assess the role that they can play in promoting the implementation of green chemistry to meet environmental and economic goals simultaneously. There are rapidly growing activities in government, industry and academia worldwide.

Also in Europe, starting from about 1998, various green chemistry initiatives were launched. In the UK the Green Chemistry Network (GCN) and the Environment, Sustainability & Energy Gateway were launched by the Royal Society of Chemistry, which also started a new journal entitled *Green Chemistry* (http://www.rsc.org/). The 10th anniversary of this journal was celebrated recently with a special issue [1]. Also in UK the CRYSTAL Faraday Partnership was created – a virtual center of excellence in green chemical technology aimed at promoting lower-cost, sustainable manufacturing, for the chemical industry. GCN also launched the Green Chemistry Center for Industry to provide competitive, tailor-made solutions, which reduce waste and environment damage to obtain better profits in chemical industry.

In Italy, the Interuniversity National Consortium "Chemistry for the Environment" (INCA) was founded in 1993 and rapidly become an internationally recognized reference center for green chemistry. The many initiatives include the periodical *School on Green Chemistry*, the organization of the first International IUPAC Conference on Green Chemistry and the Green Chemistry Publication Series. INCA is also the hub center for IGN, as indicated above.

In Germany, activities related to green chemistry are widespread throughout scientific and governmental organizations [2], as well as in other European countries such as France, Spain, Netherlands, Denmark, and so on. Many symposia and international congresses on green chemistry have been organized in these countries.

In Australia the Center for Green Chemistry was established at Monash University, and the Green Chemistry Challenge Awards by the Royal Australian Chemical Institute. In Brazil there is the "Química Verde" (Green Chemistry) program. Japan created, in 1998, the Green Chemistry (GC) Initiative, a task force consisting of representatives from Japanese chemical organizations. In 2000 the Green & Sustainable Chemistry Network (GSCN) was launched. Many other activities on green chemistry have been implemented worldwide and for reasons of space cannot be cited here. However, this short overview evidences the numerous activities concerning green chemistry starting from the first reporting of the phrase "green chemistry" by Paul Anastas (EPA) in 1991.

To further strengthen this concept, Figure 1.1 reports the number of publications per year, over the last 15 years, reporting the concept "green chemistry" (*SciFinder*: as entered, journals, English only). The exponential growth of activity is clear; indeed, the actual number of publications is higher because, often, the concept of green

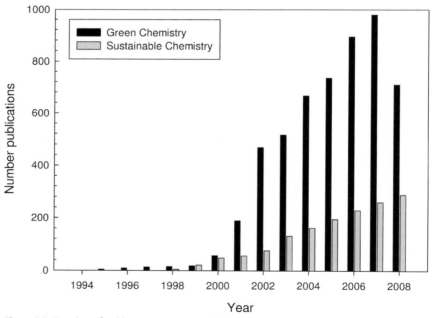

Figure 1.1 Number of publications per year in the last 15 years reporting the concept "Green Chemistry" (*SciFinder*: as entered, journals, English only) or the concept "Sustainable Chemistry." Values for 2008 extrapolated to full year from data available in July 2008.

chemistry is indicated differently and/or is simply not reported, even if the research falls under this area.

Figure 1.1 also reports the number of publications per year using the concept "Sustainable Chemistry," which is often considered as synonymous with green chemistry.

1.1.1
Green versus Sustainable Chemistry

About a decade ago, Hutzinger [3] discussed the topic of "green versus sustainable" chemistry. The controversial debate was a response to discussions within the Federation of European Chemical Societies, Division for Chemistry and the Environment, the only truly multinational (European) Environmental Chemistry Section. Hutzinger [3] argued that whilst most scientists use English as the *lingua franca* of science, cultural-sociological factors giving different meanings to terms and words should also be considered. "Green," for instance, in a cultural context simply means something different in countries such as the USA and Germany.

As pointed out before, in the USA and UK "green chemistry" has become synonymous with chemical industrial processes that avoid (toxic) by-products as much as possible: the greening of industry. The following sections analyze in more detail the principles and some examples of this green chemistry. In many countries, concepts such as "green accounts," "green procurement" and "green taxes" are accepted.

In Germany, the term "green" in many cases elicits political-sociological feelings – for instance, fear of production shutdowns in chemical industry – and general opposition to genetic engineering projects and atomic energy by members of the Green Party. There has been strong opposition to the term "green chemistry" by IUPAC, OECD, CEFIC and GDCh headed by the German speaking members. In addition, "green" was abandoned by the European Commission in the section of the fifth Framework program, likewise headed by the German speaking members.

Furthermore, the underlying meaning of the terms "green chemistry" and "sustainable chemistry" is different. Sustainable chemistry is the maintenance and continuation of an ecological-sound development, whereas green chemistry focuses on the design, manufacture, and use of chemicals and chemical processes that have little or no pollution potential or environmental risk and are both economically and technologically feasible. In Europe, apart from in the UK, the term sustainable chemistry is now preferred over green chemistry, but this practice is extending worldwide, as pointed out in Figure 1.1, which shows that the use of the term sustainable chemistry is expanding.

We could add that sustainable chemistry is not synonymous with green chemistry, for example, of a chemistry having a lower impact on environment and human health, but goes beyond the latter concepts, seeing chemistry as part of an integrated vision where chemistry, sustainability and innovation are three key components for the future of our society [4, 5].

1.1.2
Sustainability through Chemistry and the F³-Factory

A key driving force promoting this change from green to sustainable chemistry vision is the European Technology Platform on Sustainable Chemistry (ETP SusChem) [6] promoted by Cefic (European Chemical Industry Council) and EuropaBio (European Association for Bioindustries). To implement this vision, however, requires not only a concerted strategy of the stakeholders of the sector (objective of the cited platform), but also the progressive introduction of radical new approaches to chemical industrial production that foster innovation in chemical industry as the key factor to meeting sustainability. Two relevant concepts are:

1. down-size and integrate (chemical) production to minimize transport and storage, avoid large plants and concentration in one single location;
2. integrate processes (multi-step reactions – catalysis, reaction & separation).

Down-sizing chemical production, for example, break-down of scale economy though a modular design, is a new challenge for production of chemicals. To reach this objective, the development of many tools is necessary, ranging from micro-reactors and process intensification, integration of reaction-separations, to new ways to supply energy (from heat to photo, electro, microwave, etc.), new catalysis and nanotechnology, smart products (on-line microsensors), and so on. The advantages are to make production compatible with environmental sustainability (self-cleaning capacity), reduce investment (accelerate the introduction of new processes) and facilitate the introduction of cleaner production to non-chemical areas.

Realizing new processes for small-scale modular production also requires the development of new solutions for direct syntheses that avoid multistep reactions even in cascade mode.

The three main documents prepared by the European Technology Platform on Sustainable Chemistry (Vision Paper, Strategic Research Agenda, and Implementation Action Plan; they can be download free from cited ETP SusChem web site) evidence how *sustainability through chemistry* (probably a better definition than sustainable chemistry) requires a far more concerted and integrated effort than only a greener chemistry, which, nevertheless, is a core part of this vision.

These concepts are an integral part of the strategy for the future F³-Factory (future, fast, flexible), the visionary idea promoted by ETP SusChem for the future of chemical production (Figure 1.2). Future sustainable F³ chemicals production will combine a much broader range of production scales with interlinking technologies and logistics. An important strategy to meeting the F³ challenge is process intensification. This is a strategic and interdisciplinary approach employing different tools (such as micro-reaction technology and modularization) to improve processes holistically. A core technology to achieving process intensification is a new design of catalysts (e.g., to develop catalysts suitable for micro-reactor technology), evidencing how advances in catalysis are strictly related to innovation and sustainability in chemical production.

Faster, more flexible and environmentally benign chemical production are the three base elements for the F³-Factory visionary project. Rapid response to market

Figure 1.2 Vision of the F^3-Factory (future, fast, flexible).
Source: elaborated by ETP SusChem (http://www.suschem.org).

demand, an ability to handle a diverse product portfolio and improved sustainability are the key characteristics of tomorrow's chemical production facilities. The goal is to produce chemical products and intermediates that have minimal environmental impact through the use of highly eco-efficient, scalable and adaptive process that have smaller physical and ecological footprints.

One of the major drivers for this change is fierce competition from emerging countries, where large quantity production is cheaper and potentially more flexible. New competitive business models should be thus based on novel manufacturing technologies and a holistic approach to process development, where sustainability through chemistry, safety and respect of the environment are the driver for innovation and not the elements slowing down the development, the cultural model that has often dominated the business in the past.

The synthetic processes of traditional chemical industry will be strongly affected by new opportunities offered by biotechnology and process intensification (such as micro-reaction technology). The flexible integration of inherently safe process technologies with small holding volumes will be a key to success for future products.

The F^3-Factory is not only a crucial step and a prerequisite for successful and competitive future processes in Europe and worldwide, but is also the basis to implementing a new sustainable industrial chemistry. A major objective will be a substantial drop in capital expenditure for new plant and/or for retrofit of high-performance intensified devices into existing infrastructure.

Achieving these objectives requires that the whole chain for production processes is revised in an holistic approach that enables new products by new processing technologies and production concepts, encompass full lifecycles to optimize the whole production chain, and minimizes the use of resources and improves eco-efficiency, and delivers demonstration plants and a technology platform for future processes.

Integrated process units and combined unit operations should be linked with advanced process modeling tools, in-line monitoring, model-based process management and advanced process control to form centers of excellence for fast process development.

Sustainability through chemistry, for example, a sustainable industrial chemistry, is thus an approach that starts from green chemistry concepts and goes on to a vision for the future sustainability of society. This is clearly a process in evolution and this book aims to contribute to this goal by presenting, on one hand, some of the concepts and tools necessary to reach this scope and, on the other hand, some examples of interesting case histories, written from researchers operating in companies, and which could be used to better understand with practical cases the opportunities and problems in developing this new sustainable industrial chemistry.

This evolution from green to sustainable chemistry parallels the change in the concept of sustainability that has occurred in the last few years. The original concept of sustainability [7] emphasized the needs to combine social objectives (health, quality of life, employment) to the management of scarce resources (energy and raw materials) and the preservation of the natural bases for life, for example, the need to adopt all actions such as cleaner processes, recycle waste, reduce pollutant emissions necessary to preserve biodiversification. The actual concept of sustainability is broader and takes into account that sustainability is also the engine for innovation. In fact, the fast modifying socio-economic and geo-strategic context requires a societal change and adaptation to put the capacity of innovation at the core of competitiveness and economics.

1.1.3
Role of Catalysis

Emphasis throughout this book is given to catalysis, because it is a driver for sustainability and societal challenges [8]. Catalysis plays a critical role in promoting the feasibility, eco-efficiency and economics of over 90% of chemical processes. However, catalysis is also one of the critical enabling factors for sustainability in the specific field of chemical processes. In the mobility sector, without the introduction of catalytic converters starting from the 1960s, it was not possible to counterbalance the increasing level of emissions of pollutants such as NO_x, CO and hydrocarbons associated with the worldwide expansion of the car market. In particular, without the introduction of catalytic converters it was not possible to reduce the total NOx emissions below values sustainable for the society, for example, to limit issues such as smog formation, acid rain and the increase of diseases and allergies due to pollutants.

The achievement of a sustainable mobility was thus linked to the capacity to develop and consecutively further improve the catalytic converters for emissions from cars, buses and trucks. Several challenges remain in this area, particularly related to the need to find more effective catalysts for the reduction of NOx in the presence of oxygen (a very relevant problem to address the increasing share of diesel engines), better catalytic filters for particulate and catalysts more active at low temperature (to reduce cold start emissions). Catalysis was thus one of the enabling factors for a critical societal need such as mobility, and will continue to play a critical role in this area, because further R&D is needed. Regulations on CO_2 emissions from vehicles under discussion currently also poses new challenges for catalysis, because new requirements for catalytic converters derive from the necessary changes in engines to meet these limits.

In chemical and refinery productions, catalysts are not simply components functional to the process economy, for example, which allow improvements in yields/productivity, and reduce process costs (longer catalyst life, milder reaction conditions, reduction of separation and environmental costs). A new catalytic process is an opportunity to gain market share or to enter into new markets. New regulations (REACH in Europe, in particular), which impose the need to record the production process of the chemicals, will further change the use of catalysis from being a tool to achieve process targets to be part of the strategic vision of the companies. Not only the quality of a chemical product will be important in the future but also the quality of their production chain. It will thus no longer be possible to produce in remote areas where the environmental criteria or controls are less severe. The environmental performance indexes, from energy efficiency to greenhouse gases and pollution emission factors (and other factors which may be estimated from life cycle assessments) will be the key parameters for evaluation in a society, which should include in the production cost also the use of the limited natural resources. The actual dramatic lack of some raw materials is already a clear signal of the need for progress in this direction. The increasing concerns over greenhouse gas emissions is another signal.

The lack of energy and natural resources are two main factors that have determined a drastic change in the priorities for society and thus for chemistry in the last few years. The progressive revolution in chemical and fuel areas deriving from the socio-political pressure of new uses in biomass and in general terms of renewable resources [9] has largely changed R&D priorities. Handling biomass is far more complex than oil and converting biomass into a concentrated and easy transportable form of energy is more difficult than for the equivalent based on oil derivatives. Biomass utilization requires therefore an intense research effort. Many process steps require the development of novel catalysts and processes, starting from the need for efficient and stable solid catalysts for vegetable oil transesterification, to solid catalysts for cracking, hydrolysis or selective depolymerization of (hemi)cellulose and lignin, and to new enzymes for fermentation to products other from ethanol. The passage from first- to the second-generation biofuels requires intensified research on new catalysts.

A recent study by the US Department of Energy (DoE) [10] "Catalysis for Energy" indicated the following three priority research directions for advanced catalysis science for energy applications:

1. Advanced catalysts for the conversion of heavy fossil energy feedstocks.
2. Advanced catalysts for conversion of biologically derived feedstocks and specifically the deconstruction and catalytic conversion to fuels of lignocellulosic biomass.

3. Advanced catalysts for the photo- and electro-driven conversion of carbon dioxide and water.

Fuel cells, due to their higher efficiency in the conversion of chemical into electrical energy with respect to thermo-mechanical cycles, are another major area of R&D that has emerged in the last decade. Their effective use, however, still requires an intense effort to develop new materials and catalysts. Many relevant contributions from catalysis (increase in efficiency of the chemical to electrical energy conversion and the stability of operations, reduce costs of electrocatalysts) are necessary to make a step forward in the application of fuel cells out of niche areas. This objective also requires the development of efficient fuel cells fuelled directly with non-toxic liquid chemicals (ethanol, in particular, but also other chemicals such as ethylene glycol are possible). Together with improvement in other fuel cell components (membranes, in particular), ethanol direct fuel cells require the development of new more active and stable electrocatalysts.

Fuel cells should also be considered as an element of a broader area in which catalysis is used in combination with electrons to perform selective reactions. For example, it is possible to feed waste streams from agro-food production to an electrochemical device essentially analogous to a fuel cell to produce at the same time electrical energy and chemicals [8]. This approach is interesting in SMEs (Small Medium Enterprises) for using wastewater or by-product solutions derived from agro-food production. A limiting factor is the need to develop new nanostructured electrocatalysts, because conventional fuel cell electrodes have limited effectiveness and they are tailored for total oxidation. The challenge is to develop new electrocatalysts that do not break the $C-C$ bond, have a high activity to make the process industrially feasible (e.g., have close to 100% Faradaic efficiency, and current densities of about $100-150\,mW\,cm^{-2}$ at temperatures lower than 90 °C) and are stable in the strong basic medium required to use anion-exchange membranes. Target products are low-molecular weight oxygenates of industrial valuable interest. Recent results [11] have shown that it is possible to oxidize selectively ethanol, glycerol, 1,2- and 1,3-propanediol and ethylene glycol to the corresponding (di)carboxylic acids, hydroxy- or keto-acids, in DAFC-type cells (polymeric membrane fuel cells fed with alcohols).

Catalytic chemistry with fuel cells may be thus considered part of the general effort towards new delocalized chemical productions, because this approach is especially suited for SMEs.

1.1.4
Sustainable Industrial Chemistry

The above few examples (more are discussed later in this book) show that a very rapid change in priorities, methodologies and issues has occurred in chemistry over the last few years, driven by the fast evolving socio-economical context. There is thus the need to re-consider chemistry in the light of these changes, in addition to the motivations discussed before, for example, to re-address the topic from the point of view of sustainable industrial chemistry, the aim of this book. The present book is in agreement with the philosophy of the journal *ChemSusChem*, published by Wiley-VCH Verlag (Weinheim, Germany).

Although recently, various books have been published on green catalysis [12–16], often also with focus on catalysis, we consider this further book necessary focused in particular on highlighting the new vision for sustainable industrial chemistry, but complemented by a series of industrial examples that could be used either for educational purposes or as case histories.

The present chapter introduces the basic concepts of green and sustainable chemistry and engineering as a background for subsequent chapters. An outlook for future needs is also provided with a short analysis of the key priorities identified in the cited documents of the ETP SusChem.

1.2
Principles of Green Chemistry, Sustainable Chemistry and Risk

The previous section outlined briefly the historical development of the Green Chemistry Movement, which started in the early 1990s as a means of encouraging industry and academia to use chemistry for pollution prevention. More specifically, the green chemistry mission was: "To promote innovative chemical technologies that reduce or eliminate the use or generation of hazardous substances in the design, manufacture and use of chemical products." Practical motivations, however, were to counteract the negative image of chemistry, deriving from several factors:

- The limited attention to impact on the environment and human health given by companies (apart few of them) during the period of fast development of industrial chemical productions (approximately the 1960–1980s).

- The progressive increase in the size of chemical plants (due to scale economy) and concentration of many plants in the same locations (due to the integration of chemical production and need of common utilities, services, etc.), which led to an amplification of local impact on environment.

- Major chemical accidents that reinforced this poor public image [17–19]. Examples are (i) the Bhopal accident in 1984, in which 3000 people were killed and more than 40 000 injured and (ii) the grounding of the Exxon Valdez [20] in the Prince William Sound in Alaska in 1989, which still affects the marine ecosystem nearly 20 years later.

Figure 1.3 reports the results of a poll made by the UK Chemical Industries Association to analyze the general public's view of the chemical industry by (MORI survey) [21, 22]. Although the survey was limited to UK, it is indicative of the general perception of the chemical industry by the public [23]. A constant decline, with a maximum of unfavorable opinions around the end of the 1980s, is evident.

The practical consequences were a rapid increase in environmental legislation, and public opposition to building new plants, and also a decline in the number of young people interested in R&D in chemistry, for example, the number of applicants

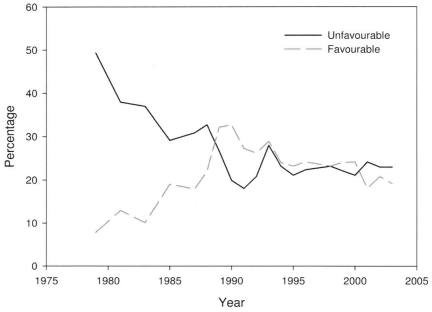

Figure 1.3 Chemical industry favorability. Source: elaborated from Lancaster [21].

reading chemistry at university had been falling steadily for several years during the cited period. Therefore, an effective push from both academia and the chemical industry was needed to give chemistry a different image.

This was the general cultural context of the green chemistry movement, which explains both the rapid increase in interest (Figure 1.1) and the limitations of the approach. Often, the term green chemistry was adopted to indicate with a new term, apparently more attractive and ecologically-sound, the natural evolution of R&D in the field. We could cite, as an example, the industrial process for the synthesis of maleic anhydride from butane as a substitute for the old process starting from benzene. This industrial process was introduced about 15 years before the concept of green chemistry. Although the process could be read in terms of green chemistry principles (Table 1.1), the effective motivations were industrial [5]. A series of other examples of industrial motivation versus principles of green chemistry in the development of new processes have been reported by Centi and Perathoner [5].

There is, thus, often no contrast between industrial objectives and the principles of green chemistry, but the right motivations in R&D should be clearly pointed out (Table 1.2). Voluntary programs for reducing environmental impact are present in several chemical companies, even if the driving force for process selection has for long been economics. With the rising costs of technologies to reduce emissions to levels compatible with legislation limits, a change of philosophy from end-of-pipe pollutant elimination to avoid waste formation is a natural trend. However, the introduction of new processes is costly and the effective introduction of cleaner industrial processes has been limited in the last decade.

Table 1.2 Comparison of industrial motivation and the principles of green chemistry in the industrial synthesis of maleic anhydride from *n*-butane. Source: adapted from Centi and Perathoner [5].

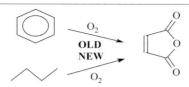

Green chemistry principle	Industrial motivation
Atom economy	The loss of two carbon atoms (as starting from benzene) is avoided. Thus better yield by weight
Simple and safe process	Easy maleic anhydride recovery and separation, due to the reduced by-products
No waste	Only by-products CO_x and CH_3COOH (minor amount)
Avoid toxic chemicals or solvents	Toxicity aspects related to use of benzene were avoided

An effective increase in the rate of introduction of new cleaner processes could derive only from a change of perspective, in which "greener technologies" are not only seen as a strategy to improve the image of chemistry but as a novel business strategy for innovation. This is the concept of sustainable industrial chemistry versus green chemistry. It uses green chemistry concepts and integrates them into a larger vision to create an effective strategy for sustainability though chemistry. Green and sustainable chemistry are thus not synonymous, as indicated in the cited books [12–16], but instead green chemistry is the core around which a new strategy for chemistry should be built up.

In addition, we may observe that the principles for green chemistry (see later) are of general validity, but their implementation was often questionable and/or with limited impact. We note, for example, that no relationship could be seen between the growing of publications on "green chemistry" (Figure 1.1.) and public perception of importance of chemistry (Figure 1.3). This is more evidence as to why a further step is necessary, for example, to pass from green to sustainable industrial chemistry.

In conjunction with the American Chemical Society, the EPA has developed a set of 12 guiding principles for green chemistry [24]. These principles can be summarized as being concerned with ensuring that:

- the maximum amounts of reagents are converted into useful products (atom economy);
- production of waste is minimized through reaction design;
- non-hazardous raw materials and products are used and produced wherever possible;
- processes are designed to be inherently safe;
- greater consideration is given to use of renewable feedstocks;
- processes are designed to be energy efficient.

Table 1.3 Principles of green chemistry [24, 25].

1	Waste prevention is better than treatment or clean-up
2	Chemical synthesis should maximize the incorporation of all starting materials
3	Chemical synthesis ideally should use and generate non-hazardous substances
4	Chemical products should be designed to be nontoxic
5	Catalysts are superior to reagents
6	The use of auxiliaries should be minimized
7	Energy demands in chemical syntheses should be minimized
8	Raw materials increasingly should be renewable
9	Derivations should be minimized
10	Chemical products should break down into innocuous products
11	Chemical processes require better control
12	Substances should have minimum potential for accidents

Table 1.3 lists all twelve principles. The underlying objective of these principles is that green chemistry encompasses much more of the concepts of sustainability than simply preventing pollution. However, a more detailed reading evidences that these principles focus mainly on the reaction (especially on organic syntheses) rather than on the industrial processes, even if relevance is given to the design for energy efficiency and the use of renewable feedstocks.

The concept of green chemistry is defined as "the utilization of a set of principles that reduces or eliminates the use or generation of hazardous substances in the design, manufacture and application of chemical products" [24, 25].

1.2.1
Sustainable Risk: Reflections Arising from the Bhopal Accident

The focus of green chemistry on the reaction more than on the process and its industrial feasibility, even considering the attention given to aspects such as green chemistry and engineering [16], indicates that aspects related to the safety of a chemical or manufacturing process, measured in terms of risk, are not given enough consideration. In its simplest form risk can be expressed as:

$$\text{Risk} = \text{Hazard} \times \text{Exposure}$$

To date, legislation has sought to minimize risk by limiting exposure to chemicals, that is, controlling use, handling, treatment and disposal. Green chemistry seeks to improve safety by minimizing hazard and so shifts the risk from circumstantial to intrinsic factors. Methods of providing this inherent safety in chemical processes are the generation of hazardous intermediates *in situ*, avoiding storage of large quantities of chemicals on site and avoiding the use of dangerous chemicals altogether.

The example of the well-known Bhopal accident (in 1984) is instructive and is a good lesson [26–28]. Bhopal is located in North Central India. It is a very old town in a picturesque lakeside setting and was a tourist center. It is the capital of Madhya Pradesh and industry was encouraged to go there as part of a policy of bringing industry to less developed states. The annual rent was $40 per acre and thus Union Carbide took the decision in 1970 to build a plant to produce the pesticide SEVIN, a DDT substitute (brand name for Carbaryl-1 – naphthyl methylcarbamate). The initiative was greatly welcomed and was cited at that time as part of India's Green Revolution.

SEVIN was manufactured from carbon monoxide (CO), monomethylamine (MMA), chlorine (Cl_2), alpha-naphthol (AN) – the first two imported by truck and the latter two made on site. The process route was as follows:

$$CO + Cl_2 \rightarrow COCl_2 \,(Phosgene) \tag{1.1}$$

$$COCl_2 + MMA \rightarrow MMC + MIC \tag{1.2}$$

MIC is stored in three $\sim$50 m^3 tanks.

MIC + AN $\rightarrow$ Carbaryl(SEVIN):

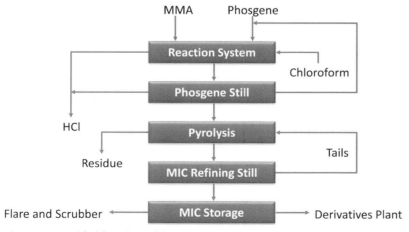

$$\tag{1.3}$$

Figure 1.4 shows a simplified flow chart of the process. An intermediate in the processes is methyl isocyanate (MIC), which was stored in a series of tanks partially underground and equipped with cooling systems and a series of safety control devices (Figure 1.5). In fact, MIC is a highly toxic (maximum exposure: TLV-TWA, during an 8-hour period is 20 parts per billion), flammable gas that has a boiling point near to ambient temperature and gives a runaway reaction with water unless chilled below 11 °C. Table 1.4 gives the list of MIC safeguards.

Figure 1.4 Simplified flow chart of the Bhopal Union Carbide process.

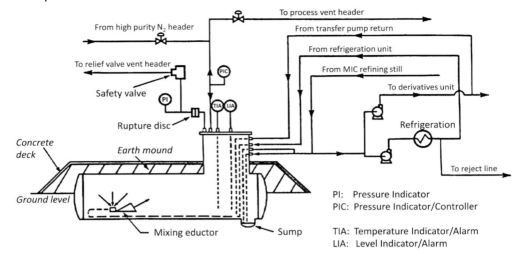

Figure 1.5 Methyl isocyanate (MIC storage) tank used in the Bhopal Union Carbide process. Source: adapted from Reference [31].

Safeguards may be equipment items or procedures designed to prevent the initiating event, limit or terminate the propagation, or mitigate the outcome. In fact, accidents are normally characterized by a sequence of events leading from the initiating event, propagation of the accident and realization of the undesired outcome.

Active safeguards are those that require human procedures or mechanical initiation to operate (e.g., work permit procedures, scrubber caustic circulation). Passive safeguards are those that do not require any initiation (e.g., concrete fireproofing, elevated vent stack for dispersion).

Both active and passive safeguards can be defeated through inadequate safety management systems. A safety management system is the most efficient way to

Table 1.4 Safeguard devices for the methyl isocyanate (MIC) tank employed in the Bhopal Union Carbide process.

Safeguards	Type
Mounded/insulated MIC Tanks	Passive
Refrigeration below reaction initiation temperature	Active
Refrigeration uses non-aqueous refrigerant (Freon)	Active
Corrosion protection (cathodic) to prevent water ingress	Active
Rigorous water isolation procedures (slip blinds)	Active
Nitrogen padding gas used for MIC transfer not pumped	Active
Relief valve and rupture disk	Passive
Vent gas scrubber with continuous caustic circulation	Active
Elevated flare	Active + Passive
Water curtain around MIC Tanks	Active

allocate resources for safety, since it not only improves working conditions but also positively influences employees' attitudes and behavior with regards safety, consequently improving the safety climate [29]. The Bhopal accident was crucial for widespread adoption of more rigorous process safety procedures and management systems [30]. Process safety is a comprehensive, systematic approach encompassing the proactive identification, evaluation and mitigation or prevention of chemical releases that could occur as a result of failures in process, procedures or equipment.

We may note in Table 1.4 that the vent gas scrubber was an active safeguard, while a passive type (as well as a change from active to passive of the other safeguards) would probably have avoided the accident, as discussed later.

In December 1984 water leaked into a storage tank and reacted to cause an increase in pressure that subsequently blew the tank. Around 41 tonnes of methyl isocyanate was released into the atmosphere and rained back down on the local population. It killed 3800 people instantly and several thousand others through long-term exposure effects. Several hundred thousand people have suffered permanent disabilities.

To understand the motivations of this accident, within the frame of the general context of green/sustainable chemistry, it is necessary to clarify first the industrial context of the accident, and in particular the plant problems that are the effective precursors to disaster. The first observation is that the alpha-naphthol plant was shut down and thus SEVIN production was no longer profitable, and the plant run intermittently. This situation determined a series of critical issues:

- minimum maintenance,
- safety procedures simplified for small jobs,
- refrigeration unit shut down and Freon sold,
- scrubber circulation stopped,
- manning cut to 600 and morale low,
- slip blinding no longer mandatory during washing,
- high temperature alarm shut-off as T now >11 °C,
- relief valve and process vent headers joined (for maintenance),
- emergency flare line corroded, disconnected.

The chemistry causing the accident is well established. Forty-one tonnes of MIC in storage reacted with 500 to 900 kg water plus contaminants. The resultant exothermic reaction between MIC and water caused an increase of temperature that reached 200–250 °C. As a consequence, the tank pressure rose to 200 + psig (14 + bar); the tank was designed for 70 psig (4 bar). The MIC tank thus overheated was over-pressured and vented through the scrubber, which was out of operation. This caused an elevated discharge of a massive quantity of MIC (approximately 25 tonnes). The source of water was clearly the determining element for the accident. The following observations could be made:

- filters were flushed using high-pressure water;
- drain line from filter was blocked, operator observed no flow to drain;

- flushing continued despite blockage;
- high pressure could cause valve leak; this forced water into the relief header.

Figure 1.6 shows the probable route of entrance of water into tank number 610 containing the MIC. The admission of water to the MIC tank was thus caused by a series of concomitant events:

- relief valve (RV) and process vent headers were joined by a jumper pipe, no blinds;
- MIC tank was not pressurized;
- head of water sufficient for flow;
- slow initial reaction between MIC and H_2O allowed enough water to enter into the tank.

Notably, no universally accepted cause exists [31], and sabotage, for example, whereby somebody deliberately connected a water hose to piping that directly entered into the storage tank, is still supported by some authors. However, this theory would require (i) an intimate knowledge of piping around the tank, where to physically make the correct connection, (ii) removal of a pressure indicator and then (iii) the re-attachment of piping fittings. This theory is thus unlikely.

When water started to react with MIC, no control of the temperature/pressure started, because it was late at night and the operational staff was reduced to a minimum, and the MIC tank's alarms had not worked for 4 years. The gas leakage followed, for the first 30 min, approximately the inverse route of water entrance (Figure 1.6), except that which reached the atmosphere through the vent collection system (VCS). However, the flare tower and the vent gas scrubber (Figure 1.7) had been out of service for 5 months before the disaster. After the first 30 min, the rupture disk bursts and this increased the rate of release of MIC.

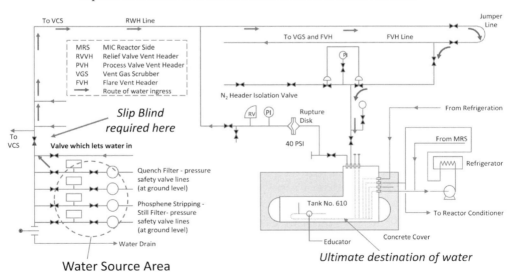

Figure 1.6 Probable route of entrance of water into the MIC tank in the Bhopal Union Carbide process.

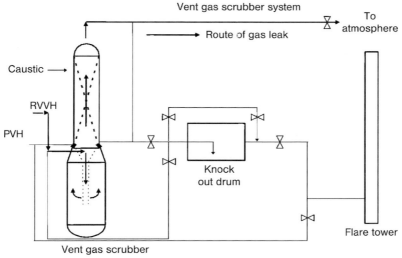

Figure 1.7 Schematic diagram of the emergency relief effluent treatment system that includes a scrubber and flare tower in series in the Bhopal Union Carbide process.

Owing to a poor design, the maximum pressure that could be handled by the NaOH scrubber was only one-quarter of that developed after the first 30 min, and thus, even if working, the scrubber was not effective in preventing the accident [32]. Finally, a further cause of the release was that the refrigeration system, designed to inhibit the volatilization of MIC, had been left idle – the MIC was kept at 20 °C, not the 4.5 °C advised by the manual, to reduce energy costs, and to allow the use of the coolant elsewhere.

Other causes that contributed to the accident include (i) lack of slip-blind plates that would have prevented water from pipes entering into the MIC tanks via faulty valves, and (ii) use of carbon-steel valves, despite the fact that they corrode when exposed to acid (the leaking carbon-steel valve that allowed water to enter the MIC tanks was not earlier repaired as this was too expensive).

However, neither of above is a root cause. The root causes were management decisions:

- to neglect to repair the flare system;
- to place a scrubber system on stand-by to save on operating expenses;
- to remove coolant from the refrigeration system used to cool the MIC storage tank.

Additional root causes made the accident more severe:

- inadequate emergency planning and community awareness;
- lack of awareness of the potential impact of MIC on the community by the people operating the plant;
- lack of communication with community officials before and during the accident;

- inadequate community planning, allowing a large population to live near a hazardous manufacturing plant. This situation was not unusual in the chemical industry in the early 1980s, and one major impact of Bhopal was to warn all chemical plants about the importance of these considerations in the siting and operation of facilities.

It may be also noted that (i) between 1981 and 1984 six accidents with phosgene or MIC occurred, (ii) a 1982 audit was critical of the MIC tank and instrumentation and (iii) in 1984 a warning of a potential runaway reaction hazard was given. Therefore, clues to this accident were available before it happened.

The roots of this accident extend even further back. Optimistic market-size expectations led to an oversized plant – and therefore MIC storage capacity oversized by a factor of three. Failure of state and local government to control the shantytown growth near the plant meant a considerably larger impacted population was unable to take shelter.

1.2.2
Risk Assessment and Sustainable versus Green Chemistry

The above paragraph shows that the reasons and dimensions of the accident at Bhopal are connected to many factors, but principally to a lack of management responsibility in plant management and procedures of maintenance of partially decommissioned plants, which were oversized due to poor market-size projections, and to a lack of control by the authorities.

Sheldon, in an editorial on the first year of activity of the journal *Green Chemistry* [33], citing a conference of Guy Ourisson (President of the French Academy of Science), invoked the terms black chemistry and red chemistry to describe two shortcomings of many traditional processes:

"Black chemistry stands for waste = pollution and conjures up images of the industrial revolution, black smoking chimneys and Blake's *dark satanic mills*. Red chemistry, on the other hand, denotes danger and evokes associations with incidents such as Bhopal and Seveso. Many chemical processes in use today are black or red, or both. Hence, the goal of the chemical industry is, or should be, the replacement of red and/or black chemistry with green alternatives."

The example of the Bhopal accident shows that instead it is not a problem of bad versus good (green) chemistry but of correct design and management of risky processes. When the risk (intended in its larger meaning, which includes risk for environment, health and process safety) is less than the benefits (for society, which includes economic access to goods produced directly by the process or though further consecutive steps), the risk is acceptable, even if it is better to further minimize it. A sustainable risk, which implies the accounting for sustainability and the use of tools such as life-cycle assessment (LCA) for a correct analysis of the risk, which will be discussed later, is what differentiates the concept of sustainable industrial chemistry from green chemistry.

Minimizing risk means reducing both the magnitude of the possible event and its frequency, by proper design of both the engineering and chemistry of the process, but which could be only effective by proper risk assessment. For example, as shown later, there is the possibility of substituting phosgene with other chemicals, which are less toxic or risky. This solution was discussed in several papers and indicated as the necessary route for a green chemistry [34–45].

However, the minimization of the risk, both in terms of environment and process safety, could be equally reached by adopting an on-demand synthesis of phosgene and MIC [46]. This is the approach preferred industrially and evidences that the same goal could be reached by a different philosophy, other than the substitution of chemicals indicated by green chemistry principles.

1.2.3
Inherently Safer Process Design

In general, there are two approaches to design safety into a chemical process – either handling the hazards with engineering and management controls or eliminating them altogether [47]. If a process hazard cannot be eliminated, it may still be possible to reduce its potential impact sufficiently as far it is not capable of causing major injury or damage. Whenever feasible, it is desirable to eliminate or minimize process hazards. Engineering and management controls have been effective in reducing risk to very low levels in the chemical process industries, as demonstrated by the industries' excellent safety record. However, no engineering or management system can ever be perfect, and the failure of these systems can result in major accidents. Moreover, the original installation and ongoing operations and maintenance of these safety and risk-management systems is often expensive. If the manufacturing technology can be changed to eliminate the hazards, the result will be a safer and a more cost-effective plant design.

We refer to a process that eliminates or minimizes hazards as "inherently safer," because the safety basis of the design is inherent in the process chemistry and operations, rather than coming from added safety equipment and procedures. The process designer is challenged to change the process to eliminate hazards, rather than to develop add-on barriers to protect people, the environment, and property. This is best accomplished early in the product and process design cycle, but it is never too late to apply these concepts. This concept is discussed more in detail in the next section.

There are four major strategies for inherently safer process design [48–50]:

1. Minimize the size of process equipment.
2. Substitute a less-hazardous substance or process step.
3. Moderate storage or processing conditions.
4. Simplify process and plant design.

As mentioned before, scaling-down chemical processes and making them modular is an essential element for a new vision of sustainable industrial chemistry.

This is also fully in line with the concept of process intensification (see below) and the strategy of minimizing the size of process equipment for a inherently safer process design.

Reducing the size of chemical process industry (CPI) equipment generally improves safety by reducing both the quantity of hazardous material that can be released in case of loss of containment and the potential energy contained in the equipment. This energy may derive from high temperature, high pressure or heat of reaction.

There are many opportunities to minimize the inventory of hazardous materials in a CPI plant without fundamental changes in process technology. Following the accident in Bhopal, India, in 1984, most CPI firms reviewed their operations to identify opportunities to reduce quantities of toxic and flammable materials on-hand. These companies did not rebuild their plants using a different technology, or make dramatic changes to the process equipment, as both solutions were too costly. Instead, they carefully evaluated existing equipment and operations, and identified changes that would allow them to run with a reduced inventory of hazardous materials. Some of the possibilities are:

- **Storage**: The need for storing large quantities of raw materials and intermediates was often simply adopted to make "easier" the operations of a plant, for example, more flexibility in ordering raw materials, secure from fluctuations and transportation delays. However, when the associated risks are fully considered, it is often worth significantly reducing the quantities on-hand and devoting additional resources to ensure a reliable supply. Modern inventory-control systems, improved communications with raw materials suppliers and transport companies, and strategic alliances with raw materials manufacturers also allow plants to function with a smaller stock of hazardous substances. Often, large storage tanks for hazardous in-process intermediates are used. This storage decouples sections of a plant from each other. Parts can continue to run, either filling or emptying these tanks, while a unit at the facility is shut down for maintenance or operating problems. However, by reducing plant shutdowns it is possible to greatly reduce or eliminate such storage.

- **Piping**: When designing piping for hazardous materials, one should minimize the "inventory" (holdup) in the system. Piping should be large enough to transport the quantity of material required, but not greater. The quantity in pipes can be also minimized by using the material as a gas rather than a liquid. The Dow Chemical Exposure Index [51] is a tool to measure inherent safety with regards to potential toxic exposure risks. For example, the risk related to a pipe carrying liquid chlorine from a storage area to a manufacturing building, where chlorine should be vaporized and fed to a process, could be reduced by installing the vaporizer in the storage area. This reduces the inventory of chlorine in the pipe by a factor of over ten.

- **Process intensification** means using significantly smaller equipment. Examples include novel reactors, intense mixing devices, heat- and mass-transfer designs that provide high surface-area per unit of volume, equipment that performs one or more unit operations, and alternative ways of delivering energy to processing equipment – for example, via ultrasound, microwaves, laser beams or simple electromagnetic radiation. These technologies can increase the rate of physical and chemical processes, allowing high productivity from a small volume of material. A small, highly efficient plant can be expected to be cheaper and more cost-effective, but also can reduce the magnitude of potential accidents. If the plant is small enough, the maximum possible accident may not pose any significant hazard. This will reduce the safety equipment, emergency alarms and interlocks, and other layers of protection required to manage the risk. Installation and ongoing operation of this safety equipment is often a major expense; if it can be eliminated, there will be additional cost savings. Safety needs not necessarily mean spending money. Safer can also be cheaper, if a small, efficient inherently safer process can be developed.

- **On-demand (on-site) synthesis:** The use of novel equipment/technologies for process intensification avoids the storage/transport of hazardous chemicals. They could be generated on-site and immediately converted into final product. In the Bhopal process, the introduction of on-demand synthesis of methyl isocyanate (MIC) has reduced the total inventory of less than 10 kg of MIC, for example, which is 0.024% of the amount released during the accident [48, 49]. Phosgene, the other hazardous chemical present in the Bhopal process, could be also produced on-demand. A continuous tubular reactor has been developed to make this chemical on-demand available for immediate consumption by batch processing vessels [52]. One plant using the new design contains 70 kg of gaseous phosgene, compared to an inventory of 25 000 kg of the liquid in equipment and storage in the old facility. As the new unit is quite small it is also possible to provide secondary containment for the phosgene, further enhancing the process safety [53].

1.2.4
On-Demand Synthesis and Process Minimization

Novel developments in micro-reactor technology offer new possibilities to combine process intensification, safer operations and on-demand production. For example, in the phosgene on-demand synthesis it is possible to use a silicon micropacked-bed reactor to achieve complete conversion of chlorine for both a 2:1 $CO:Cl_2$ feed at 4.5 $cm^3 min^{-1}$ and a 1:1 feed at 8 $cm^3 min^{-1}$ [54]. The latter gives a projected productivity of approximately 100 $kg yr^{-1}$ from a ten-channel microreactor, with the opportunity to produce significant quantities by operating many reactors in parallel. The increased heat and mass transfer inherent at the submillimeter reactor length scale provides a larger degree of safety, control and suppression of gradients with respect to those present in macro-scale systems.

   Phosgene is widely used as a chemical intermediate for the production of isocyanates used in polyurethane foams and in the synthesis of pharmaceuticals and pesticides [55]. Processes using phosgene require specialized cylinder storage, environmental enclosures, pipelines, fixtures under negative pressure and significant preventative maintenance. Moreover, phosgene is under various transportation restrictions. As a consequence, most phosgene is consumed at the point of production today. Micro-chemical systems stand to provide an opportunity for flexible point-of-use manufacturing of chemicals such as phosgene. Banks of reactors can be turned on or off as needed to maintain close to zero the storage. Single reactor failures would lead to extremely small chemical release.

The cited micro-fabricated silicon packed-bed reactor for phosgene on-demand production (Figure 1.8) was made out of single crystal silicon with standard micro-fabrication processes developed for integrated circuits and MEMS. The geometry is defined using photolithography and created with silicon etching. The reactor consists of a 20 mm long, 625 mm wide, 300 mm deep reaction channel (3.75 mL volume) capped by Pyrex. Figure 1.8b shows a scanning electron micrograph (SEM) of the inlet where the flow is split among several interleaved channels (25 mm wide) that meet at the entrance of the reaction channel. Perpendicular to the inlet channels are 400-mm wide loading channels used to deliver catalyst particles to the reactor. Catalyst is loaded by placing a vacuum at the exit of the reactor and drawing in particles through the loading channels. At the outlet of the reaction chamber, a series of posts with 25 mm gaps acts as a filter to retain the catalyst bed (Figure 1.8c).

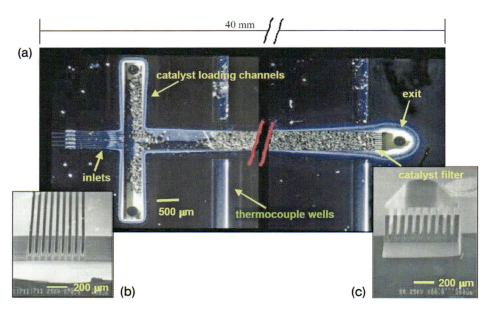

Figure 1.8 Microfabricated silicon packed-bed reactor for phosgene on-demand production. (a) Top-view of reactor partially loaded with 60-mm activated carbon particles – the reactor channel is 20 mm long, and the image is spliced to fit the 20 mm reaction channel by omitting the long channel midsection; (b) SEM of the 25 mm wide interleaved inlets; (c) SEM of the catalyst filter structure. Source: adapted from [53].

There are also four 325-mm wide channels perpendicular to the reaction channel along its length for holding thermocouples. Access ports for flow come from underneath at the inlet (not shown in Figure 1.8), the reactor exit and at the ends of the catalyst loading channels.

There has been great interest recently in micro-reactor technology (as discussed also later in this book) as a new efficient tool for process intensification and risk reduction [56–58]. They will be one of the key tools for decentralized chemical production. Actual chemical production is still at the stage where the large, centralized production facility dominates. Although not for all processes, we believe that this will change, and that small, dedicated plants producing on-demand specific materials at the site where the material is needed will be an important component of the future chemical production industry (CPI). Changes in this direction in other industries have been largely driven by the need for flexibility and convenience, along with the development of economically competitive, small, distributed systems.

The CPI has an additional factor that will drive the change to small, distributed facilities – the greatly reduced inventory of hazardous materials of an inherently safer, small plant. These plants will be highly automated and largely unstaffed, and use commonly available and relatively nonhazardous raw materials. They will contain small quantities of hazardous material and energy, and will not be capable of causing a major incident. The *scale economy* leading to large, centralized plants should be substituted with the *smart economy* of building small, cheap plants that can be set down where they are needed, used and then taken away or recycled. In addition, this modular design would greatly reduce the costs in developing new process (increasing the production would only need an increase in the number of reaction/separation modules) and will significantly boost the introduction of new processes/technologies (reduced cost of investment). This future vision of the CPI has, for example, been described in articles by Benson and Ponton [59], Ponton [60], Hendershot [47] and also us as well [5]. However, the increasing energy and raw materials costs, global competiveness, and social pressure on safety and environment protection, has forced chemical companies to re-consider their strategy of development. For this reason, this concept of modular-design and small/distributed production is now a key part of the vision for the future of chemical industry indicated in the documents prepared by the European Technology Platform of Sustainable Chemistry [6], which derives from a joint effort of the main chemical companies in Europe and the scientific community.

Long term, it will even be possible to manufacture many hazardous materials in plants as small as a silicon chip, using large numbers of "plants on a chip" in parallel to produce the quantity of material required [61, 62]. It is unlikely that the large, world-scale petrochemical plant will ever completely disappear. However, distributed manufacture of chemicals in small plants will become an important part of the industry, in particular for the improved safety through distributed manufacturing of hazardous chemicals. There are many relevant examples besides those cited previously of phosgene and methyl isocyanate synthesis, for example HCN [63] and aqueous peracetic acid [64] on-demand syntheses.

1.2.5
Replacement of Hazardous Chemicals and Risk Reduction

The replacement of hazardous chemicals with benign and inherently safer alternatives indicated by green chemistry principles is certainly a valuable measure, because a hazardous chemical that is no longer present can no longer be involved in an accident. However, from industrial point of view there are often alternative solution that reduce the risk to a sustainable level at a much lower cost than that required by completely changing the production.

The problem of cost is a key element for an effective introduction of cleaner and safer technologies. Green chemistry approaches [12, 13, 24, 25, 65, 66] often stress the need for replacement of a hazardous ingredient in chemical synthesis or process, with respect to alternatives such as on-site/on-demand production of high-risk compounds, and reduced reliance on those hazardous chemicals that cannot be replaced. However, a sustainable industrial technology, for example, which reaches the optimum in the triangle of process economy, safety of operation and environment protection, is preferable to a "greener" technology that may have a lower impact on the environment or better intrinsically safer operations, but which will be not implemented due to too higher costs in a global economy.

Sustainability is step-by-step progressive development with a moving frontier. It is necessary to consider and teach this concept; a realistic approach will lead to a faster implementation of better industrial technologies, with an effective reduction of risks and waste, and improved use of resources. Owing to the large cost of investment, substitution of chemicals is often not cost-effective with respect to alternative engineering solutions, even if it should be discussed case-by-case. There are, however, often discrepancies between the emphasis given in scientific research to the issue of substituting hazard chemicals and industrial solutions to reduce risk to a sustainable level. We can further discuss this issue by returning to the problem of the Bhopal synthesis of SEVIN.

After the Bhopal accident, new regulations were introduced in almost all countries to limit the amount of methyl isocyanate (MIC) that may be stored in a plant. The EU allows a maximum of half a tonne storage on site – around 67 tonnes were stored at the Bhopal facility. The use of MIC could be avoided by changing the order of synthesis, as illustrated in the alternative route reported in Figure 1.9 [67].

Avoiding the use of MIC removes a significant amount of the hazard associated with the process and as a result gives an inherently safer process. However, the alternative process still uses phosgene, which is extremely toxic. A process that avoids phosgene would provide further intrinsic safety.

1.2.6
Replacement of Hazardous Chemicals: the Case of DMC

Dimethyl carbonate (DMC) is a versatile compound that is an attractive alternative to phosgene [68, 69] and which could be synthesized in a eco-friendly process by catalytic oxidative carbonylation of methanol with oxygen (Enichem, Italy [70] and

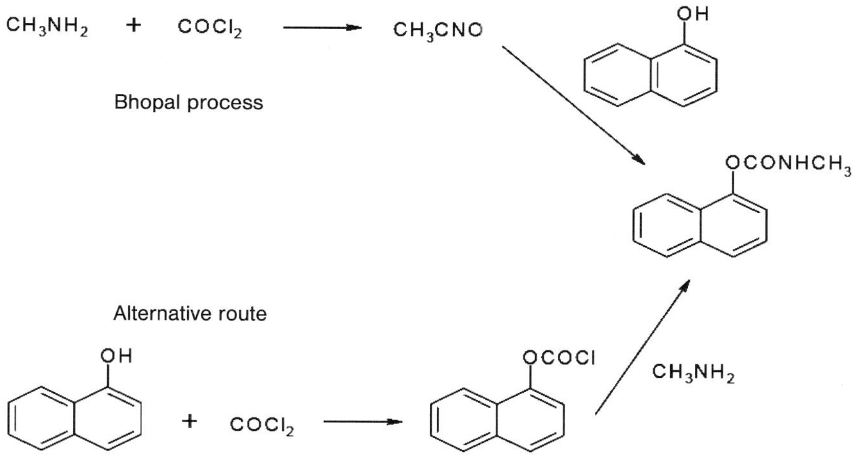

Figure 1.9 Bhopal and alternative routes to *N*-methyl 2-naphthyl carbamate [67].

UBE, Japan [71] processes). DMC can act as a substitute for phosgene and toxic methylating agents like dimethyl sulfate and methyl chloride, and can also find use as a safe solvent and as a possible emission-reducing, high-oxygen-containing additive for fuels [69, 72]. The topic of phosgene substitution is of high relevance because its high reactivity has been known since the beginning of the chemical industry. However, owing to the high reactivity and toxicity its utilization is increasingly burdened by growing safety measures to be adopted during the production, transportation, storage and use and by growing waste disposal costs to be faced. The reasons for phosgene substitution stem not only from considerations relating to its high toxicity, but also to the fact that its production and use involves chlorine as a raw material and results in the generation of large amounts of halogenated by-products. The formation of HCl and chlorine salts as by-products gives rise to contaminated aqueous streams that are difficult to dispose of, or the credited value of the by-produced HCl may render uneconomical its recovery, purification and re-use. Moreover, reactions involving phosgene often require the use of halogenated solvents, like CH_2Cl_2 and chloro- or *o*-dichlorobenzene, which are likely to raise environmental problems.

Phosgene ranks highly among the industrially produced chemicals. Although its production output is most exclusively captive and, therefore, only approximate production statistics are available, a yearly worldwide production of about $5–6t\,yr^{-1}$ can be estimated. The main uses of phosgene are in the production of isocyanates and polycarbonates. The production of isocyanates represents the major output. Of exceeding importance is the production of di- and polyisocyanates, both as commodities, like TDI (toluene diisocyanate) and MDI (diphenylethane diisocyanate), and as specialties, like the aliphatic isocyanates (HDI – 1,6-hexane diisocyanate, IPDI – isophorone diisocyanate and HMDI – dicyclohexane diisocyanate), for the production of polyurethanes. Monofunctional isocyanates like methyl isocyanate, cyclohexyl isocyanate and aryl or phenyl isocyanate are used in lower amounts for agrochemicals and pharmaceutical products.

The production of polycarbonates, mostly the aromatic polycarbonates derived from bisphenol A, is the second largest area of phosgene usage and is probably the most important growing area. It accounts for about $1.5 \, t \, yr^{-1}$ of polycarbonates, corresponding to a yearly phosgene consumption over $0.6 \, t \, y^{-1}$.

DMC is classified as a non-toxic and environmentally compatible chemical [69]. In addition, the photochemical ozone creation potential of DMC is the lowest among common VOCs (2.5; ethylene = 100). The areas in which DMC acts, or can act, as a potential phosgene substitute correspond to the main areas of phosgene industrial exploitation, that is, production of aromatic polycarbonates and isocyanates, leading the production of these important chemicals out of the chlorine cycle.

The traditional production of DMC involves phosgene:

$$COCl_2 + 2CH_3OH \rightarrow (CH_3O)_2CO + 2HCl \tag{1.4}$$

Therefore, any possible use of DMC as substitute of phosgene should be based on a different synthesis of DMC, not involving phosgene. Non-phosgene alternative routes for DMC production, basically, have relied on the reaction of methanol with carbon monoxide (oxidative carbonylation) or with carbon dioxide (direct carboxylation with CO_2, or indirect carboxylation, using urea or alkylene carbonates as CO_2 carriers) (Figure 1.10) [72].

Oxidative carbonylation of methanol to DMC, which takes place in the presence of suitable catalysts, has been developed industrially by EniChem (later Polimeri Europa). Carbonylation/transesterification of ethylene oxide to DMC via ethylene carbonate is also an attractive route. However, this route is burdened by the complexity of the two-step process, the co-production of ethylene glycol (even if it

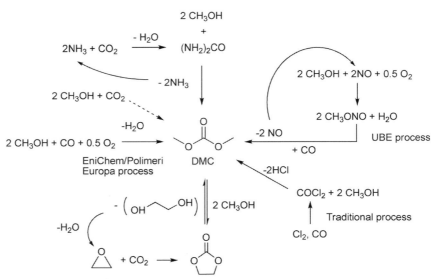

Figure 1.10 DMC synthesis routes.

could be recycled) and the use of toxic and risky ethylene oxide, which is presented as carbon-friendly because it allows the use of CO_2 instead of CO.

The UBE process was also developed on a commercial scale, in Japan, and uses methyl nitrite as intermediate for the gas-phase palladium-catalyzed ($PdCl_2/CuCl_2$ on active carbon) carbonylation to DMC.

Copper compounds, besides being the most widely used co-catalysts for palladium re-oxidation, are themselves active in DMC formation. Exploiting the catalytic properties of CuCl, EniChem developed its DMC production process of one-step oxy-carbonylation of methanol. This process has operated industrially since 1983. The single step is carried out in the liquid phase in a continuous reactor fed with CH_3OH, CO and O_2. Reaction conditions are in the range of 120–140 °C and 2–4 MPa. The $CO:O_2$ ratio is kept outside the explosion limits by the use of a large excess of CO and the adopted high oxygen conversion per pass. As depicted in Figure 1.11, the reactor-evaporator concept is adopted: the catalyst is kept inside the reactor, where the products are vaporized, mainly taking advantage of the heat of reaction ($\Delta H_r = -74\,\text{kcal mol}^{-1}$), and removed from the reaction system together with the excess gas leaving the reactor [73]. This design allows the use of high catalyst concentrations and simplifies catalyst separation from the products. Quite high DMC productivity (up to $250\,\text{g L}^{-1}\text{h}^{-1}$) is achieved under optimized reaction conditions.

The use of CuCl as a catalyst affords minimization of by-products, high purity of the product and practically endless catalyst life. The only co-products are water and CO_2, which are produced in substantial amounts. By adopting a suitable process, the co-produced CO_2 can be re-utilized as a carbon source in the CO generation. All these features characterize the presented DMC production process as a clean technology.

Since a halide free, non-corrosive catalyst for DMC production would be a further process improvement, alternative catalytic systems have been investigated. Cobalt(II) complexes with N,O ligands, such as carboxylates, acetylacetonates and Schiff bases, have been shown to produce DMC with a good reaction rate and selectivity [74].

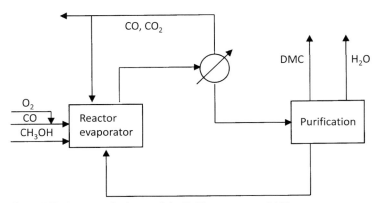

Figure 1.11 Conceptual scheme of the EniChem one-step DMC production process. Source: Rivetti [72].

Table 1.5 Comparison between DMC- and phosgene- or dimethyl sulfate (DMS)-based reactions. Source: Tundo [75].

Phosgene or DMS	DMC
Dangerous reagent	Harmless reagent
Use of solvent	No solvent
Waste water treatment	No waste water
NaOH consumption	The base is catalytic
By-products: NaCl, Na$_2$SO$_4$	By-products: MeOH, CO$_2$
Exothermic	Slightly or not exothermic

DMC is thus considered a prototype example of a green reagent, since it is nontoxic, made by a clean process, is biodegradable and it reacts in the presence of a catalytic amount of base, thereby avoiding the formation of undesirable inorganic salts as by-products [75–77]. Table 1.5 shows the major environmental benefits of DMC-based procedures. DMC Enichem/Polimeri Europa technology has been licensed to (i) General Electric Japan for a DMC/DPC unit at Chiba (DMC unit was 11.7 kt yr^{-1} capacity and started up in 1993) and (ii) General Electric España for a DMC/DPC unit at Cartagena (DMC unit was 48.3 kt yr^{-1} capacity and started up in 1998; in 2004 the total capacity has been increased to 96.6 kt yr^{-1} with the start up of a second unit).

DMC has been proven to perform advantageously as a substitute for phosgene in several reactions. A non-phosgene process for the melt polymerization production of aromatic polycarbonates has been established commercially [69, 72]:

(1.5)

This process also avoids the use of methylene chloride as a solvent and the co-production of NaCl salt. Another well-established application of DMC in the field of polycarbonates relates to the production of poly[diethyleneglycol bis(allylcarbonate)], a thermosetting resin used in the production of optical glasses and lenses. The non-phosgene process involves the intermediate formation of diallyl carbonate from DMC – whereas the traditional process was based on the use of diethyleneglycol bis(chloroformate) that in turn was obtained from phosgene – and allows high flexibility in terms of customer-tailored products.

The non-phosgene production of isocyanates takes place through thermolysis of the corresponding carbamate. The carbamate synthesis may involve several alternative possible ways, such as the reaction of a nitro-compound with CO, or the reaction of an amine with CO and O_2, with urea and alcohol, or with a carbonic ester. Among these routes, the reaction of DMC, or DPC (diphenyl carbonate), with aliphatic amines is a very efficient way to produce carbamates.

A non-phosgene process for the production of methyl isocyanate, starting from methylamine and diphenyl carbonate as raw materials, has been established by EniChem/Polimeri Europa, resulting in the commercialization of two production units in the USA (1988) and China (1994) [78].

Recently, a comparative evaluation of dimethyl carbonate versus methyl iodide, dimethyl sulfate and methanol as methylating agents has been made in terms of green chemistry metrics [79]. These provide a quantitative comparisons based on measurable metrics able to account for several aspects of a given chemical transformation, including: (i) the economic viability, (ii) the global mass flow and the waste products and (iii) the toxicological and eco-toxicological profiles of all the chemical species involved (reagents, solvents, catalysts and products) [80–83].

Figure 1.12 summarizes the result of this comparative evaluation of DMC as green methylating agent. The assessment was based on atom economy (AE) and mass index (MI) for three model transformations: O-methylation of phenol, the mono-C-methylation of phenylacetonitrile and the mono-N-methylation of aniline. In terms of chemical and toxicological properties, DMC shows the lower toxicity index (LD_{50}, e.g., lethal dose for 50% of rats in toxicology experiments) and irritating properties, but costs about twice as much as methanol.

The atom economy (AE, as percentage) was calculated considering the mass-balance of a process related to its stoichiometric equation, that is, the percentage of atoms of the reagent that end up in the product:

$$AE = \frac{MW \text{ product}}{\Sigma MW \text{ of all reagents used}} \times 100$$

where MW is the mass weight in $g\,mol^{-1}$. To include the chemical yield and the selectivity towards the desired product, as well as the mass of all reagents, solvents, catalysts, and so on, used in the examined reactions a more all-encompassing metric, the mass index (MI), could be used. All values are expressed by weight (kg).

$$MI = \frac{\Sigma \text{ reagents} + \text{catalysts} + \text{solvents} + \text{etc.}}{\text{Desired product}}$$

The atom economy generally follows the trend (Figure 1.12a):

$$MeOH \gg DMC \geq DMS > MeI$$

Two factors account for this behavior: (i) for methanol, 47% of its mass is incorporated in the final products, more than twice as much as the other reagents (DMC 16%, DMS 12%, or 24% when both methyl groups are incorporated, MeI 11%) and (ii) methanol and DMC require catalytic base or zeolites, as opposed to DMS and MeI. MeOH and DMC offer similar low values of MI (on average, in the range 3–5.5), which are better than those achievable with DMS and MeI (Figure 1.18b).

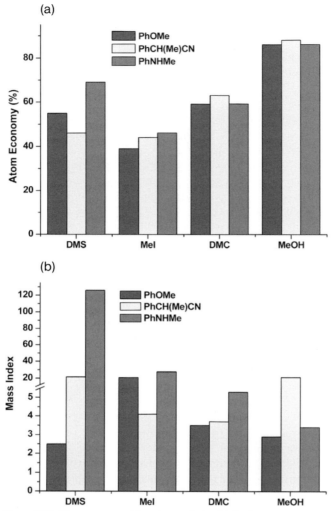

Figure 1.12 Atom economy (a) and mass index (b) for the reaction of phenol, phenylacetonitrile and aniline with different methylating agents. Source: Selva and Perosa [79].

DMC thus yields very favorable mass indexes (in the range 3–6), indicating a significant decrease of the overall flow of materials (reagents, catalysts, solvents, etc.) and thereby providing safer greener catalytic reactions with no waste.

One conclusion that may be derived from these studies is that DMC is an ideal reactant and that no reasons could exist to still use phosgene. As mentioned above, the process of clean DMC production has been commercialized for over 15 years, an apparently long time and long enough for a substantial substitution of phosgene by DMC. It is thus interesting to look at the market for phosgene in comparison with DMC.

The United States, Western Europe and Asia are currently the major producing and consuming regions for phosgene – primarily consumed captively to manufacture *p,p'*-methylene diphenyl diisocyanate (MDI), toluene diisocyanate (TDI) and polycarbonate resins. In 2006, global production/demand was estimated at over 7 million metric tons. Demand for phosgene grew by about 3.25% per year in the period 2001–2006, while it was about 6.4% per year in the period 1997–2002. About 75–80% of global phosgene is consumed for isocyanates, 18% for polycarbonates and about 5% for other fine chemicals. Fine chemical applications are further broken down to 50% for intermediates, 25% for agrochemicals, 20% for pharmaceuticals and 5% for monomers and coloring agents.

By contrast, total global dimethyl carbonate capacity was about $170\,000\,t\,y^{-1}$ in 2002 and output and consumption were both about 90 000 tonnes. Production was concentrated in Western Europe, the USA and Japan and capacity in these regions accounts for about 70% of the total. Consumption was in polycarbonate synthesis (about 50 000 tonnes, accounting for 56.1% of the total), pharmaceutical production (about 20 000 tonnes, accounting for 22.5%), pesticide production (about 7000 tonnes, accounting for 7.9%) and other sectors (about 12 000 tonnes, accounting for 13.5%).

With respect to the global production of polycarbonate, three companies (GE, Cartagena, Spain; Bayer, Antwerp, Belgium; and Asahi Kasei, Taiwan in 2002) use non-phosgene based manufacturing units, with a market share of 12% of the polycarbonate produced by this phosgene-free technology. This market share increased to ca. 20% in 2007.

We may thus conclude that more than 15 years after the introduction of clean processes of DMC synthesis and the large amount of advertising by the scientific community, which still continues, about the use of DMC as clean and safer reactant as a replacement for phosgene, the market penetration of DMC is still quite limited. The reasons for this are several, some general, as discussed in the following section, and some more specific. These include, first, the already noted observation that a safer use of phosgene is possible. There are two main options:

- **On-demand (or on-site) production**. Over 99% of produced phosgene is not transported and is consumed on-site to avoid risk of transport. New legislations limit the amount of phosgene that can be stored on-site, and on-demand production is spreading. Davy Process Technology – DPT (Switzerland) offers modular phosgene generators with production ranging from 3 to $10\,000\,kg\,h^{-1}$ [55]. These modular generators produce phosgene from CO and Cl_2 over carbon-supported catalysts (carbon itself is active or may by doped with 0.1–2% of active metal, in particular to reduce the formation of CCl_4 side product to less that 150 ppm). Figure 1.13 shows the process flow of the phosgene generation section of a phosgene generator from DPT [55]. It consists of two sections, a phosgene generator (Figure 1.13) and a safety absorption module. Note that for safety the reactors are located in a secondary containment and all the lines and systems could be vented with an inert gas. Novartis Crop Protection Inc. (Monthey, Switzerland) has developed an intrinsically safe equipment for the on-demand manufacture of phosgene [84]. Furthermore, confinement in a double envelope of the phosgene

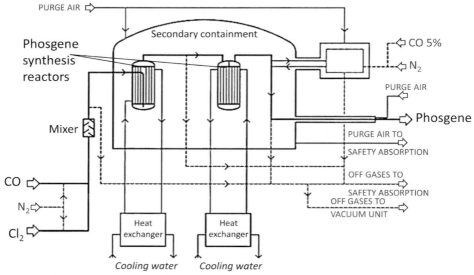

Figure 1.13 Process flow of the phosgene generation section of a phosgene generator of Davy Process Technology. Source: adapted from Cotarca and Eckert [55].

production, supply and utilization equipment makes it possible to collect any leakage with ultimate destruction of the phosgene in specific installations. Chemical Design, Inc. (US) (http://www.chemicaldesign.com/Phosgene.htm) has also designed and built phosgene plants ranging from 0.5 to over 160 tons per day of high purity phosgene. State-of-the-art bellow seal valves are used to virtually eliminate emissions. They use a compact, skid mounted phosgene reactor design that allows the entire reaction system to be installed inside a controlled building that acts as secondary containment. The complete plant allows phosgene to be safely produced on-site, on demand, thereby eliminating transportation and storage concerns.

- **Use of a safer phosgene source**. Triphosgene is used as a phosgene source. It may be used in pre-packaged cartridge for on-demand production of phosgene by triphosgene catalytic depolymerization. Laboratory generators for on-demand production of phosgene are available. In cooperation with Buss ChemTech, Sigma-Aldrich offers a safe and reliable phosgene generation kit, giving simple access to small quantities of high purity, gaseous phosgene exactly when needed, while no transport and storage of liquid phosgene is necessary. The generator converts safe triphosgene into phosgene on demand using a patented catalyst [85]. Phosgene generation can be stopped at any time. A total containment approach eliminates the risk that phosgene can reach the environment.

Phosgene substitution is thus an emblematic case for sustainable industrial chemistry and how this question should be considered in view of a rational risk assessment more than on generic principles. Phosgene is still central to the chemistry of pharmaceutical, polyurethanes and polycarbonates. This is a very large market and still about eight million tons of phosgene are used industrially worldwide. New uses have also been discovered, from the synthesis of high purity synthetic diamonds to the production of the nutritive sweetener aspartame; it was also used as fuel in molecular motors. The book *Phosgenations – A Handbook* [55] discusses in detail novel and old uses of phosgene (see Sections 1.4 and 1.5, in particular).

However, phosgene is clearly highly toxic (threshold limit value – TLV of 0.1 ppm), but acrolein, for example, has the same TLV and is produced in quantities of several millions worldwide. Acrolein is also produced at barbecue parties by roasting foods, without provoking health alarms. Clearly, a low TLV implies the adoption of special safety procedures and limited storage. On-demand production and other safety procedures, such as those discussed above, are the solution to minimizing the risk to a sustainable level.

The substitution of phosgene with other chemicals (DMC, in particular) should be thus weighted between intrinsic (yield, reactivity, handling, work-up) and extrinsic (safety, toxicity, environmental impact) criteria [55]. Modern technologies for industrial chemical production allow proper and safe operation with toxic chemicals. Their sustainability (or green content) in terms of effective impact on environment and safety of workers is not necessarily lower than their substitution. It is only a problem of economy. More toxic feedstocks means higher costs in safety devices and thus alternative, less toxic reactants are implemented when the overall cost is lower. However, in terms of risk assessment, a more toxic reactant could be equally used, when the appropriate measures are adopted. This is why phosgene is still largely used.

The example of Bhopal teaches that the background for the accident is a poor design (actual processes no longer have this problem) but the reason is a poor management. The human factor is a critical element for all industries, but particularly for chemical ones. It is wrong to give the image of a clean and safe ("green") chemistry in opposition to the "red" and/or "black" chemistry cited in Section 1.2.2. All chemical production should be sustainable, but unavoidably they handle risky substances, not only in terms of intrinsic toxicity but also of explosive/flammable nature and so on. There are proper procedures and technologies to operate safely, and their progress is in continuous evolution. However, these need skilled operators, high-quality management, continuous maintenance and training. Under these conditions risk could be sustainable, and therefore exporting chemical production to regions of cheap labor could be a problem, owing to weakness in education and sensibility to risk.

1.2.7
Final Remarks on Sustainable Risk

We end this section with a further question arising from the discussion in the previous section. Should the slow substitution of phosgene by DMC be associated only with the slow turnover of chemical processes? No. There are various examples of faster changes of technology even in large-scale chemical productions, while in small-scale applications, for example, production of fine chemicals turnover is typically fast. The problem is that the substitution of chemical processes by more sustainable ones is a complex issue. It clearly involves a techno-feasibility and environmental assessment. The use of toxic reactants implies the use of special safety procedures and monitoring devices, and the production of waste, particularly that which is difficult to treat, determines additional costs of raw materials and disposal treatments. They are thus an integral part of a techno-economic assessment.

The problem is the boundary limit of the techno-economical assessment, for example, which costs are effectively considered (see also later discussion regarding life-cycle assessment). This is a moving boundary that should be determined from the best-available-technology (BAT) and the related legislative limits on emissions. However, a more advanced concept is to consider the chemical process as a component of the environment and set the local legislative limits on emissions to values that do not decrease the biodiversity in the specific area where the process is localized. There are many problems in implementing this concept which introduces the idea that emissions from chemical production (and in general from all industrial and human activities) should have a value connected to the capacity of the environment to sustain the life (biodiversity).

However, global competiveness and market should be considered. This is the real barrier for a more sustainable industrial chemistry. Therefore, the real step forward could derive only when the product value of chemical compounds also includes components related to the process of production, its impact on environment and safety of operations. We live in a global world, not only economically but also environmentally, where the impact of industrial production and human activities is no longer on a local scale. The value of products should thus not be related only to the "local" cost of production, which hides part of the effective costs (for the environment, for example, or for society, when risk is too high). It is thus necessary to adopt transparent procedures where cost is not only determined from the market but also includes the production procedures. This concept of traceability of chemical products is one of the concepts around which the new REACH legislation was built. The next section discusses in more detail this legislation, because it is an important component for the sustainability of chemistry and industrial chemical production not only inside Europe.

1.3
Sustainable Chemical Production and REACH

REACH is a new European Community Regulation on chemicals and their safe use (EC 1907/2006) (http://www.reachlegislation.com, http://ec.europa.eu/env-ironment/chemicals/reach/reach_intro.htm). It deals with the Registration, Evaluation, Authorisation and Restriction of Chemical substances. The new law entered into force on 1 June 2007. The aim of REACH is to improve the protection of human health and the environment through the better and earlier identification of the intrinsic properties of chemical substances. At the same time, the innovative capability and competitiveness of the chemicals industry should be enhanced. The

Table 1.6 REACH timeline.

Date	Action
June 2007	New law entered into force
June 2008	European Chemicals Agency to become operational
2008–2010	The first phase of registrations. This will apply to substances supplied at 1000 tonne or more, as well as some other priority high-risk substances
By 2013	The second phase of registrations, to be completed 6 years after REACH comes into force, and will apply to substances supplied at 100 tonne or more
By 2018	The final phase of registrations for substances supplied at 1 tonne or more

benefits of the REACH system will arrive gradually, as more and more substances are phased into REACH (Table 1.6).

Under REACH, manufacturers, importers and downstream users are required to demonstrate that the manufacture/import/use of a substance does not adversely affect human health and that risks are adequately controlled. Information on chemical properties and safe uses of chemicals will be communicated up and down the supply chain.

The main elements characterizing REACH are the following:

- **Registration**. Each producer and importer of chemicals in volumes of 1 tonne or more per year and per producer/importer – around 30 000 substances – will have to register them with a new EU Chemicals Agency, submitting information on properties, uses and safe ways of handling them. The producers and importers will also have to pass the safety information to "downstream users" – manufacturers that use these chemicals in their production processes – so that they know how to use the substances without creating risks for their workers, the end consumers and the environment.

- **Evaluation**. Through evaluation, public authorities will look in more detail at registration dossiers and at substances of concern. They can request more information if necessary. At this stage, they will also scrutinize all proposals for animal testing to limit it to the absolute minimum. REACH makes data sharing on animal test results compulsory and prescribes the use of alternative methods wherever possible.

- **Authorisation**. Use-specific authorisation will be required for chemicals that cause cancer, mutations or problems with reproduction, or that accumulate in our bodies and the environment. Authorisation will be granted only to companies that can show that the risks are adequately controlled or if social and economic benefits outweigh the risks where no suitable alternative substances or technologies are available. This will encourage substitution – the replacement of such dangerous chemicals with safer alternatives.

- **Restrictions**. The use of certain dangerous substances at EU level could be still restricted, but REACH will introduce clearer procedures and allow decisions to be taken more quickly than currently. The provisions for restrictions will act as a safety net.

The REACH Regulation gives greater responsibility to industry to manage the risks from chemicals and to provide safety information on the substances. Manufacturers and importers will be required to gather information on the properties of their chemical substances, which will allow their safe handling, and to register the information in a central database run by the European Chemicals Agency (ECHA, fully operational on June 2008) in Helsinki. The Agency will act as the central point in the REACH system: it will manage the databases necessary to operate the system, co-ordinate the in-depth evaluation of suspicious chemicals and run a public database in which consumers and professionals can find hazard information.

The Regulation also calls for the progressive substitution of the most dangerous chemicals when suitable alternatives have been identified. The main legislative texts of REACH are the following:

- Regulation (EC) No 1907/2006 is the central act of the new European chemicals policy. It is often referred to as the "REACH Regulation."

- Directive 2006/121/EC contains technical adaptations of Directive 67/548/EEC that are necessary in the light of the new REACH Regulation (Directive 67/548 concerns the classification, packaging and labeling of dangerous substances and applies in parallel with REACH).

The latter Directive originates from the need of an internationally-harmonized approach for classification and labeling to ensure the safe use, transport and disposal of chemicals. The new system, which is called "Globally Harmonized System of Classification and Labeling of Chemicals" (GHS, a United Nations system) addresses classification of chemicals by types of hazard and proposes harmonized hazard communication elements, including labels and safety data sheets. It aims to ensure that information on physical hazards and toxicity from chemicals is available to enhance the protection of human health and the environment during the handling, transport and use of these chemicals. GHS also provides a basis for harmonization of rules and regulations on chemicals at national, regional and worldwide level, which is an important factor for trade facilitation. The GHS aims to identify hazardous chemicals, and to inform users about these hazards through standard symbols and phrases on packaging labels and through safety data sheets (SDSs).

1.3.1
How does REACH Works

The REACH legislation is by far the largest legislative project adopted by the EU in recent years. It replaces 40 legislative texts and creates a single EU-wide system for the management of chemicals produced in Europe or imported into Europe.

Before REACH, there was a general lack of knowledge regarding 99% of the chemicals (around 100 000 substances) that were placed on the market before 1981. Prior to that date, no stringent health and safety tests were needed to market chemicals.

REACH is based on the idea that industry itself is best placed to ensure that the chemicals put on the market in the EU do not adversely affect human health or the environment. This requires that industry has certain knowledge of the properties of its substances and manages potential risks. REACH creates a single system for both "existing" and "new" substances; substances are now described as *non-phase-in* substances (i.e., those not produced or marketed prior to the entry into force of REACH) and *phase-in* substances. Its basic elements are described below:

- All substances are covered by the REACH Regulation unless they are explicitly exempted from its scope.
  - REACH is very wide in its scope, covering all substances, whether manufactured, imported, used as intermediates or placed on the market, be that on their own, in preparations or in articles, unless they are radioactive, subjected to customs supervision or are non-isolated intermediates. Waste is specifically exempted. Food that meets the definition of a substance, on its own or in a preparation, will be subject to REACH, but such substances are largely exempted from Registration, Evaluation and Authorisation.

- Manufacturers and importers of chemicals should obtain relevant information on their substances and to use that data to manage them safely.
  - To reduce testing on vertebrate animals, data sharing is required for studies on such animals.
  - For other tests, data sharing is required on request by other registrants.
  - Better information on hazards and risks and how to manage them safely will be passed down and up the supply chain.
  - Downstream users are brought into the system.

- Evaluation is undertaken by the agency for testing proposals made by industry or to check compliance with the registration requirements. The agency co-ordinates substance evaluation by the authorities to investigate chemicals with perceived risks. This assessment may be used later to prepare proposals for restrictions or authorisation.

- Substances with properties of very high concern will be made subject to authorisation; the agency will publish a list containing such candidate substances. Applicants will have to demonstrate that risks associated with uses of these substances are adequately controlled or that the socio-economic benefits of their use outweigh the risks.
  - Applicants must also analyze whether there are safer suitable alternative substances or technologies. If there are, they must prepare substitution plans, if not they should provide information on research and development activities, if appropriate.
  - The European Commission may amend or withdraw any authorisation on review if suitable substitutes become available.

- The restrictions provide a procedure to regulate as to whether the manufacture, placing on the market or use of certain dangerous substances shall be either subject to conditions or prohibited.
  - Restrictions act as a safety net to manage EC wide risks that are otherwise not adequately controlled.
  - A classification and labeling inventory of dangerous substances will help to promote the agreement within industry on the classification of a substance. For some substances of high concern a EC wide harmonization of classification by the authorities should be made.

1.3.2
REACH and Sustainable Industrial Chemistry

The REACH legislation will improve the life of all citizens because it concerns more than 30 000 substances currently used in everyday products. At the same time, REACH will provide important opportunities for European industry. This is true for the chemicals industry itself. But REACH will also provide important opportunities to other industries that look to the chemicals industry as a driver of innovation and as key to resolving critical challenges, such as higher energy efficiency and combating climate change.

There are several beneficial characteristics of REACH for both companies and society [86–88]:

- closure of data gaps, for example, missing knowledge on hazards is no longer an advantage;
- equal data requirements for new and existing chemicals;
- authorisation of the most dangerous chemicals (CMR, PBT, vPvB) keeps them under control;
- substitution of most hazardous substances encouraged;
- improved confidence of consumers by better information on hazards and risks.

REACH plays a relevant role for innovation:

- encourages market entrance of new chemicals by lower data requirements up to a production volume of $10\,t\,yr^{-1}$;
- no competitive advantage for existing substances because of equal requirements;
- use of substances for R & D purposes facilitated;
- improved knowledge about chemicals, allowing better predictability of their risks;
- downstream users receive improved information to help find innovative solutions;
- incentives for manufacturers through improved knowledge about their uses and exposure patterns.

On a long-term view REACH legislation represents the way to introduce traceability of chemical products, for example, to include in the products and their chain of production the information on the production method and their impact on the environment. Notwithstanding the obvious strong resistance and difficulties in implementation, this is the only real possibility to include the cost for environment and society in the production cost and to avoid the excuse of global competitiveness to introduce new cleaner and sustainable chemical processes and technologies. Note that this concept is quite distant from the concepts of "green taxes," which have shown great limits.

Including the method of production and its impact on environment and health on the trading cost of chemical products will be a relevant boost to the introduction of new advanced processes/technologies and thus for innovation and sustainability. The chemical industry is a high-tech sector in which competiveness has long been dominated by the capacity of companies to be innovative. The percentage of annual budget dedicated to R&D was thus nearly twice that of the mean value in the manufacture sector. This picture has changed in the last one–two decades, in parallel with the change of the overall market and the situation of chemical production in the world.

However, new sustainability issues require R&D to be put back at the core of chemical industry. The first element to move in this direction is the development of a longer term vision for the chemical industry, which contrasts the very short term view (few years) planning of several companies and the continuous restructuring of the chemical industry in the last one–two decades. One of the positive signals in this direction is the cited European Technology Platform on Sustainable Chemistry and the related prepared document [6], which plans to define the scenario for R&D for the next 20 years.

REACH is another important step towards sustainability of chemicals, because:

- Information on hazardous properties of chemicals is generated.
- Information upstream and downstream the supply chain will be improved (enhanced dialogue).
- Most hazardous chemicals must be authorised and will be under control.
- Chemicals can only be marketed if it is proven that their identified use is safe.

1.3.3
Safety and Sustainability of Chemicals

A chemical substance is safe for a certain use if it is demonstrated that it poses no risk when taking into consideration risk reduction measures that reduce exposure of man or environment [86–88]. However, this is not enough to indicate that it is also a sustainable chemical. A sustainability target consists of the development of inherently safe chemicals (ISCs), for example, without risks for human health and the environment, even without specific exposure control.

We may distinguish two aspects for ISC: those used at workplaces (in particular, in SMEs), where in principle only substances not classified as dangerous for human health should be used, and those released to the environment, which should be not persistent and bioaccumulative or not persistent and highly mobile.

Table 1.7 Inherent safety of chemicals. Source: adapted from Steinhäuser [86].

Unsustainable	Sustainable
CMR properties	No irreversible and chronic effects
Respiratory sensitizers	Low acute (eco) toxicity
Extreme acute (eco) toxicity	Low persistence
PBTs/vPvBs	No bioaccumulation
High persistence and mobility	Low spatial range

Table 1.7 summarizes some of the characteristics for unsustainable versus sustainable chemical substances. The combination of persistence with bioaccumulation is of great concern, because:

- bioaccumulation enhances probability of toxic effects;
- persistence causes irreversibility of environmental exposure;
- long-term adverse effects are unpredictable.

The combination of persistence with mobility is also of great concern, because:

- mobility enhances probability of exposure to large areas;
- persistence causes irreversibility of environmental exposure;
- long-term adverse effects are unpredictable.

Sustainable chemicals should be short-range chemicals, that is, in which their (i) spatial range is low (mobility), (ii) temporal range is low (persistence) and (iii) effects are not irreversible. Clearly, not all chemicals that should be used can have the above characteristics because, for example, fuels must be inflammable, pesticides must be toxic (even with controlled properties) and reactive reagents must be aggressive/corrosive. The challenge is thus a careful control of the properties ("just needed") and hazardous properties should be not linked to not-necessary functionalities.

However, it is necessary to go beyond inherent safety sustainable chemicals, because they should be produced with low resource demand (energy, feedstock, auxiliaries), high yield and low discharges of sewage and waste. Three steps towards sustainability of chemicals should be evidenced:

1. **Safety:** A chemical is safe if it is demonstrated that it poses no risk, taking into consideration risk reduction measures.
2. **Inherent safety:** Chemicals unlikely to pose a risk for human health and environment – even without specific exposure control – due to the lack of hazardous properties.
3. **Sustainable chemicals:** in addition low energy and resource demand.

Sustainability in chemistry is not only a qualitative target, but also a quantitative one, because chemical production has increased since 1930 from 1 million to 500 million tonnes per year in 2000. As a consequence, measures needed to reduce chemicals' use in various branches are very important. One of these actions consists in selling services instead of substances. "Chemical leasing" may reduce chemical consumption up to 30–50%. Chemical leasing (ChL) is an initiative of the United Nations Industrial Development Organisation (UNIDO) (http://www.-chemicalleasing.com). The idea is to promote sustainable management of chemicals and close the material cycles between suppliers and users of chemicals ("closing the loops").

Traditionally, chemicals are sold to a customer, who becomes the owner of the substances and therefore responsible for its use and disposal. Their suppliers have a clear economic interest in increasing the amount of chemicals sold, which is usually related to a negative release to the environment. Compared to this approach, the concept of ChL is much more service-oriented. In this business model the customer pays for the benefits obtained from the chemical, not for the substance itself. Consequently, the economic success of the supplier is no longer linked with product turnover. Chemical consumption becomes a cost rather than a revenue factor for the chemicals supplier. The supplier will thus try to optimize the use of the chemical and improve the conditions for recycling to reduce the amount consumed, which again reduces the environmental pollution.

Against this background ChL can be seen as a key element of sustainable chemicals management systems. The application of ChL models brings economic advantages for all partners involved, provides concrete solutions for efficient chemicals management and ways to reduce negative releases to the environment. Since chemical products provide a broad variety of services such as "cleaning," "coating," "coloring" and "greasing" the ChL model is applicable in a multitude of industry sectors.

When applying ChL business models, the producer does not just provide the chemical but also his know-how on how to reduce the consumption of chemicals and how to optimize the conditions of use. While in the traditional model the responsibility of the producer ends when the chemical is sold, in ChL business models the producer remains responsible for the chemical during its whole life cycle, including its use and disposal.

1.4
International Chemicals Policy and Sustainability

Sustainable development was originally defined in the Brundtland report "Our Common Future" in 1987 as follows: "Sustainable Development should meet the needs of the present without compromising the ability of future generations to meet their own needs." Sustainable Development balances three principal requirements:

- the needs of society (the social objective);
- the efficient management of scarce resources (the economic objective);
- the need to reduce the load on the eco-system to maintain the natural basis for life (the environmental objective).

Although this original definition has evolved (see, for example, the United Nations Commission on Sustainable Development – CSD: http://www.un.org/esa/desa/

aboutus/dsd.html), and new definitions have been proposed to consider, for example, dynamic models that describe the inter-relationships of environmental, economic and social variable [89], the background concepts remain valid. One of the central aspects is the need for cooperation on a worldwide scale, and for a sustainable industrial chemistry the definition of international chemicals policies is of fundamental importance.

At the United Nations (UNs) level, the commitment to sound management of chemicals was renewed in an action plan on chemicals, agreed by Heads of State at the Johannesburg World Summit on Sustainable Development in September 2002. This plan – welcomed by the global chemical industry – provides that by 2020, chemicals be used and produced in ways that lead to minimization of significant effects on human health and the environment. The approach is based on sound risk assessment and risk management, and follows the precautionary approach. Following this, two initiatives were developed in parallel by UNEP (United Nations Environmental Program): the Marrakech process and the Strategic Approach to International Chemicals Management (SAICM).

From chemical industry side, the commitment to sustainable development has resulted in a wide range of programs and initiatives. The most known are the Responsible Care Global Charter, the Global Product Strategy, the Long Range Research Initiative and the HPV program. These programs and initiatives being global are managed by ICCA (International Council of Chemical Association; http://www.icca-chem.org). ICCA is the worldwide voice of the chemical industry, representing chemical manufacturers and producers all over the world. It accounts for more than 75% of chemical manufacturing operations with a production exceeding US$ 1.6 trillion annually. ICCA has a central role in the exchange of information within the international industry, and in the development of position statements on matters of policy. It is also the main channel of communication between the industry and various international organizations that are concerned with health, environment and trade-related issues, including the United Nations Environment Program (UNEP), the World Trade Organization (WTO) and the Organisation for Economic Co-operation & Development (OECD).

Through the above programs and initiatives the chemical industry is contributing to the SAICM objectives as well as to the goal of the Marrakech process to achieve sustainable consumption and production (SCP). Achieving the objectives of SAICM is essential to improving public confidence in the safe management of chemicals and to further promote the benefits of chemistry. SAICM will provide the framework for future international chemicals management globally and will influence the direction of national regulatory systems until 2020.

Figure 1.14 shows a roadmap of governmental initiatives and regulations and industry programs and initiatives. The Rio and Johannesburg Summits (in 1992 and 2002, respectively) are the Earth Summits organized by United Nations to discuss the growing environmental and development problems facing the planet. In Rio the 178 governments attending the Earth Summit signed up to *Agenda 21*, an ambitious global action plan for achieving sustainable development. This document set out a long-term vision for balancing economic and social needs

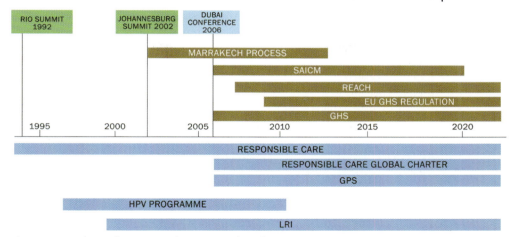

Figure 1.14 Roadmap of governmental initiatives and regulations and industry programs and initiatives. Source: European Federation of Chemical Industry – Cefic – web site: www.cefic.org.

with the capacity of the earth's natural resources. In the immediate aftermath of Rio, governments, NGOs (non-governmental organizations) and other stakeholders joined forces to implement the plan. There was a real belief that global leaders were on their way to tackling issues such as poverty eradication, social injustice and environmental degradation. However, a decade on, it had become clear that the vision and commitment shown at the Rio Summit did not last. While some real progress was made – for instance with the convention on climate change and other national and regional initiatives – many of the actions agreed have been still not implemented. The movement towards a more sustainable world has been slower than many expected.

In 2002, the international community met in Johannesburg to once again take up the challenge of sustainable development. The World Summit on Sustainable Development was one of the largest and most important international gatherings ever held on the subject.

The International Conference on Chemicals Management (Dubai, 2006) was a critical turn-out point for strategies in chemical production. During the conference the chemical industry made a public commitment to enhance chemical safety throughout the value chain. At the same time, the Responsible Care Global Charter and the Global Product Strategy (GPS) were launched. Related programs and initiatives include the Long-range Research Initiative (LRI), the HPV Chemicals program and the SusChem platform.

The Marrakech process is a global process to support regional and national initiatives to promote the shift towards Sustainable Consumption and Production (SCP) patterns. In the European Union, the Marrakech process is leading the development of an EU SCP Action Plan that entails amongst other things the Integrated Product Policy (IPP) and Green Public Procurement (GPP).

The International Conference on Chemicals Management (ICCM) organized by UNEP (United Nation's Environmental Program) in Dubai (2006) led to the "Dubai Declaration on the Strategic Approach to International Chemicals Management (SAICM)." At the same time, the Johannesburg Summit called for technical and financial assistance for developing countries to strengthen their capacity for the sound management of chemicals and hazardous waste, as well as for the national implementation of a new globally harmonized system of classification and labeling of chemicals (GHS).

In the European Union the Marrakech process has led to a renewed European Sustainable Development Strategy, which identifies Sustainable Consumption and Production (SCP) as one of the key objectives to be achieved in the context of the European Union's commitment of sustainable development. SCP aims to "promote sustainable consumption and production by addressing social and economic development within the carrying capacity of ecosystems and decoupling economic growth from environmental degradation."

The two key pillars in the context of SCP Action Plan in Europe are the Green Public Procurement (GPP) and the Integrated Product Policy (IPP). One of the main pillars of European SCP Policy is GPP, which means that "public purchasers take account of environmental factors when buying products or services."

Another key pillar is the Integrated Product Policy (IPP). IPP aims to reduce the environmental impact of products. Relevant initiatives include pilot projects, the creation of an EU Platform on life cycle assessment (LCA), and product prioritization.

The Strategic Approach to International Chemicals Management (SAICM) is a non-binding agreement launched in 2003 and aimed to improve the sound management of chemicals throughout their entire life cycle. More than 100 countries represented by their environment and health ministers have adopted this international approach to stimulate the safe production, transport, storage, use and disposal of chemicals. The objectives of the Global Plan of Action are (i) risk reduction, (ii) knowledge and information, (iii) governance, (iv) capacity-building and technical cooperation, and (v) stopping illegal international traffic.

Worldwide hazard communication of chemicals requires a change from the existing systems to a more harmonized one. This gave rise to the elaboration of a "Globally Harmonized System of Classification and Labeling of Chemicals" (GHS) by the United Nations Economic Commission for Europe (UNECE, 2002). This system proposes harmonized hazard communication elements, including labels and safety data sheets.

Regarding the trade of chemicals, the following conventions should be finally remembered:

- The "Basel Convention," a multilateral agreement to reduce the movement of hazardous wastes between nations, and specifically to prevent transfer of hazardous wastes from developed to less developed countries (LDCs). The convention entered into force in 1992.

- The "Rotterdam Convention," a multilateral agreement that became legally binding on its parties in 2004 to promote shared responsibilities in relation to the international trade of certain hazardous chemicals and to contribute to the environmentally sound use of those chemicals by facilitating exchange of information about their characteristics. The convention creates legally binding obligations for the implementation of the Prior Informed Consent (PIC) procedure.

1.5
Sustainable Chemistry and Inherently Safer Design

Process safety, a discipline that focuses on the prevention of fires, explosions and accidental chemical releases at chemical process facilities, is a key element for a sustainable industrial chemistry, as indicated in the previous sections. There are three key elements for process safety: behavior, system and process.

Thorough and effective analyses of workplace incidents are critical components of a comprehensive safety management system. Yet, many incident analysis processes (i.e., accident investigations) fall short. They frequently fail to identify and resolve the real root causes of injuries, process incidents and near misses. Because the true root causes of incidents are within the system, the system must change to prevent the incident from happening again.

Process safety differs from the traditional approach to accident prevention [90]:

- there is greater concern with accidents that arise out of the technology;
- attention is given to foreseeing hazards and taking action before accidents occur;
- accidents that cause damage to plant and loss of profit but do not injure anyone should be considered in addition to those that do cause injury.

In general, there is greater emphasis on a systematic rather than a trial-and error approach, particularly on methods that identify hazards and estimate their probability and consequences. The term loss prevention can be applied in any industry, but is widely used in the process industries (particularly chemical industries), where it usually means the same as process safety.

Chemical plants, and other industrial facilities, may contain large quantities of hazardous materials. The materials may be hazardous due to toxicity, reactivity, flammability or explosivity. A chemical plant may also contain large amounts of energy – the energy is required either to process the materials or is contained in the materials themselves. An accident occurs when control of this material or energy is lost. An accident is defined as an unplanned event leading to undesired consequences. The consequences might include injury to people, damage to the environment, or loss of inventory and production, or damage to equipment.

The practices to process safety have been progressively changed with time, as briefly shown in Table 1.8. Process safety does not depend only on human errors or faults in equipments but on the whole system management. From the end of the 1970s there has thus been a large effort to develop risk assessment techniques and systematic approaches, as well as suitable strategies for successful process safety management (PSM). In the 1980s, the Bhopal accident pushed the chemical industry

Table 1.8 Process safety milestone practices.

Period	Type approach	Practice
Pre-1930s	Behavior	Identify who caused the loss and punish the guilty
Pre-1970s	Process	Find breakdown in, and fix man–machine interface
1970s, 1980s	Management systems	Development of risk assessment techniques and systematic approaches
1980s +	Comprehensive	Performance-, risk-based standards, regulations; sustainable and inherent designs

towards a further step for a more comprehensive approach, for example, to an inherent safer design.

Inherently safer design of chemical processes involves the use of smaller quantities of hazardous materials, the use of less hazardous materials, the use of alternative reaction routes or process conditions to reduce the risk of runaway exothermic reactions, fires, explosions and/or the generation or release of toxic materials.

Notably, in some cases changes made to improve the environment have resulted in inherently less safe designs. For example, the collection of vent discharge gases for incineration or for absorption on carbon beds has resulted in explosions when the composition of the gases in the vent system has entered the flammable range.

Chemical process safety strategies can be grouped into four categories [91]:

1. **Inherent**: when the safety features are built into the process, not added on; for example, replacement of an oil-based paint in a combustible solvent with a latex paint in a water carrier.

2. **Passive**: for example, safety features that do not require action by any device – they perform their intended function simply because they exist; for example, a blast resistant concrete bunker for an explosives plant, or a containment dike around a hazardous material storage tank.

3. **Active**: for example, safety shutdown systems to prevent accidents (e.g., a high level alarm in a tank shuts automatic feed valves) or to mitigate the effects of accidents (e.g., a sprinkler system to extinguish a fire in a building). Active systems require detection of a hazardous condition and some kind of action to prevent or mitigate the accident. Multiple active elements involve typically a sensor (detect hazardous condition), a logic device (decide what to do) and a control element (implement action).

4. **Procedural**: or operating procedures, for example, operator response to alarms, emergency response procedures, safety rules and standard procedures, training. An example is a confined space entry procedure.

In general, inherent and passive strategies are the most robust and reliable, but elements of all strategies will be required for a comprehensive process safety management program when all hazards of a process and plant are considered.

Table 1.9 Examples of process risk management strategies. Source: adapted from Mannan [90].

Risk management strategy category	Example	Comments
Inherent	An atmospheric pressure reaction using nonvolatile solvents that is incapable of generating any pressure in the event of a runaway reaction	No potential for overpressure
Passive	A reaction capable of generating 22 kPa pressure in case of a runaway, carried out in reactor which may operate up to 36 kPa	The reactor can contain the runaway reaction, but 2 kPa pressure is risky and reactor could fail due to a defect, corrosion, physical damage or other cause
Active	A reaction capable of generating 22 kPa, realized in a reactor with a 1 kPa high-pressure interlock to stop reactant feeds and a properly sized 3 kPa rupture disc discharging to an effluent treatment system	The interlock could fail to stop the reaction in time, and the rupture disk could be plugged or improperly installed, resulting in reactor failure in case of a runaway reaction. The effluent treatment system could fail to prevent a hazardous release
Procedural	The same reactor described in example 3 above, but without the 1 kPa high-pressure interlock. Instead, the operator is instructed to monitor the reactor pressure and stop the reactant feeds if the pressure exceeds 3 kPa.	There is a potential for human error, the operator failing to monitor the reactor pressure, or failing to stop the reactant feeds in time to prevent a runaway reaction

Table 1.9 gives some examples of process risk management strategies. Note, however, that these examples refer only to the categorization of the risk management strategy with respect to the hazard of high pressure due to a runaway reaction. The processes described may involve trade-offs with other risks arising from other hazards. For example, the non-volatile solvent in the first example may be extremely toxic, and the solvent in the remaining examples may be water. Decisions on process design must be based on a thorough evaluation of all the hazards involved.

Table 1.9 refers to a batch chemical reactor as an example. The hazard of concern is a runaway reaction causing high temperature and pressure and potential reactor rupture. The preferable (inherent) approach is to develop a chemistry that is not exothermic, or mildly exothermic, for example, where the maximum adiabatic exothermic temperature is lower than the boiling point of all ingredients and onset temperature of any decomposition or other reactions. In a passive approach, the maximum adiabatic pressure of a reaction is lower than the maximum reactor pressure design. The hazard (pressure) still exists, but is passively contained by the pressure vessel. In an active strategy the maximum adiabatic pressure for 100% reaction is higher than the reactor design pressure, but an active control is present, for example, progressive introduction of the limiting reactant with temperature control to limit potential energy from reaction. In addition, high temperature and pressure interlocks to stop feed and apply emergency cooling are used and emergency relief systems are provided. In the procedural approach, the automatic devices are substituted by a trained operator to observe temperature, stop feeds and apply cooling, if the temperature exceeds critical operating limit.

There are various techniques to achieve classical risk reduction, but generally these approaches to safety are mostly an afterthought in the design. They may use a safety review or process hazards analysis (PHA), such as a hazard and operability study (HAZOP) or a "what if?/checklist study", merely as a project check instead of a preemptive hazards reduction tool. If these studies are carried out at the latter stages of engineering or during construction there is a natural tendency to avoid expensive redesign or rework. In the inherently safer design, elimination or significant reduction of the process hazards occurs during the design by adopting suitable approaches, which fall into the following categories:

- **Minimize**: Significantly reduce the quantity of hazardous material or energy in the system, or eliminate the hazard entirely if possible. It is necessary to use small quantities of hazardous substances or energy in (i) storage, (ii) intermediate storage, (iii) piping and (iv) process equipment, as discussed in the previous sections. The benefits are to reduce the consequence of incident (explosion, fire, toxic material release), and improve the effectiveness and feasibility of other protective systems (e.g. secondary containment, reactor dump or quench systems). Process intensification (see below) is also a way to reach this objective.

- **Substitute**: Replace a hazardous material with a less hazardous substance, or a hazardous chemistry with a less hazardous chemistry. Examples are water-based coatings and paints in place of solvent-based alternatives. They reduce fire hazard, are less toxic, have a better smell and lower VOC (volatile organic compound) emissions, and reduce hazards for end user and also for the manufacturer. Safer use and better sustainability thus go in the same direction. Another example is substitution of chemicals used for refrigeration. Initially, ammonia, light hydrocarbons and sulfur dioxide were used. They were later substituted by inherently safer alternatives, for example, CFCs (chloro-fluoro-carbons). However, in around the 1980s, CFCs were discovered to be active in stratospheric ozone destruction and thus were later banned (Montreal Protocol entered into force in 1989). CFCs were initially substituted by HCFCs, where not all the C-H bonds in alkanes were substituted by C-X bonds (X is an halogen group), but the phasing out of also these chemicals is programmed. New substitutes for hydrofluorocarbons (HCFC) should be thus developed, but their impact should be also minimized both by severe regulations on their disposal and by re-design of refrigerators to minimize the quantity of flammable hydrocarbons. Currently, in home refrigerators as little as 120 grams of hydrocarbon refrigerant is used. This example shows that substitution of chemicals sometimes is not a simple problem. In fact, this is one of the critical points in REACH legislation discussed in the previous section.

Substitution of an hazardous chemical is often an even more complex problem, in particular regarding the trade-off between inherently safer design and sustainable chemistry. Several examples are discussed in subsequent chapters. We thus limit our discussion here to a few aspects. Up until around the 1960s the Reppe process was employed for of synthesis of acrylic esters:

$$CH \equiv CH + CO + ROH \xrightarrow[HCl]{Ni(CO)_4} CH_2 = CHCO_2R \qquad (1.6)$$

It was substituted by the new process of oxidation of propylene to acrylic acid via acrolein using heterogeneous Bi-molybdate based catalysts followed by acid-catalyzed reaction of acrolein with the alcohol:

$$
\begin{aligned}
CH &= CHCH_3 + 1.5\,O_2 \xrightarrow{Solid\ Cat.} CH_2 = CHCHO \\
CH_2 &= CHCHO \xrightarrow{Solid\ Cat.} CH_2 = CHCOOH + H_2O \qquad (1.7) \\
CH_2 &= CHCOOH + ROH \xrightarrow{H^+} CH_2 = CHCO_2R + H_2O
\end{aligned}
$$

Although substitution was motivated by the availability at that time of propylene and lower cost of the process, it was also a significant improvement in terms of safety, because acetylene is flammable and extremely reactive, carbon monoxide is also toxic and flammable, nickel carbonyl catalysts are toxic, environmentally hazardous (heavy metals), and carcinogenic, and anhydrous HCl (used in the reaction) is toxic and corrosive. However, the new process from propylene cannot be considered inherently safer. Hazards are primarily due to the flammability of reactants, corrosivity of the sulfuric acid catalyst for the esterification step (new solid acids have eliminated this hazard, as discussed in subsequent chapters), small amounts of acrolein as a transient intermediate in the oxidation step, and reactivity hazard for the monomer product.

- **Moderate:** Reduce the hazards of a process by handling materials in a less hazardous form, or under less hazardous conditions, for example at lower temperatures and pressures. Dilution is one of the key words. Aqueous ammonia should be used instead of anhydrous NH_3. Aqueous HCl in place of anhydrous HCl. Sulfuric acid in place of oleum. Figure 1.15 shows an example of the relevant effects observed for the concentration of ammonia measured in air as a function of distance from the place of rupture of a tank containing anhydrous or diluted ammonia solution. Less severe processing conditions are also another keyword. The use of improved catalysts is a critical element in reaching this objective and will be discussed extensively in the following chapters.

- **Simplify:** Eliminating unnecessary complexity to make plants more "user friendly" and less prone to human error and incorrect operation. In the previous section we emphasized how an objective of sustainable chemistry is the development of novel solutions to reduce complexity of chemical processes, which also allows a better control and an improvement of safety.

One way to simplify processes is to eliminate equipment, by combining reaction and separation. The use of membranes is discussed in Chapter 4. Another relevant example is reactive distillation. Figure 1.16 compares the traditional methyl acetate process with that based on reactive distillation (Eastman Chemical) [97–99]. Eastman

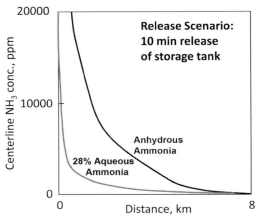

Figure 1.15 Concentration of ammonia measured in air as a function of the distance from the place of the rupture of a tank containing anhydrous or diluted ammonia solution.

Chemical Co.'s methyl acetate reactive distillation process and processes for the synthesis of fuel ethers are classic success stories in reactive distillation. Improvements for the Eastman process are very high: five-times lower investment and five-times lower energy use than the traditional process. However, combining reaction and distillation is not always advantageous and in some cases it may not even be feasible. The methyl acetate process based on reactive distillation has fewer vessels, pumps, flanges, valves, piping and instruments. This is an advantage also in terms of safety and maintenance. However, a reactive distillation column itself is more complex (multiple unit operations occur within one vessel) and thus more difficult to control and operate. It is thus not possible to make unique conclusions.

The concept of inherently safer design was first proposed by Kletz, who developed a set of specific design principles for the chemical industry [92] (see also [50]), but it has been publicized and promoted later by many technologists from petrochemical and chemical companies such as Dow, Rohm and Haas, ExxonMobil, and many others. A relevant source of information is the book *Inherently Safer Chemical Processes: A Life Cycle Approach* [49].

For inherent safety, while prevention, detection and mitigation are all considered, the emphasis should be on prevention. For example, moving the proposed location of a flammable liquid storage tank away from a public fence line may greatly reduce the consequences of a release and may reduce or eliminate the costs of providing the added protection system required if it is not. Inherent safety includes the consideration of more than just design features of a process. Inherent safety principles include human factors, in particular those related to the design and operating conditions. Finding an error-likely situation, such as controls being too difficult to access or too complicated, and working to reduce the clutter and confusion or to improve the accessibility to reduce the chance of a human error is an example of inherent safety in action.

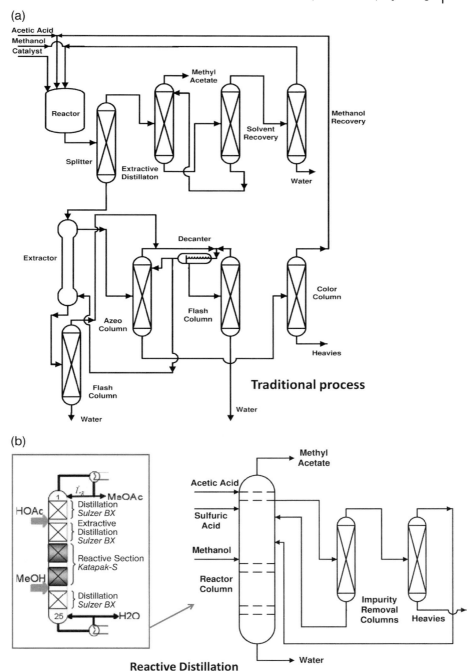

Figure 1.16 Traditional (a) versus reactive (b) distillation process for the synthesis of methyl acetate.

Inherently safer design includes the following strategies and approaches:

1. **Hazard elimination**
   - Concept – eliminate hazards as a first priority rather than accepting them and mitigating them as a risk reduction strategy once they exist.
   - Potential methods:
     - eliminate the hazardous materials;
     - substitute with a non-hazardous material;
     - discontinue the operations.

2. **Consequence reduction**
   - Concept – where hazards cannot be completely eliminated, find less hazardous solutions to accomplish the same design objective by focusing on the consequences.
   - Potential methods:
     - reduce the quantity of the hazardous material;
     - provide a curbed area with a drain to contain and evacuate a spill and produce a smaller pool area of a spill;
     - separate the operation by adequate spacing to reduce exposure to adjacent operations and personnel.

3. **Likelihood reduction**
   - Concept – where hazards cannot be completely eliminated and after consideration of consequence reduction, consider ways to reduce the likelihood of events occurring.
   - Potential methods:
     - reduce the potential for human error through simplicity of design;
     - control ignition sources;
     - provide redundant alarms.

Inherently safer design is thus an intrinsic part of the sustainable chemistry and engineering [93]. It focuses on "safe" accidents, for example, immediate consequences of single events (fires, explosions, immediate effects of toxic material release) and includes considerations of chemistry as well as engineering issues such as siting, transportation and detailed equipment design. Several methodologies are relevant for an inherently safer design and implementation of the above strategies [92–96]:

- **Intensification:** The most widely used method of an inherently safer design is intensification. This involves the use of minimal amounts of hazardous materials so that a major emergency is not created even if all the plant contents are released. For example, hazardous reactants, such as phosgene, are often generated as required in an adjacent plant so that the actual amount in the pipeline is kept to an absolute minimum. Intensified designs are also available for reactors, liquid–vapor contacting equipment, heat exchangers, mixers, scrubbers, dryers, heat pumps and so on. "Inherently safer" designed equipment is smaller than conventional equipment, often cheaper, as well as being safer because less add-on protective equipment is needed.

- **Substitution:** If intensification cannot be achieved an alternative is substitution with safer materials. Nonflammable or less flammable, less toxic solvents, refrigerants or heat transfer materials should be used instead of flammable or toxic ones. For example, some ethylene oxide plants use hundreds of tonnes of boiling paraffin to cool reaction tubes and this represents a bigger hazard than the mixture of ethylene and oxygen in the tubes. Modern plants now use water for cooling instead of paraffin. Care must be taken to avoid the introduction of new hazards and risks as a result of substitution. For example, to protect the environment and prevent damage to the ozone layer, chlorofluorocarbon refrigerants are being replaced by liquefied petroleum gas and ammonia. However, this change can cause additional fire, health and safety risks, if not properly controlled.

- **Alternative reaction routes:** In addition to safer chemicals, it is also possible to reduce the risks associated with manufacture introducing changes in the reaction routes. We discussed extensively this case above with reference to the Bhopal accident. For the design of alternative reaction routes it is important to consider, in turn, reactants, catalysts, solvents, intermediates and the compatibility of all materials used. For example, in a particular process, acetone was used as a solvent. However, due to the heat of reaction, the uncontrolled addition of one of the reactants, or the loss of cooling, could led to a vigorous boiling of the mixture. The consequence is the possible over-pressurization of the reactor and loss of containment. The simple replacement of acetone with toluene, which has a higher boiling point, eliminated this hazard.

- **Modified storage arrangements:** If the above alternatives to achieve inherently safer processes cannot be applied and large quantities of hazardous material are still needed, then it should be handled in the least hazardous form or in minimum quantities. Consequently, large quantities of ammonia, chlorine and liquefied petroleum gas are now usually stored as refrigerated liquids, at low pressure below their boiling point rather than under pressure at atmospheric temperature. If a leak occurs in such circumstances the driving force is low and the evaporation rate is comparatively small. The inventories of toxic or flammable materials that are not manufactured on site can be reduced significantly from hundreds of tonnes to tens of tonnes if reliable and regular supplies can be delivered, perhaps on a daily basis or on a just-in-time basis. In these circumstances a release of such materials, although still involving a comparatively large quantity, reduces the potential to cause injury or damage. This approach does not necessarily involve modifications of existing plants, though smaller storage tanks would prevent inadvertent increases in the inventories in the future.

- **Energy limitation:** Consideration should also be given to limit the amount of energy available in the manufacturing process. For example, it is better to prevent overheating by limiting the temperature of the heat exchange fluid than to rely on interlocks that may fail or be disconnected.

- **Simplicity:** The final method to achieve inherently safer processes is to consider simplicity. Simple plants are inherently safer, because less equipment can fail and fewer opportunities for human error exist.

To develop inherently safer processes along the lines reported above, several formal review procedures are available that can be applied to a new chemical process or during the review of an existing process. These procedures include hazard and operability (HAZOP) studies and life cycle assessment (LCA) Studies. HAZOP studies are normally carried out late in the design when detailed diagrams are available and it is then too late to make significant changes. However, similar detailed studies should be made at the beginning of a project when decisions are made about products, routes, and so on. Figure 1.17 gives a flow diagram identifying the stages where inherently safety issues could be addressed within the process life cycle, starting with conceptual research and development through plant decommissioning and environmental fate.

It is at the conceptual stage in the process development where inherent chemical and process safety benefits can be best introduced. It becomes progressively more difficult to achieve such benefits in the later stages of process development.

In conclusion, inherent safety is achievable through safer processes, improved engineering and stringent safety procedures. Inherent safety will be cost-effective in the medium and long terms, even if higher costs may be incurred in the short term. It should be pursued to safeguard the health and safety of all involved, including consumers, and to protect the environment.

1.6
A Vision and Roadmap for Sustainability Through Chemistry

To realize and implement a sustainable industrial chemistry requires not only the development of suitable methodologies and tools, and a legislative and social framework, as shown in this chapter, but also a vision and roadmap for sustainability through chemistry that is shared between all the stakeholders, from companies and research institutions to NGOs and civil/environmental associations. In fact, while industry has made great progress in adopting processes and chemistry with sustainable characteristics, a further decisive step towards a true sustainability requires long-term planning of R&D activities, for example, to establish a roadmap that can overcome the typical short-term R&D planning that has characterized the last decade. In fact, future progress requires technical and political challenges to be addressed. Industrial biotechnology appears to be a particularly fruitful area for building industrial sustainability, but new problems are posed. Radical improvements can also be achieved by new process design, including new reactor configurations, integration of operations both within and between enterprises and through a focus on recycling and reusing materials [100].

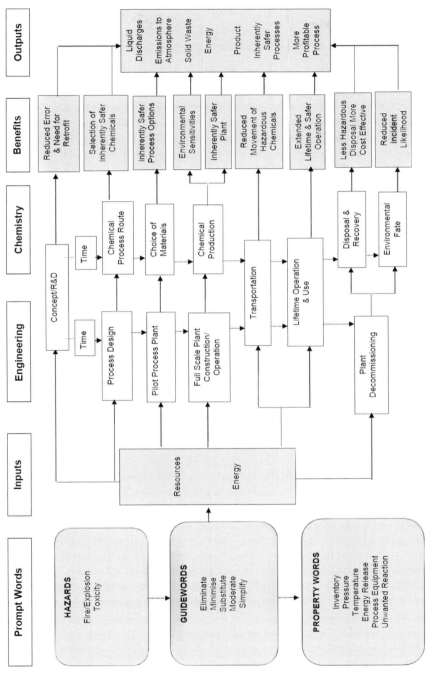

Figure 1.17 Inherent safety life cycle. Source: Wrightson [94].

Many barriers exist to limit the implementation of radical improvements to industrial performance: incremental debottlenecking of plant and expansion is preferred in times of slow growth and a 'first-of-a-kind' process faces a high investment hurdle. The economic and regulatory environment needs to support investment in new technologies that demonstrates sustainable advantage but carry also higher technical risk than conventional operations.

To overcome these barriers and foster innovation and sustainability through chemistry requires a more synergetic integration of fundamental and applied research, and to effectively coordinate the effort along common priorities. One of the problems affecting the chemical industry in the last one–two decades has been the continuous re-organization and short-term view of the objectives that has caused a progressive worsening of the innovation index of the chemical industry; the index is defined as the number of new products/processes introduced on the market versus the number of products/processes already presented for more than five years. At the same time, new problems and issues have arisen: the increase in the cost of energy and raw materials, the double-digit economic growth in Asia and Latino/ America, the increasing social pressure for sustainability, the need to use biore-sources, and so on. It is thus necessary to reconsider the market strategies for development in chemical companies and put again innovation at the heart of chemical production. Defining a roadmap and an alliance to implement this roadmap is the first step in this direction.

These were the motivations for developing in Europe a Technology Platform for Sustainable Chemistry, which was promoted by Cefic (European Chemical Industry Council, http://www.cefic.be) and EuropaBio (The European Association for Bioin-dustries, http://www.europabio.org), but which involved also the active participation of many researchers and managers from academy, companies and other public or private organizations. The results of this effort are three documents: the initial Vision document, the Strategic Research Agenda (SRA) and the Implementation Action Plan (IAP) [6]. Although a roadmap exercise has also been made in other countries, these documents have probably seen the largest active participation in their prep-aration. They represents thus a genuinely shared effort to define priorities and planning actions for R&D in chemistry. Notwithstanding some limits and possible disagreement on some specific aspects, it is useful to summarize here the main aspects of these documents, as a good and reliable basis to define the needs and priorities to develop an industrial sustainable chemistry.

The original structure of the Technology Platform for Sustainable Chemistry (SusChem, http://www.suschem.org), which is then reflected in the cited document, was an organization in three main technology-oriented sections: (i) Industrial Biotechnology, (ii) Materials Technology and (iii) Reaction & Process Design. In addition, a horizontal section was also present. The actual structure has changed, reflecting the need for reorganization of objectives after completing the preparation of the cited documents.

While the SRA document was structured along the three sections indicated above, the IAP document was instead organized along eight areas (bio-based economy, energy, healthcare, information and communication technologies, nanotechnology, sustainable quality of life, sustainable product and process design, and transport) which reflected the structure of the seventh European Framework Program (http://cordis.europa.eu/fp7/home_en.html) in support of R&D in Europe. The SRA describes which science questions need to be answered to accomplish the vision, will define themes for future research and will contain technology roadmaps for those thematic areas aimed at providing 'fertile ground' for subsequent commercial development and exploitation in Europe. The IAP defines how the research themes, as identified in the SRA, are to be implemented and how the innovation framework conditions in Europe need to be altered to enable or accelerate innovation to directly promote the competitiveness of the EU chemical industry and optimize the benefits for all stakeholders.

The IAP also contains an introduction to three visionary projects that will practically demonstrate the benefits and impact on daily life of the sustainable industrial chemistry: (i) the Smart Energy Home, (ii) the Integrated Biorefinery, and (iii) the F^3 Factory; the latter has already been discussed in this chapter.

For conciseness, it is not possible to discuss in depth all the research themes and priorities identified, and the issues that need to be addressed in the short-, medium- and long-term to realize the full potential of research and innovation. We thus outline here only the general aspects, with references given to cited documents for details.

1.6.1
Bio-Based Economy

Industrial biotechnology is a key technology to realizing the knowledge-based bioeconomy by transforming the knowledge of life sciences into new, sustainable, eco-efficient and competitive products. This includes an optimized combination of the biotechnology processes with classical and new biochemical processes – especially in the chemical, materials and biofuels sectors. The following three topics have been identified as being of major importance to facilitate the development of a bio-based economy and industrial biotechnology:

1. Biocatalysis – novel and improved enzymes and processes. Biocatalysis focuses on two aspects: (i) the discovery and improvement of novel selective biocatalysts suitable for industrial use and (ii) the development of a systematic process design technology for a quick and reliable selection of new and clean high-performance manufacturing process configurations. The aim of research in biocatalysis is to: (i) employ nature's toolkit to enable cleaner, safer and more cost-efficient processes, (ii) address the increasing need for selectivity, stability and efficiency using enzymes as catalysts, (iii) enable novel chemo-enzymatic processes through the discovery, evolution and/or design of enzymes and (iv) solve reaction and process problems through the search for novel biocatalytic functions and the selection of new high-performance process configurations. We may cite that

Table 1.10 Case studies that demonstrate sustainability. Source: adapted from Kamm *et al.* [101].

Selected case studies	Environmental impact			Economic
	Energy efficiency	Raw materials	CO$_2$ emissions	Production costs
Vitamin B2 (BASF)	+	++	+	+
Antibiotic cephalexin (DSM)	++	++	+	+
Scouring enzyme (Novozymes)	+	+	0	+
NatureWorks (Cargill Dow)	+	++	++	0
Sorona (DuPont)	+	++	+	+

several environmental studies in which the impact of replacing chemical synthesis with biotech routes has demonstrated the benefits of industrial biotechnology [101]. In particular, two reports – one authored by the OECD [102], which includes 21 case studies on the impact of biotechnology on the environment, the other by a consortium of companies, industry associations, the Öko-Institut, and McKinsey [103] – have demonstrated clearly that industrial biotech can help to create jobs, boost profits and benefit the environment (Table 1.10). The German chemical company BASF was able to adopt biotech processes to transform the production of vitamin B2. Traditionally, its synthesis requires a complicated eight-step chemical process, but biotech reduces it to just one step. Soy oil is fed to a mould and vitamin B2 is recovered as yellow crystals directly from the fermentation process. This has cut production costs by 40% and reduced CO$_2$ emissions by 30% and waste by 95%. The antibiotic cephalexin has been produced on an industrial scale by the Dutch chemicals firm DSM for several years. Metabolic pathway engineering helps to establish a bio-route that reduce substantially the number of steps needed in the process. The biotech process uses 65% less energy, 65% less input chemicals, is water-based and generates less waste. In total, the variable costs of the process decreases to nearly half. Novozymes, a Danish biotech company, produces enzymes for the scouring process in the textile industry. Scouring, which removes the brown, non-cellulose parts of cotton, traditionally requires a harsh alkaline chemical solution. Use of enzymes not only reduces discharges into the water by 60% but reduces also energy costs by a quarter. Environmental and economic benefits go hand in hand: the new process is also 20% cheaper than the chemical treatment. Cargill Dow's bio-polymer PLA made from corn requires 25 to 55% fewer fossil resources than the conventional polymers against which it competes. With the help of biomass and potentially other forms of renewable energy for processing, the joint venture between Cargill and Dow Chemical believes PLA could even become a net carbon sink. In the near future DuPont's Sorona polymer will be based on propanediol (PDO) produced by fermentation, in collaboration with Tate and Lyle. This is estimated to reduce greenhouse gas emissions by approximately 40%.

2. Developing the next generation of high efficiency fermentation processes, including novel and improved production of microorganisms/hosts. Fermentation processes are commonly used today to manufacture numerous products; however, major technological improvements are needed to increase competitiveness. Current bottlenecks include low volumetric productivity and low yield of the microorganisms under non-optimal fermentation conditions in bioreactors. Another setback is the limited understanding of cellular behavior in bioreactor surroundings. Therefore, the major aims are to (i) enhance existing or new microorganisms to reach optimum production capacities under industrial conditions, (ii) develop analytical tools for monitoring the events in the bioreactor and mathematical models to control better processes and to improve their understanding for strain optimization and (iii) improve fermentation process engineering through better bioreactors and downstream processing.

3. Process eco-efficiency and integration: the biorefinery concept. Since it produces multiple products a biorefinery maximizes the value derived from the complex biomass feedstock. It relies on the best use and valorization of feedstock, optimization and integration of processes for a better efficiency, optimization of inputs (water, energy, etc.) and waste recycling/treatment. The main focus points of research are: (i) improving biorefining technologies, (ii) integrating the products into existing value chains and (iii) establishing strategies and business models for sustainability and competitiveness.

Regarding biorefinery, the recently published book *Biorefineries – Industrial Processes and Products* [101] provides an excellent overview of the status quo and future directions in this area.

Biotech also plays a critical role in rekindling chemical innovation. At a time of increasing competition from Asia in established products and the subsequent commoditization and strong price decline, chemical companies are once again looking at innovation as a key source of differentiation, as commented in previous sections. The importance of stimulating innovation can be seen by looking at the introduction of new polymers. During the twentieth century the development of fossil-fuel-based polymers increased steadily through to the post-war period, stimulated by the abundance and low cost of basic petrochemicals. It has, however, declined dramatically since 1960. Innovation in the traditional polymer industry today is mainly related to the application and blending of these polymers, rather than to the invention of new ones (Figure 1.18). Just as low-cost petrochemical building blocks such as ethylene, propylene and butadiene became available with the introduction of crackers in the 1930s it is necessary now to introduce new bio-based building blocks. These include lactic acid, which can be polymerized to the biopolymer PLA (polylactic acid). PLA has started to replace polyester because of its competitive cost and new applications. Lactic acid can also be processed into chiral drugs, acrylic acid, propylene glycol, food additives, and more. Other examples of innovation abound – Cargill is exploring the potential of 3-hydroxyproprionic acid as a new building block; BASF is looking into new chemistry around the simple organic molecule succinic acid; and DuPont will use cheap propandiol (PDO) as a monomer for its Sorona polymer.

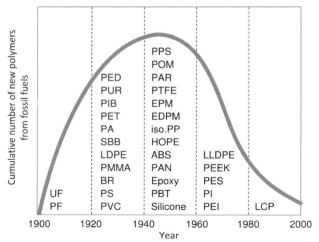

Figure 1.18 Polymer innovation based on fossil-fuel building blocks. Source: adapted from Kamm *et al.* [101].

1.6.2
Energy

A sustainable, safe and efficient energy supply is crucial for every country's economy. There is a critical need to rethink energy supply and use, since existing energy resources are limited both in volume and geographical distribution in the light of exploding global energy requirements. The problem of energy is complex and multifaceted. In terms of sustainable chemistry, there are three critical aspects: (i) developing alternative energy sources, (ii) saving energy by reducing energy loss through the smart application of materials and technologies and (iii) ensuring the safer and better storage of energy through innovative ways of use and transport.

The diversification of energy sources tailored to the requirements and resources of each country using nature's renewable resources such as the sun (photovoltaics), wind power, geothermal energy and biomass is a definite requirement. If solar cells are chosen to provide an alternative to fossil fuels, significant research work is needed: (i) to develop new routes for the production of crystalline silicon, (ii) in the development of amorphous silicon hybrid materials that could result in enhanced efficiencies, (iii) for further development of thin-layer technology, (iv) in concerted efforts for cheaper and more stable dyes, (v) in improving the efficiency of the dye-sensitized cells and (vi) in process development to deliver enhanced device performances, ensure sustainability and reduce production costs on an industrial scale.

However, to overcome present barriers, sustainable energy vectors should be developed in a harmonized way, taking into account all possibilities and related technologies, and not be limited to one or few sources. For example, in the transportation sector, several technologies present great potential: biomass conversion into biofuels, hydrogen fuel cells, hybrid engines and the exploration of metal nanoparticles as a fuel source. In the specific area of fuels from biomass, the production of liquid fuels, to be competitive, needs cheap and reliable sources of renewable raw materials and efficient production processes. The attention is on organic fuels such as methanol, ethanol, butanol and their derivatives, ETBE, MTBE, which can be produced by fermentation or gas to liquid technologies. Other points of attention are biodiesel and biogas, which can provide interesting biomass-based alternatives to diesel and LPG.

Bio-fuels should be not in competition with food. Therefore, new technologies need to be developed to efficiently convert cellulosic, fiber or wood-based, waste biomass into fermentable sugars. Similarly, to make biodiesel competitive as a transport fuel, efforts should be directed to diversify the use of raw materials and to improve the processes while making them more economic by developing added-value uses for by-products such as glycerol. Catalysis plays a critical role in achieving these objectives [9].

More efficient use and conservation of energy is possible, for example, by using OLEDs for lighting or novel nanofoams for insulation. Portable technologies require novel materials for the storage of energy, such as supercapacitors or new batteries. Higher specific energy, shorter charging times and long cycle life batteries need to be attained to keep pace with the increasingly demanding needs of personal electronic equipment. Lithium ion technology is currently the most promising for rechargeable batteries. Here, safety is key for the development of new systems (e.g., hybrid vehicles, energy networks, electronic devices).

Nanomaterials are expected to have a huge impact on active material design, due in part to their better resistance to structural strains and improved kinetics whatever the electrochemical system used (lead-acid, Li-ion or Ni-MH).

The envisaged hydrogen economy requires an efficient and safe hydrogen storage system. Hybrid organic-inorganic materials are a promising family of materials that includes dispersions of inorganic nanoparticles in organic (polymer) matrices; the converse, porous metal organic framework materials; pure nanoporous polymers; and organic macromolecules into mesoporous oxides such as silicas.

Supercapacitors containing electronically conducting polymer electrodes are of great interest, in particular for hybrid and electric vehicles due to their potential in the storage of large amounts of energy in a small volume. The largest hurdles towards marketable products are material development, production scale up and quality control.

1.6.3
Healthcare

The challenges for the next century in healthcare are (i) an ever increasing number of patients with allergic, inheritable or contagious diseases and cancers, (ii) demographic trends and (iii) the exploding costs of healthcare.

Nanotechnology has the potential to revolutionize medical technologies and therapies, by providing the tools to cope with these challenges. The health sector will greatly benefit from the new products and technologies provided by materials science: treatments, medical devices and delivery systems. Drug delivery, whether targeted or smart, may be achieved through the novel design of materials, in the form of coatings or formulations. These materials enhance the binding of the active ingredient/pharmaceutical to receptor sites or delivery/entry into specific cell types (e.g., cancer).

Functionalized (magnetic) metal nanoparticles with suitable polyfunctional organic groups or polymer chains are able to link to specific biomolecules or receptor sites. Different nanostructured inorganic matrices are suitable as bone tissue substitutes and, furthermore, have the potential for controlled drug delivery, specific to bone pathologies. Currently, there are several problems with the application of nanoparticles in delivery systems, for example, their tendency to aggregate during storage or under physiological conditions. Understanding and preventing aggregation is thus extremely important.

Exploiting nanoparticulate formulations with nanoscale functionalities for new applications will have an immense impact on healthcare and quality of life. This will be achieved through effective drug applications, faster and non-invasive medical diagnostics, and through treatments. An important task will be to establish appropriate safety standards for these new products and formulations. Many of these actions require extensive research to be realized, particularly in the understanding of how structure–property relationships affect the interactions between biological cells/organs and pharmaceuticals, how sensors work in the body, and how to interface biological systems with signal transporting materials/devices.

Note that the focus here was on materials for healthcare, because this is the area that needs to be fostered from recent developments in science. The identification and design of new active pharmaceutical ingredients is clearly equally important, but innovation in this area is already a driver.

1.6.4
Information and Communication Technologies

Information and communication technology (ICT) is a fast-moving consumer sector with a strong need for new materials and products to meet demands. The adoption of newly developed architectures and materials will lead to fundamental changes in the concepts and design paradigms of integrated circuits. Integration of nanostructures in device materials is thus a priority. Several new nanostructured materials, including carbon nanotubes, and silicon and germanium nanowires, are of potential interest for future device applications. Note that this problem of advanced nanoarchitecture of materials is critical not only for ICT but also in several other sectors (e.g., novel materials for energy). However, the mechanisms that control nanostructure, composition and size are not yet understood or mastered.

The role of catalyst composition and structure, in conjunction with temperature and gas composition (and other factors), in the production of, for example, carbon nanotubes is not understood in determining the structure. Understanding the mechanisms of how to control growth, structure, composition and orientation on predefined templates will be vital.

New methodologies of preparation are also vital to go beyond metal-oxide-semiconductor (CMOS) technology. Already, the semiconductor industry is investigating some very interesting areas such as spintronics, quantum computing and optical computing. Priorities will thus be as follows: (i) the application and introduction of new materials (polymers, nanomaterials, etc.), (ii) the control, production and integration of these new materials into devices with respect to cost and (iii) energy consumption.

Important facets are the understanding of interfaces between various materials, whether organic or inorganic; the use of self-assembly and patterning techniques in production; and the identification of new semiconductors, for example, carbon nanotubes. Furthermore, materials for information storage have also been addressed, such as holographic or new forms of switches (e.g., DNA, picosecond).

The need to fully characterize interfaces between nanomaterials integrated in an IC chip is of critical importance for their controlled production. Here, novel metrology techniques will be required to characterize the electronic and structural interactions at nanometer scale interfaces.

Studying the interaction forces between molecules and interfaces by means of atomic force microscopy and fluorescence-based technologies on the nanometer scale will provide valuable information on interfaces. Investigating surface functionalization processes with organic and biomolecular moieties, as well as self-assembly at surfaces, will enable insights into the basic properties of the interfaces.

1.6.5
Nanotechnology

Nanotechnology is enabling new developments in material science, while providing innovations for industries such as construction, information and communications, healthcare, energy, transportation and security. The sustainable development of nanomaterials, including their potential for environmental protection and the appropriate assessment of possible risks, will contribute to sustainable economic growth.

The discovery of new materials with tailored properties and developed processing are the rate-limiting steps in new business development for many industries. The development of novel nanomaterials is necessary: their synthesis and production, their analysis and their design by computer modeling. A priority topic is synthesis, surface chemistry and processing. Innovations in materials technology and the design of advanced materials with tailored macroscopic properties based on their molecular structure are prerequisite for innovations in many industries. Interesting applications are nanostructured biosensors, smart packaging with enhanced barrier properties and functional surfaces.

The demands of tomorrow's technology translate directly into increasingly stringent demands on the chemicals and materials involved, for instance their intrinsic properties, costs, processing and fabrication, their benign health and environmental attributes, and their recyclability when focusing on eco-efficiency. There is a pressing industrial need to better understand complex physical, chemical and biological phenomena relevant to the mastering and processing of multifunctional and eco-efficient materials.

Another priority is the understanding of structure–property relationships, particularly at the nanometer scale. To understand the phenomena arising at this scale and to gain the ability to control structure, and integrate new properties related to a reduction of the material size, it is necessary to understand the structure–property relationship, including the control of materials' shape and size. There is a need to be able to analyze individual components, but also to analyze the properties of the whole system at the macroscopic level. Computational modeling will lead the way in providing insights into nanophenomena and thus assist the design of new materials.

1.6.6
Sustainable Quality of Life

Sustainable industrial chemistry core goal is to improve the quality of life of citizens by applying new technologies and at the same time making citizens' lifestyles more sustainable, by achieving a lower environmental impact by consuming less energy, using fewer resources and reducing emissions. The focus revolves around the home – from energy collection via photovoltaic cells, novel insulation and lighting technologies, to eco-efficient processes. Particular attention is paid to the actions for waste treatment, for example, water purification and recycling to maintain this precious resource. The cited SusChem visionary project, the Smart Energy Home, will demonstrate how these technologies integrate in a home that produces, instead of consumes, energy and which minimizes the use of other resources such as water.

Reaching this objective requires the development of new energy-efficient construction materials, smart appliances, alternative energy systems in the home environment, insulation and sustainable products for the consumer.

1.6.7
Sustainable Product and Process Design

The building blocks for sustainable product and process design are engineering, catalysis and chemical synthesis to achieve intensified, more eco-efficient, environmentally benign and competitive processes and production technologies.

Five main areas need to be developed to implement innovative process technologies and knowledge-based design as well as plant operation methodologies in chemical, pharmaceutical and biotechnological production:

1. Diversification of the feedstock base.
2. Innovative eco-efficient processes and synthetic pathways.
3. Knowledge-based manufacturing concepts for targeted and tailored products.
4. Implementation and integration of intensified process technologies.
5. Life cycle analysis.

Development in these areas will also provide opportunities for new business models, utilizing various feedstocks and directly targeting specific product end-use properties. A holistic approach is depicted to rationally integrate innovative chemistry and engineering developments and to develop the most viable option regarding the whole process.

The following research activities specifically aim to innovate processes and chemical/pharmaceutical manufacturing excellence in Europe:

- Innovative eco-efficient processes and synthetic pathways. This activity aims to achieve more eco-efficient chemical syntheses and corresponding processes with high resource efficiency and reduced amounts of waste. Some examples addressed are (i) specific transformations of functional groups, (ii) the utilization of highly selective multifunctional catalysts, (iii) the increased use of benign and easy-to-handle oxidants and (iv) alternative solvents. Targeted process technologies are the development of integrated reactive and hybrid separations and the utilization of non-conventional forms and sources of energy. There is clearly some coincidence with the topics discussed above regarding the green chemistry approach, but it may be noted that the general context and philosophy is different, further stressing the core concept of this chapter of the need to pass from a green chemistry to a sustainable industrial chemistry approach.

- Knowledge-based manufacturing concepts for targeted and tailored products. It is necessary to master product's end use properties through process design. Intensified product engineering requires extending the capabilities of continuous processing, in particular those of highly viscous and/or solid-containing process fluids with the ultimate goal of intelligent, self-adapting process devices. Other requirements are advanced formulation technologies, high-throughput tools for formulation engineering and process systems engineering techniques, as well as reliable scale-up methods for microencapsulation and the production of fine particles and nanoformulations.

1.6.8
Transport

Chemistry plays an important role in realizing sustainable mobility and developing vehicles with improved eco-efficiency and recyclability, while increasing the safety and mobility of citizens. Key priorities are:

- sustainable management of materials, including the development of elastomeric products and materials to catalyze the decomposition of COx, NOx and SOx;
- development of new assembly technologies;
- onboard H_2 production;
- production of fuels containing low sulfur and aromatic components.

In the area of sustainable materials management, the need for lightweight but strong materials, which combine high performance with reduced environmental impact, should be noted. Reducing weight can be accomplished by developing innovative thermoplastic products for structural parts, leading to lower fuel consumption: (i) tough thermoplastic foams, (ii) organic filler-based thermoplastic composites and (iii) polymers with improved thermal resistance for further metal replacement (body panels).

More efficient airbag systems need to be developed to bring the best degree of protection together with weight reduction. Fibers offering a better cost–performance compromise could replace metal chords in tires, resulting in a significant weight reduction, thus producing less rolling resistance and reducing fuel consumption.

Alternative body assembly technologies – gluing versus welding and reversible assembly – need to be pursued to complement the move away from rigid assembly processes. The use of adhesives in the assembly process gives also the potential to reverse the process cheaply and efficiently. This can easily be done by dissolving the glue, particularly at the end of the lifecycle. Research should focus on high-throughput formulations and testing techniques as well as on the development of nondestructive analytical tools.

Chemistry also is central for the development of materials and technologies for traffic management sensors (collision avoidance, night vision), instant diagnostics, constant repairing materials (coatings), and silent cars and roads.

1.6.9
Risk Assessment and Management Strategies

The introduction of emerging technologies requires reconsideration of the overall approach to risk assessment and management in the early stages of development. This is needed to ensure that the objectives for growth and innovation, and the protection of health and environment, are not compromised.

Especially in the light of long-term investments in innovation, a reliable consistent risk-based framework is essential on the regulatory side and in the public debate. The goals are the following:

- Identification and development of a reliable consistent risk management framework for the introduction of emerging technologies – addressing the critical points throughout product use to allow faster and more effective innovation.
- Capture and dissemination of integrated approaches for safety assessment within the context of regulatory decision processes for emerging/breakthrough technologies.
- Evaluation of how the gap between risk and perceived risk can be closed for new technologies.

The work should aim to define and characterize (i) risk, (ii) perceived risk, (iii) risk management options and (iv) risk communication options to meet stakeholder concerns. Key activities to reach these scopes are:

- intelligent risk management strategies;
- integrated assessment and acceptance;
- safety assessment of new technologies.

1.7
Conclusions

The central concept of this chapter is the idea that sustainable industrial chemistry is more than green chemistry or engineering, and not an alternative politically-correct definition, as is usually reported in the literature and cited in lectures. Sustainable industrial chemistry is an holistic approach that has as its final aim the realization of sustainability through chemistry. It includes green chemistry and engineering concepts, and inherently safer design as well, but in a general framework and vision centered on the balance between economic growth and development, environment preservation and society promotion (health protection and quality of life), where innovation plays the driving role to realize this balance.

We have discussed a series of examples and aspects, such as the problem of risk and sustainability assessment, tools and principles for a sustainable industrial development (in particular, the issue of scaling-down and intensification of chemical processes, and the role of catalysis), and problems and opportunities in substituting chemical and processes (also in the view of REACH legislation, and of the international chemicals policy on sustainability). These topics are expanded in the following chapters, while the final section on industrial case histories for sustainable chemical processes provides further hints on these aspects.

The final section of this chapter has been dedicated to a concise discussion of a vision and roadmap for sustainability through chemistry, taken from the implementation action plan prepared by the European Technology Platform on Sustainable Chemistry. This plan evidences the pervasive role of chemistry for our society and thus how a next step is necessary in R&D to really realize a sustainable world through chemistry innovation.

References

1 Sheldon, R.A. (2008) *Green Chem.*, **10**, 359.

2 Förster, A. (2004) *Green Chem.*, **6**, G33.

3 Hutzinger, O. (1999) *Environ. Sci. & Pollut. Res.*, **6** (3), 123.

4 Centi, G. (2008) *ChemSusChem*, **1**, 7.

5 Centi, G. and Perathoner, S. (2003) *Catal. Today*, **77**, 287.

6 The European Technology Platform for Sustainable Chemistry, http://www. suschem.org (accessed in 2009).

7 UN World Commission On Environment and Development (headed by G.H. Brundtland), (1987) *Our Common Future*, Oxford University Press, Oxford.

8 Centi, G. and Perathoner, S. (2008) *Catal. Today*, **138**, 69. (Opening lecture at 5th European Conference on Catalysis, Ottrott (France), September 12–16th, 2007).

9 Centi, G. and van Santen, R.A. (eds) (2007) *Catalysis for Renewables*, Wiley-VCH Verlag, Weinheim.

10 Bell, A.T., Gates, B.C. and Ray, D. (2008) *Basic Research Needs: Catalysis for Energy (PNNL-17214)*, U.S. Department of Energy.

11 Bert, P., Bianchini, C., Giambastiani, G., Marchionni, A., Tampucci, A. and Vizza, F. (2007) It Patent. FI2007A000078.

12 Sheldon, R.A., Arends, I. and Hanefeld, U. (2007) *Green Chemistry and Catalysis*, Wiley-VCH Verlag, Weinheim.

13 Clark, J. and Macquarrie, D. (eds) (2002) *Handbook of Green Chemistry and Technology*, Blackwell Science, Oxford, UK.

14 Rothenberg, G. (2008) *Catalysis – Concepts and Green Applications*, Wiley-VCH Verlag, Weinheim.

15 Tundo, P. and Esposito, V. (eds) (2008) *Green Chemical Reactions*, Springer, Dordrecht. (Proceedings of the NATO Advanced Study Institute on New Organic Chemistry Reactions and Methodologies for Green Production Lecce Italy 29 Oct–10 Nov 2006).

16 Doble, M. and Kruthiventi, A.K. (2007) *Green Chemistry and Engineering*, Elsevier Science & Technology Books, Amsterdam.

17 de Souza Porto, M.F. and de Freitas, C.M. (1996) *Risk Anal.*, **16**, 19.

18 Gunster, D.G., Bonnevie, N.L., Gillis, C.A. and Wenning, R.J. (1993) *Ecotoxicol. Environ. Safety*, **25**, 2002.

19 Wenning, R.J. (1993) *Ecotoxicol. Environ. Safety*, **25**, 202.

20 Maki, A.W. (1991) *Environ. Sci. Technol.*, **25**, 24.

21 Lancaster, M.(1st Nov. 2004) *Responding to the Green Challenge*, UK Chemical Industries Association (GC&C).

22 MORI (1999) *The Public Image of the Chemical Industry. Research study conducted for the Chemical Industries Association*, MORI, London.

23 CEFIC (2000) *CEFIC Pan European Survey 2000. Image of the Chemical Industry Summary*, CEFIC, Brussels.

24 Anastas, P.T. and Warner, J.C. (1998) *Green Chemistry, Theory and Practice*, Oxford University Press, Oxford.

25 Clark, J.H. (2002) in *Handbook of Green Chemistry and Technology* (eds J.H. Clark and D. Macquarrie), Blackwell Science, Oxford, UK, p. 6.

26 Gupta, J.P. (2002) *J. Loss Prevention in the Process Industries*, **15**, 1.

27 Allen, B. (2000) *Green Chem.*, **2**, G56.

28 Fischer, M.J. (1996) *J. Risk and Uncertainty*, **12**, 257.

29 Fernández-Muñiz, B., Montes-Peóna, J.M. and Vázquez-Ordás, C.J. (2007) *J. Loss Prevention in the Process Industries*, **20**, 52.

30 Hood, E. (2004) *Env. Health Perspectives*, **112**, A352.

31 Willey, R.J., Hendershot, D.C. and Berger, S. (2006) The Accident in Bhopal: Observations 20 Years Later. Presented at American Institute of Chemical Engineers 2006 Spring National Meeting, 40th Annual Loss Prevention Symposium, Orlando, FL, USA, April 24–26.

32 Weir, D. (1987) *The Bhopal Syndrome: Pesticides, Environment, and Health*, Sierra Club Books, San Francisco, USA.

33 Sheldon, R. (2000) *Green Chem.*, **2**, G1.

34 Fukuoka, S., Tojo, M., Hachiya, H., Aminaka, M. and Hasegawa, K. (2007) *Polym. J. (Tokyo, Japan)*, **39**, 91.

35 Gong, J., Ma, X. and Wang, S. (2006) *Appl. Catal., A: General*, **316**, 1.

36 Wang, X.-K., Yan, S.-R., Cao, Y., Fan, K.-N., He, H.-Y., Kang, M.-Q. and Peng, S.-Y. (2004) *Chin. J. Chem.*, **22**, 782.

37 Kim, W.B., Joshi, U.A. and Lee, J.S. (2004) *Ind. & Eng. Chem. Res.*, **43**, 1897.

38 Fukuoka, S., Kawamura, M., Komiya, K., Tojo, M., Hachiya, H., Hasegawa, K., Aminaka, M., Okamoto, H., Fukawa, I. and Konno, S. (2003) *Green Chem.*, **5**, 497.

39 Shi, F., Deng, Y., SiMa, T., Peng, J., Gu, Y. and Qiao, B. (2003) *Angew. Chemie, Int. Ed.*, **42**, 3257.

40 Chaturvedi, D. and Ray, S. (2006) *Monatsh Chem.*, **137**, 127.

41 Ballivet-Tkatchenko, D., Camy, S. and Condoret, J.S. (2005) in *Environmental Chemistry* (eds E. Lichtfouse, J. Schwarzbauer, and R. Didier), Springer, Berlin, p. 541.

42 Rivetti, F. (2002) DGMK Tagungsbericht 2002. 2002-4 (Proceedings of the DGMK-Conference "Chances for Innovative Processes at the Interface between Refining and Petrochemistry, Deutsche

Wissenschaftliche Gesellschaft für Erdöl, Erdgas und Kohle, p. 53.

43 Bigi, F., Maggi, R. and Sartori, G. (2000) *Green Chem.*, **2**, 140.

44 Aresta, M. and Quaranta, E. (1997) *Chemtech*, **27**, 32.

45 Ono, Y. (1997) *Catal. Today*, **35**, 15.

46 Osterwalder, U. (1966) *Symposium Papers – Institution of Chemical Engineers, North Western Branch 1996 (5, Batch Processing III)*, **6**, Institution of Chemical Engineers, North Western Branch, US, p. 1.

47 Hendershot, D.C. (2000) *Chem. Eng. Progress*, **96**, 35.

48 Center for Chemical Process Safety (CCPS) (1993) *Guidelines for Engineering Design for Process Safety*, Wiley-VCH Verlag, Weinheim.

49 Bollinger, R.E., Clark, D.G., Dowell, R.M. JIII, Ewbank, R.M., Hendershot, D.C., Lutz, W.K., Meszaros, S., Park, D.E., Wixom, E.D. and Crowl, D.A. (eds) (1996) *Inherently Safer Chemical Processes: A Life Cycle Approach*, The Center for Chemical Process Safety (CCPS) of the American Institute of Chemical Engineers (AlChE), New York, (re-published by Wiley-VCH Verlag, Weinheim, 2005).

50 Kletz, T.A. (1998) *Process Plants: A Handbook for Inherently Safer Design*, CRC Taylor & Francis, Boca Raton, FL.

51 Dow Chemical, Co. (1994) *Dow's Chemical Exposure Index Guide*, 1st edn, AIChE, New York.

52 Osterwalder, U. (1996) Continuous process to fit batch operation: safe phosgene production on demand. Symposium Paper (Batch Processing 111), Inst. Chem. Eng. North West. Branch (Rugby, Warwickshire, UK), IChemE (Institution of Chemical Engineers), Vol. 6, p.1.

53 Delseth, R. (1998) *Chemia*, **52**, 698.

54 Ajmera, S.K., Losey, M.W., Jensen, K.F. and Schmidt, M.A. (2001) *AIChE J.*, **47**, 1639.

55 Cotarca, L. and Eckert, H. (2004) *Phosgenations – A Handbook*, Wiley-VCH Verlag, Weinheim.

56 Wörz, O., Jäckel, K.-P., Richter, Th. and Wolf, A. (2001) *Chem. Eng. & Techn.*, **24**, 138.

57 Ehrfeld, W., Hessel, V. and Löwe, H. (2004) *Microreactors*, Wiley-VCH Verlag, Weinheim.

58 Mills, P.L., Quiram, D.J. and Ryley, J.F. (2007) *Chem. Eng. Sci.*, **62**, 607.

59 Benson, R.S. and Ponton, J.W. (1993) *Trans. IChemE*, **71**, 160.

60 Ponton, J.W. (1996) The disposable batch plant. Proceedings 5th World Congress of Chemical Engineering, Vol. II, San Diego, CA, AIChE, New York, July 14–18, p.1119.

61 DeWitt, S.H. (1999) *Curr. Opin. Chem. Biol.*, **3**, 350.

62 Lowe, H. and Ehrfeld, W. (1999) *Electrochim. Acta*, **44**, 3679.

63 Koch, T.A., Krause, K.R. and Mehdizadeh, M.E. (1997) *Process Saf. Prog.*, **16**, 23.

64 Vineyard, M.K., Moison, R.L., Budde, F.E. and Walton, J.R. (2006) U.S. Patent 173209 (assigned to Peragen Systems Inc., US).

65 Anastas, P.T. and Zimmerman, J.B. (2003) *Environ. Sci. Technol.*, **37**, 95.

66 Anastas, P.T. and Kirchhoff, M.M. (2002) *Acc. Chem. Res.*, **35**, 686.

67 Lancaster, M. (2002) *Green Chemistry: An Introductory Text*, Royal Society of Chemistry, Cambridge, p. 242.

68 Tundo, P. and Selva, M. (2002) *Acc. Chem. Res.*, **35**, 706.

69 Rivetti, F. (2000) in *Green Chemistry: Challenging Perspectives* (eds P. Tundo and P. Anastas), Oxford University Press, Oxford, p. 201.

70 Romano, U., Rivetti, F. and Di Muzio, N. (1979) US Patent 4,318,862 (assigned to Enichem, Italy).

71 Nisihra, K., Mizutare, K. and Tanaka, S. (1991) EP Patent Appl. 425 197 (assigned to UBE Industries, Japan).

72 Rivetti, F. (2000) *C. R. Acad. Sci. Paris, Ser. IIc, Chim: Chem.*, **3**, 497.

73 Di Muzio, N., Fusi, C., Rivetti, F. and Sasselli, G. (1991) EP Patent 460 732 (assigned to to EniChem SpA., Italy).

74 Delledonne, D., Rivetti, F. and Romano, U. (1995) *J. Organomet. Chem.*, **488**, C15.

75 Tundo, P. (2001) *Pure Appl. Chem.*, **73**, 1117.

76 Shaik, A.-A. and Sivaram, S. (1996) *Chem. Rev.*, **96**, 951.

77 Tundo, P. and Selva, M. (2002) *Acc. Chem. Res.*, **35**, 706.

78 Rivetti, F., Mizia, F., Garone, G. and Romano, U. (1987) US Patent 4 659 845 (assigned to Enichem Synthesis, Italy).

79 Selva, M. and Perosa, A. (2008) *Green Chem.*, **10**, 457.

80 Tucker, J.L. (2006) *Org. Process Res. Dev.*, **10**, 315.

81 Winterton, N. (2001) *Green Chem.*, **3**, G73.

82 Curzons, A.D. and Constable, D.J.C. (2002) *Green Chem.*, **4**, 521.

83 Eissen, M., Hungerbühler, K., Dirks, S. and Metzger, J. (2003) *Green Chem*, **5**, G25.

84 Delseth, R. (1998) *Chimia*, **52**, 698.

85 Eckert, H. and Forster, B. (1987) *Angew. Chem. Int. Ed. Engl.*, **26**, 894.

86 Steinhäuser, K.G. (2007) Sustainable chemicals in the product chain. Presented at EU-Workshop on Sustainable Chemistry – Implementation of a Scientific Concept in Policy and Economy, 15–16th May, Berlin, Germany.

87 Steinhäuser, K.G., Richter, S. and Penning, J. (2004) *Green Chem.*, **6**, G41.

88 Steinhäuser, K.G., Greiner, P., Richter, S., Penning, J. and Angrick, M. (2004) *Environ Sci & Pollut Res.*, **11**, 281.

89 Rassafi, A.A., Poorzahedy, H. and Vaziri, M. (2005) *Sust. Dev.*, **14**, 62.

90 Mannan, S. (2005) *Lees' Loss Prevention in the Process Industries*, Elsevier Butterworth-Heinemann, Oxford, UK.

91 Hendershot, D.C. and Berger, S. (2006) Inherently Safer Design and Chemical Plant Security and Safety. Presented at the United States Senate Environment and Public Works Committee, Washington, DC, USA, June 21.

92 Kletz, T.A. (1978) *Chem. Ind.*, 287–292, 6.

93 Hendershot, D.C. (2003) Green chemistry, green engineering, and inherently safer design. Presented at 7th Annual Green Chemistry and Engineering Conference, Washington, DC, USA, June 23–26.

94 Wrightson, I.(Chairman of the Working Party of the RSC Environment, Health and Safety Committee – EHSC (2007) EHSC Note on Inherently Safer Chemical Processes, The Royal Society of Chemistry, Cambridge, UK.

95 Hendershot, D.C. (2006) *Process Safety Progr.*, **25** (2), 98.

96 Hendershot, D.C., Sussman, J.A., Winkler, G.E. and Dill, G.L. (2006) *Process Safety Progr.*, **25** (1), 52.

97 Tang, Y.T., Hsiao-Ping, H. and I-Lung, C. (2003) *J. Chem. Eng. Jpn.*, **36**, 1352.

98 Jan Harmsen, G. (2007) *Chem. Eng. Process*, **46**, 774.

99 Malone, M.F. and Doherty, M.F. (2000) *Ind. Eng. Chem. Res.*, **39**, 3953.

100 Jenck, J.F., Agterberg, F. and Droescher, M.J. (2004) *Green Chem.*, **6**, 544.

101 Kamm, B., Gruber, P.R. and Kamm, M. (2006) *Biorefineries – Industrial Processes and Products*, Wiley-VCH Verlag, Weinheim.

102 Organisation for Economic Co-Operation and Development (OECD) (2001) *The Application of Biotechnology to Industrial Sustainability*, OECD Pub., Paris.

103 EuropaBio (2003) *White Biotechnology: Gateway to a More Sustainable Future*, EuropaBio Pub., Brussels.

2
Methods and Tools of Sustainable Industrial Chemistry: Catalysis

Gabriele Centi and Siglinda Perathoner

2.1
Introduction

Chapter 1 introduced the concept that industrial sustainability should be defined as the continuous innovation, improvement and use of clean and safer technologies to reduce pollution levels, consumption of resources and improve safety and quality of work, while maintaining industrial competiveness. In practical terms, industrial sustainability means employing technologies and know-how to use less material and energy, maximizing renewable resources as inputs, minimizing generation of pollutants or harmful waste during product manufacture and use, and producing recyclable or biodegradable products [1].

This is a company internal objective that is also driven by external factors, between which social pressure and legislation from one side and market competiveness driven by innovation from the other side are the driving elements. Owing to global competiveness, the effectiveness of these driving forces to foster innovation is related to the capacity to introduce in the product value elements that account for how the product (and related chain of production) impact on the environment, as well risks, are minimized. Chapter 5 discusses these aspects and the modalities to quantify sustainability of chemical production.

The present chapter and Chapter 3 will instead focus on the discussion of general aspects of the other relevant components to achieving sustainability: the availability of new technologies and processes that allow us to reduce the process and product impact on the eco-system (including also products that could be easily recycled or be biodegradable and, therefore, less toxic), minimize the use of resources (which includes also their more rational use, e.g., introducing alternative raw materials), and improve process and product safety.

Various books on green or sustainable chemistry have discussed these aspects [2–15]. Many review papers have discussed general [15–24] or more specific aspects of green/sustainable chemistry and engineering such as monitoring strategies [25], microwaves in green and sustainable chemistry [26], polycarbonate production as

Sustainable Industrial Processes. Edited by F. Cavani, G. Centi, S. Perathoner, and F. Trifiró
Copyright © 2009 WILEY-VCH Verlag GmbH & Co. KGaA, Weinheim
ISBN: 978-3-527-31552-9

an industrial example of green and sustainable industrial chemistry in practice [27], heterogeneous metal catalysts for green and sustainable chemistry [28], selective oxidation of butanes as an example of green/sustainable chemistry [29], life-cycle evaluation of chemical processing plants [30], greener nanosynthesis [31], gold catalysts for sustainable chemistry [32], ionic liquids for sustainable chemical products [33], solvent-free and highly concentrated reactions for a green chemistry approach to asymmetric catalysis [34] and solid acids for green chemistry [35]. These are only few examples to evidence the large variety of topics discussed in the broad area of sustainable/green chemistry, because even when limited to reviews there is an exponential growth of publications in this topic, as indicated in Chapter 1.

In addition to these review papers, issues of *Chemical Reviews* [36] and *Accounts of Chemical Research* [37] have been dedicated to green chemistry, and there are the journals *Green Chemistry* (editorial for 10 years activity [38]), *Green Chemistry Letters and Reviews* (opening editorial of the first issue [39]), and *ChemSusChem* (opening editorial of the first issue [40] of this new international journal of chemistry and sustainability, energy and materials).

A large variety of books and review papers is thus available, but often they either do not employ a comprehensive approach or are not consistent with the approach of sustainable industrial chemistry that is the core of this book. In addition, for educational purposes, it is useful to overview the main methods and tools available to develop a novel and innovative sustainable industrial chemistry. This chapter is dedicated to catalysis as key tool for sustainability and process intensification.

2.2
Catalysis as Enabling Factor of Sustainable Chemical Production

A catalyst, in its simpler definition, is a substance that enables a chemical reaction to proceed at an usually faster rate or under different conditions (such as at a lower temperature) than otherwise possible. The catalyst interacts with the reagents and intermediates of reaction but is regenerated to the initial state during the reaction cycle. The turnover number (TON) indicates how many cycles a single catalytic center could perform the reaction cycle without being deactivated. In complex reactions with multiple possible products (the usual case in chemistry), the catalyst enables us to maximize a specific reaction pathway and thus to provide a selective synthesis.

From this definition, the role of catalysis to enable sustainable chemical processes is evident:

- it allows us to avoid stoichiometric reactions, thus reducing waste and solvents;

- it allows us to intensify the reaction, thus making it possible to operate with smaller reactors and often with continuous instead of discontinuous reactors (thereby reducing the amount of waste and improving safety, besides improving the economics), and typically under less severe reaction conditions (lower temperature and pressure), thus reducing energy consumption;

- it allows us to maximize the desired product, thus reducing costs and the environmental impact of separation, and decreasing the formation of by-products and waste.

The key role of catalysis for industrial chemical production and sustainability need not to be further demonstrated. We may simply cite that over 90% of chemical industrial processes use a catalyst. About 15–20% of the economies of the developed Western nations rely directly or indirectly on catalytic processes. Anastas *et al.* [41] have defined catalysis as a foundational pillar of green chemistry and in an interview indicated that "the most overarching concept in green chemistry is catalysis" [42]. The relationship between catalysis and sustainable (green) chemistry has been discussed by Centi and Perathoner [43]. The books *Green Chemistry and Catalysis* [2] and *Catalysis–Concepts and Green Applications* [5] are specifically dedicated to this subject. A review on designing catalysts for clean technology, green chemistry and sustainable development has been published by Thomas and Raja [44] and the role of catalysis for sustainable technologies has been highlighted by Poliakoff and Licence [45].

Catalysis is also one of the key enabling factors for the European Technology Platform of Sustainable Chemistry (www.suschem.org). A specific coordination of European innovation-driven research in the field of catalysis and sustainable chemistry has been made by the ACENET ERA-NET network (www.acenet.net) and a European Network of Excellence (IDECAT, www.idecat.org) has been dedicated to the role of catalysis for sustainable production and energy. In Japan the Green and Sustainable Chemistry Network (www.gscn.net) dedicates much attention to catalysis. For example, the 7th Green and Sustainable Chemistry Awards given in 2008 were dedicated to the development of (i) an environmentally benign THF polymerization process utilizing solid acid catalysis (various researchers of Mitsubishi Chemical), (ii) advanced molecular transformation of alkenes by low-valent ruthenium catalysts (T. Kondo) and (iii) selective oxidation systems by fine control of metal oxide cluster catalysts (F. Mizuno). Several of the awards given by the UK Royal Chemistry Society Green Chemistry Network (www.rsc.org/chemsoc/gcn) and US Environmental Protection Agency Green Chemistry Program (www.epa.gov/gcc) are dedicated to catalysis. Most contributions of a recent special topic issue of *Pure and Applied Chemistry* [46] on green-sustainable chemistry are also related to catalysis.

These examples indicate the general agreement in identifying catalysis as one of the major enabling factors toward sustainable chemical production. The economic value of catalysis has also to be noted. The world catalyst market will reach $12.3 billion in 2010, driven by growing demand in the chemical, polymer and refining industries for more energy efficient processes and products. Polymer catalysts will grow rapidly, while chemical synthesis types will remain dominant. Frost & Sullivan (http://www.chemicals.frost.com) evidenced that in Europe the emission control catalyst market earned revenues of $2500 million in 2006 and estimates these to reach $3484.2 million in 2013 [47]. The chemical processing catalyst market earned revenues of $1020 million in 2006 and expects to reach $1277.3 million in 2013.

They also observed [48] that in Europe the catalyst market will be driven mainly from the various legislative standards that stipulate permissible quantities of contaminants/pollutants from automotive and other industrial plants (for instance,

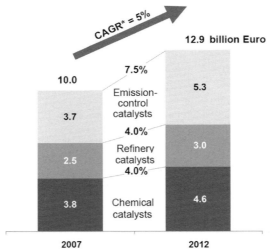

Figure 2.1 Estimated trend of catalyst market in markets in billion euro. CAGR: Compound annual growth rate. Source: adapted from Smith [49].

the Clean Air for Europe – CAFE). This has led to compulsory application of catalysts by almost every industry to minimize the volumes of contaminants discharged. In addition, notwithstanding the saturation in the chemical processing catalyst market due to the high fuel costs in Europe, which have compelled many petrochemical manufacturers to explore alternatives in regions offering lower fuel costs such as Middle East, the need to set up new innovative production plants in Europe will further push the role of catalysts in chemical processes.

BASF, the first global catalyst supplier with about 22% of global market, has estimated that the growth rates in the period 2007–2012 will be about 3% for the global chemical market, while about 5% for the catalysts market (excluding precious metals) (Figure 2.1) [49]. The high catalyst demand for the chemical market is driven by global trends: (i) increasing energy costs (more efficient usage of crude oil, gas, coal and chemical feedstocks, and raw material change), (ii) tightening clean air regulations (transportation, industry) and (iii) emerging technologies (nanotechnology).

It should be also considered that catalyst performance drives manufacturing costs and catalyst performance determines overall production plant investment. On average, each euro invested in catalyst results in about €250 of chemical product values. We may thus conclude that catalysis is not only a driver for economy of chemical production but also a key to sustainability [50].

Traditionally, catalysts are divided into homogeneous and heterogeneous catalysts, biocatalysts (enzymes), photocatalysts, and electro-catalysts. In order of number and volume of chemical processes, most use heterogeneous catalysts, that is, in a phase different (typically a solid phase) than that of the reagents and products. We will thus discuss the use of solid catalysts to promote sustainability of chemical processes in more detail. However, the importance of homogeneous catalysis is increasing and relevant examples and outlooks are also discussed.

Biocatalysis, in relation to white biotechnologies for chemical processes, is also an active growing area, driven by the progresses made in gene technology and the increasing use of bioresources as raw materials. Photocatalysis has long been confined mainly to the field of photodegradation of pollutants in water and air, but the need for a major use of solar energy has increased the interest of photocatalytic routes for the production of H_2 from water or bio-derived products (bioethanol, etc.), and in converting CO_2. Electro-catalysis has also received renewed interest from active research on fuel cells and new nanomaterials. This has extended interest in electro-catalysis to other applications, such as the combined chemical transformation of carbohydrates (in waste streams) into chemicals and electrical energy.

Catalysis is thus a driver for sustainability and societal challenges [51] and for a sustainable energy [52, 53]. New demand for applications (e.g., the area of biorefineries [54, 55]) and new advances in both the ability to control catalyst characteristics through nanotechnologies [56, 57] and to understand catalytic reactions [58–62] have greatly renewed the interest in catalysis and changed the research topics and approaches with respect to few years ago. We could thus conclude that catalysis is not only a key element for the sustainability of chemical processes but also that the recent advances in this area have further enhanced its critical role.

2.3
Homogeneous Catalysis and the Role of Multiphase Operations

Several large-scale industrial processes use homogeneous catalysts [e.g., hydroformylation, hydrocyanation (DuPont), ethene-oligomerization (SHOP), acetic acid (Eastman Kodak), acetic acid anhydride (Tennessee-Eastman), acetaldehyde (Wacker) and terephthalic acid (Amoco)] as well as smaller scale applications [e.g., metolachlor (Novartis), citronellal (Takasago), indenoxide (Merck) and glycidol (ARCO, SIPSY)].

However, homogenous catalysts are still far less used on an industrial scale than heterogeneous catalysts, despite the great advantages in terms of selectivity and possibility of fine tuning of performance (e.g., by changing the ligands), for two main reasons: (i) cost of catalyst recovery and (ii) low turnover per reactor volume, for example, productivity. The latter aspect is often not considered.

In terms of process cost and sustainability (energy consumption, waste formation), the productivity per volume of reactor is a key factor, because fixed costs (reactor and separation) depend on this parameter, as well as running costs (energy, utilities, etc.). The need for a solvent, and often of relatively diluted reaction conditions using homogeneous catalysts, notwithstanding good turnover numbers (TON) of the catalyst, is the critical element that has often prevented commercialization of homogeneous catalysis processes, in addition to catalyst cost (more than the catalyst itself, the cost of catalyst make-up per unit of production, which includes, for example, the loss of catalyst with time on stream in continuous processes or during consecutive batches in discontinuous syntheses) and catalyst recovery/recycle.

Homogeneous catalysts, in addition, often pose problems of safety of preparation, which also increase the costs.

Other aspects that are important in determining the overall process economics include the typical more limited range of temperature of operations using homogeneous than heterogeneous catalysts, the higher pressure, the need for special materials for the reactor in some cases (corrosive medium, a problem often not considered in academic studies, but critical for industrial development and safety of operations), and the limitations in mass- and heat-transfer. Various industrial processes operate in the liquid phase using solid catalysts. Therefore, the limited number of processes using homogeneous catalysis do not derive from the use of liquid phase operations, but instead from the reactor downstream costs, and the lower productivity per unit of reactor volume when using homogeneous catalysts.

As discussed later regarding the role of multiphase operations for homogeneous catalysis, research has been focused mainly on the issue of recyclability of homogeneous catalysts. Relevant progress has been made in this aspect, but the other problem, which could be indicated as intensification of homogeneous catalysis processes, is the issue that should be solved, at least for medium-large scale industrial productions.

In fine and specialty chemicals production the process cost is a less relevant aspect; on the other hand, the time taken to realize the industrial production is typically the critical factor. This aspect, together with the limited resources dedicated for R&D, determine the preference in companies for multipurpose catalysts with respect to optimized, but more specific, catalysts. This applies also to the process itself, where simpler, not optimized, batch reactors are preferred to better, but less flexible, more complex operations. This is one of the key aspects to consider in evaluating the use of multiphase operations in the synthesis of this class of chemicals.

In contrast, homogeneous catalysts are certainly superior to heterogeneous (solid) catalysts in terms of possible fine tuning of performances, through a proper design of the ligands, and also thanks to the general better level of understanding on a molecular level with respect to solid catalysts. Therefore, the selectivity that could be obtained with homogeneous catalysts is typically superior to that of heterogeneous catalysts. In terms of sustainability of chemical processes, the pros of homogeneous catalysts are thus relevant to a possible lower formation of by-products, but the cons are the higher energy consumption and amount of waste.

The following discussion shows that the industrial examples of multiphase homogeneous catalytic processes are those for which the limitations discussed above have been overcome. It is always necessary to have clearly in mind these limitations, otherwise it cannot be explained why, despite intense research in this area and the several claims of advantages in multiphase homogeneous catalysis operations, the number of industrial applications is still quite limited.

We conclude here by indicating that heterogeneous catalysts are still preferable in developing sustainable chemical processes, but that the notable recent progress in multiphase operations to allow improved catalyst separation, recovery and recycle of homogeneous catalysts [63–67] indicates that the use of homogeneous catalysis in industrial processes will certainly increase in the near future.

2.3.1
Multiphase Operations: General Aspects

The use of multiphase systems to avoid difficulties in separation and recycle, while maintaining the advantages of homogeneous over heterogeneous catalysis (typically higher specific activity, e.g., relative to metal content, higher selectivity, possibility of fine tuning and better understanding at a molecular level) is an area of great relevance for a sustainable industrial chemistry. Also in the area of biocatalysis, much progress has been made recently in the industrialization of biotechnology processes through the use of multiphase systems [68–72]. The use of biocatalysis for industrial synthetic chemistry is on the verge of significant growth. Biocatalytic processes can now be carried out in organic solvents as well as aqueous environments, so that apolar organic compounds as well as water-soluble compounds can be modified selectively and efficiently with enzymes and biocatalytically active cells. As the use of biocatalysis for industrial chemical synthesis becomes easier, several chemical companies have begun to increase significantly the number and sophistication of biocatalytic processes used in their synthesis operations.

There are various possible approaches for multiphase operation of homogeneous catalysis, to improve their usability and recycle: processes with organic/organic, organic/aqueous, or "fluorous" solvent pairs (solvent combinations), non-aqueous ionic solvents, supercritical fluids, and systems with soluble polymers. Figure 2.2 reports a general scheme of the possibilities for homogeneous catalysis.

Table 2.1 gives a comparative overview of the pro and cons of the various options in processes with multiphasic operations. On going from left to right in the table the industrial relevance of the processes decreases. The RCH/RP oxo process and the Shell Higher Olefin Process (SHOP) discussed later belong to the first class (aqueous biphase) and second class (organic biphase), respectively.

2.3.2
Aqueous Biphase Operations

An excellent example of sustainable chemical processes using homogeneous catalysts is Ruhrchemie/Rhône-Poulenc's (RCH/RPs) oxo process, which is a prototype of an aqueous biphasic process (Figure 2.3) [65, 66]. The first commercial oxo plant using the RCH/RPs biphasic process went on stream in 1984 and has produced over 5 million tons of *n*-butanyraldehyde (as well as less than 4% iso-butyraldehyde).

The oxo-active $HRh(CO)_4$ catalyst is modified by trisulfonated triphenylphosphine (TPPTS) ligands. Owing to the solubility of the Rh(I) complex in water and its insolubility in the oxo products, the oxo unit is essentially reduced to a continuous batch reactor followed by a phase separator (decanter) and a stripping column. Propene and syngas are added to the stirred, noncorrosive catalyst solution in the reactor. After reaction the crude aldehyde passes to the decanter and, while being degassed, is thus separated into the aqueous catalyst solution and the organic aldehyde phase. The catalyst solution exchanges heat and produces process steam in the heat exchanger is replaced by the same amount of water dissolved in the crude

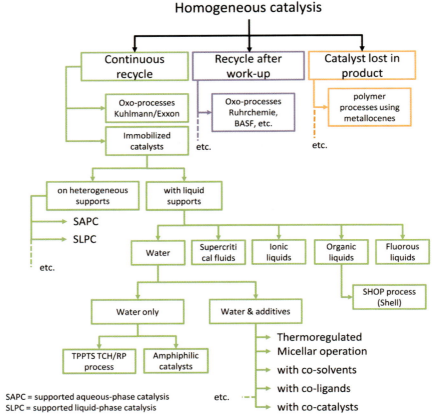

Figure 2.2 General scheme of the possibilities for homogeneous catalysis to improve in terms of usability and recycling. Source: adapted from Cornils *et al.* [63].

aldehyde and is returned to the oxo reactor. The crude aldehyde passes through a stripping column in which it is treated with counter-current fresh syngas. No side-reactions occur to decrease the selectivity or yield of the crude aldehyde, since the aldehyde is stripped out in the absence of the oxo catalyst – a distinctive feature of RCH/RPs process.

The crude aldehyde is fractionally distilled into *n*- and isobutanal in a conventional aldehyde distillation unit. The reboiler of this "n/iso column" is designed as a heat-absorbing falling film evaporator incorporated in the oxo reactor, thus providing a neat, efficient method of recovering heat by transferring the heat of reaction in the reactor to cold *n*-butanal, which subsequently heats the n/iso column. The preferred hydroformylation temperature is 110–130 °C and is therefore used for the production of process steam. Whereas other oxo processes are steam "importers," the RCH/RP process including the distillation of n-/isobutanol exports steam. No special pre-treatment or even purification steps are necessary for the catalyst. This reduces the environmental burden still further.

Table 2.1 Pro and cons of the various options in processes with multiphasic operation. Source: adapted from Cornils *et al.* [63].

	Aqueous biphase	Organic biphase	Ionic liquids		sc fluids	Fluorous biphase	Polymer supported
Solvents							
Availability	High	High	Medium	High		Low	Low
Costs	Low	Medium	High	Low		High	High
Flammability	None	High	High	None[a]		None	High
Thermal stability	High	Medium	Medium[c]	High		High[b]	Low
Toxicity	None	Various	Various[c]	None		Various	Various
Ligands							
Availability	High	Various	Low to medium				
Price	Low	Various	High				Various
Catalysts							
Variability	Medium	High	Medium				Unknown
Activity	High	High	High				Unknown
Service life	High	High	Various				Unknown
Recyclability	Proven	Proven	Probable				Unknown
Process							
Variability reactants	Limited	Limited	Unknown/limited				Limited
Sensitivity poisons	Low	Unknown	Unknown				Unknown
Eco-impact	None	None	Medium to high				Low
Recyclability	Proven	Proven	Probable				Unknown
Scale-up	Proven	Proven	Probable to unknown				Unknown

sc = supercritical.
[a]Valid only for scCO$_2$ and scH$_2$O.
[b]Except for atmospheric ozone depletion potential of perfluorinated compounds.
[c]Except for corrosive effects and potential toxicity.

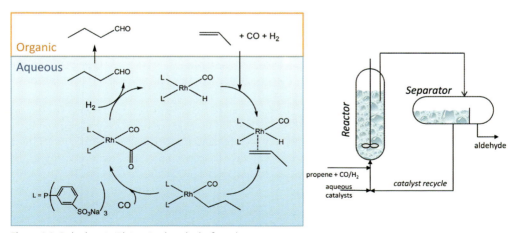

Figure 2.3 Ruhrchemie/Rhône-Poulenc hydroformylation process. Source: adapted from Sheldon *et al.* [2].

The high selectivity toward the sum of C_4 products is a special feature of a two-phase operation and results from the availability of water during hydroformylation. The high selectivity toward C_4 products (a maximum of 1% of higher-boiling components, "heavy ends", are formed relative to butanals) makes fractional distillation after aldehyde distillation unnecessary, reduces expenditure and thus also minimizes the environmental load. The manufacture of the by-products becomes part of the 2-EH (2-ethylhexanal) process since the heavy ends consist mainly of 2-ethyl-3-hydroxyhexanal, which, during downstream processing, is also converted into 2-EH. This (and the avoidance of butyl formates) is the reason for the considerable simplification of the process flow diagram compared with other process variants (Figure 2.4).

The fundamental advance represented by the RCH/RP process in terms of conservation of resources and minimization of environmental pollution can be demonstrated by various criteria. In terms of E-factor (defined in Chapter 3) it decreases for about 0.6–0.9 for the conventional oxo processes (cobalt catalyst) to

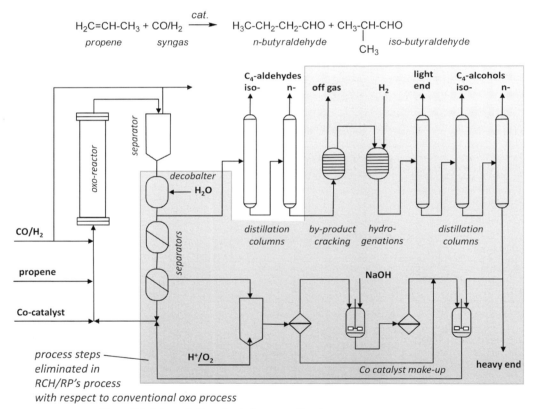

Figure 2.4 Schematic flow-sheet of a conventional oxo process, with an indication of the parts that are eliminated in the RCH/RPs process. Source: adapted from Cornils and Herrmann [65].

below 0.1 in the RCH/RP process. Process simplification is well demonstrated in Figure 2.4. Owing to milder reaction conditions (in particular, lower pressure), a drastic reduction of energy consumption is obtained. The volume of wastewater from the new process is 70-times lower than that from the Co high-pressure process. Table 2.2 summarizes other aspects of this comparison.

The use of water as solvent also significantly reduces the risks, with respect to the Co high-pressure process, because the possible leaking of highly flammable, metallic carbonyl is reduced.

The aqueous biphasic process thus shows several advantages over the Co high-pressure process, even if the latter is still commercially implemented in various plants [73]. The motivations are essentially related to the cost of the catalyst and of its recycle.

Another limit of aqueous processes is that the performances greatly depend on substrate water solubility. Very low rates are achieved using longer chain alkenes. However, various approaches to overcoming the low rates of hydroformylation of long-chain alkenes in aqueous biphasic systems have been proposed. Some of these, such as the use of microemulsions [74] or pH dependent solubility [75], have provided improvements, often at the expense of more complex separation processes. Perhaps the most promising new approaches involve the introduction of new reactor designs where improved mixing allows much better rates of mass transport between the three phases (water, organic and gas) [76]. An interesting approach is to use a temperature dependent phase, that is, the catalyst is soluble in the organic phase under the reaction conditions, but soluble in the water phase on cooling [77].

Aqueous, two-phase catalysis is also utilized industrially in several other processes apart from hydroformylation. The hydrodimerization of butadiene and water, a

Table 2.2 Environmentally advantages of the RCH/RP process in comparison with Co high-pressure oxo-synthesis of iso-butyraldehyde. Source: adapted from Cornils and Herrmann [65].

	Co high-pressure process	RCH/RP Rh low-pressure process
Selectivities		
– Toward C_4 products (%)	93	>99
– Toward C_4 aldehydes (%)	86	99
Products other than n-butyraldehyde (%)	31	<5
n/i Ratio	80:20	93–97:7–3
Waste water volume	70	1
Energy consumption figures		
– Steam	82	6.5 (steam exporter)
– Power	>2	1
– Syngas compression	1.7	1
Reaction conditions		
– Pressure (bar)	300	<50
– Temperature (°C)	150	120

telomerization variant yielding 1-octanol or 1,9-nonanediol, is carried out at a capacity of 5000 tonnes per annum by the Kuraray Corporation in Japan. They use the process as a step to produce nonanediamine for heat-resistant polyamide that is used in electronic, electrical and automotive parts [78].

Homogeneous, aqueous two-phase catalysis is also of industrial interest for the production of the important intermediate phenylacetic acid (PAA), which is used in perfume and pesticides syntheses. The previous process (benzyl chloride to benzyl cyanide with hydrolysis of the latter) suffered from the formation of large amounts of salt (1400 kg per kg of PAA). The new carbonylation method reduces the amount of salt by 60% and makes use of the great cost difference between $-CN$ and $-CO$ [79–81]. Finally, the Suzuki coupling of aryl halides and arylboronic acids, substituting Pd/TPPMS with Pd/TPPTS catalysts, should be also mentioned.

Aqueous biphasic systems (ABS) or aqueous two-phase systems are clean alternatives for traditional organic–water solvent extraction systems. ABS are formed when two polymers, one polymer and one kosmotropic salt, or two salts (one chaotropic salt and the other a kosmotropic salt) are mixed together at appropriate concentrations or at a particular temperature. The two phases are mostly composed of water and non-volatile components, thus eliminating volatile organic compounds. Solutes are defined as kosmotropic if they contribute to the stability and structure of water–water interactions. Kosmotropes (e.g., sulfate, phosphate, Mg^{2+} and Li^+) cause water molecules to favorably interact, which also stabilizes intermolecular interactions in macromolecules such as proteins. A chaotropic salt (e.g., guanidinium chloride and lithium perchlorate) instead disrupts the three-dimensional structure in macromolecules such as proteins, DNA, or RNA and denatures them. Chaotropic agents interfere with stabilizing intramolecular interactions mediated by non-covalent forces such as hydrogen bonds, van der Waals forces and hydrophobic effects.

ABSs have been used for many years in biotechnological applications as denaturing and benign separation media [82]. Recently, they have been used for metal ion separations, environmental remediation, metallurgical applications and as reaction media.

2.3.3
Organic Biphase Operations

The SHOP process (linear α-alkene via ethene oligomerization) was the first commercial catalytic process to benefit from two-phase non-aqueous liquid/liquid technology [83, 84]. In this special case two immiscible organic phases are used to separate the catalyst from the products formed, with the more or less pure products forming the upper phase. The first commercial plant was built at Geismar (LA, US) in 1977 with an initial capacity of about 200 000 tons per year, which was then progressively increased to over 900 000 tons per year. A second plant was constructed in Stanlow (UK) to reach a total capacity of over one million tons per year. The present market for alpha olefins is nearly four million tons. The global leaders in the alpha olefin business are CPChem, Ineos and Shell. Each operates its own process, which provides a different product spectrum. Three others, with smaller capacities – Idemitsu, Mitsubishi, and Nizhnekamskneftekhim – utilize a similar process technology. The alpha olefins' business is very complex as major producers serve a broad range of chemical industry segments from polyethylene commoners (C_4–C_8) through synthetic lubricants (C_{10}) and detergent intermediates (C_{12}–C_{14}) to oilfield chemicals, paper sizing agents (C_{16}–C_{18}), lubricant additives (C_{20+}) and wax rheological modifiers (C_{24+}). In addition there are a myriad of fine and performance chemical intermediates. By 2004, several of the producers added capacity, and there will be two new producers via ethylene oligomerization – SABIC, producing a full range of alpha-olefins, and Q-Chem, producing hexene.

Shell operates the Shell Higher Olefin Process (SHOP), combining an alpha olefin process catalyzed using a ligand modified nickel based system with a metathesis step to produce a broad product distribution of C_4–C_{10} materials, selected higher fractions and internal olefin fractions dedicated to captive detergent intermediates production. In addition to competitive ethene oligomerization processes, alpha olefins are also produced by Fischer–Tropsch synthesis (Sasol). This process is economically competitive. The carbon number distribution of alpha olefin produced could be narrowed down to the C_4–C_{10} range by zeolitic or zirconium/aluminum catalysts.

Figure 2.5 reports the scheme of the SHOP process and the simplified flow-sheet of the oligomerization biphasic process that uses 1,4-butanediol as the catalyst phase and a nickel catalyst modified with a diol soluble phosphine, R_2PCH_2COOH [85]. The nickel concentration in the catalyst system is in the range 0.001–0.005 mol.% (approx. 10–50 ppm). The rate of the reaction depends on the rate of addition of the catalyst. The process operates at 80–120 °C and 1500 psi ethylene. A high partial pressure of ethene is required to obtain good reaction rates and high product linearity. The linear α-alkenes produced are obtained in a Schulz–Flory-type distribution with up to 99% linearity and 96–98% terminal alkenes over the whole range from C_4 to C_{30+}.

While ethylene is highly soluble in butanediol, the higher olefins phase separates from the catalyst phase. The heat of reaction is removed by water-cooled heat exchangers between the reactors. In a high-pressure separator the insoluble products and the catalyst solution as well as unreacted ethene are separated. The catalyst solution is fed back into the oligomerization reactor. Washing of the oligomers by fresh solvent in a second separator removes traces of the catalyst. This improves product quality and the catalyst utilization, because traces of remaining catalyst in the product can yield insoluble polyethene during upstream processing, resulting in fouling of process equipment.

Further processing of the produced α-alkenes (Figure 2.5a) involves separation into the desired product fractions in a series of distillation columns. First the lower C_4–C_{10} α-alkenes are stripped off. In a heavy-ends column the C_{20+} α-alkenes are removed from the desired C_{12}–C_{20} α-alkenes. Finally, the middle-range products meeting the market needs are separated into the desired cuts and blends. This process scheme allows a great flexibility in the production, which is one of the major advantages.

The SHOP process, together with the previously discussed RCH/RP oxo process, are the two successful industrial processes cited as examples of the contribution

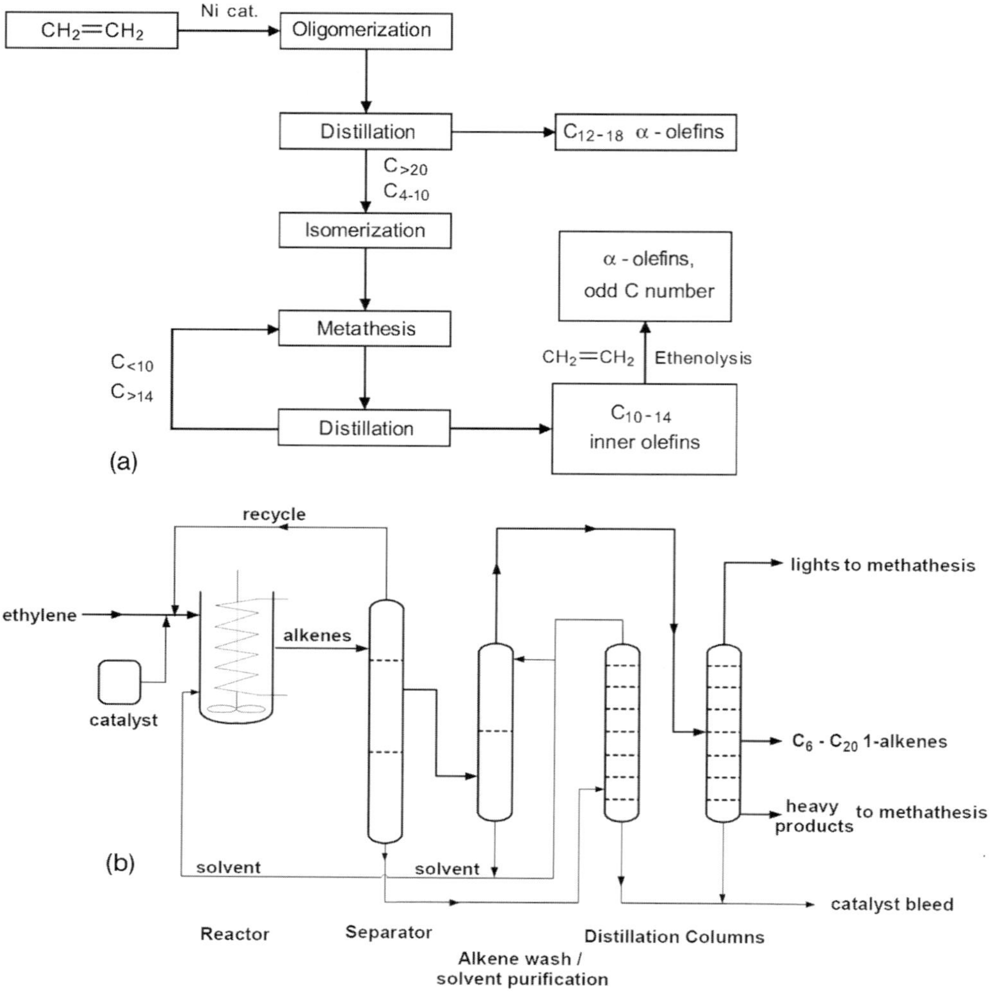

Figure 2.5 (a) Scheme of the SHOP process; (b) flow-sheet of the oligomerization biphasic process. Source: adapted from Hagen [60].

of biphasic homogeneous catalysis to sustainable chemistry [86–89]. While the use of biphasic systems has improved the recyclability of homogeneous catalysts, we could observe that only very limited industrial examples exist. A large variety of other liquid–liquid biphasic systems are under active investigation, which are summarized in Figures 2.2 and 2.6. Solvent properties of the products govern the selection of the catalyst phase.

However, notwithstanding expectations, none of the alternative strategies for catalyst/product separation in homogeneous catalysis, besides the cited examples and a few more, have yet reached the point where it can be commercialized [64].

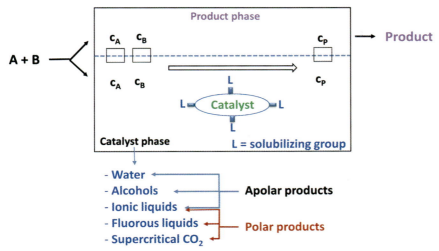

Figure 2.6 General scheme of liquid–liquid biphase catalysis and how the solvent properties of the products govern the selection of the catalyst phase. Source: adapted from Cornils *et al.* [63].

2.3.4
Catalysts on Soluble Supports

Catalysts on soluble supports, such as dendrimers and soluble polymers, provide alternative supports to solids, which have the advantage that access to the catalytically active sites is not restricted. Dendrimers are large tree-like, soluble molecules with a globular shape (Figure 2.7), making them suitable for ultrafiltration where the solvent and reaction product(s) pass through while the dendrimer is retained. The main problem in these cases is not the catalysis – reactions with high rates and selectivities have been reported – but rather the separation, which relies on nano- or ultrafiltration. One way of circumventing these problems is to support dendrimers on solid beads so that flow systems or simple filtration can be used. Because of the nature of the dendrimers, all their active sites are on the surface and so they are all accessible from the liquid phase even if they are supported on an insoluble support. This approach has been applied only in a very restricted number of reactions up to now.

For dendrimers and soluble polymers, there are two main problems: leaching of metal from the catalyst and lack of retention of the dendrimer or polymer by the membrane. Leaching could perhaps be addressed by improving catalyst design, such as using bidentate coordination and covalent attachment of the ligand, whilst the retention issue is potentially more difficult. The macromolecules could leak through the membrane either because the size of the channels is varied and some are large enough to allow the macromolecules to pass through or because of imperfections such as pinholes.

Although membranes have been successfully developed for ultrafiltration of biological systems in water (removal of whole cells), research on membrane materials

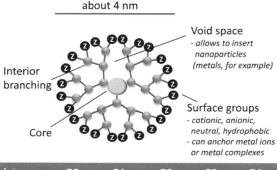

about 4 nm

Void space
- allows to insert nanoparticles (metals, for example)

Interior branching

Core

Surface groups
- cationic, anionic, neutral, hydrophobic
- can anchor metal ions or metal complexes

Nomenclature	G0	G1	G2	G3	G4
Nr. surface groups	3	6	12	24	48
Diameter, nm	1.4	1.9	2.6	3.6	4.4

Figure 2.7 General structure and nomenclature of dendrimers. Source: adapted from Dendritic Nanotechnologies Inc. (http://dnanotech.com).

for use with organic solvents and under gas pressure is less developed. Preventing membrane fouling is also an area where more information is required. Cross flow arrangement (hollow fibers) is preferable to dead-end filtration and also allows continuous (loop) operation.

2.3.5
Fluorous Liquids

Fluorous biphasic catalysis is another active area in multiphasic homogeneous catalysis. The term fluorous was introduced [90] as the analogue to the term aqueous, to emphasize the fact that one of the phases of a biphase system is richer in fluorocarbons than the other. Fluorous biphase systems can be used in catalytic chemical transformations by immobilizing catalysts in the fluorous phase. A fluorous catalyst system consists of a fluorous phase containing a preferentially fluorous-soluble catalyst and a second product phase, which may be any organic or inorganic solvent with limited solubility in the fluorous phase (Figure 2.8a).

Conventional homogeneous catalysts can be made fluorous-soluble by incorporating fluorocarbon moieties to their structure in appropriate size and number. The most effective fluorocarbon moieties are linear or branched perfluoroalkyl chains with high carbon numbers that may contain other heteroatoms. The perfluoroaryl groups offer dipole–dipole interactions, making perfluoroaryl-containing catalysts soluble in common organic solvents and therefore less compatible with fluorous biphase systems. The most effective fluorous solvents are perfluorinated alkanes, perfluorinated dialkyl ethers and perfluorinated trialkylamines. Their remarkable chemical inertness, thermal stability and nonflammability, coupled with their unusual physical properties, make them particularly attractive for catalyst immobilization. These materials have low toxicity by oral ingestion, inhalation or intraper-

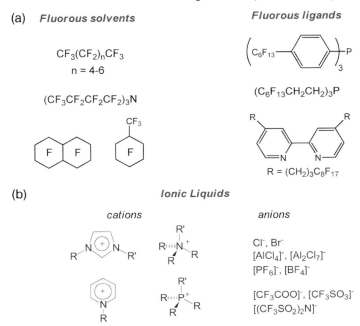

Figure 2.8 (a) Examples of fluorous solvents and ligands;
(b) typical cations and anions used for the formation of ionic
liquids. Source: adapted from Sheldon *et al.* [2] and Cole-Hamilton
and Tooze [64].

itoneal injection, but their thermal degradation can produce toxic decomposition products.

A fluorous biphase reaction could proceed either in the fluorous phase or at the interface of the two phases, depending on the solubilities of the substrates in the fluorous phase. When the solubilities of the substrates in the fluorous phase are very low, the chemical reaction may still occur at the interface, or appropriate phase-transfer agents may be added to facilitate the reaction. A fluorous biphase system might become a one-phase system by increasing the temperature [90]. Thus, a fluorous catalyst could combine the advantages of one-phase catalysis with biphasic product separation by running the reaction at higher temperatures and separating the products at lower temperatures. Alternatively, the temperature-dependent solubilities of solid fluorous catalysts in liquid substrates or in conventional solvents containing the substrates could eliminate the need for fluorous solvents [91].

Fluorous catalysts are best suited for converting apolar substrates into products of higher polarity, as the partition coefficients of the substrates and products will be higher and lower, respectively, in the fluorous phase.

It is generally considered that the major obstacle to the commercialization of reactions employing the fluorous biphasic concept is the cost of the fluorous solvent and the cost of the ligand, which must contain very large amounts of fluorine to retain the catalyst within the fluorous phase. Cost may be reduced by a proper selection of the solvent and ligand, but often the unsolved problem of progressive ligand degradation still exists. In addition, notably, the environmental persistence and very high global warming potential of fluorinated compounds may make their licensing for wide scale use problematic.

One advantage of fluorous ligands, which also has implications for reducing the cost, is that such electron-withdrawing fluorous ligands (called ponytails) increase the stability constants for complex formation and favor the linear product more than less electron-withdrawing ligands. This means that high selectivity towards the linear product can be obtained at low ligand:metal ratio (low ligand loading).

2.3.6
Ionic Liquids

Ionic liquids (IL) [92] are a major topic in the area of multiphase homogeneous catalysis. ILs are characterized by the following three criteria: (i) consist entirely of ions, (ii) have melting points below 100 °C and (iii) exhibit no detectable vapor pressure below the temperature of their thermal decomposition. As a consequence of these properties most ions forming ionic liquids display low charge densities, resulting in a low intermolecular interaction. Figure 2.8b displays some of the most common ions used so far for the formation of ionic liquids.

Most of these ILs are now commercially available, a fact that contributes to the development of their applications. The most commonly used ILs are probably those with BF_4^- or PF_6^- anions. For example, [BMIM][PF$_6$] (BMIM: 1-butyl-3-methyl-imidazolium chloride) has often been used as a solvent for catalytic reactions for a combination of reasons, including its ease of preparation and purification, its hydrophobicity and the possibility of separating a large range of reaction products by decantation without the need of addition of an extracting solvent. But despite the widespread use of these salts, the PF_6^- anion, and to a lesser extent the BF_4^- anion, are known to decompose. In the presence of water, they give HF and phosphoric acid [93]. Alternative anions, more stable toward hydrolysis, have been introduced in which, for example, the fluorine is bonded to carbon as in the $CF_3SO_3^-$ and $CF_3CO_2^-$ anions.

While the nonvolatile character of IL is a major advantage over conventional solvents, it is a limitation due to the impossibility of purifying ILs by distillation. Purification can be thus a significant challenge.

The range of homogeneous reactions that has been transposed into ILs is probably wider than into scCO$_2$ or perfluorinated solvents due to the great versatility of ILs. However, most of these reactions are limited to laboratory- or bench-scale with just a few examples of pilot-scale. A relevant industrial example is the Difasol process, which can be seen as an extension of the Dimersol family of processes developed by IFP [94]:

- Dimersol G converts a C3 cut into a gasoline effluent having excellent octane blending properties. In this way, the dimerization reaction increases the yield of high-RON gasoline that can be obtained from a cracker.

- Dimersol E is used to upgrade $C_2 + C_3$ fuel gas. Co-oligomerization of ethylene and propene leads to a gasoline stream very similar to the Dimersol G product. Mixed butenes are also obtained with Dimersol E (from ethylene dimerization). They can be used in paraffinic alkylation or to make propene through a subsequent cross-metathesis reaction with ethylene.

- The Dimersol X process has been developed to produce octenes as raw material for isononanol manufacture via oxo reaction.

The dimidiation process is commonly operated solvent-free with the active catalyst, a cationic nickel complex of the general form $[LNiCH_2R'][AlCl_4]$ (where L is PR_3). The catalyst is soluble in aromatic and halogenated hydrocarbon solvents, and shows higher catalytic activity in solution. Although this process is still used widely, the separation of products from the catalyst is a major problem and leads to increased operational costs and environmental impact. Chauvin (winner of the Nobel Prize in chemistry in 2005) and coworkers at IFP in France [95–97] reasoned that chloroaluminate ILs would be good solvents for the nickel catalyst, and discovered that by using a ternary ionic liquid system ([bmim]Cl–AlCl$_3$–EtAlCl$_2$) (bmimIbutyl-3-methylimidazolium) it is possible to form the active catalyst from a $NiCl_2L_2$ precursor and that, most importantly, the ionic liquid solvent stabilizes the active nickel species.

Using the ionic liquid catalyst, the Dimersol reaction can be performed as a two-phase liquid–liquid process at atmospheric pressure at between -15 and $5\,^\circ$C. Under these conditions, alkenes are immersed with activities well in excess of that found in both solvent-free and conventional solvent systems. The products of the reaction are not soluble in the ionic liquid, and form a second less-dense phase that can be separated easily. The nickel catalyst remains selectively dissolved in the ionic liquid phase, which permits both simple extraction of pure products and efficient recycling of the liquid catalyst phase. In addition to the ease of product/catalyst separation, the key benefits obtained using the ionic liquid solvent are the increased activity of the catalyst (1250 kg of propene dimerized per 1 g of Ni catalyst), better selectivity to desirable dimers (rather than higher oligomers) and the efficient use of valuable catalysts through simple recycling of the ionic liquid.

This process, using the ionic liquid solvent system, has been commercialized by IFP, as the Difasol process. In this process butene is dimerized in a continuous two-phase procedure with high conversion of olefin and high selectivity to the dimer (Figure 2.9). Catalyst consumption is divided by a factor of about ten and a higher yield of dimers is obtained. Most important, the Difasol system can be retro-fitted into existing Dimersol plants to give improved yields, lower catalyst consumption and associated costs and environmental benefits.

The Difasol reaction involves a mechanically stirred reactor and settlers. An injection of fresh catalyst components is defined to compensate the detrimental effects of accidental impurities present in the feed and slight carryover of the catalyst. Mixing of the solvent phase with the organic phase ensures advantageous butene conversion. However, importantly, the stirring power combined with a high

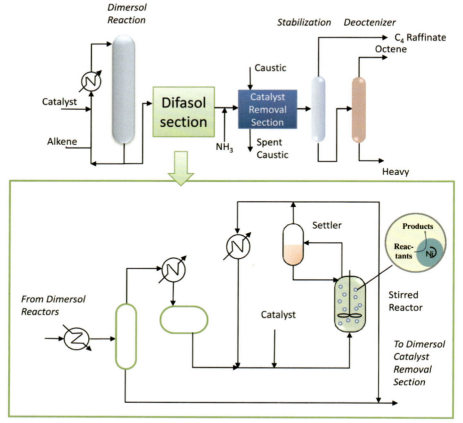

Figure 2.9 Schematic flow-sheet of the Dimersol + Difasol package reaction section scheme.

ratio of IL can result in a longer decantation time. Instead of one decanter, two distinct settling zones can be used. One settler has a moderate residence time and returns the ionic phase to the reactor while the organic phase is recirculated via a pump through a heat exchanger. The other settler has a more limited residence time and sends the product phase to the neutralization section.

The Difasol catalyst is concentrated and operates in the ionic phase, or perhaps at the phase boundary. The reaction volume is therefore much lower than in the conventional one-phase Dimersol process, where the catalyst concentration is very low. Notwithstanding the large academic effort spent on ILs as alternative solvents replacing classical solvents in existing processes, with over 2000 papers published in 2006, and claims that the age of IL has arrived [98, 99], the Difasol is still the only commercial process that uses an IL as solvent for homogeneous catalysis. However, ILs could be advantageously used also in a different way, as shown in BASFs BASIL (Biphasic Acid Scavenging utilizing Ionic Liquids) process.

Many organic chemical processes, such as esterifications, produce by-product acids that must be scavenged to prevent decomposition of the primary reaction

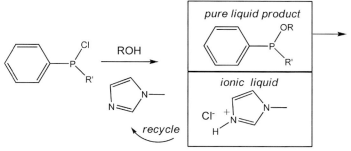

Figure 2.10 Scheme of the chemistry in the BASIL process.

product, or to prevent unwanted side-reactions. To address this problem, tertiary amines, such as triethylamine, are typically added to the reaction mixture, resulting in the production of a solid ammonium salt. This solid exhibits several problems, including reduction of heat transfer and reaction rates, reduction of yields and solid separation (Figure 2.10).

The BASIL process economically avoids the problems resulting from solids generation by making use of ionic liquids to scavenge acids. Instead of using a tertiary amine, 1-alkylimidazole is used to scavenge the produced acids. As the imidazole reacts with the acid, an alkylimidazolium salt is formed that is an ionic liquid at the reaction temperature. As a liquid, the alkylimidazolium salt can be easily removed by a liquid–liquid phase separation.

1-Alkylimidazole acts also as a nucleophilic catalyst, thereby improving reaction rates, and increasing yields and selectivities. The IL liquid in the BASIL process thus acts as co-catalysts and scavenger for by-products. The space–time yield was increased from $8 \, \mathrm{kg \, m^{-3} \, h^{-1}}$ using the old process (without ionic liquid) to $690\,000 \, \mathrm{kg \, m^{-3} \, h^{-1}}$ using the BASIL process: an 80 000-fold increase in productivity.

The main advantage of ILs compared to using water is thus the possibility of tuning their properties, to dissolve the substrate, even if it is of relatively low polarity, so that high reaction rates can be achieved. The main disadvantage, in a chemical sense, is that the IL may be soluble to some extent in the organic phase, thus leading to leaching of both IL and catalyst. The cost of the IL means that such losses must be kept to an absolute minimum. ILs that dissolve the organic substrate, but not the product, would be the ideal answer to this problem.

Another relevant problem is that, often, heavier products remain in the IL, and thus their continuous removal would be required.

In the Eastman process for 2,5-dihydrofuran production (Figure 2.11) [100], the oligomers formed in the process are highly polar and insoluble in alkanes. The IL $[P(oct)_3C_{18}H_{37}]I$ and the Lewis acid catalyst $[Sn(oct)_3]I$, which are non-toxic $(LD_{50} > 2 \, \mathrm{g \, kg^{-1}}$ for each), non-flammable (flammability 1) and non-corrosive (340 stainless steel is used for the reactor), have been designed to be soluble in heptanes. Part of the IL containing the oligomers, which is recycled from the distillation unit to the reactor, is purged to a counter-current extractor where the catalyst and the IL are extracted using heptanes. They are then stripped from the

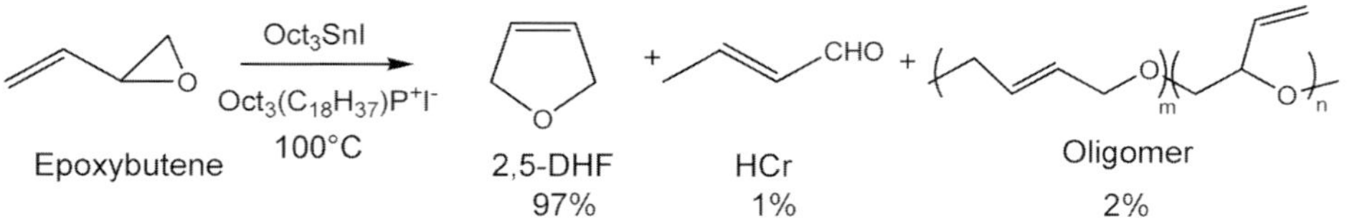

Figure 2.11 Reaction scheme for the isomerization of epoxybutene to 2,5-dihydrofuran (Eastman Chemicals).

heptanes and returned to the reactor whilst the oligomers, now free from the catalyst and IL, are incinerated.

The Eastman process has operated in Texas (USA) with three continuous, stirred-tank reactors, a wiped-film evaporator, a distillation train and a continuous, counter-current, liquid–liquid extractor for recovery of the catalysts since 1996 (1400 metric tons per year capacity, e.g., semiworks facility), but full commercial capacity has not yet been reached. The main problem is still the formation of oligomers. Another answer to the problem of IL loss to the organic phase consists in the extraction of the products from the IL using a supercritical fluid [101], but operational costs are high.

Other problems with ILs in addition to their cost, which is falling, are the following, even if the class of IL is very wide and thus the following comments could be not valid for all IL:

- environmental persistence, especially important when they are lost with the heavies purge;
- lack of flammability, which means that destruction by incineration is difficult;
- slight but significant toxicity [102, 103];
- corrosive properties [104].

Ranke *et al.* [105], in discussing the eco-toxicity of IL, evidenced the still great uncertainty of the risk in using these solvents. A recent study of the effects of several ILs and traditional organic solvents on the growth of a microalga (*Selenastrum capricornutum*) showed that the toxicities of the ILs were two to four orders of magnitude higher than those of organic solvents, although ILs are usually considered as green solvents [106]. Wells and Coombe [107] have also investigated the ecotoxicological tests of ILs, finding that several of them have very high toxicity towards freshwater algae and the freshwater invertebrate *Daphnia magna* ($\sim 10^4$–10^6 times more ecotoxic than methanol for the worse cases). In addition, these compounds show high levels of toxicity to the microorganisms responsible for biodegradation in the environment. They concluded that these materials have a significant potential to damage the aquatic ecosystem if released into water. ILs (imidazolium-based IL, in particular) also have an eco-toxic effect when dispersed on the soil [108].

As indicated, IL is a broad class of solvents and it is not correct to derive unique conclusions. However, it is also not possible to present ILs as "green solvents" [109], or to claim as sustainable a chemical process because it uses ILs as solvents. They can be corrosive, flammable or toxic. Their impact on aquatic ecosystems, due to their medium to high solubility in water, is another critical element. Their nonvolatile nature could be a factor for a lower impact on the environment and human health, but it is not the only element to consider in assessing the sustainability of a chemical process.

2.3.7
Supercritical Solvents

A supercritical fluid (SCF) is any compound above its critical point, which is the maximum in both temperature and pressure at which a gas and liquid can coexist. Above the critical point, isothermic compression yields a continuous increase in density without condensation to a liquid state. All substances theoretically have a critical point, but many experience thermal degradation well before reaching it.

Figure 2.12 is the classic pressure–temperature (PT) representation of the phase changes of a pure component. There are three primary phases of pure components: solid liquid, and vapor; solid–solid transitions, liquid crystal phases, and so on, are also possible but will not be considered here. The solid lines represent the sublimation curve (solid → vapor), the vapor pressure curve (liquid → vapor) and the melting curve (solid → liquid) of the pure component. The triangle represents the triple point, at which a solid, liquid and vapor coexist in equilibrium. The circle represents the pure component critical point, where the "supercritical" region begins.

Carbon dioxide is the most employed substance by far in supercritical fluid processes (scCO$_2$). Like water, carbon dioxide is an environmentally attractive solvent. The ability to control its solvating power by simple swings in temperature and pressure makes it an ideal medium for homogeneous catalysis. The main problem compared with conventional systems is that such swings, especially in pressure, require costly recompression and care must be taken to control the temperature whilst the pressure swing is occurring so that mixed liquid and gas phases do not form.

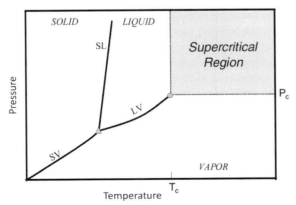

Figure 2.12 Pressure–temperature diagram of the phase transitions of a pure substance, illustrating the critical region; the circle represents the critical point and the triangle the triple point. Source: adapted from Scurto [110].

$scCO_2$ is largely used to process food (extraction or fractionation), but other applications, such as the fluoropolymer synthesis by DuPont, hydrogenation or alkylation by Thomas Swan, coatings by Union Carbide, and polyurethane processing by Crain Industries, are still in development [111]. The application of supercritical fluids (SCFs) as reaction media with homogeneous catalysts has been mainly investigated on a laboratory scale.

In the area of multiphase homogeneous catalysis, $scCO_2$ could be useful to solve a specific problem. In fact, any method for the immobilization of organometallic catalysts faces the paradoxical requirements for intimate contact of reagents and catalyst during the reaction, and maximum ease of separation during the consecutive separation stage. Separation can be facilitated by introducing phase boundaries and confined spaces, but this approach will often create additional mass-transfer barriers, thus reducing turnover rates and/or selectivities. This problem is illustrated by the existing aqueous biphasic hydroformylation system, which cannot be applied to long-chain olefins because of their low solubility in the catalyst-containing aqueous phase. Furthermore, catalyst leaching and cross-contamination between the substrate and catalyst phases are crucial factors for practical implementation of some biphasic approaches.

The properties of $scCO_2$ can be beneficial in this context. The gas-like mass-transfer properties, for example, can facilitate exchange between catalyst and substrate phases. Cross-contamination of CO_2 into the catalyst phase is not a problem if the catalyst is chemically inert toward CO_2. Indeed, this is often an important requirement for efficient mixing during the reaction, as the CO_2 will generally act as the mobile phase bringing reactants to, and separating products from, the catalyst. The relatively poor solvent power of $scCO_2$ for most organometallic compounds generally leads to reduced leaching compared with classical organic solvents. Most importantly, the design of SCF-based continuous-flow systems that resemble gas-phase heterogeneous catalytic processes is a very attractive new approach to reaction engineering of organometallic catalysis.

Three different approaches to catalyst immobilization involving SCFs can be distinguished. First, the tunable solvent properties of the SCF are used to control the solubility of the organometallic catalyst in the reaction medium with no additional support or solvent. In certain cases, these systems operate under truly monophasic supercritical conditions during the reaction stage. This method is referred to as "catalysis and extraction using supercritical solutions" (CESS). The second approach is liquid/supercritical multiphase catalysis, where the traditional counterpart is in fact triphasic (liquid–liquid–gas), if gaseous reagents are involved. Finally, the organometallic catalyst can be anchored to a solid organic or inorganic support, which is then contacted with the supercritical reaction medium.

Most examples so far have concentrated on $scCO_2$ as the phase containing substrates and/or products, corresponding to the mobile phase in continuous flow operation. More recently, the reverse situation, where the catalyst is retained in the $scCO_2$ phase, has also found increasing interest. These systems have been referred to as 'inverted' biphasic catalysis.

Using the CESS approach, the rates and selectivities for the hydroformylation of 1-octene and other long-chain alkenes can be commercially attractive [64]. An attractive aspect of the CESS process is that the formation of heavy aldehyde dimers will be reduced due to the low aldehyde concentration. In addition, those that eventually form will precipitate during the reaction and thus their separation is easy.

Biphasic systems in which the $scCO_2$ simply acts as a transport vector, when the catalyst is dissolved in a separate liquid phase, mean not only that catalyst retention can be improved but also that fewer pressure swings are required. The main disadvantage of these systems is that the solvent [water or poly(ethylene glycol)] can show some solubility in $scCO_2$, thus complicating the separation. Ionic liquids allow excellent, although not as yet perfect, catalyst retention and no loss of solvent [101]. Heavy products will be reduced because both the ionic liquid and the $scCO_2$ will act as diluents, even if they will eventually build up in the ionic liquid phase.

2.3.8
Supported Liquid Films

The discussion in the previous sections has evidenced that the use of biphasic systems has solved, at least in various cases, the problem of homogeneous catalyst recovery and recycle, but there still exists the problem of the cost of recycle and especially of reaction rate per volume of reactor, which derives in large part from mass- and heat-transfer limitations, but also from the low amount of catalytic centers per volume of reactor necessary to avoid side reactions and maintain a high selectivity, and/or limit catalyst deactivation or loss. These aspects often emerge only during the scaling-up and industrialization of the reaction and this is one of the reasons why many interesting reactions at the laboratory scale fail in commercialization.

An interesting approach to overcome these limits and thus combine the advantages of homogeneous and heterogeneous catalysis is that of supported liquid phase catalysts (SLPC or SLP). In SLPC the organometallic complex active components are dissolved in a small quantity of liquid phase dispersed in the form of an isle or film on the surface of supports. A SLPC has been applied successfully for several chemical transformations [113], particularly in the Wacker-type ethylene oxidation to acetaldehyde and vinyl acetate production by ethylene acetoxylation [114], and in other reactions catalyzed by Pd-complexes such as the Heck reaction [115].

Notably, there are other examples of old commercial catalysts, such as copper chloride-based Deacon catalysts used for the oxychlorination of ethylene, and alkali-promoted vanadium catalysts for sulfur dioxide oxidation in sulfuric acid manufacturing, that under the working conditions inside the reactor form a pseudo-melted surface layer [116], even if the concept is different from that of SLPC. Kieselguhr-supported phosphoric acid catalysts, used in the oligomerization of alkenes and alkylation of aromatics for gasoline and polymer production as well as in the hydration of alkenes to alcohols, are also characterized by a thin phosphoric acid film during operations. Although substituted by zeolite or other solid acid catalysts, these kieselguhr-supported phosphoric acid catalysts are still commercially used, due to their low cost of production. Supported trifluoromethanesulfonic acid (triflic acid, TFMS) is another example of supported liquid-phase catalyst for acid-catalyzed reactions. TFMS supported on TiO_2 for the synthesis of flavone and chromone derivatives [117] or on amorphous silica for alkylation of isobutane with n-butenes to yield high-octane gasoline components [118] are two representative applications. Other super-acids can be also supported, for example 1,1,2,2-tetrafluoroethanesulfonic acid (TFESA), which can be used in a range of acid-catalyzed applications, (aromatic alkylation, acylation of arenes, isomerization, oligomerization and the Fries rearrangement) [119]. Notably, these supported liquid acid catalysts are presented as an example of "green" chemistry because they allow the development of safer and more cost-effective acid catalysts (lower loadings and reaction temperatures, increased selectivity and less by-products, shorter reaction times and higher throughput with respect to acids commonly used in processes such as sulfuric or hydrofluoric acid, and aluminium chloride) [119]. These supported liquid-film acid catalysts do not contain an organometallic catalyst as in SLPC, but the general conceptual aspects are quite similar.

Supported ionic liquid catalysis is one of the main examples of SLPC adopted [120] to take advantage of ionic liquid properties without the drawbacks evidenced in Section 2.3.6. The viability of this concept has been confirmed by several studies that have successfully confined various ionic phases to the surface of support materials and explored their potential catalytic applications. Although most of the evaluated supports were silica based, several studies have focused on polymeric materials, including membranes. These materials were prepared by using two different immobilization approaches. The first involves the covalent attachment of ionic liquids to the support surface whereas the second simply deposits the ionic liquid phases containing catalytically active species on the surface of the support.

Supported ionic liquid containing dissolved Rh-complexes have been applied successfully by ExxonMobil researchers as hydroformylation catalysis (Figure 2.13a) [121]. Rhodium complexes in silica-supported ionic liquid phase were observed to show also excellent activity and selectivity towards acetyl products in methanol carbonylation [122] (Figure 2.13b). A dimeric Cr(salen) catalyst has been successfully immobilized in a silica-supported ionic liquid, demonstrating very high selectivity and good reactivity for asymmetric ring-opening reactions of epoxides [123]. The heterogenization of the Cr(salen) catalyst offers the possibility of use in a continuous-flow reactor as an alternative for the homogeneous reaction. Supported ionic liquids also offer good opportunities for asymmetric catalysis [124]. SLPC has been successfully applied in the transesterification reaction for the synthesis of biodiesel from vegetable oils [125]. In the same reaction (production of biodiesel from soybean oil) enzymes (*Pseudomonas cepacia* lipase) dissolved in a supported ionic liquid [1-n-butyl-3-methylimidazolium bis(trifluoromethylsulfonyl)imide] provides an interesting alternative [126]. These examples show the versatility of the possible applications of this concept from large-scale applications to novel catalysts for fine chemical production.

It must be also noted that supported ionic liquid phase (SILP) catalysis can also be successfully combined with supercritical fluids. Cole-Hamilton *et al.* [127] have reported recently high activity (rates up to $800\,h^{-1}$), stable performances ($>40\,h$) and minimum rhodium leaching (0.5 ppm) in the hydroformylation of 1-octene using a system that involves flowing the substrate, reacting gases and products dissolved in

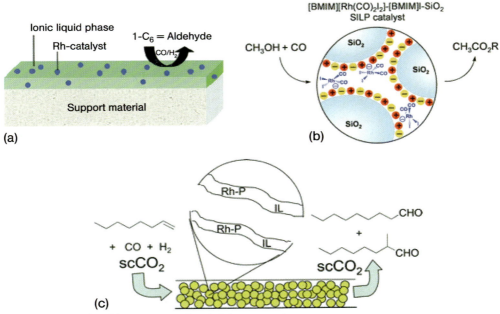

Figure 2.13 Examples of three concepts of catalysis in supported ionic liquids: (a) hydroformylation catalysis [121]; (b) methanol carbonylation [122]; (c) supported ionic liquid phase catalysis combined with scCO₂ [127].

supercritical CO_2 (scCO₂) over a fixed bed supported ionic liquid phase catalyst (Figure 2.13c).

SILP systems have proven to be interesting not only for catalysis but also in separation technologies [128]. In particular, the use of supported ionic liquids can facilitate selective transport of substrates across membranes. Supported liquid membranes (SLMs) have the advantage of liquid phase diffusivities, which are higher than those observed in polymers and grant proportionally higher permeabilities. The use of a supported ionic liquid, due to their stability and negligible vapor pressure, allow us to overcome the lack of stability caused by volatilization of the transport liquid. SLMs have been applied, for example, in the selective separation of aromatic hydrocarbons [129] and CO_2 separation [130, 131].

A further aspect to be considered is that supported ILs can also host chemicals, for example, complexing agents, which could further enhance the selectivity in the separation. This concept opens up the possible use of ILs supported on a membrane, particularly ceramic type nano-membranes (e.g., an alumina nanoporous membrane obtained by anodic oxidation) to develop novel systems that can combine catalysis and separation. The concept is shown in Figure 2.14.

Supported aqueous-phase catalysis (SAPCs) offer a second relevant possibility for heterogenization of homogeneous catalysts [132, 133]. Initially, water was used as the hydrophilic liquid and these catalysts are therefore denoted as SAPCs [134].

**Alumina nano-porous membrane obtained
by anodic oxidation**

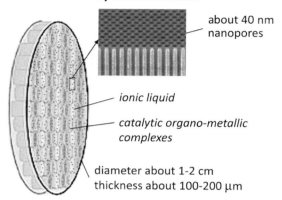

about 40 nm
nanopores

ionic liquid

*catalytic organo-metallic
complexes*

diameter about 1-2 cm
thickness about 100-200 μm

Figure 2.14 Concept of using ionic liquids (hosting organometallic catalytic complexes) supported on an alumina nano-membrane to develop novel systems that can combine catalysis and separation.

Subsequently, the concept was expanded to other hydrophilic liquids such as ethylene glycol and glycerol [135]. SAPC consists of adsorbing on the surface of an hydrophilic solid a thin film of water containing the catalyst precursor. The catalyst should contain hydrophilic ligands, which allow its dissolution in the water film and thus its anchorage on the support. The hydrophilicity of the ligands and the support creates interaction energies sufficient to maintain the immobilization. The metal atom is oriented toward the organic phase, with the catalytic reaction taking place efficiently at the aqueous–organic interface.

SAPC can perform a broad spectrum of reactions such as hydroformylation, hydrogenation and oxidation, for the synthesis of bulk and fine chemicals, pharmaceuticals and their intermediates. Rhodium complexes are the most extensively used, but complexes of ruthenium, platinum, palladium, cobalt, molybdenum and copper have also been employed [63–65]. Owing to interfacial reactions, one of the main advantages of SAPC upon biphasic catalysis is that the solubility of the reactant in the catalytic aqueous-phase does not limit the performance of the supported aqueous phase catalysts.

However, despite promising results, SAPC has not yet been used for large commercial production. In base petrochemistry the main target is the hydroformylation of higher olefins, particularly in continuous fixed-bed reactor operations. They will allow both much higher catalyst support loading, up to 60%, and better reaction performance than slurry reactors. The other field of application will be in fine and pharmaceutical chemistry, for example, for the selective hydrogenation of α,β-unsaturated aldehydes such as retinal [136] and for the production of the commercially anti-inflammatory agent naproxen [137]. Enzyme-based SAP catalysts, using porous glass beads and the enzymes polyphenol oxidase and horseradish peroxidase have also been studied [53]. These SAP catalysts were active in the reaction of phenol with O_2 or H_2O_2, respectively.

Finally, we mention supported molten metal catalysis (SMMC), in which molten metal catalysts are dispersed as nanodroplets or as thin film on the surface of porous supports. Supported salt melts provide a well-defined volume, accessible to few reactant components, with a surface that is dynamically restructuring to give access to metal cations. The supported molten salt forms a thin layer on the top of the support that is stable up to high temperatures (600 °C). Usually, the whole surface is covered, but micro- and small meso-pores are preferentially filled. Such catalysts possess very interesting properties for the oxidative dehydrogenation of light alkanes [138].

For example, alkali and alkaline earth metal chloride supported on Dy_2O_3/MgO (MD) (the eutectic mixtures decreases the melting point of pure LiCl to as low as 366 °C) were shown to have excellent properties in the oxidative dehydrogenation of ethane [139]. The maximum ethene selectivity (over 80%) was found to be directly correlated with the melting point of the eutectic melt on the catalyst support.

Also in this case, the concept could be extended and viewed as the inorganic counterpart of an ionic liquid. It is thus possible to realize a very broad range of supported liquid films between the two extremes of pure inorganic (molten salts) and organic (ionic liquids) liquid phases, with a consequent large range of possible temperatures of operation and stability. These thin liquid films could also host, in addition to metallo-organic complexes and enzymes (in low temperature applications), other co-catalysts such as porphyrins or functionalized calixarenes, or metal anions such as Keggin units selectively dissolved and enriched on the surface of the melts.

A further possible extension regards the possibility to optimize the design of the hosting metal-oxide support, and to structure the surface in terms of nanoreactors that host the liquid film (Figure 2.15) [140]. By anodic oxidation and other techniques it is possible to prepare various metal oxides showing a nanostructure characterized by an ordered array of 1D elements, such as nanotubes. The possibilities offered for catalysis by this design in the case of TiO_2 materials have been discussed recently.

By using the concept of nanoconfined liquids it is possible to improve both the stability of supported liquid films (a critical problem in all catalysts previously discussed) and the catalytic performances, because:

- the liquid is hosted in a well-defined environment, allowing for easy heat/mass transfer (unlike from conventional high surface area oxides where micro- and smaller meso-pores are completely filled, and thus only the pore mouth can be catalytically active);

- consecutive reactions can be limited due to faster desorption to the gas phase;

- more complex materials can be explored (Figure 2.15), where the catalytic elements are dissolved in the liquid and/or anchored at the surface of metal-oxide nano-reactors. In addition, the metal oxide can also play a co-catalytic role.

The nanoreactor can thus provide a confined 3D environment for optimal integration between heterogeneous, homogeneous and biocatalysis. By combining advances in knowledge in homo-, hetero- and biocatalysis it is thus possible to design

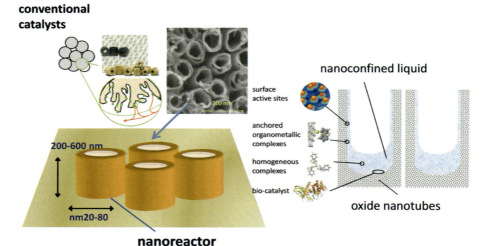

Figure 2.15 Opportunities for catalysis offered by the concept of nanoreactor and confined liquids supported on a nanostructured oxide. Source: Centi and Perathoner [140].

a new generation of catalytic materials that really can boost the implementation of novel sustainable industrial chemical processes.

2.3.9
Conclusions on Multiphase Homogeneous Catalysis for Sustainable Processes

The topic of multiphase homogeneous catalysis is a very active R&D area, but apart from a few commercially operating processes, such as Ruhrchemie/Rhône-Poulenc and SHOP, and some others still at the demonstration stage, the exploitability for novel sustainable chemical processes has still to be demonstrated. Actually, a limited variety of reactants suitable for the aqueous-phase mode are known. The same is true for the organic/organic variant, due to the inadequate number of solvent pairs that act suitably. Sensitivity to poisons, with the exception of the aqueous-phase process, is often still an issue. There are growing studies on fluorous and ionic liquids, but a lack of basic knowledge is still present. The same is true for scCO$_2$ operations. Therefore, this is an area that should still grow in terms of fundamental knowledge before conclusions can be made about exploitability.

However, we should raise some concerns regarding the sustainability of multiphase homogeneous catalytic processes. They are presented as one of the relevant directions for green chemistry. The suitability of water as a process solvent is certainly environmentally benign, and in the case of the RCH/RPs process clear advantages in terms of process simplification and reduction of energy consumption (Table 2.2) are demonstrated. However, the process has not found wider applicability and the impact on the environment, in terms of wastewater and other emissions, has to be assessed.

The use of biphasic organic solvents could improve the performances and catalyst recyclability, but in principle it is not a true sustainable process in consideration of the type of solvents used. Processes based on supercritical carbon dioxide (or scH_2O) could also in principle be good examples of sustainable processes, but when the catalysts or ligands require, for example, perfluoro "ponytails" [112] the sustainability is questionable. The same concerns apply to fluorous ligands: partially fluorinated or perfluorinated solvents are very stable, inert materials and the uncertainty about their effect on atmospheric chemistry would probably prevent the authorities from giving approval for commercial realizations: ozone depletion or the greenhouse warming potential of fluorine compounds would once more raise questions about the fate of even traces of losses during any further processing steps (e.g., recycling, work-up, etc.).

Ionic liquids are indicated as "green" solvents because of their negligible vapor pressure. However, major concerns exist regarding their impact and persistence in the environment, raising several doubts about whether they could really be presented as an advance towards sustainable chemical production.

We could thus conclude that this is a worthy area for R&D, but the sustainability of proposed processes still presents more doubts than certainties. However, new developments, such as that discussed for supported liquid films, open up very interesting perspectives, although intense R&D is still necessary.

2.4
Bio- and Bioinspired-Catalysts

"White biotechnology" (biotechnology applied to industrial processes) and biocatalysis are a key element to improving traditional chemical technology. They can, in principle, reduce pollution and waste, decrease the use of energy, raw materials and water, lead to better quality food products, and create new materials and biofuels from waste. In fact, the ability of enzymes to catalyze organic reactions in the moderate pH range of 4–9 at reasonable temperatures (usually 10–50 °C), and without extremes in pressure or the addition of metals, can provide an environmentally acceptable method for many reactions that otherwise require highly acidic or alkaline environments, high energy inputs for heating or toxic metal catalysts. However, the drawbacks of biocatalysis are underestimated. These consist of problems of mixing and mass/heat transfer (and related energy costs), the cost of separation (even if recent advances in integrating biocatalysis with membranes have reduced these costs), the high sensitivity to small changes in operational parameters (need of continuous fine control), the problems of wastewater, the short lifetimes of biocatalysts, the high cost of producing many enzymes and co-factors for biocatalyst applications, and so on.

In other words, biocatalysis should be not considered by itself but integrated in the whole bio-process (Figure 2.16a). Substantially increased emphasis on biocatalyst development is an important goal for chemistry-related industries, even if biocatalysts cannot reach their potential without a concerted effort on the parallel development of other components of bioprocessing, as well as an integration with

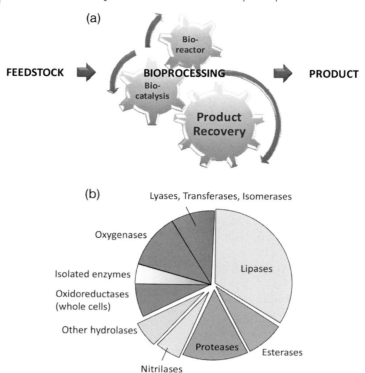

Figure 2.16 (a) Key elements in bioprocessing; (b) frequency of use of enzymes in biotransformations. Source: adapted from Faber [145].

chemo-catalysis in cascade reactions [142]. There is also the need for further knowledge on the following aspects:

- development of a tool box of biocatalysts, that is, biocatalysts that can catalyze a broader range of reactions and have greater versatility than now;
- increased temperature stability, activity and solvent compatibility;
- development of molecular modeling to permit rapid de novo design of new enzymes and in general creation of better tools for new biocatalyst development;
- improvement of the limited knowledge of enzyme/biocatalyst mechanisms;
- improvement of the poor understanding of metabolic pathways for secondary metabolites, including pathway interactions;
- reduction of the high production costs for many enzymes and co-factors.

2.4.1
Industrial Uses of Biocatalysis

The earliest industrial application of biocatalysts began without a scientific foundation: in 1874 Christian Hansen founded the first biocatalysts company, which supplied enzymes for cheese production. Now, over 130 years after the founding of this enzyme business and significant progresses in scientific understanding [143], the world annual turnover of biocatalysts is worth well over US$2 billion. More than 130 industrial biotransformation processes had been launched by 2002 [144], but the classes of enzymes utilized is small. Their relative distribution is reported in Figure 2.16b [145].

Reactions catalyzed by various types of hydrolases (lipases, esterases, proteases, nitrilases and nitrile hydratases, epoxide hydrolases, phosphatases and kinases, dehalogenases and halohydrin epoxidases, glycosidases) are predominant among biotransformations. The lack of sensitive cofactors, which have to be recycled, makes them particularly attractive for organic synthesis. Consequently, they account for about two-thirds of all reactions reported. In particular, reactions involving the cleavage (or formation) of an amide or ester bond are easiest to perform by using lipases, esterases and proteases. Other types of hydrolysis reactions involving phosphate esters, epoxides, organo-halogens and nitriles are still hampered by a restricted availability of enzymes, even if they have great synthetic potential. A second relevant class of enzymes is that able to make redox reactions (dehydrogenases, oxygenation, peroxidation). Table 2.3 provides examples of industrial applications of enzymes [146].

Table 2.3 Examples of industrial enzymatic processes.
Source: adapted from Adamczak and Krishna [146].

Enzyme	Main industrial application
Lipases	Interesterification of triacylglycerols, chiral synthesis
Glutaminase	Flavor-enhancing enzyme
Glycosynthases	Oligosaccharide synthesis
Epoxide hydrolases	Resolution of aryl and substituted alicyclic epoxides. Enantioselective catalyst
Monooxygenases, oxidizing enzymes	Oxidation, Baeyer–Villiger oxidation
Amylases	Pharmaceutical and fine-chemical industries
Oxidoreductases	Oxidation/reduction reactions
Lipoxygenases	Biosynthesis of inflammatory mediators
Nitrilase family	Different reactions

2.4.2
Advantages and Limits of Biocatalysis and Trends in Research

Table 2.4 summarizes the strengths and weaknesses of the most widely used enzymes [145]. However, the field of biocatalysis is undergoing rapid development, and it can be expected that some of the limitations will be overcome in the near future by using novel (genetically engineered) enzymes and through the development of novel techniques. Examples of recent developments of potential industrial bioprocesses include:

Table 2.4 Pros and cons of biotransformations according to enzyme types. Source: adapted from Faber [145].

Enzyme type	Reaction catalyzed	Strength	Weakness
Lipase	Ester-hydrolysis, -formation, -aminolysis	Many stable enzymes organic solvents	Low predictability for stereoselectivity
Esterase	Ester-hydrolysis, -formation	Pig liver esterase, proteases	Few esterases organic solvents
Protease	Ester, amide hydrolysis, ester aminolysis, peptide synthesis	Many stable proteases	No D-proteases
Nitrilase, nitrile hydratase	Nitrile hydrolysis	Chemo- and regio-selectivity	Enantioselectivity, no commercial enzyme
Epoxide hydrolase phosphatase/ kinase haloalkane dehalogenase	Hydrolysis of epoxides, phosphate esters, haloalkanes	Chemocatalysis fails	Few enzymes
Glycosidase	Oligosaccharide formation	Anomeric selectivity	Regio- and diastereo-selectivity, limited yields
Dehydrogenase	Reduction of aldehydes + ketones	Prelog-selectivity NADH-enzymes whole cells	Anti-Prelog-selectivity, NADPH-enzymes
Enolate reductase	Reduction of enones, α,β-unsaturated esters	Whole-cell systems	Isolated enzymes
Mono-oxygenase	Hydroxylation, Baeyer–Villiger, epoxidation	Chemocatalysis fails	Multi-component sensitive NADPH-enzymes
Di-oxygenase	Dihydroxylation of aromatics	Whole-cell systems	Isolated enzymes
Peroxidase	Peroxidation, epoxidation	No cofactor	H_2O_2 sensitive
Aldolase	Aldol reaction in H_2O	Stereo-comple-mentary DHAP-enzymes	Dihydroxyacetone phosphate
Transketolase	Transketolase reaction	Non-phosphory-lated donors	Few enzymes
Hydroxynitrile lyase fumarase, aspartase	Cyanhydrine formation	R & S enzymes	Few enzymes
Fumarase, aspartase	Addition of H_2O, NH_3	Chemocatalysis fails	Narrow substrate tolerance
Glycosyl transferase	Oligosaccharide synthesis	High selectivity	Phosphorylated donors

- conversion of hydrophobic substrates such as oils and fats to produce a range of possible products for healthy food, nutritional supplements, neutraceuticals, chiral synthons, specialty chemicals, surfactants, biopolymers, and antimicrobial and physiologically active agents;

- biocatalysis in microaqueous organic media, including lipase, esterase, protease, and so on;

- epoxidation and hydroxylation of fatty acid alcohols for integration into biorefineries;

- synthesis of value-added products from carbohydrate substrates;

- bioelectrocatalysis for synthesis of chemicals, fuels and drugs;

- synthesis of chiral intermediates for drug development. Some examples are: (i) anticancer drugs (paclitaxel, orally active taxane, deoxyspergualin and antileukemic agent), (ii) antiviral drugs (BMS-186318, HIV protease inhibitor, Atzanavir, crixivan), (iii) reverse transcriptase inhibitor (Abacavir, Lobucavir), (iv) antihypertensive drugs (angiotensin converting enzyme inhibitor, captopril, monopril), (v) neutral endopeptidase inhibitors, (vi) melatonin receptor agonists, (vii) anti-Alzheimer's drugs, anti-infective drugs, and so on;

- *trans-para*-hydroxycinnamic acid from renewable resources;

- carboxylation using decarboxylases and carbon dioxide.

To overcome some of the limitations mentioned in Table 2.4, as well as exploring new enzymes, new research areas are:

- use of ionic liquids as a solvent;
- reaction under extreme temperatures (for enzymes);
- use of light energy as a driving force to proceed biocatalysis;
- catalysis by enzyme–metal combinations.

These new trends and types of applications are described in some recent books on biocatalysis [147–149].

2.4.3
Biocatalysis for the Pharmaceutical Industry

Biocatalysts are either proteins (enzymes) or, in a few cases, they may be nucleic acids (ribozymes). Their application covers a wide range, for example, food, animal feed, pharmaceuticals, fine and bulk chemicals, fibers, hygiene and environmental technology, and they are also applied for analytical purposes, especially in diagnostics and sensors.

In chemical synthesis, particularly for pharmaceutical applications, they give highly stereo- and regioselective syntheses that often are impossible using classical catalysts. Enzymatic reactions combined with enantio-resolution can generate enantiopure materials with high yields [150]. Enzymes can be extremely enantioselective, achieving more than 99% enantiomeric excess (e.e.) in reactions. They can catalyze the formation of chiral centers from prochiral substrates or selectively discriminate between enantiomers in a racemic mixture. This latter property makes enzymes effective resolving agents. A drawback is that the maximum theoretical yield of the desired enantiomer in resolutions is 50%, and half may be waste material. However, with dynamic kinetic resolution (DKR) it is possible to reach virtually 100% yield [151]. This combines enantioselective resolution and *in situ* racemization to recycle the unwanted enantiomer.

Several multi-ton industrial processes still use enzymatic resolution, often with lipases that tolerate different substrates. BASF, for example, makes a range of chiral amines by acylating racemic amines with proprietary esters. Only one enantiomer is acylated to an amide, which can be readily separated from the unreacted amine. Many fine chemicals producers also employ acylases and amidases to resolve chiral amino acids on a large scale. *l*-Acylases, for example, can resolve acyl *d,l*-amino acids by producing the *l*-amino acids and leaving the N-acyl-*l*-amino acid untouched; after separation, the latter can be racemized and returned to the reaction. *d*-Acylase forms the alternative product. Likewise, DSM and others have an amidase process that works on the same principle: *d,l*-amino acid amides are selectively hydrolyzed, and the remaining *d*-amino acid amide can be either racemized or chemically hydrolyzed.

The problem of chiral synthesis is critical in making pharmaceuticals. Researchers at GlaxoSmithKline, AstraZeneca and Pfizer have examined 128 syntheses from their own companies and found that as many as half of the drug compounds made by their process R&D groups are not only chiral but also each contains an average of two chiral centers [152]. To meet regulatory requirements, enantiomeric purities of 99.5% were found to be necessary. Biocatalysis is thus an essential tool for pharmaceutical research, and contributes to the development of more sustainable processes.

2.4.4
Biocatalysis for Sustainable Chemical Production

In Chapter 1 (Section 1.6.1, Bio-based Economy) it was noted how "white" bio-technologies contribute to sustainability. We may cite, as an additional example, that DSM's route to the antibiotic cephalexin (a combination of a fermentation and an enzymatic reaction with respect to multistep chemo-synthesis) reduces of 65% the materials and energy used, and about 50% the variable costs [153]. Other examples of biocatalysis and "white" biotechnologies used industrially by DSM are summarized in Table 2.5.

Figure 2.17a reports in more detail the process simplification possible by biocatalysis in the case of cephalexin synthesis [154]. Figure 2.17b shows the results of a life cycle analysis (Chapter 5) of the old chemical route versus the new white biotech route [154]. The significant improvement in the sustainability of the new process is clearly evidenced.

BASF's vitamin B2 process (a one-step fermentation process as opposed to an eight-step chemical synthesis) is another example of sustainable processes by bioroutes: it reduces of 40% the environmental impacts and the costs. Novozymes'

Table 2.5 Biocatalysis and "white" biotechnology processes used by DSM for the production of fine chemicals, anti-infectives and vitamins – commercial products ranging from 100 to 100 000 tpa. Source: adapted from Wubbolts [154].

Product	Process
Citric acid	Fermentation
Vitamin C	Biotransformation: sorbitol → sorbose
Vitamin C	Biotransformation: sorbitol → KGA
Vitamin B2	Fermentation: metabolic engineering
Aspartame	Biocatalysis: protease
Aspartic acid	Biocatalysis: ammonia lyase
D-p-Hydroxyphenylglycine	Biocatalysis: hydantoinase/carbamoylase
6-APA (penicillins)	Fermentation and biocatalysis: acylase
7-ADCA (cephalosporins)	Fermentation and biocatalysis: metabolic engineering, acylase
Cephalexin	Biocatalysis: acylase
Agro application	Biocatalysis: hydroxynitrile lyase

enzymes for the scouring process (removal of the brown, non-cellulose parts of cotton) in the textile industry reduces by about 25% the primary energy demand and by 60% the emissions to water with respect to treatment with hot alkaline solution, with a cost reduction of about 20%.

Another interesting example from DSM is the synthesis of monomers for novel polyamides. In contrast to chemical routes, an enzyme (nitrilase) performs single hydrolysis of dinitrile to a mono-acid that can then be hydrogenated to novel cyclic polyamide (co)monomers (Figure 2.18). This is also an example of integration between bio- and chemo-catalysis.

Another example of new possible bioroutes in the manufacture of large-scale chemicals is the synthesis of adipic acid, an intermediate in the synthesis of Nylon-6,6, polyurethane, lubricants and plasticizers. Figure 2.19 compares the proposed bioroute with the commercial route of adipic acid production, where the use of nitric acid as oxidant gives rise to large amounts of N_2O as by-product (even if all current plants have a system for N_2O abatement). The new bioroute, which received the 1998 Academic Award in Green Chemistry from the US EPA (Karen M. Draths and John W. Frost of Michigan State University) converts glucose into cis,cis-muconic acid using a single, genetically engineered microbe. This novel biosynthetic pathway was assembled by isolating and amplifying the expression of genes from different microbes, including *Klebsiella pneumoniae*, *Acinetobacter calcoaceticus* and *Escherichia coli*. The cis,cis-muconic acid, which accumulates extracellularly, is hydrogenated to afford adipic acid.

Although still too costly, the new bioroute starts from renewable materials and allows a significant reduction in process complexity. Draths-Frost has also shown that using a single, genetically engineered microbe it is possible to catalyze the conversion of glucose into catechol.

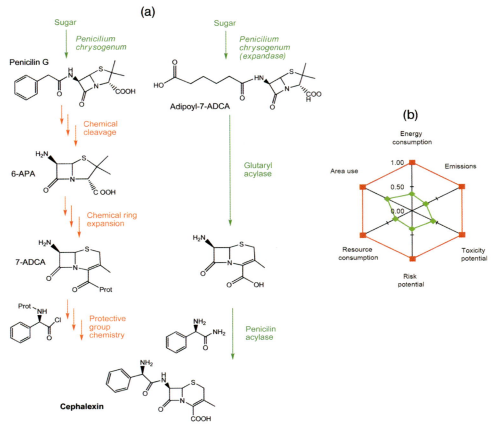

(a)

Figure 2.17 (a) Comparison of chemo- versus bio-route in the synthesis of cephalexin; (b) results of LCA analysis of the comparison of the two routes. Source: adapted from Wubbolts [154].

Figure 2.18 Synthesis of novel monomers by integration of bio- and chemo-catalysis. Source: adapted from Wubbolts [154].

Another interesting example is Mitsubishi's new route for the biocatalytic production of acrylamide [155], an intermediate produced in over 100 000 tons per year. The biocatalytic route (Figure 2.20a) is somewhat simpler than the chemical process (Cu catalyst). It is based on immobilized whole cells of *Rh. rhodocrous J1* and operates under mild conditions (5 °C), that is, no polymerization inhibitors

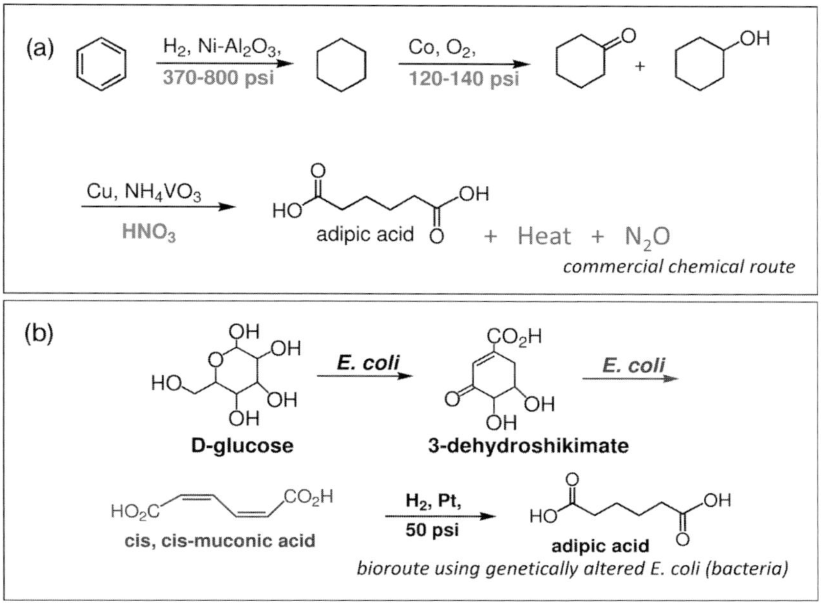

Figure 2.19 Comparison of commercial (a) and proposed bioroute (b) for the synthesis of adipic acid.

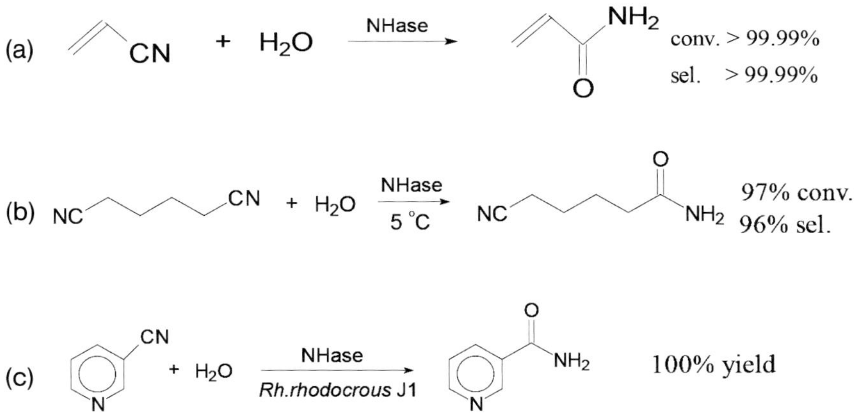

Figure 2.20 (a) Mitsubishi's new route for the biocatalytic production of acrylamide; (b) and (c) DuPont and Lonza bioroutes, respectively, for the hydrolysis of nitriles.

are needed. The reaction rates are quite good ($>400 \, g \, L^{-1} h^{-1}$) and product quality is high.

Other industrially relevant examples are the use of enzymes for the hydrolysis of nitriles. Figure 2.20b and c reports applications of DuPont and Lonza, respectively.

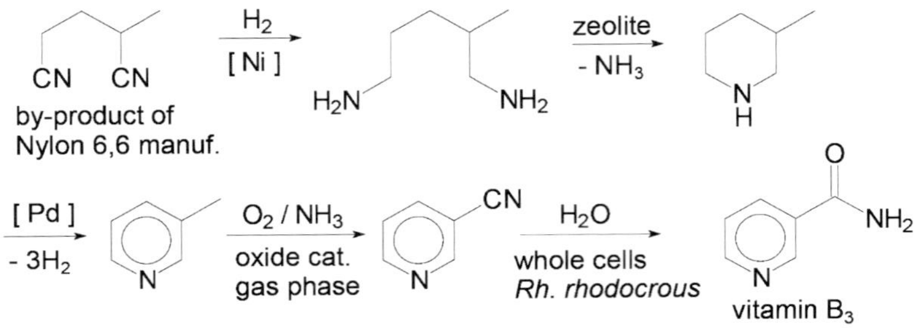

Figure 2.21 Reaction scheme of the Lonza process for the synthesis of vitamin B3 (niacinamide).

The first uses immobilized whole cells of *P. chlororaphis B23*. Catalyst consumption is very low (about 0.006 kg per kg-product) and the bioroute gives higher conversion and selectivity than the chemical process (operating with MnO_2 catalyst at 130 °C) [156]. The Lonza process uses *Rh. rhodocrous J1* and the yield is remarkably high [157]. This represents one step of Lonza's process for the synthesis of vitamin B3 (Figure 2.21) [158]. The first plant (capacity 5000 tons per year), Lonza's niacinamide plant, was built in Guangzhou (China). Some of the features and advantages of this process are listed below:

- high atom and carbon efficiencies (90–100%) with little waste (some CO_2);
- a catalyst is used in every step to increase efficiency of reaction and reduce energy requirements; energy recuperated from exothermic reactions;
- waste prevented by a highly selective four-stage process;
- ammonia liberated in picoline production integrated into ammoxidation;
- use of hazardous materials is avoided; low level of toxic by-products (catalytic destruction);
- benign solvents are used (water and toluene for extraction of cyanopyridine); toluene is practically 100% recycled;
- reagents used (3-picoline, water, air) are not particularly toxic (picoline $LD_{50} = 420\,\text{mg}\,\text{kg}^{-1}$).

2.4.5
Biocatalysis in Novel Polymers from Bio-Resources

The production of polymers using biocatalysis is another major area of activity [159, 160]. Celanese is developing applications for polylactic acid (PLA) polymer. Cargill Dow is supplying the polymer, which is derived from renewable sources. Lactic acid raw material is produced via fermentation of corn sugars. By substituting corn feedstocks for petroleum, PLA production uses 20–50% less fossil fuel than conventional plastics. The use of corn or other crops as the source of fermentable sugars for high-volume chemicals and polymer production is appealing, but it raises questions about costs and the ability to fuel market growth meeting long-term demand.

Table 2.6 Examples of enzyme-catalyzed polymer synthesis. Source: adapted from Uyama [160].

Enzymes	Typical polymers
Oxidoreductases	Phenolic polymers, polyanilines, vinyl polymers
Transferases	Polysaccharides, cyclic oligosaccharides, polyesters
Hydrolases	Polysaccharides, polyesters, polycarbonates, poly(amino acid)s, polyphosphates

DuPont is instead producing 1,3-propanediol (PDO) from corn sugar rather than petroleum feedstocks. PDO is a key component of DuPont's new Sorona 3GT polymer. Other examples of microbial polymers under study are polyhydroxyalkanoates and γ-poly(glutamic acid).

Studies on enzyme-catalyzed polymerization ("enzymatic polymerization") has been of increasing importance as a new trend in macromolecular science. Enzyme catalysis has provided a new synthetic strategy for useful polymers, most of which are difficult to produce by conventional chemical catalysts. Enzymatic polymerization also affords a great opportunity for use of nonpetrochemical renewable resources as starting substrates of functional polymeric materials (as shown in the industrial examples cited above).

In enzymatic polymerizations the product polymers can be obtained under mild reaction conditions without using toxic reagents. Therefore, enzymatic polymerization has great potential as an environmentally friendly synthetic process of polymeric materials, providing a good example of achieving "sustainable polymer chemistry."

Enzymes are generally classified into six groups. Table 2.6 shows typical polymers produced with catalysis by respective enzymes. The target macromolecules for enzymatic polymerization have been polysaccharides, poly(amino acid)s, polyesters, polycarbonates, phenolic polymers, poly(aniline)s, vinyl polymers, and so on.

Most of these syntheses are still at a laboratory scale; in terms of potential industrial applications the most relevant enzymatic syntheses are those of polyesters and phenolic polymers.

Many studies concerning syntheses of aliphatic polyesters by fermentation and chemical processes have been made in terms of biodegradable materials [161, 162]. Recently, another approach to synthesis of biodegradable polyesters has been developed by polymerization using lipase as catalyst [163, 164]. As enzymatic reactions are virtually reversible, the equilibrium can be controlled by appropriately selecting the reaction conditions. Based on this concept, many of the hydrolases, which are enzymes that catalyze a bond-cleavage reaction by hydrolysis, have been employed as catalyst for the reverse reaction of hydrolysis, leading to polymer production by a bond-forming reaction.

Lipase is an enzyme that catalyzes the hydrolysis of fatty acid esters normally in an aqueous environment in living systems. However, lipases are sometimes stable in organic solvents and can be used as catalyst for esterifications and transesterifications. By utilizing the catalytic specificities of lipase, functional aliphatic polyesters have been synthesized by various polymerization modes. Lipase-catalyzed polymerizations

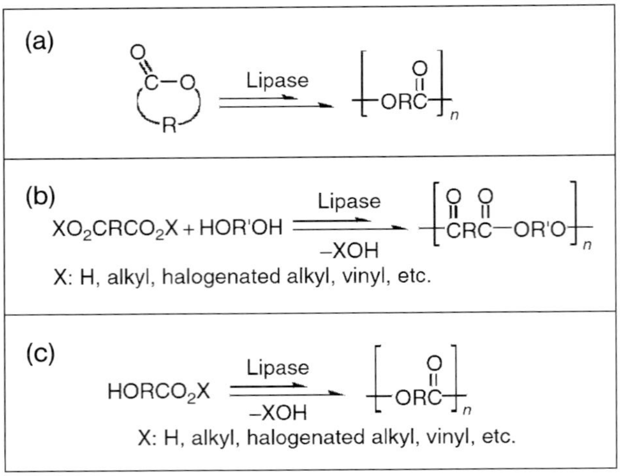

Figure 2.22 Examples of reactions of polymerization leading to polyesters catalyzed by lipase. Source: adapted from Uyama [160].

also produce polycarbonates and polyphosphates. Figure 2.22 summarizes examples of reactions of polymerization leading to polyesters catalyzed by lipase [160].

Phenol–formaldehyde resins using prepolymers such as novolaks and resols are widely used in industry. These resins show excellent toughness and thermal-resistant properties, but the general concern over the toxicity of formaldehyde has resulted in limitations on their preparation and use. Therefore, an alternative process for the synthesis of phenolic polymers that avoids the use of formaldehyde is strongly desired. The enzymatic synthesis of phenolic polymers is thus an alternative, interesting and sustainable route.

So far, several oxidoreductases, peroxidase, laccase, polyphenol oxidase (tyrosinase), and so on have been reported to catalyze oxidative polymerization of phenol derivatives; among which peroxidase is most often used [165, 166]. Peroxidase is an enzyme that catalyzes the oxidation of a donor to an oxidized donor by the action of hydrogen peroxide, liberating two water molecules. Horseradish peroxidase (HRP) is a single-chain β-type hemoprotein that catalyzes the decomposition of hydrogen peroxide at the expense of aromatic proton donors.

Peroxidase-catalyzed oxidative coupling of phenols proceeds rapidly in aqueous solutions, furnishing oligomeric compounds. However, mixtures of organic solvents and water are used to better control the quality of the polymer.

2.4.6
Progresses in Biocatalysis

Table 2.7 gives examples of novel products that can be obtained by biocatalytic routes [167]. Recent progress in research has largely increased the availability of microorganisms and enzymes for chemical products through improved methods for the isolation of new enzyme genes, and optimization of existing enzymes by

Table 2.7 Examples of chemical products obtained by biocatalytic routes. Source: adapted from Ogawa and Shimizu [167].

Product	Enzyme	Origin
D-Amino acids (CF)	D-Hydantoinase	*Pseudomonas putida* *Bacillus* sp.
	D-Decarbamoylase	*Blastobacter* sp., *Agrobacterium* sp.
L-3,4-Dihydroxyphenylalanine	β-Tyosinase	*Erwinia herbicola*
L-Serine (CF, H)	Serine hydroxy- methyltransferase	*Methylobacterium* sp.
Acrylamide (Ch) Nicotinamide (H)	Nitrile hydratase	*Rhodococcus rhodochrous*
Acrylic acid (Ch) Nicotinic acid	Nitrilase	
(2S,3R)-3-(4-Methoxyphenylglycidic acid) methyl ester	Lipase	*Serratia marcescens*
Carbacephem (H)	o-Phthalyl amidase	*Xanthobacter agilis*
Chiral epoxide	Alkene monooxygenase	*Nocardia corallina*
(R)-2-(4-Hydroxyphenoxy)propionic acid (Ch)	Hydroxylase	*Beauveria bassiana*
(S)-p-Chlorophenylethanol	Alcohol dehydrogenase	*Rhodococcus erythropolis*
Chiral 2,3-dichloro-1-propanol	Halyohydrin hydrogenhalidelyase	*Alcaligenes* sp., *Pseudomonas* sp.
(S)-1,2-Pentanediol	Alcohol dehydrogenase and reductase	*Candida parapsilosis*
D-Pantoic acid (H)	Lactonase	*Fusarium oxysporum*
Theobromine (CF)	Oxygenase	*Pseudomonas putida*
Adenosylmethionine (H)	Adenosylmethionine synthetase	*Saccharomyces sake*
Adenosylhomocysteine (H)	Adenosylhomocysteine hydrolase	*Alcaligenes faecalis*
Adenine arabinoside	Nucleoside phosphorylase	*Enterobacter aerogenes*
Ribavirine (H)	Nucleoside phosphorylase	*Erwinia carotovora*
5-Methyluridine	Nucleoside phosphorylase	*Erwinia carotovora*
Arachidonic acid	Multistep conversion	*Mortierella alpina*
Eicosapentaenoic acid	Multistep conversion	*Mortierella alpina*

directed evolution [168]. The *ad hoc* possible design of biocatalysts, due to advances in selection and mutation through recombinant DNA technology, enables production of process-compatible enzymes that can reduce the limits in industrial applications deriving from evolution-led catalyst traits [169]. This has led to the indication that "the future is now" for biocatalysis in the synthesis of pharmaceutical intermediates [170]. However, to fully realize the potential of biocatalysis for making pharmaceutical processes greener, targeted research is necessary to increase the commercial availability of key enzymes, as well as to improve enzyme stability and substrate repertoire [171].

Miniaturization in biocatalysis and fermentation is another necessary step. This will allow continuous processes with the benefits that could derive in terms of process intensification and reduction of waste. Miniature (less than 10 mL) stirred reactors and microtiter plates (MTP) have been introduced mainly with the idea of allowing high-throughput screening to speed up bioprocess development, even though they are available now also for production uses [172–174]. Notably, problems emerge with these miniature bioreactors (MBRs), such as evaporation and surface tension, which determine the performances, but which are masked in larger bioreactors.

In addition to food manufacture (bread, cheese, beer), biocatalysis is currently used mainly to produce a wide range of products in the fields of fine chemicals (e.g., amino acids, vitamins) and pharmaceuticals (e.g., derivatives of antibiotics) [175]. However, some new trends will significantly change the panorama of applications of enzymes in the very near future: (i) the fast increasing applications of biocatalysis in environmental fields, from enzymatic bioremediation to the synthesis of renewable and clean energies, as well as biochemical cleaning of "dirty" fossil fuels [176]; (ii) the increasing relevance of bioenergy (bio-diesel, -ethanol, -hydrogen, -gas) and the consequent role of biocatalysis [177]; and (iii) the increasing robustness and range of applications of enzyme-based biofuel cells [178].

For all these developments, as well as for other applications (implantable devices, sensors, drug delivery, microchips and portable power supplies), the critical factor for success will be the availability of improved methodologies for immobilization of the enzymes. Enzyme immobilization technology has been extensively studied since the 1960s [179]. Immobilization methods had evolved from the physical adsorption, entrapment and encapsulation in the porous structure of supports to chemical covalent-bonding of the ligands on the materials. Recent trends involve immobilization on nanostructured materials, due to larger specific surface areas and also to specific nanoarchitecture that provides more positions for enzyme attachment. Anchoring on nanomaterials could also provide new functionalities for the enzymatic action. Intensively successful and interesting enzyme immobilization attempts have been explored on nanomaterials, such as gold nanoparticles [180], nanoporous materials [181, 182], nanocapsules [183], and on conductive materials such as carbon nanotubes and nanocomposites [184].

This area of research is also denoted as nanobiocatalysis. A range of possible ways to stabilize enzymes in these nanostructures has been developed. Figure 2.23 reports, as an example, a cartoon of the encapsulation of an enzyme inside a silica shell (single enzyme nanoparticles, SENs) [185], and its support over (i) conductive materials (carbon or oxide semiconductor nanofibers or nanotubes) to realize biosensors or electrodes for biofuel cells and (ii) mesoporous materials (e.g., SBA-15, or MCM-41) to develop robust biocatalysts for bioremediation or chemical applications.

Biocatalysis is thus an active and promising area for the development of sustainable chemical syntheses at an industrial scale [186], as well as for several other interesting applications, from environmental bioremediation to bioenergy.

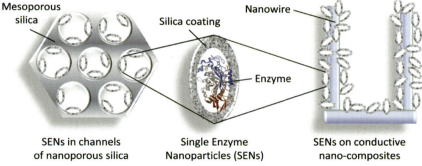

Mesoporous silica · Silica coating · Nanowire · Enzyme

| SENs in channels of nanoporous silica | Single Enzyme Nanoparticles (SENs) | SENs on conductive nano-composites |

Figure 2.23 Encapsulation of an enzyme inside a silica shell (single enzyme nanoparticles, SENs), and the support of these SENs over conductive supports or silica mesoporous material. Source: adapted from Kim *et al.* [178].

2.4.7
Biomimetic Catalysis

The development of biomimetic or bioinspired catalysts or artificial enzymes [187] has long been a challenge and holy-grail for catalysis and part of the general area of bioinspired organic chemistry [188] or biologically inspired technologies [189] that has progressed significantly in recent years. In fact, the interface between chemistry and biology has become increasingly blurred as biological processes are understood in molecular terms. This provides inspiration for the design of new molecular systems that reproduce enzymatic processes. At the same time, great progress has been made in the construction of complex and supramolecular catalysts [190]. Therefore, the gap between bio- and bioinspired catalysts will be soon bridged; indeed, good examples are already available [191].

A recent significant push on R&D in this area, arising from increasing energy problems, has stimulated bioinspired chemistry [192] in topics ranging from a better use of bioresources and conversion into fuels and chemicals to photosynthetic biomimicking catalysts for the use of solar energy in H_2 production or CO_2 conversion.

An interesting, quite active area of research in bioinspired catalysis is that of copper-catalyzed oxidation reactions [193], which is a good example of how progresses in this field could lead to more sustainable processes. In fact, the controlled oxidation of C$-$H bonds is one of the most challenging and difficult reactions in organic chemistry. Generally, it requires either stoichiometric amounts of toxic heavy metal salts or very expensive catalysts containing transition metals such as palladium, rhodium or ruthenium. Enzymatic and biomimetic oxidation catalysts, involving copper (in particular binuclear centers) as the active metal center, offer instead not only the possibility to operate under mild conditions but also to realize reactions that cannot be performed with conventional approaches, including the hydroxylation of alkanes and methanol synthesis from methane. Understanding the chemistry of copper oxidase and how to develop bioinspired copper-model-complexes that react with O_2 is the key to developing new efficient catalysts capable of aerobic oxidation of hydrocarbons [194]. Recent progress on bioinspired models of copper proteins (in particular of catechol oxidase) has been reviewed [195]. In addition, Thomas has overviewed recent developments in the modeling of the copper(II)-phenoxyl entity of the metalloenzyme galactose oxidase (GO) [196].

A second interesting area of bioinspired catalysts is that of metalloporphyrins as models for cytochrome P450 chemistry and oxidation catalysis [197]. Since their discovery in the 1960s, cytochrome P450-dependent monooxygenases have very much attracted the interest of chemists due to their ability to catalyze the selective transfer of an oxygen atom from O_2 to diverse substrates, including alkanes. The high-valent iron(IV)-oxo intermediate, formed by the reductive activation of molecular oxygen via peroxo-iron(III) and hydroperoxy-iron(III) intermediates by cytochrome P450, is responsible for the *in vivo* oxidation of drugs and xenobiotics. Porphyrins are based on 16-atom rings containing four nitrogen atoms; they are of perfect size to bind nearly all metal ions. Heme proteins (which contain iron porphyrins) are ubiquitous in nature and serve many roles, including O_2 storage and transport (myoglobin and hemoglobin), electron transport (cytochromes *b* and *c*) and O_2 activation and utilization (cytochrome P450 and cytochrome oxidase). Related macrocycles include the chlorophylls (which have a central magnesium ion) and pheophytins (which are metal free) in the photosynthetic apparatus of plants and bacteria as well as vitamin B12 (which contains cobalt) found in bacteria and animals.

An example of application of Fe(III) porphyrins is the hydroxylation of the anticancer drug cyclophosphamide to active metabolite 4-hydroxycyclophosphamide in yields similar or higher than those typically obtained by the action of liver enzymes *in vivo* [198]. This allows the development of novel anticancer drugs for the treatment of tumors with less toxic side effects to the patient. There are many other examples of metalloporphyrin-based systems for the synthesis of drugs or agrochemicals that mimic P450 catalyzed processes.

In terms of development, initial attention focused on the construction of relatively simple host–guest systems that were able to mimic certain aspects of enzymatic catalysis, for example, substrate binding in a cavity and conversion at a nearby catalytic center. The reactions studied were often taken from known enzymatic reactions, such as ester hydrolysis and aldehyde condensations, but Diels–Alder reactions have also led to nice examples of accelerations and selectivity changes.

One of the first examples were cyclodextrins containing metal complexes on one of the rims, which were used as catalysts for ester hydrolysis [199]. Breslow functionalized the β-cyclodextrin with two imidazole moieties (Figure 2.24). Selectively, catechol cyclic phosphate carrying a 4-*tert*-butyl group binds into the cavity of the catalyst in water solution, and is then hydrolyzed by the combined action of one imidazole ring acting as a base and the other one, protonated, acting as an acid. The reaction proceeds in a controlled specific direction, forming the ring-opening ester product 100-times faster than the non-catalyzed reaction [200].

Another area of investigation was based on nickel complexes anchored to cyclo-dextrins as an alternative to ethylene oligomerization catalysts for the Shell Higher Olefins Process (SHOP). The idea was that the cyclodextrin cavity could lead to

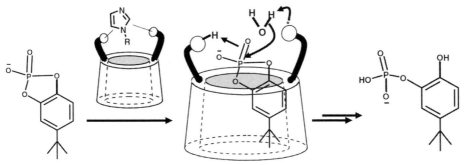

Figure 2.24 Host–guest complexation for enhanced hydrolysis.
Source: adapted from van Leeuwen [190].

a change in selectivity through the limited space in the apolar cavity dissolved in a polar solvent. Other applications of cyclodextrins made use of their size selective preference for alkene complexation in two-phase hydrogenation and Wacker oxidation [187, 201].

Later, more sophisticated supramolecular complexes capable of improved molecular recognition started to be studied. New supramolecular approaches to construct synthetic biohybrid catalysts were developed [190]. An example is the giant amphiphiles, formed by a (hydrophilic) enzyme headgroup and a synthetic apolar tail. These biohybrid amphiphilic compounds self-assemble in water to yield enzyme fibers and enzyme reaction vessels (nanoreactors [202]).

However, notwithstanding the relevant effort exerted in host–guest (supramolecular) biomimetic catalysis [190], few examples have led to relevant accelerations of reactions and improved selectivity. Noticeably, in large part the studies were based on a "the lock-and-key" idea, for example, a void host configuration that matches the configuration of the transition state. The action of enzymes is more complex and they have a dynamic and flexible configuration that not only favors and orients the binding of the substrate but also blocks the channel of access to other reactants and solvents. Thus, up to now, the large R&D effort on supramolecular catalysts has produced beautiful systems, but has made less impressive advances in terms of application. One possible approach is to use weaker bonding and go in the direction of self-assembling systems [203–205]. Self-assembly is not intrinsically simpler than covalent synthesis, but it appears to be promising.

We may conclude that bioinspired host–guest catalysis in industrial applications is still a far-off objective, even if the recognition of substrates is highly interesting and may find application in areas (e.g., sensors) other than catalysis. A further push in substrate-selective catalysis mimicking nature will derive from the use of renewable bio-feedstocks. In fact, these feedstocks are highly complex, and thus the possibility of converting a specific substrate, without the need of isolating it from the others, would industrially be highly attractive.

For example, biomimetic catalysis for hemicellulose hydrolysis in corn stover have been reported [206]. In fact, efficient and economical hydrolysis of plant cell wall polysaccharides into monomeric sugars is a significant technical hurdle in biomass processing for renewable fuels and chemicals. One possible approach to overcome this hurdle is a biomimetic approach with dicarboxylic acid catalysts mimicking the catalytic core microenvironment in natural enzymes. The field of catalysis for renewables [54] thus opens up new perspectives for biomimetic catalysis to develop novel sustainable chemical processes.

2.5
Solid Acids and Bases

The substitution of Brønsted acids (H_2SO_4, HF, HCl, p-toluenesulfonic acid), Lewis acids ($AlCl_3$, $ZnCl_2$, BF_3) or bases (NaOH, KOH, and so on) in liquid-phase homogeneous systems with solid acids or bases, to avoid the neutralization of unreacted acids/bases at the end of reaction, which leads to the generation of inorganic salts ending up in aqueous waste streams, is probably the best known successful industrial example of green (sustainable) chemistry.

Many industrial processes are based on acid/base catalysis (over 130). Examples include alkylation, etherification, cracking, dehydration, condensation, hydration, oligomerizations, esterification, isomerization and disproportionation. The dimensions of the processes range from very large scale in the field of refinery (thousand tons per day) to very small productions in fine and specialty chemical industries. In the latter case, acids and bases are often used in stoichiometric quantities, leading thus to large amounts of waste.

2.5.1
Classes of Solid Acid/Base Catalysis

The variety of solid materials used as catalysts is large and includes the following classes:

- *Zeolites and zeotype materials* [207]. These include, for example, crystalline aluminosilicates that are made up of corner-sharing SiO_4 and AlO_4 tetrahedra, and consist of a regular system of pores (channels) and cavities (cages) with diameters of molecular dimensions (0.3 to 1.4 nm). Some 136 different structures for zeolites have been synthesized, but the theoretical number is higher. Although several different natural zeolites are known, those used industrially are typical of synthesis under hydrothermal conditions (basic or HF medium) and in the presence or absence of suitable templating agents. Analogous structures containing TO_4 tetrahedra composed of Si, Al or P as well as other main group and transition elements, for example, B, Ga, Fe, Ge, Ti, V, Cr, Mn and Co, have also been synthesized and are generically referred to as zeotypes. They include, for example, AlPOs, SAPOs and MeAPOs [208]. Brønsted acidity in zeolite materials derives from the substitution of Si^{4+} with trivalent Al^{3+} (or other trivalent ions such as B^{3+}, Ga^{3+}, etc.). For charge compensation a negatively charged oxygen atom

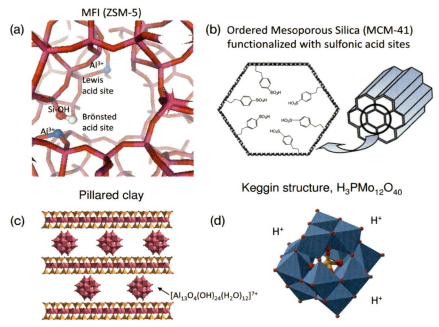

MFI (ZSM-5)

(a)

Al³⁺
Lewis
acid site

Si-OH
Brönsted
acid site
Al³⁺

(b) Ordered Mesoporous Silica (MCM-41)
functionalized with sulfonic acid sites

Pillared clay

(c)

$[Al_{13}O_4(OH)_{24}(H_2O)_{12}]^{7+}$

Keggin structure, $H_3PMo_{12}O_{40}$

(d)

H⁺

H⁺

H⁺

Figure 2.25 Examples of solid acid catalysts: (a) zeolite MFI
(ZSM-5); (b) functionalized ordered mesoporous silica
(MCM-41); (c) pillared clay; (d) a heteropolyacid ($H_3PMo_{12}O_{40}$).

forms, which should be compensated by an extraframework cation, an alkaline
metal or a proton (Figure 2.25a). The Brønsted acidity (in the case of proton)
derives from the stability of the associated SiO_4^- moiety, which depends on the
concentration of Al (or other trivalent ions) sites, and the structure of the zeolite.
Upon high temperature treatment, dehydroxylation of these Brønsted sites occurs
with the creation of Lewis acidity. Weaker acid sites are also present in zeolites,
due to defects (hydroxyl nests) and incomplete surface saturation. Zeotype AlPOs
materials, instead, in principle are not acidic (apart from defects), but acidity could
be generated by substitution (SAPOs and MeAPOs, where S indicates Si and Me is
a transition metal). The protons in the zeolite can be exchanged with metal ions,
giving rise to numerous possible catalysts containing either acid or redox centers.
Some of zeolites (e.g., ZSM-22) have a layered structure and, by a swelling
procedure, could be delayered, giving rise to a new family of high-surface acid
materials (ITQ-2 from MCM-22), which are characterized by a local structure
analogous to that of the parent zeolite, but without an ordered porous structure.
Basicity in zeolites can be created by introduction of alkaline- or alkaline-earth ions.
Notwithstanding the large number of possible structures, less than ten are actually
industrially utilized, the principals of which are MFI (ZSM-5, silicalite), MOR
(mordenite), FAU (faujasite), LTA (zeolite A), MEL (ZSM-11) and BETA (Beta).
They are different in terms of ring size, presence of monodimensional or

interconnected channels and type of cavities. Both these aspects determine the effective diffusivity of reactants and products within the channels, and constrains in the dimensions of possible transition states within the porous structure (shape selectivity).

- *Mesoporous materials* [209, 210]. Ordered mesoporous (alumino)silicates have been synthesized with the help of surfactant micelle templates. They are characterized by straight channels with a diameter ranging from 1–2 nm (MCM-41) to 5–10 nm (SBA-15). They are amorphous, unlike zeolites and zeotype materials. The inner channels are characterized by weak acidity (silanol groups), but acid sites can be created either by recrystallization (to form locally nanozeolites) or by introduction of organic groups during the synthesis that may be eventually further transformed later. Post-synthesis functionalization by grafting organic moieties is also possible (Figure 2.25b). In this way, either strong acid or strong basic groups lying inside the channels can be created.

- *Acid clays and analogous layered materials* [211]. Clays are layered (alumino)silicates in which the basic building blocks – SiO_4 tetrahedra and MO_6 octahedra (M = Al^{3+}, Mg^{2+}, Fe^{3+}, etc.) – polymerize to form two-dimensional sheets. One of the most commonly used clays is montmorillonite, in which each layer is composed of an octahedral sheet sandwiched between two tetrahedral silicate sheets. Typically, the octahedral sheet consists of oxygens attached to Al^{3+} and some lower valence cations such as Mg^{2+}. The overall layer has a net negative charge that is compensated by hydrated cations occupying the interlamellar spaces. The more electronegative is the cation, the stronger is the acidity. Both the amount and strength of Brønsted and Lewis acid sites can be enhanced by cation exchange or treatment with a mineral acid. Clays or acid-treated clays are also effective supports for Lewis acids such as $ZnCl_2$ or $FeCl_3$. The stability of these materials can be increased by developing so-called pillared clays (PILCs), in which the layered structure is intercalated with pillaring agents, the most common of which are inorganic polyoxocations such as $[Al_{13}O_4(OH)_{24}(H_2O)_{12}]^{7+}$ (Figure 2.25c). Pillaring with Al_{13} provides an interlamellar space of about 0.8 nm, which remains after drying. Layered perovskites are another class of materials that are also attractive for their acid properties, especially upon substitution. A different type of layered material is that of anionic clays, the best known of which are hydrotalcites [212]. Hydrotalcite is a natural mineral of ideal formula $Mg_6Al_2(OH)_{16}CO_3 \cdot 4H_2O$, having a structure similar to brucite, $Mg(OH)_2$. In hydrotalcites the Mg cations are partially replaced with Al^{3+} and the resulting positive charge is compensated by anions, typically carbonates, in the interlamellar space between the brucite-like sheets. After removal of carbonate anions, Brønsted base sites (OH^-) in the interlamellar space are created. Activated hydrotalcites prepared in this way have been used as solid base catalysts. The Mg and Al ions can be substituted to create also in this case a family of materials with tunable characteristics.

- *Mixed oxides* [213, 214]. These may be either crystalline or amorphous, and can be prepared with a large variety of surface characteristics and porosity, and which may be further functionalized with inorganic or organic complexes, for fine tuning of the properties; examples of interesting mixed oxides used in acid reactions are phosphates and sulfated zirconia. Recent advances include the possibility of controlling their nanoarchitecture for further tuning of properties.

- *Heteropolyacids* [215]. These are formed from a polyoxometalate anion (e.g., twelve MoO_6 groups in a octahedral coordination forming a cage around a central atom of P, in the most common Keggin structure, $XM_{12}O_{40}{}^{n-}$) (Figure 2.25d). They are widely used as homogeneous and heterogeneous catalysts, particularly those based on the Keggin structure, due to their qualities such as good thermal stability, high acidity and high oxidizing ability.

- *Hybrid inorganic-organic materials.* These are materials with an ordered pore structure, characterized by alternating inorganic and organic groups (in an ordered manner) [216]. They differ from hybrid inorganic–organic mesoporous silicates with uniform channel structures [217], where reactive and passive organic groups can be incorporated in the porous solids by grafting methods or by co-condensation under surfactant control. Periodic mesoporous organosilicas (PMOs) also belong to this class. In all these materials, various organic and organometallic groups may be integrated into the framework, creating materials with novel, tunable properties.

- *Ion-exchange resins* [216]. Polymeric ion-exchange resins are commercially used in a wide range of industrially important transformations, from alkylation, transalkylation, isomerization, oligomerization, acylation, esterification and nitration. The two main classes of ion-exchange resins are centered on styrene-based sulfonic acids (Amberlyst and Dow type resins), which show very high activity in the areas of esterification and etherification, and the perfluorosulfonic acid-based catalysts, including the recently developed Nafion resin/silica nanocomposites. These show very high activity in the area of linear alkyl benzene formation, isomerization and some selective acylation reactions.

The number and variety of solids with acid/base properties is thus extremely large, not only in terms of acid/base characteristics but also in terms of properties (from thermal to mechanical) and the possibility of incorporating other functional groups to realize multifunctional properties.

Figure 2.26 shows an example of multistep cascade reactions on multifunctional layered materials [211]. In particular, it is an example of tandem deprotection–aldol reaction with different steps, either acid and base catalyzed [219]. The overall yield, which was observed using Ti^{4+}-montmorillonite (Ti^{4+}-mont) and hydrotalcite (HT), was 93%. On using the single catalysts alone or in combination with soluble acids or bases, the worst performances were obtained. The key aspect is the possibility of localizing strong acid sites inside the layers of a montmorillonite by inserting Ti^{4+} ions, and using as base Brønsted sites those present on the external surface of the hydrotalcite particles. Owing to the localization of the active Ti^{4+} acid sites inside the narrow interlayer of the montmorillonite, the interaction with the base

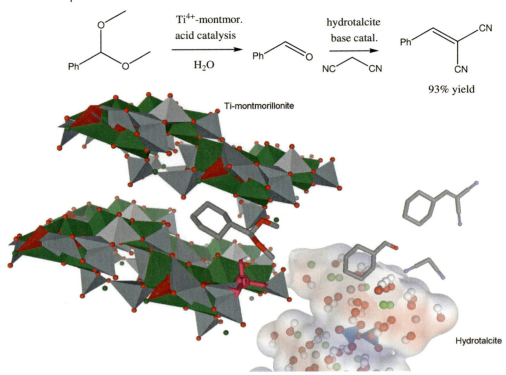

Figure 2.26 Cascade reactions requiring both strong acid and basic Brønsted sites: reaction scheme and pictorial representation of tandem deprotection–aldol reaction with different acids and basic steps using a combination of Ti^{4+}-montmorillonite $(Ti^{4+}$-mont) and hydrotalcite (HT). Elaborated from the results of Kaneda *et al.* [219]. Source: adapted from Centi and Perathoner [211].

sites, localized on the surface of the large HT particles, is hindered. Otherwise, it would be impossible to realize in a single step a reaction requiring both acid and base sites, because they will self-neutralize.

Another interesting example is the sequential coupling of the transesterification of cyclic carbonates with the selective N-methylation of anilines catalyzed by faujasites [220]. Anilines ($RC_6H_4NH_2$; R = H, p-MeO, p-Me; p-Cl and p-NO$_2$) react with a mixture of ethylene carbonate and methanol at 180 °C in the presence of alkali metal exchanged faujasites – preferably of the X-type – to give the corresponding N,N-dimethyl derivatives ($RC_6H_4NMe_2$) in yields of up to 98%. The reaction takes place through two sequential transformations, both catalyzed by faujasites: first transesterification of ethylene carbonate with MeOH to yield dimethyl carbonate, followed by the selective N-methylation of anilines by dimethyl carbonate. The overall process is highly chemoselective since the competitive reactions between the anilines and the cyclic carbonates is efficiently ruled out. Figure 2.27 shows a pictorial

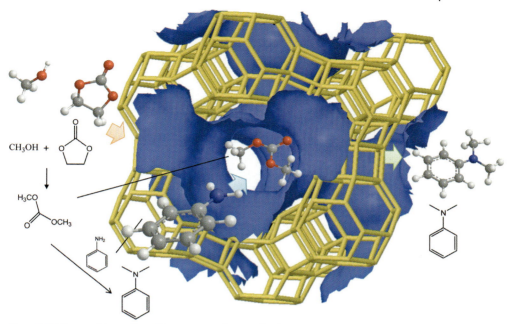

CH_3OH +

H_3CO OCH_3

NH_2

Figure 2.27 Sequential coupling of the transesterification of cyclic carbonates with the selective N-methylation of anilines catalyzed by faujasites [220].

representation of this reaction. In FAU (zeolite-X) zeolites the surface of the channel system is highlighted.

These examples show that the use of solid catalysts offers not only the possibility of avoiding the formation of waste with respect to homogeneous catalysis but also offers new possibilities to establish new sustainable processes. In fact, realizing, in one step, complex multistep reactions means a reduction in costs, amount of waste and risks of operations. Therefore, it is a primary objective for a sustainable chemistry.

2.5.2
Alkylation with Solid Acid Catalysts

The industrial alkylation of aromatics with olefins is one of the major examples of development of environmentally friendly processes with solid acid catalysts [221, 222]. The principal products obtained are ethylbenzene (EB), cumene (CUM), p-diethylbenzene, p-diisopropylbenzene, C_{10}–C_{14}linear alkylbenzenes (LAB) and cymene. Figure 2.28 summarizes several aromatic alkylations industrially applied for the preparation of important chemical intermediates [222]. These reactions include the most important aromatic substrates, benzene, toluene and xylene, and different olefins. They also include two different kinds of alkylation: electrophilic alkylation on the aromatic ring catalyzed by acids and side-chain alkylation catalyzed by bases. In terms of production volume, acid-catalyzed alkylations are by far the most

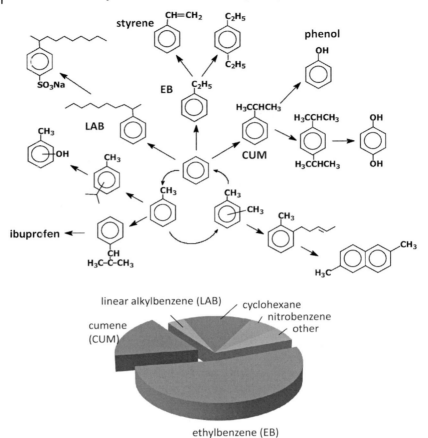

Figure 2.28 Industrial aromatic alkylations and breakdown of benzene used in the chemical industry. Source: adapted from Perego and Ingallina [221].

important. The estimated world overall demand for benzene in 2008 is around 45 million tons. Alkylbenzene derivatives account for about 75% of the total benzene production (Figure 2.28).

Transalkylation is often combined with alkylation to convert low valued by-products such as polyalkylbenzenes into their monosubstituted homologues, globally improving the efficiency of the process. Older technologies available to perform these alkylations/transalkylations are based on catalysts that have drawbacks. Often such catalysts are strong mineral acids or Lewis acids (e.g., HF, H_2SO_4, $AlCl_3$). These acids are highly toxic and corrosive.

They are dangerous to handle and to transport as they corrode storage and disposal containers. In addition, because the reaction products are mixed with acids, the separation at the end of the reaction is often a difficult and energy consuming process. Very frequently at the end of the reaction these acids are neutralized and therefore the corresponding salts have to be disposed of. Considerable effort was thus made in substituting these acids with solid acid catalysts, which are safer to handle and do not produce waste liquids.

Note, however, that liquid acids are still largely used in refinery and petrochemical processes. For example, HF alkylation (for isobutane alkylation with light olefins) is still among the top-ten refining processes licensed by UOP, with over 100 units installed worldwide. However, UOP introduced from 2002 the Alkylene process, which uses a liquid phase riser reactor with a solid acid catalyst for the isobutane alkylation. However, HF alkylation remains the best economic choice [223], notwithstanding environmental and corrosion problems. Also in this case, the conventional process has been improved, for example by HF aerosol vapor suppression. Other aspects of isobutane alkylation have been reviewed by Hommeltoft [224].

In aromatic alkylation with olefins, the solid acid catalyst based process has instead largely substituted the homogeneous acid catalysis process. This evidences that the change of substrate (isobutane vs. aromatic) could change completely the applicability of one technology with respect to another.

The alkylation of benzene with ethylene is an electrophilic substitution on the aromatic ring. Alkylation reactions are commonly considered as proceeding via carbenium-ion-type mechanisms. On a Brønsted acid site ethylene is protonated to form the active species. The latter can follow two major routes:

- It can react with benzene, producing EB, which can later undergo other reactions producing mainly diethyl- and triethylbenzene (DEB and TEB).
- It can react with another ethylene molecule, producing a C_4 species that can undergo further transformation by alkylation, oligomerization, isomerization or cracking to give other alkylbenzenes and olefins.

To a very small extent EB undergoes alkylation to diphenylethane. The DEB and TEB can be easily recovered by transalkylation with benzene to EB, so they can be considered useful products. Conversely, the formation of olefins and other alkylbenzenes heavily affects the efficiency of the process by increasing the specific consumption of ethylene and benzene and reducing the EB quality.

In the traditional process, developed since the 1930s, alkylation is performed by reacting benzene and ethylene in the presence of a Friedel–Crafts catalyst (i.e., $AlCl_3$–HCl) under mild conditions. Table 2.8 shows the advantages and disadvantages of this process. Starting from the mid-1960s different zeolite-based

Table 2.8 Advantages and disadvantages of the alkylation of benzene with ethylene catalyzed by $AlCl_3$. Source: adapted from Perego and Ingallina [221].

Pros	Cons
Single step (alkylation + transalkylation)	Corrosion
Low benzene/ethylene	Acidic waste
Mild conditions (160 °C, 0.7 MPa)	Chlorinated impurities

catalysts have been evaluated extensively in the alkylation. However, the first industrial alkylation process based on a zeolite catalyst (ZSM-5) started only in 1976. Further improvements were obtained by introducing (i) the liquid-phase alkylation and (ii) a separate step of transalkylation. The liquid phase has the advantage of better thermal control and longer catalyst life, which allows off-site catalyst regeneration and therefore easier control of pollution. To do this it was necessary to move from medium-pore zeolites like ZSM-5 to large-pore zeolites such as Beta and Y. The introduction of new zeolite catalysts together with process up-grading produced significant yield improvements and, especially, lower formation of trans-alkylate [225]. The performances greatly depend on the zeolites structure, and the Si/Al ratio. Upon decreasing the Si/Al ratio, an increase in performance was observed. The two preferable zeolites are MCM-22, developed by Mobil, and Beta, developed by ENI group. For similar Si/Al, Beta performs a little better than MCM-22. The latter performance could be improved by delamination to form ITQ-2.

Several industrial units have been revamped or realized with new zeolite catalysts operating in the liquid-phase. MCM-22 has been successfully applied in a new process called EBMax and licensed by Mobil/Raytheon since 1995. Zeolite Beta has been successfully employed by Polimeri Europa (formerly EniChem) in the revamp-ing of an existing EB unit in 2002. An energy-saving process for EB production based on catalytic distillation (CDTECH EB) also uses a modified zeolite Beta as catalyst.

Figure 2.29a shows a typical EBMax plant flow diagram [226]. The alkylation reactor is maintained in the liquid phase and uses multiple catalyst beds and ethylene injections. The ethylene conversion is essentially 100% in the alkylation reactors, and the reactors operate nearly adiabatically. The exothermic heat of reaction is recovered and used to generate steam, heat reactor feed streams, or as heat duty in the distillation columns. The transalkylation reactor can be in either the vapor or liquid phase, but the latter shows improved energy efficiency.

The alkylation and transalkylation reactor effluent streams are sent to the distil-lation section, which consists primarily of three fractionation columns. The first column is a benzene column, which separates unconverted benzene into the overhead stream for recycling to the reactors. The benzene column bottom stream feeds the EB column. The EB column recovers the EB product in the overhead stream, and the bottom stream of the EB column feeds the PEB column where PEB is fractionated overhead and recycled to the transalkylation reactor. The bottom stream of the PEB column is removed as a residue stream and is generally used as fuel in an integrated styrene complex.

With respect to the older Monsanto/Lummus AlCl$_3$-catalyzed EB process the advantages are the lower costs and risks (i) in the handling and disposal of the aluminum chloride catalyst and waste and (ii) for equipment and piping corrosion and fouling (major equipment pieces needed to be replaced on a regular schedule because of corrosion).

The CDTECH EB process (Figure 2.29b) is based on a mixed liquid–vapor phase alkylation reactor section. The design of a commercial plant is similar to the liquid phase technologies except for the design of the alkylation reactor, which combines catalytic reaction with distillation into a single operation. The reactor consists of two

(a)

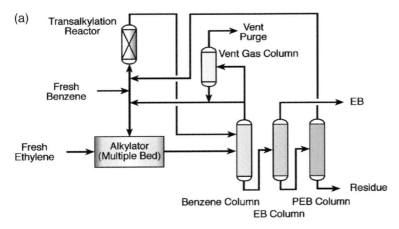

(b)

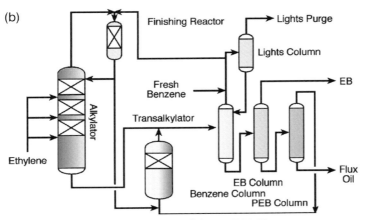

Figure 2.29 (a) Mobil/Badger's EBMax and (b) CDTECH EB processes for ethylbenzene (EB) production by benzene alkylation. Source: adapted from Woodle [226].

main sections – a catalytic distillation section and a standard distillation section. Benzene is fed to the top of the alkylation reactor and ethylene is fed as a vapor below the catalytic distillation section, creating a counter-current flow of the alkylation reactants through the catalytic distillation section. Throughout the catalytic distillation section, a vapor–liquid equilibrium is established with ethylene largely concentrated in the vapor phase. The ethylene that dissolves into the liquid phase rapidly alkylates benzene on the catalyst active sites to produce EB. The rapid reaction of ethylene in the liquid phase creates a driving force for additional ethylene to dissolve into the liquid phase where the alkylation reaction occurs on the catalyst active sites. The exothermic heat of reaction creates the vaporization necessary to effect the distillation. In the lower section of the alkylation reactor, standard distillation occurs and the reactor bottom stream contains primarily EB, PEB and other high-boiling by-products. Since the alkylator operates in a mixed vapor–liquid phase it can utilize dilute ethylene feeds, for example offgas from a fluid catalytic cracking plant or dilute ethylene from a steam cracker plant.

Of around 70 EB units operating in the world (in 2004), 24% were still based on $AlCl_3$–HCl. The rest are based on zeolite catalysts: 40% in the gas phase and 36% in the liquid phase [222]. This highlight that, notwithstanding the clear advantages both in terms of economy and environmental impact of the process using solid catalysts with respect to that based on $AlCl_3$, more than 20 years from the initial introduction one-quarter of the plants still use the old technology.

This point evidences the slow turnover in changing technologies to more sustainable ones, even in the case of evident economic advantages. When these aspects are less relevant, such as in the case of the process cited above of isobutane alkylation, the turnover is even lower. In the field of fine and specialty chemicals production, where the fixed costs are much lower, the rate of introduction of the novel, more sustainable processes, could be faster, but it is contrasted with the lower economic incentives, due to lower production volumes. In refinery/base petrochemistry, the product volumes justify the introduction of new processes, but the problem instead is the large cost of construction (and sometimes also revamping) of the plants in a period where uncertain economics, due to a global market, disincentives new investments. This is the dilemma for sustainable chemical processes.

2.5.3
Synthesis of Cumene

The other major process in the synthesis of alkylbenzenes is the manufacture of cumene [228]. The first alkylation plants used H_2SO_4 as the catalyst. However, around the 1960s the problem of waste using H_2SO_4 was overcome by employing supported phosphoric acid (SPA) on kieselguhr. This catalyst is still largely used for the production of cumene. However, owing to release of the acid, corrosion problems frequently arise. Besides, owing to the formation of organic residues on the catalyst surface, at the end of the life cycle this catalyst cannot be regenerated. The evolution of the catalytic systems for cumene production is similar to that described for EB. However, unlike with EB, the research efforts took longer to reach an industrial application. This time lag probably depends on the inability of ZSM-5, the first catalyst of EB Mobil–Badger process, to satisfactorily catalyze the cumene synthesis in the gas-phase. ZSM-5, in fact, produces an elevated cumene isomerization to *n*-propylbenzene, possibly due to the high temperature necessary to overcome the diffusion constraints in its ten-membered ring pores. In addition, a quite rapid decay was observed due to the higher tendency of propylene to produce oligomers with respect to ethylene. On the other hand, ZSM-5 shows poor catalytic activity in liquid-phase alkylation, probably because of diffusion constraints.

It was thus necessary to move to 12-membered ring zeolites (e.g., Y, mordenite, ZSM-12, Omega, Beta and MCM-22), before obtaining good results with liquid-phase operations (Figure 2.30). Based on these zeolites, new commercial processes or

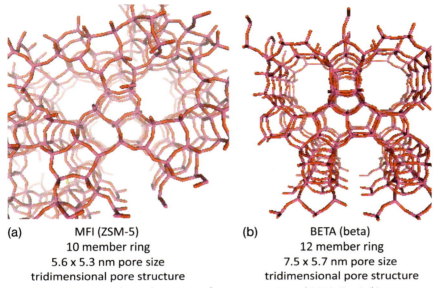

(a)   MFI (ZSM-5)   (b)   BETA (beta)
          10 member ring   12 member ring
         5.6 x 5.3 nm pore size   7.5 x 5.7 nm pore size
       tridimensional pore structure   tridimensional pore structure

Figure 2.30 Structure and pore dimensions of MFI (ZSM-5) (a) and BETA (Beta) (b).

industrial test runs were announced in the 1990s by Dow-Kellogg, Mobil-Raytheon, CDTech, EniChem and UOP [227].

The reaction pathway of benzene alkylation with propylene catalyzed by acids is very similar to that already reported for EB. The main difference is represented by the tendency of cumene to isomerize to *n*-propylbenzene, which is thermodynamically more stable at increased temperature. Also, cumene can undergo further alkylation to diisopropylbenzene (DIPB), which could be recovered by transalkylation with benzene to give cumene. The transalkylation reaction requires a higher temperature than the related alkylation. In addition, not all of the alkylation catalysts are suitable for transalkylation. Beta or dealuminated mordenite are suitable catalysts for transalkylation. The first industrial demonstrations of cumene technologies based on zeolite catalysts were started-up in 1996 by Mobil-Raytheon, EniChem and UOP, independently. In 2001, worldwide, 14 cumene units were already operating with zeolite catalysts. Around 98% of cumene is used to produce phenol and expected world production of cumene in 2008 is around 9 million tons. For cumene, among the 40 units in the world (2004), 14 cumene plants were in operation with zeolite catalysts [222]. Today over 70% of cumene plants use a zeolite as the catalyst.

This example shows that the problem of avoiding liquid acid waste was already solved for cumene synthesis using supported phosphoric acid, but the main problems were (i) the loss of H_3PO_4, which also causes problems of corrosion and deactivation, (ii) the impossibility of regeneration of the catalyst due to the type of carbon-species that accumulate on it and (iii) the formation of relatively high amounts of diisopropylbenzene (3.0–3.5%), which, even if converted by transalkylation, increases the cost of the process. However, their cost of production was much lower than that of large-pore zeolites and thus, initially, the change of technology (favored by the possibility of revamping and increasing in the process productivity) was relatively slow. Only by lowering the production costs of zeolites, by optimizing the production and cost of templates, did the zeolite-based process become preferable.

This example evidences that the change to more sustainable technologies is a complex problem in which various factors –not only the availability of new technologies – determine the possibility of success.

2.5.4
Friedel–Crafts Acylation

Another well known example of successful application of Beta zeolite is the substitution of $AlCl_3$ for Friedel–Crafts acylation. This reaction is an important industrial process, used for the preparation of various pharmaceuticals, agrochemicals and other chemical products, since it allows us to form a new carbon–carbon bond onto an aromatic ring. Friedel–Crafts acylations generally require more than one equivalent of, for example, $AlCl_3$ or BF_3. This is due to the strong complexation of the Lewis acid by the ketone product.

Rhône-Poulenc (now Rhodia) developed up to commercial scale an alternative process based on zeolite Beta [229, 230] using acetic anhydride as reactant (Table 2.9). The original process used acetyl chloride in combination with 1.1 equivalents of $AlCl_3$ in a chlorinated hydrocarbon solvent, and generated 4.5 kg of aqueous effluent, containing $AlCl_3$, HCl, solvent residues and acetic acid, per kg of product. The

Table 2.9 Comparison of the Friedel–Crafts acylation of anisole using $AlCl_3$ (liquid phase) and H-Beta (heterogeneous). Source: adapted from Sheldon et al. [2].

Using AlCl₃	Using H-Beta
Liquid phase using chlorinated hydrocarbon as solvent	Gas-phase
85–95% yield (hydrolysis of products)	>95% yield, higher purity (no hydrolysis and chlorinated by-products)
Complex separation (12 unit operations)	Lower separation costs (only 3 unit operations)
4.5 kg aqueous effluent per kg	0.035 kg aqueous effluent per kg

catalytic route using H-Beta instead avoids the production of HCl in both the acylation and in the synthesis of acetyl chloride. It generates only 0.035 kg of aqueous effluents per kg of product. Furthermore, a product of higher purity is obtained, in higher yield (>95% vs. 85–95%), the catalyst is recyclable and the number of unit operations is reduced from twelve to a few.

Alternative catalysts for this reaction are polymer-supported alkyl sulfonic acids [231], even if they show lower performances than the zeolite. Two of the problems in the reactions are the need to vaporize the reactant and the periodic regeneration of the rapidly deactivating zeolite catalysts. It was thus proposed recently that continuous catalytic Friedel–Crafts acylation can be performed in the biphasic medium of an ionic liquid and supercritical carbon dioxide [232].

There are many other examples of useful applications of zeolite and mesoporous materials, as well as other solid acid catalysts, for developing sustainable chemical processes. Several have been reviewed by Corma [233] and other authors [234–236].

2.5.5
Synthesis of Methylenedianiline

A less well-known example, but which further exemplifies the potential of solid acid catalysts in improving process sustainability, is the industrial production of methylenedianiline (MDA), used as an intermediate for the production of high-tech polyurethanes. Figure 2.31a shows the industrial process currently used. It is a multistep process with various drawbacks:

- production of a large quantity (about 2 moles for mole of MDA) of salt contaminated with aromatic compounds, which need special treatment to be disposed of;
- consumption of a relevant quantity of chemicals;
- handling of acids and caustics with related safety and corrosion problems.

Figure 2.31b shows instead the process that could be performed in two stages using acid zeolites. There is no salt production, no chemical waste and no corrosion problem. Various solid acids have been investigated: (i) fluorinated graphite, which gives large amounts of by-products, for example, N-methylated; (ii) intermetallic compounds, which give a low yield in the intermediate aminal conversion; (iii) ion-exchange resins, which give low productivity and are not regenerable; and (iv) clays, which show low water resistance and instability after regenerations [238]. Very interesting results have instead been shown by Beta zeolite, which gives (i) complete aminal conversion, (ii) low N-methylated content (<0.5%), (iii) relevant water resistance (up to 3% in the feed), (iv) high productivity (>260-g MDA per g-catalyst), (v) catalyst regenerability by thermal treatment and/ or rejuvenation and (vi) is active at high temperature ($\geq 150\,^{\circ}$C) [239]. By silylation treatment of Beta zeolite it was possible to further significantly increase the ratio of the 4,4′ MDA/(2,4′ + 2,2′ MDA) isomers from about 2.5 to nearly 6 at 99.9% conversion, due both to the elimination of acid sites on the external surface of zeolite crystals and to modification of channel accessibility, which increases also the para selectivity.

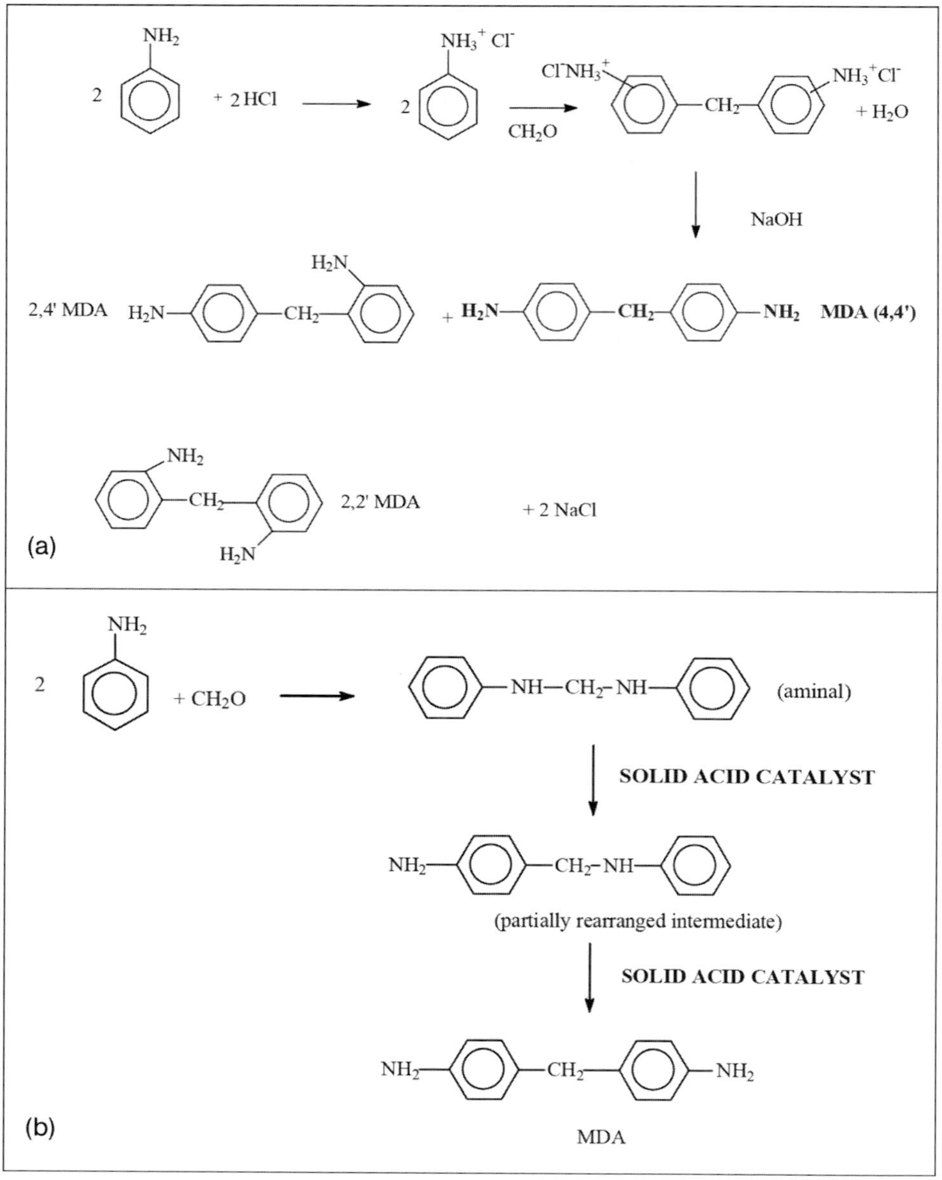

Figure 2.31 Synthesis of methylenedianiline (MDA). Comparison of the commercial process (a) with a new one based on the use of solid acid catalysts (b). Source: adapted from de Angelis et al. [237].

Amorphous aluminosilicates with controlled porosity in the region of micro-pores (ERS-8, SA) and meso-pores (MCM-41, HMS and MSA) have also been investigated [240]. Mesoporous MSA has shown similar performances of H-Beta in terms of MDA yields, but has a lower catalyst life.

This example evidences how the combination of shape selectivity and controlled acid properties still offers large industrial opportunities in improving current processes.

2.5.6
Synthesis of Caprolactam

Not only strong acid sites but also relatively weak acid sites could be utilized in the development of improved industrial processes. One of the most significative ex-amples is the new Sumitomo process for caprolactam manufacture, which combines a first step of ammoximation (originally developed by the Eni group [241–243]) with a second step of Beckmann rearrangement based on the use of silicalite-1 (Figure 2.32). The first step operates in the liquid phase with ammonia and H_2O_2 as the reactants and titanium-silicalite (TS-1) as the catalyst. TS-1 is a zeolite, developed by Eni, having a structure that belongs to the same MFI family as ZSM-5, but in which Al is absent (acid sites are detrimental for selectivity) and substituted by tetravalent Ti ions, which can activate H_2O_2 and give selective reactions of oxidation (Figure 2.33; see also Chapter 6 on propene oxide for further aspects).

Silicalite-1 is a high Si zeolite with an MFI structure and in which the weak Brønsted acid sites derive from the presence of defects [244, 245]. The active sites of the catalyst are silanol nests (very weak acidity) located close to the pore mouth of

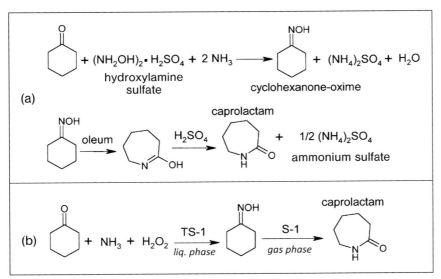

Figure 2.32 Commercial (a) versus Sumitomo (b) process for caprolactam manufacture.

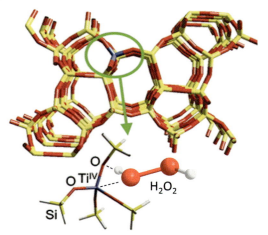

Figure 2.33 Model of titanium-silicalite (TS-1) and the activation of H_2O_2.

the zeolite [246] (Figure 2.34). Stronger acid sites catalyze the polymerization and build-up of carbonaceous species on the catalyst.

The conventional process (Figure 2.32a) involves the reaction of cyclohexanone with hydroxylamine sulfate (or another salt), producing cyclohexanone oxime that is subjected to the Beckmann rearrangement in the presence of stoichiometric amounts of sulfuric acid or oleum. The overall process generates about 4.5 kg of ammonium sulfate per kg of caprolactam, divided roughly equally over the two steps. The Sumitomo process (Figure 2.32b) instead produces virtually no waste and allows caprolactam to be obtained in >98% yield (based on cyclohexanone; 93% based on H_2O_2).

The process was commercialized by Sumitomo in 2003 at its Ehime plant in Japan. The process is sometimes called "green nylon-6", because caprolactam is the monomer for this large volume polymer. The main limit in the process is the cost of H_2O_2. Some alternative solutions are possible that can reduce largely the environmental impact of the process, but none have been commercialized. Toray in Japan has bypassed the need for the cyclohexanone or oximation steps by commercializing a photochemical process to convert cyclohexane into cyclohexanone oxime in the presence of nitrosyl chloride and hydrogen chloride. This process provides substantial capital cost savings, with the elimination of both cyclohexanone, hydroxylamine and oximation plants. However, the process requires access to low-cost power to be truly cost effective. Large-scale photochemical reactors are difficult to design and require constant cleaning to remove tar-like reaction residues.

DSM, working initially with DuPont and then later with Shell, have developed a process using butadiene and carbon monoxide feedstocks to make caprolactam without ammonium sulfate production in the mid-1990s. Called Altam, the process employs four steps – carbonylation, hydroformylation, reductive amination and cyclization. DSM claims cost reductions of 25–30%, simplified plant operations and lower energy consumption, but the process never reached commercial scale.

(a)

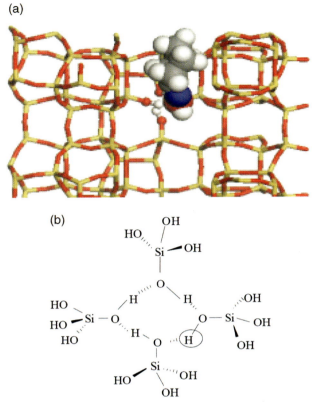

(b)

Figure 2.34 (a) Coordination of cyclohexanone oxime on the active site; (b) schematic representation of hydroxyl nests. Source: adapted from Ishida *et al.* [246].

In the late 1990s, BASF and DuPont investigated the feasibility of investing in a butadiene to adiponitrile/HMDA/caprolactam process in China. Rhodia, meanwhile, developed its own alternative approach to caprolactam manufacture called Capucine. Both the BASF and Rhodia processes involve the hydrogenation of adiponitrile to make 6-aminocapronitrile with hexamethylenediamine (HMDA) as co-product, using different operating conditions and catalysts. Adiponitrile can be manufactured from butadiene and hydrogen cyanide, and by electrolysis from acrylonitrile, the latter being produced catalytically from propene by ammoxidation. The conversion of 6-aminocapronitrile is highly selective and uses the weak acido-base characteristics of alumina to catalyze the reaction. The strength of the process is that adiponitrile is the intermediate for the production of nylon 6,6 obtained by polymerization with adipic acid. Therefore, it could be possible to obtain a flexible production of nylon 6 and nylon 6,6 from this intermediate.

To complete the analysis of caprolactam process, it should be also considered how the cyclohexanone is produced, starting from the raw material, which is benzene,

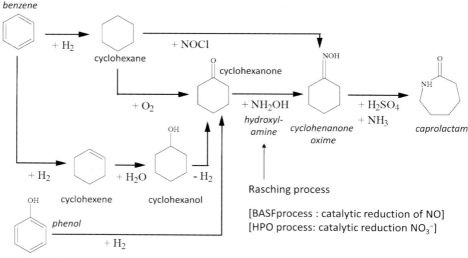

benzene

cyclohexane

cyclohexanone

cyclohexanone oxime

caprolactam

+ H₂

+ O₂

+ NOCl

+ NH₂OH
hydroxyl-
amine

+ H₂SO₄
+ NH₃

NOH

cyclohenanone
oxime

OH

+ H₂

cyclohexene

+ H₂O

cyclohexanol

- H₂

Rasching process

[BASFprocess : catalytic reduction of NO]
[HPO process: catalytic reduction NO₃⁻]

phenol

OH

+ H₂

Figure 2.35 Reaction scheme of current routes of caprolactam manufacturing process.

or eventually phenol. Figure 2.35 shows the complete picture of the reaction scheme in caprolactam production.

Caprolactam (world production of which is about 5 million tons) is mostly produced from benzene through three intermediates: cyclohexane, cyclohexanone and cyclohexanone oxime. Cyclohexanone is mainly produced by oxidation of cyclohexane with air, but a small part of it is obtained by hydrogenation of phenol. It can be also produced through selective hydrogenation of benzene to cyclohexene, subsequent hydration of cyclohexene and dehydrogenation of cyclohexanol. The route via cyclohexene has been commercialized by the Asahi Chemical Company in Japan for adipic acid manufacturing, but the process has not yet been applied for caprolactam production.

Hydroxylamine (NH_2OH) is also very important as a raw material for the production of cyclohexanone oxime. Commercially, it is mainly produced by the Raschig synthesis: aqueous ammonium nitrite is reduced by HSO_4^-/SO_2 at $0\,°C$ to yield a hydroxylamido-N,N-disulfate anion, which can be hydrolyzed to give $(NH_3OH)_2SO_4$. Solid NH_2OH can be collected by treatment with liquid ammonia. Ammonium sulfate is insoluble in liquid ammonia and is removed by filtration; the liquid ammonia is evaporated to give the desired product. The Rashig step is responsible for over half of the ammonium sulfate generation in the complete caprolactam commercial synthesis (Figure 2.32a). Alternative methods for hydroxylamine production are the BASF process (catalytic hydrogenation of NO) and the DSM HPO process (catalytic hydrogenation of NO_3^- anion). The former process produces hydroxylamine hydrogensulfate by hydrogenation of NO in dilute sulfuric acid aqueous media with the aid of a Pt on carbon catalyst. The DSM process is based on nitrate hydrogenation in a buffer solution composed of phosphoric acid and ammonia aqueous solution with the aid of a Pd catalyst.

The three hydroxylamine preparation processes produce the following amount of ammonium sulfate:

- *Raschig*, 2.8 kg per kg of caprolactam.
- *BASF*, 1 kg per kg of caprolactam.
- *DSM*, 0 kg per kg of caprolactam.

The hydroxylamine then reacts with cyclohexanone to form the cyclohexanone oxime. Conversion of cyclohexanone oxime into caprolactam involves further generation of 1.6 kg ammonium sulfate per kg of caprolactam. In terms of ammonium sulfate, the cyclohexanone oxime produced through the DSM method of synthesis is thus perfectly comparable with the method involving H_2O_2/NH_3 using TS-1 catalyst (developed by ENI group and then part of Sumimoto process; Figure 2.32b). Therefore, up to the synthesis of cyclohexanone oxime, which accounts for nearly 65% of total amount of ammonium sulfate generation, the DSM route is equivalent to Sumitomo route.

Note that in all textbooks of green chemistry the new Sumitomo process of caprolactam synthesis is given as an example, because it eliminates ammonium sulfate generation, while, usually, the DSM route is not considered. In reality, ammonium sulfate generation is only one aspect of the impact on the environment. The DSM route produces large amounts of wastewater in comparison to the ammoximation route using H_2O_2/NH_3 and involves more steps. The main difference is whether nitrate ion solutions are available or, instead, relatively cheap H_2O_2 (e.g., by direct synthesis from H_2 and O_2) could be available.

Note also that the Toray company in Japan produces cyclohexanone oxime directly from cyclohexane by the PNC (photonitrozation of cyclohexanone oxime) process, where cyclohexane is reacted with nitrosyl chloride (NOCl) under UV irradiation. The HCl eliminated in the Beckmann rearrangement process is recovered and recycled into the NOCl production. Also, this process does not produce ammonium sulfate, but it is still too costly.

Comparing the routes shown in Figure 2.35 with those shown in Figure 2.36, the main evident difference is that in one case the starting raw material is cyclohexane while in the other it is propene or butadiene. A critical factor would thus be the availability and cost of these raw materials. We may observe that the route from propene or butadiene is also completely catalytic and does not produce ammonium

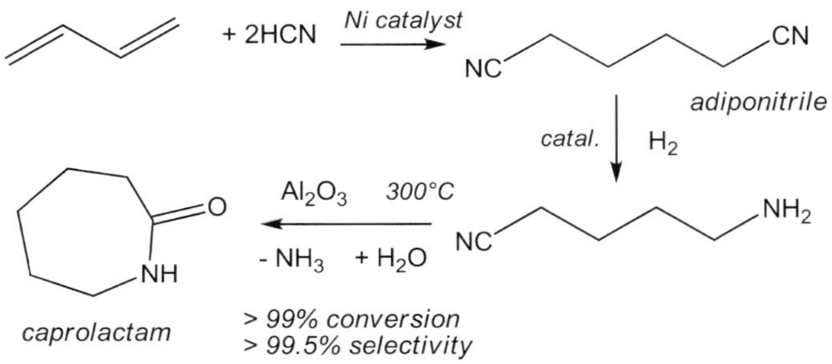

Figure 2.36 Alternative process of caprolactam synthesis.

sulfate. In terms of environmental impact this would be preferable to the actual route from cyclohexane. A critical problem, in fact, is the step of cyclohexane oxidation, which gives rise to many kinds of by-products (carboxylic acids such as adipic acid, hydroxycapric acid, glutaric acid, succinic acid, tartaric acid, oxalic acid, acetic acid, formic acid; esters; caprolactone, cyclohexyl adipate; COx). The total yield of by-products is about 20 mol.%.

In terms of sustainability, the process starting from propene would be preferable, since it avoids the risks connected with the use of HCN in the butadiene route, even if produced on demand. However, the butadiene route to produce adiponitrile (ADN) is more cost-effective, owing to the need to use an electrochemical reaction for acrylonitrile dimerization. The problem of cost, however, is highly dependent on several factors, including sensitivity to natural gas prices (which influences butadiene cost), the market for acrylonitrile, and so on. The acrylonitrile route is used by Solutia, BASF and Asahi Kasei. New plants to make caprolactam, using ADN as intermediate, are under construction in Asia.

The environmental impact of the cyclohexane oxidation could also be reduced. An alternative is to start from benzene and make a selective hydrogenation to form cyclohexene. Ru-based supported catalysts working in the liquid phase and in the presence of a co-catalysts such as Zn (Asahi Chemical Industry process) are selective in the reaction, with yields up to about 60% [247], but with cyclohexane as the main by-product. Cyclohexene is hydrated in the liquid phase with an MFI zeolite as catalyst at moderate temperature (100–130 °C). This reaction is very selective (>99%). This route was primarily developed for the synthesis of adipic acid, but could be used also to reduce the number of products and separation costs in the production of cyclohexanone.

Finally, caprolactam could also be obtained by recycling polyamide-6 (PA-6) and polyamide-6,6 (PA-6,6) waste. Rhodia recycles about 30 000 tons annually of PA-6 wastes in three different places in Europe. PA-6 production waste as well as used pure PA-6 wastes (fish nets and pure PA-6 fabrics) are used as starting products. PA-6,6 is also recycled. DSM and AlliedSignal opened a pilot plant in Richmond, Virginia, in 1997, where PA-6 carpets are depolymerized. The technology involves chemical processing of complete carpets without an expensive mechanical separation of fibers from the other carpet components.

In conclusion, a correct analysis of the sustainability in caprolactam manufacture should consider the full scheme of production and the alternatives, including the critical steps to be further developed.

2.5.7
Green Traffic Fuels

Traffic fuels account for about one-third of oil use and thus development of improved and more sustainable traffic fuels is a priority worldwide. We will not discuss this problem here, but instead take only two examples to evidence that not only zeolite or mesoporous materials are used as solid acid catalysts but that other solid acid catalysts (ion-exchange resins in these specific cases) are also widely applied [218].

To comply with the changing quality of the feedstock (more and more heavier oil is used), the changing market in traffic fuel (type and quality) and evolving environmental regulations, refineries have significantly changed their configuration in recent decades. Refinery configurations depend on what crude oil quality, product mix and quality, as well as environmental, safety, economic or other constraints were specified with its design. Although each refinery has a specific configuration, all refineries are constituted by a series of operations that grouped into six functional areas: separation, conversion or cracking, combination, reformulation, treating and other specialty or support operations.

Combination processes (Table 2.10) link two light gaseous streams to form a larger higher valued fuel product. At least one of the gas streams used in the combination process is a reactive olefin hydrocarbon molecule produced via fluid catalytic cracking, coking or outside petrochemical operation. The major combination processes are alkylation, etherification and polymerization.

Alkylation combines the lighter FCC products, for example, olefins, usually butylene or a mixture of butylenes and propylene, with a non-olefin, to produce a higher octane gasoline stream. Some aspects of this process, and in particular the still large use of hydrofluoric acid (HF) or sulfuric acid (H_2SO_4) as the acid catalysts, and related environmental/safety problems have been discussed in the previous section. In alkylation a typical reference reaction is that between isobutane and *n*-butenes to give isooctane (2,2,4-trimethylpentane), which has RON and MON numbers of 100. RON (research octane number) is a measure of the resistance of gasoline and other fuels to autoignition and is determined by running the fuel in a test engine with a variable compression ratio under controlled conditions and comparing the results with those for mixtures of iso-octane and *n*-heptane. MON (motor octane number) is a better measure of how the fuel behaves when under load.

Table 2.10 Combination technologies in a refinery.

Process technology	Typical products slates	Typical product characteristics
Alkylation	Combines FCC[a] low octane reactive molecules (olefins) with a less reactive light component to produce high value, high octane, gasoline component	Increases octane yield over either HF or H_2SO_4 catalysts
Etherification	Combines FCC[a] reactive olefins with alcohols (i.e., methanol, ethanol)	Products MTBE[b], ETBE[c] are high octane clean gasoline components
Polymerization	Combines two light olefins into high octane gasoline component	Improves gasoline yield and octane number. The products have lower octane value than products from etherification or alkylation units

[a] FCC: fluid catalytic cracking, a main refinery operation that uses heavier oil components to produce lighter components.
[b] MTBE: methyl *tertiary*-butyl ether.
[c] ETBE: ethyl *tertiary*-butyl ether.

Etherification combines lighter products from the FCC unit (various light olefins) with alcohols (methanol, ethanol) through a low temperature and pressure process, resulting in an oxygenated compound called an ether, such as methyl *tertiary*-butyl ether (MTBE), ethyl *tertiary*-butyl ether (ETBE) and tertiary amyl methyl ether (TAME). The most common ether, MTBE, is a gasoline component with high octane number and other characteristics valuable in clean gasoline production.

Polymerization also combines two light olefins to produce a high-octane gasoline component. The process employs a fixed catalyst bed at low temperatures and pressure. The process is relatively inexpensive, but the product is less desirable than alkylates or ethers. The polymerization products consist primarily of olefins that are unstable in gasoline (gum forming).

A problem present in the refinery is that, due to its fast transport in water and low biodegradability, MTBE addition to gasoline pool has been banned in some countries (from 2003 in California). MTBE is formed by acid-catalyzed reaction of isobutene with methanol. Other alcohols could be used to form different oxygenated additives, as discussed below, but the alternative is to use isobutene for conversion into another high octane number component such as isooctane, which could substitute in part the need of the alkylation process and related environmental/safety problems.

Figure 2.37a reports the reaction scheme for isooctane production by isobutene dimerization. Various isooctane processes are commercially available and are different in terms of reaction conditions and type of catalysts:

- *CDIsoether*, CD Tech and Snamprogetti.
- *InAlk*, UOP.
- *NExOCTANE*, Neste Oil Oy.
- *Selectopol*, IFP.
- *SP-Isoether*, Snamprogetti.

As an example Figure 2.37b reports the schematic flow-sheet of the NExOCTANE (high selectivity dimerization process) combined with the NExSAT (hydrogenation) processes (http://www.nestejacobs.com). The first plant (260 000 tons per year) started operations in Edmonton, Canada in 2002.

Catalysts active in the dimerization are: (i) strong cation-exchange resins, (ii) solid supported phosphoric acid, (iii) zeolites, (iv) amorphous SiO_2/Al_2O_3 and (v) supported oxides. Resins give the best results and are mainly used in the cited processes. Different organics could be used as solvents.

The ion-exchange resin is composed of a polymeric skeletal structure (e.g., copolymer of styrene and divinylbenzene), onto which ionic groups (e.g., $-SO_3^-$) are attached. The charges of the ionic groups are neutralized by counter ions, which can move freely within the polymer skeleton (e.g., H^+). Figure 2.38 shows schematically the structure of an ion-exchange resin. In applications such as acid catalysts, they are typically used as sulfonated copolymer of styrene and divinylbenzene. The maximum temperature of operation is 120–150 °C, and pore volume and pore size distribution depend on the polarity of the reaction mixture.

Ion-exchange resins also find application in the etherification reaction to produce alternative oxygenates to MTBE. Fuel ethers are preferred to alcohols for the gasoline pool because they (i) give lower emissions, (ii) are less polar, (iii) have a lower volatility,

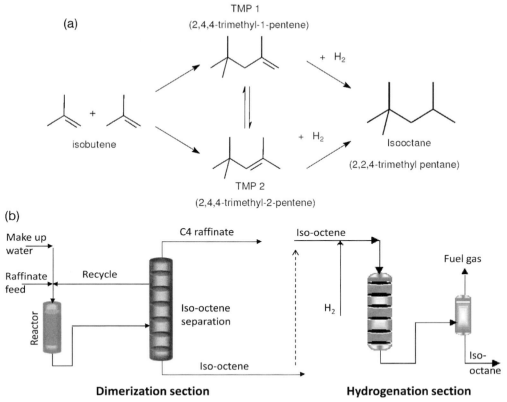

Dimerization section   **Hydrogenation section**

Figure 2.37 (a) Reaction network for isooctane production by
isobutene dimerization; (b) schematic flow-sheet of the
NEXOCTANE/NExSAT process of Neste Oil.

(iv) have a higher MON, (v) have a smaller water solubility and (vi) do not give phase
separation with H_2O. Etherification reactions of interest are the following:

- isobutene + MeOH → MTBE (MTBE = 2-methoxy-2-methylpropane),
- isobutene + EtOH → ETBE (ETBE = 2-ethoxy-2-methylpropane),
- isoamylenes + MeOH → TAME (TAME = 2-methoxy-2-methylbutane),
- isoamylenes + EtOH → TAEE (TAEE = 2-ethoxy-2-methylbutane).

Typical etherification process conditions are: (i) liquid phase ($p < 18\,bar$), (ii)
temperature $<100\,°C$ and (iii) use of a strong cation-exchange resins as catalysts.
It is an exothermic reactions with thermodynamic limitations.

Amberlyst Catalysts (Rohm & Haas) are among the most common resins used in
etherification. They are strongly acidic macroreticular catalysts based on styrene
divinilbenzene copolymers. In particular, the Amberlyst 35Wet gives the highest
performances and lifetime and was specifically developed for MTBE, ETBE and
TAME production.

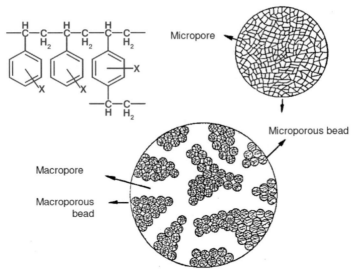

Figure 2.38 Structure of an ion-exchange resin.

2.5.8
Solid Base Catalysts

Previous sections have shown that catalysis by solid acids has received much attention due to its importance in petroleum refining and petrochemical processes. Conversely, relatively few studies have focused on catalysis by bases, even if acid and base are paired concepts. Base catalysts, however, play a decisive role in several reactions essential for fine-chemical syntheses [248–251]. Solid-base catalysts have many advantages over liquid bases. Examples of successful reactions include isomerization, aldol condensation, Knoevenagel condensation, Michael condensation, oxidation and Si−C bond formation. Various reviews have discussed catalysis by solid bases [248–255].

Table 2.11 lists the principal types of solid base catalysts. We should remember, however, that "base catalyst" is a relative definition and thus the materials listed in Table 2.11 do not necessarily function as a base in all cases. Some of these materials may act as an acid if the reactants are strongly basic. The terms, acid and base, should be used according to the function. The materials may be called solid base catalysts only if acting as a base toward the reactants by abstraction of a proton (Brønsted base) or by donation of an electron pair (Lewis base) to form anionic intermediates that undergo catalytic cycles.

We can broadly distinguish two main classes [248]:

- strong bases, such as KF/Al_2O_3, alkali metal compounds supported on Al_2O_3, and KNH_2/Al_2O_3;

Table 2.11 Principal types of heterogeneous basic catalysts. Source: adapted from Hattori [251].

Type	Class materials
Single metal oxides	Alkaline earth and alkali metal oxides (MgO, CaO, SrO, BaO) Rare earth oxides (La_2O_3, YbO_2) Transition metal oxides (ThO_2, ZrO_2, ZnO, TiO_2)
Micro- and meso-porous materials	Alkali ion-exchanged zeolites and mesoporous materials (Cs-exchanged zeolite, MgO/SBA-15) Alkali ion-added zeolites Organic functionalized mesoporous silica (MCM-41 functionalized with amino groups)
Supported alkali metal ions	Alkali metal ions on alumina (KF/Al_2O_3, Na/NaOH/Al_2O_3) Alkali metal ions on silica, zirconia, carbon Alkali metal on alkaline earth oxide (Na/MgO)
Clay minerals and mixed oxides	Hydrotalcite (MgO-Al_2O_3 and related materials, e.g., calcined and rehydrated hydrotalcite, and materials obtained by Mg or Al substitution) Chrysotile Sepiolite (magnesium silicate) Mixed oxides (MgO-TiO_2)
Non-oxides	KF supported on alumina Oxynitride (silicon oxynitride – SiON, aluminophosphate oxynitride – AlPON, zirconophosphate oxynitride – ZrPON) Lanthanide imide and nitride on zeolite Modified natural phosphate (NP) (calcined $NaNO_3$/NP)

- modest or weakly basic materials, such as zeolites, hydrotalcites (as synthesized, mixed oxides derived from them, and reconstructed hydrotalcites), oxynitrides and mesoporous silica functionalized with amino groups.

The use of these materials in a range of reactions [isomerization of alkenes and alkynes, C—C bond formation, aldol condensation, Knoevenagel condensation, nitroaldol reactions, Michael addition, conjugate addition of alcohols, nucleophilic addition of phenylacetylene, nucleophilic ring opening of epoxides, oxidation reactions, Si—C bond formation, Pudovik reaction (P—C bond formation) and synthesis of heterocycles] have been discussed in detail by Ono [248], as well as in the other cited reviews. We will thus discuss here only selected examples.

A survey of industrial applications indicated ten processes catalyzed by solid bases and 14 by solid acid–base bifunctional catalysts with respect to 103 catalyzed by solid acids. Table 2.12 gives a list of industrial examples.

2.5.8.1 Hydrotalcites

Hydrotalcites (HT) belongs to the class of anionic clay minerals, also known as layered double hydroxides (LDHs). They are probably one of the best known and used solid basic catalyst [212]. Their properties in catalytic organic reactions have been discussed extensively by Jacobs *et al.* [256]. More recent aspects have been analyzed briefly also

Table 2.12 Examples of industrial processes using solid base catalysts. Source: adapted from Hattori [251].

Type of process	Reaction	Catalyst
Alkylation	Alkylation of phenol with methanol	MgO
	Alkenylation of o-xylene with butadiene	Na/K$_2$CO$_3$
	Alkylation of cumene with ethylene	Na/KOH/Al$_2$O$_3$
Isomerization	Isomerization of safrole to isosafrole, of 2,3-dimethyl-1-butene and of 3,5-vinylbicyclo[2.2.1]heptene	Na/NaOH/Al$_2$O$_3$
	Isomerization of 1,2-propadiene to propyne	K$_2$O/Al$_2$O$_3$
Dehydration/ condensation	Dehydration of 1-hexylethanol and of iso-butyraldehyde to iso-butylisobutyrate	ZrO$_2$
	Dehydration of propylamine-2-ol	ZrO$_2$-KOH
	Dehydrotrimerization of iso-butyraldehyde	BaO-CaO
Esterification	Esterification of ethylene oxide with alcohol	Hydrotalcite
Miscellaneous	Thiols from alcohols with hydrogen sulfide	Alkali/Al$_2$O$_3$
	Cyclization of imine with sulfur dioxide	Cs-zeolite

by Centi and Perathoner [211] and Kannan [257]. Structural, surface and reactivity aspects of hydrotalcite materials are discussed in a book edited by Rives [258].

The structure of HT closely resembles that of brucite, Mg(OH)$_2$. In the latter structure, Mg^{2+} cations are octahedrally coordinated by hydroxyl groups, which are edge shared to form stacked layers. Compared to brucite, in HT some of the Mg^{2+} cations have been replaced by Al^{3+}, resulting in positively charged cation layers; the charge is compensated by anions situated together with water molecules in between the brucite-like layers, as schematically represented in Figure 2.39 [212].

The general formula of a hydrotalcite is:

$$[M^{2+}_{1-x}M^{3+}_x(OH)_2][A^{n-}]_{x/n} \cdot yH_2O$$

where M^{2+} and M^{3+} represent divalent and trivalent cations in the octahedral sites within the hydroxyl layers, x is the ratio M^{3+}/(M^{2+} + M^{3+}), with a value varying in

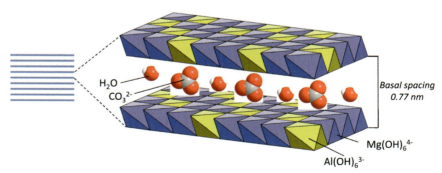

Figure 2.39 Schematic representation of a hydrotalcite structure (Mg$_6$Al$_2$(OH)$_{16}$CO$_3$·4H$_2$O).

the range 0.17–0.50, and A is an exchangeable interlayer anion. The M^{2+} and M^{3+} cations must have ionic radii that are not too different from 0.65 Å (characteristic of Mg^{2+}) to form a stable structure of hydrotalcite. In naturally occurring hydrotalcites, carbonate is the interlayer anion. However, the number of counterbalancing ions is essentially unlimited, and HT intercalated by various simple inorganics, polyoxometalates, complexes as well as organic anions have been synthesized. Therefore, it is possible to prepare tailor-made materials for specific applications by changing the cationic and anionic compositions of hydrotalcites.

The unique basic properties of HT make them very useful for catalytic purposes. However, notably, what often is indicated as a hydrotalcite is not really this compound with a layered structure. During thermal treatments a HT transforms first into an amorphous oxide and then, at higher temperatures, into a crystalline spinel-like oxide. Therefore, when the sample is calcined above 300 °C it is instead present as an amorphous oxide or as a crystalline spinel-like oxide. HT derived oxides, however, have the peculiar characteristic of reconstruction during catalytic reactions, returning to a structure resembling the starting HT structure (memory effect), which often is quite relevant in determining the catalytic performance [259–261].

Recent patents for HT use cover their application in fine chemicals productions, either as base or multifunctional catalysts. Figure 2.26 gives an example of the latter. HT also finds application for polymerization and as polymer additives, as well as with other catalytic reactions, particularly in the area of environmental catalysis.

Although HTs could be relatively easy prepared on a lab-scale by co-precipitation, they are commercially available in large amounts for industrial uses, for example by Süd-Chemie, which is one of the main industrial suppliers. An interesting example of application of HTs is the condensation and selective hydrogenation of acetone to methyl isobutyl ketone (MIBK).

MIBK, after methyl methacrylate and bisphenol-A, is the third largest tonnage product obtained from acetone. The world production of MIBK is about 2.5 millions tons per year. It is used as a solvent for cellulose- and resin-based coating systems, and as a raw material in the production of rubber antioxidants.

MIBK is produced from acetone and H_2 in three reaction steps (Figure 2.40). In the conventional process, the first step is the self-condensation of acetone to diacetone

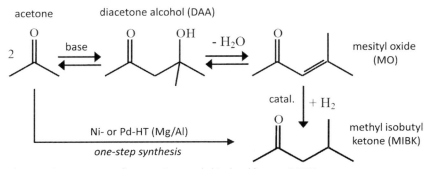

Figure 2.40 Conversion of acetone into methyl isobutyl ketone (MIBK).

alcohol (DAA), which is typically catalyzed by dilute sodium hydroxide; the DAA is then separated by distillation from the reaction mixture to undergo dehydration to form mesityl oxide (MO), catalyzed by phosphoric or sulfuric acid. The last step involves the selective catalytic hydrogenation of the C=C bond of MO to give MIBK, in either the gas- or liquid-phase. The reaction is thus complex and energetically costly (distillation step), with low yields (the first two steps are limited by the thermodynamic equilibrium) and has a relevant impact on the environment, related to the use of aqueous caustic (NaOH) in the first step and acid solution in the second step. A one-step direct catalytic reaction is thus attractive, but requires a multi-functional catalyst able to perform the three steps. Various catalysts have been investigated, such as Pd on ion-exchange resin, Pt on NaX or CsX zeolites, Pd or Pt on ZSM-5, $Pd/Nb_2O_5/SiO_2$, Cu on MgO, Pd on zirconium phosphate, Ni-γ-Al_2O_3 or Pd or Ni on MgO. Major problems were the poor selectivity and catalyst deactivation.

Figueras *et al.* ([262] and references therein) were among the first to evaluate the possible use of hydrotalcite-based materials for the one-step conversion of acetone into MIBK. HT possess both basic and dehydration properties and thus by intro-duction of an hydrogenation function it is possible to perform the reaction in a single step. Subsequently, various other research groups [263, 264] investigated this reaction on HT-derived materials, and in particular by partial substitution of Mg in the HT structure with Ni to introduce a hydrogenation activity, or supporting Pd particles for the same function. However, in general, maximum yields were not higher than 30–40%, due to the relatively low activity of HT. It was therefore necessary to increase the reaction temperature, resulting in an increase in side reactions.

To overcome the low yields, HT can be supported on carbon nanofibers (CNF), which offer a high geometrical surface area without microporosity together with optimal electron conductivity properties (Figure 2.41) [265]. Transmission electron images showed that Mg–Al hydrotalcite platelets with a lateral size of 20 nm could be deposited on carbon nanofibers and the resulting supported catalyst exhibited a specific activity in the condensation of acetone four-times higher than that of unsupported hydrotalcites, owing to the higher number of active edge sites [265, 266].

Note, however, that rehydration of Mg–Al hydrotalcites in the liquid phase using ultrasound or a high stirring speed [267] leads to nanoplatelets with surface areas of $400 \, m^2 \, g^{-1}$, displaying catalytic activities in aldol condensations up to eight-times higher than the best catalytic system reported in the literature. There are thus alternative methods to increase the performances of HT materials.

Another reaction of interest, in which a bifunctional catalyst (HT plus an hydro-genation function) works well is the conversion of *n*-butyraldehyde into 2-ethylhex-enal. Aldol and related condensation reactions such as Knoevenagel and Clai-sen–Schmidt condensations are also widely used in fine chemicals and specialty chemicals, for example, flavors and fragrances, industries.

An example is the base-catalyzed aldol condensation of citral and acetone (Figure 2.42a), producing pseudoionones (6,10-dimethyl-3,5,9-undeca-trien-2-one), which is utilized to obtain higher value products from essentials oils (citral), since

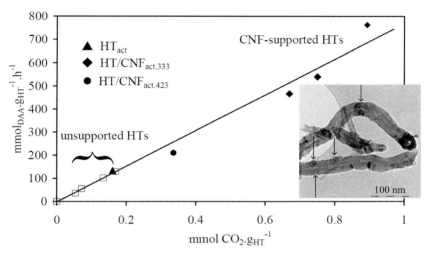

Figure 2.41 Initial activity in the condensation of acetone at 273 K versus CO_2 adsorption over unsupported activated HT and HT/CNFs (carbon nanofibers) activated at 333 and 423 K. Inset: TEM image of HT/CNFs$_{333}$. Source: adapted from de Jong et al. [265, 266].

pseudoionones are intermediates in the synthesis of ionones. The latter are extensively used in the perfume and soap industries and as precursor in the synthesis of vitamin A (β-ionone).

Pseudoionones are produced commercially by the aldol condensation of citral and acetone catalyzed by aqueous or ethanolic solutions of bases. However, these liquid catalysts have potential problems related to undesirable side reactions (self-condensation of citral and secondary reactions of pseudoionones), catalyst separation and disposal (the catalyst cannot be reused and causes serious damage to the environment) as well as purification steps, which are laborious and costly [268]. These homogeneous processes also require a large excess of acetone, typically 10–20 moles per mole of citral, for a satisfactory yield, and a large and costly plant is needed for commercial production.

These problems could be overcome with the use of basic solid catalysts. Among these, Mg,Al-mixed oxides/hydroxides derived from HTs show the best performances [268, 269], with selectivity over 70% at nearly full conversion.

Hydrotalcite-derived materials also show good performances in analogous reactions, such as the Claisen–Schmidt condensation of substituted 2-hydroxyacetophenones with substituted benzaldehydes, the synthetic route to flavonoids and the condensation of 2,4-dimethoxyacetophenone with p-anisaldehyde to synthesize Vesidryl, a diuretic drug [270]. Another similar class of reactions in which HT-based materials give good results are Knoevenagel condensations [271]. An example is the synthesis of citronitrile, a perfumery compound with a citrus-like odor, which can be prepared by HT-catalyzed condensation of benzylacetone with ethyl cyanoacetate, followed by hydrolysis and decarboxylation (Figure 2.42b) [272].

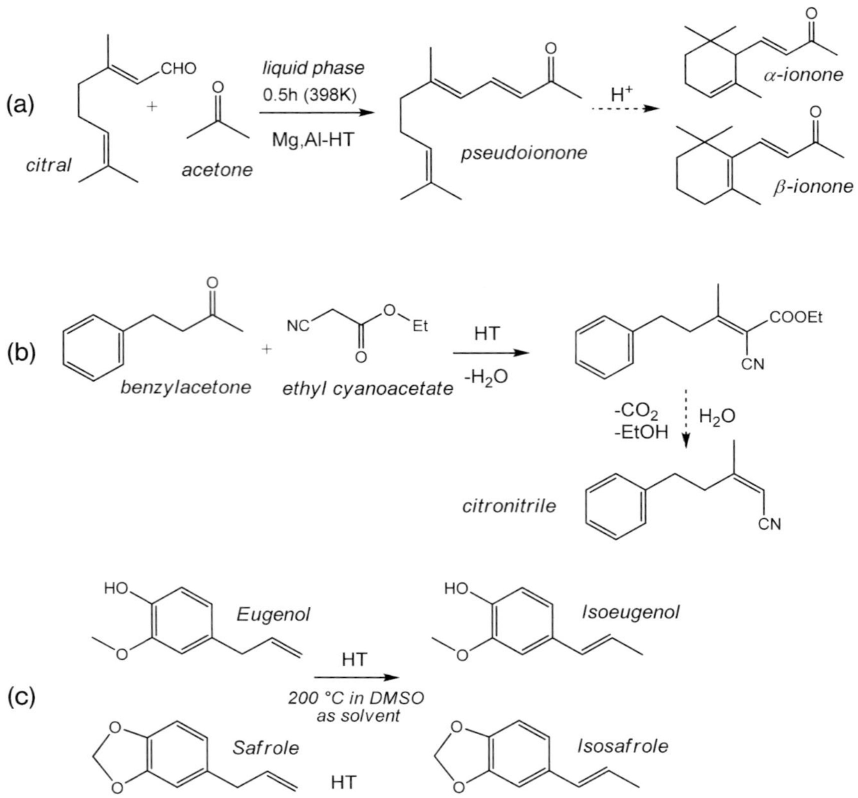

Figure 2.42 Examples of reactions catalyzed by hydrotalcite-based catalysts for the synthesis of flavors and fragrances: (a) aldol condensation; (b) Knoevenagel condensation; (c) double-bond isomerization.

Various other reactions are catalyzed by hydrotalcites, such as Michael additions, cyanoethylations and alkylations of 1,3-dicarbonyl compounds [256]. Another example is selective double bond migration. Isoeugenol and isosafrole find application in pharmaceuticals and fragrances and they can be synthesized by isomerization of, respectively, eugenol and safrole (Figure 2.42c). MgAl hydrotalcites with an Mg/Al atomic ratio 6.0 showed a maximum conversion of around 98% with a cis/trans isosafrole ratio of 1/9 at 200 °C using DMSO as solvent. Brønsted basic hydroxyl groups attached to Mg^{2+} are indicated as the active centers involved in the isomerization reaction [273, 274].

Double bond migration could be catalyzed either by bases or acids. However, one of the characteristic features of solid base catalysts is the lack of C—C bond cleavage ability, resulting in higher selectivities. Solid base catalysts have another advantage: double bond migration of unsaturated compounds containing heteroatoms such as N and O proceeds smoothly over solid base catalysts. Acidic catalysts, in contrast, interact strongly with heteroatoms and become poisoned.

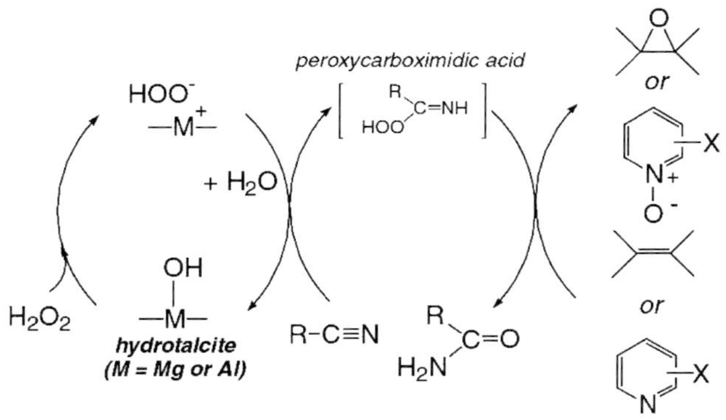

Figure 2.43 Reaction mechanism for epoxidation catalyzed by basic sites of hydrotalcite in the presence of H_2O_2 and nitriles. Sources: adapted from Kaneda *et al.* [28].

The basic sites of hydrotalcites can also activate H_2O_2 to give HOO^- anion species. In the presence of nitriles, the HOO^- can react with the nitrile to form peroxycarboximidic acid, which is able to transfer the oxygen into an olefin to produce an epoxide together with a carboxyamide [28] (Figure 2.43). Yields over 90–95% for a large variety of substrates and under mild reaction conditions (60 °C) could be obtained.

Hydrotalcites, owing to their layered structure, offer also interesting possibilities to develop multifunctional catalysts that combine the basic sites of HT with other catalytic functionalities. Numerous transition metals can be easily introduced into the brucite-like layer, interlayer space or surface by using the following characteristics:

- the cation-exchange ability of the brucite layer,
- the anion-exchange ability of the interlayer,
- surface tunable basicity,
- adsorption capacity.

Figure 2.44 shows an overview of the different possibilities, which have been explored systematically by Kaneda *et al.* [28].

The basic sites of HT (type A in Figure 2.44) are active in the Baeyer–Villiger oxidation of various carbonyl compounds, for example, the oxidation of a substrate in the presence of O_2 and an aldehyde (e.g., benzaldehyde) through the intermediate formation of a peracid. An example is the oxidation of cyclopentanone to δ-valerolactone [275]. The base property of hydrotalcites can be tuned by changing the Mg/Al ratio and the content of interlayer anion species, for example, CO_3^{2-}, Cl^- and SO_4^{2-} (type D in Figure 2.44). It is possible to observe that the yields of δ-valerolactone increase as the calorimetric heats of benzoic acid adsorption and the zeta-potential increase; the hydrotalcites with a large number of base sites gave high catalytic activities. Yields of up to 80% in δ-valerolactone were obtained.

The basic sites of hydrotalcites, the amount of which could be changed by modification of the interlayer anion species, promote the step of oxygen transfer

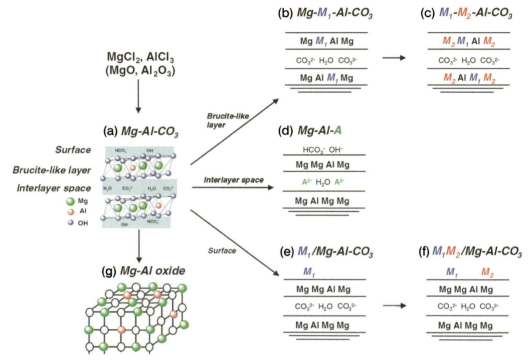

Figure 2.44 Strategy for designing highly-functionalized hydrotalcite catalysts. Source: adapted from Kaneda *et al.* [28].

from perbenzoic acid to ketone [28]. Autoxidation of benzaldehyde with O_2 produces perbenzoic acid, and the reaction of the base OH group on the hydrotalcite surface with perbenzoic acid gives a metal perbenzoate species and H_2O. Then, the perbenzoate species attacks the ketone to form a metal alkoxide intermediate, which further reacts with H_2O to give the lactone or ester accompanied by the formation of benzoic acid and the fresh hydrotalcite surface.

The performances of base hydrotalcites in the heterogeneous Baeyer–Villiger oxidation using O_2 and benzaldehyde can be significantly further improved by introducing transition metals, for example, Fe and Cu, into the brucite-like layer to form multi-metallic hydrotalcite catalysts (type B in Figure 2.44) [276]. In particular, Mg–Al–Fe–CO_3 hydrotalcites efficiently oxidize various cyclic ketones to give high yields of the corresponding lactones, while in the case of Mg–Al–Cu–CO_3 hydrotalcites, bicyclic ketones were oxidized almost quantitatively. The improved catalytic activity of multi-metallic hydrotalcites can be ascribed to the cooperative action originating from base sites and transition-metal sites. This example shows the flexibility of HT in tuning the properties to develop materials with tailored performances.

Mixed oxides derived from HT materials (type G in Figure 2.44) also possess interesting properties, and in particular show acid–base bifunctional properties that could be employed, for example, in the chemical use of CO_2 by reaction with epoxides to form five-membered cyclic carbonates [277], which are monomers for interesting polymers. The advantages of Mg–Al mixed oxides are: (i) high catalytic activity even under a CO_2 atmosphere, (ii) reusable catalysts without any toxic metals and (iii) stereospecific addition via configuration retention of epoxides.

Mg–Al oxides with a Mg/Al ratio of 5, calcined at $400\,^{\circ}C$, are the most active catalysts for the reaction of CO_2 and styrene oxide, and DMF is the best solvent. Using this Mg–Al oxide, various kinds of epoxides could be quantitatively converted into the corresponding cyclic carbonates. This addition reaction proceeds with retention of the stereochemistry of epoxides; the reaction of CO_2 with *(R)*- and *(S)*-benzyl glycidyl ether gave *(R)*- and *(S)*-4-(benzyloxymethyl)-1,3-dioxolane-2-one with >99% e.e., respectively.

Figure 2.45 shows the proposed reaction mechanism. The addition reaction is initiated by adsorption of CO_2 on the Lewis-base sites to form a carbonate species and, independently, an epoxide is coordinated on the neighboring acid site on the surface. The coordinated epoxide is ring-opened by nucleophilic attack on the carbonate species, which leads to an oxy-anion species, which yields the corresponding cyclic carbonate as a product.

A different possibility – the performance in one step of reactions that require both acid and base properties – can be realized by using a cascade reaction with two different catalysts that do not self-neutralize each other. An interesting example is given by Kaneda *et al.* [28], who investigated the combination of Ti^{4+}-exchanged montmorillonite (Ti^{4+}-mont) with large hydrotalcite particles that cannot enter the narrow interlayer space of Ti^{4+}-mont [278]. An example has been given in Figure 2.26, which reports the synthesis of benzylidenemalononitrile from

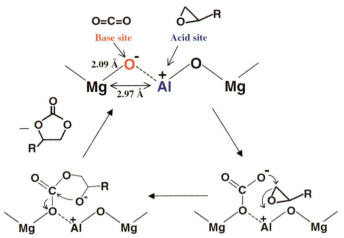

Figure 2.45 Acid–base bifunctional reaction mechanism in the synthesis of cyclic carbonates over mixed oxides obtained from a Mg-Al hydrotalcite. Source: Kaneda *et al.* [28].

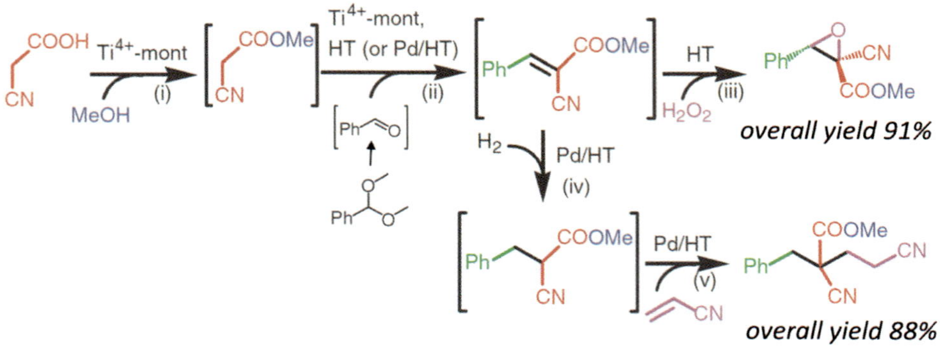

Figure 2.46 One-pot synthesis using Ti^{4+}-exchanged montmorillonite (Ti^{4+}-mont.) and Mg-Al hydrotalcite (HT). Source: adapted from Kaneda et al. [28].

malononitrile with benzaldehyde dimethyl acetal (one-pot deprotection–aldol reaction of acetals). This combination of catalysts is also active in one-pot Michael reactions and acetalizations. For example, nitromethane undergoes the Michael reaction with methyl vinyl ketone, followed by acetalization with ethane-1,2-diol to afford an 89% yield of 2-methyl-2-(3-nitropropyl)-1,3-dioxolane, instead of the less than 70% yield afforded by the conventional two-step method.

Figure 2.46 reports another example of a complex one-pot multi-stage reaction. Epoxynitrile, an intermediate for the synthesis of various heterocyclic compounds, was successfully obtained using methanol, cyanoacetic acid and hydrogen peroxide in four sequential acid and base reactions, namely, (i) esterification, (ii) deacetaliza-tion, (iii) aldol reaction and (iv) epoxidation, in a single reactor. It is also possible to perform a one-pot synthesis of glutaronitrile using the Ti^{4+}-mont and the Pd/HT catalysts. After reaction of the unsaturated nitrile under 1 atm of H$_2$, the Michael reaction with acrylonitrile occurs at the base sites of Pd/HT to afford 2-carbomethoxy-2-benzyl-glutaronitrile with an excellent overall yield.

2.5.8.2 Other Solid Bases

In the previous section, the use of HT-based materials was discussed in detail, as a good example of how it is possible to use the growing knowledge in catalysis to develop sustainable chemical processes ranging from large-volume intermediates to small-volume chemicals for the fine and specialty industry. Even if HT-materials are of much use for their good properties, flexibility in modification and low-cost of production, Tables 2.11 and 2.12 show that various other solid bases exist and are applied. Their comparison requires the utilization of standard model reactions. One that is widely used is the Knoevenagel reaction of benzaldehyde with ethyl cyanoa-cetate [279] (Figure 2.47). However, a better analysis of the base properties requires spectroscopic methods such as FTIR and UV spectroscopies [280] or temperature-programmed desorption (TPD) methods (of CO$_2$, pyrrole or other probe molecules).

Pure basic oxide (alkaline and alkaline-earth oxides) are also used industrially, but due to the difficulty in obtaining thermally-stable materials with high surface area

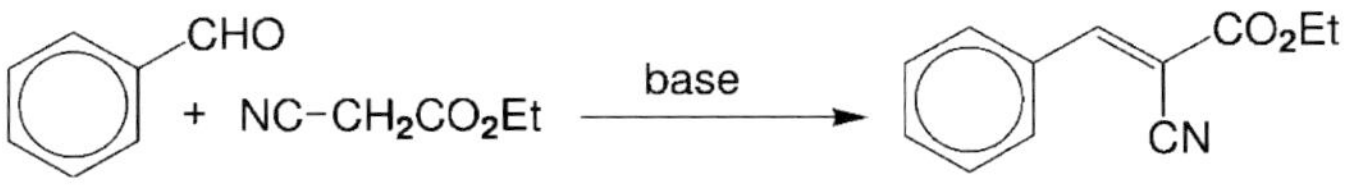

benzaldehyde   ethyl cyanoacetate

Figure 2.47 Knoevenagel reaction used as a model test to compare the properties of solid base catalysts.

they are often supported. Oxides such as alumina, silica and zirconia are often used, but one problem could be either ion diffusion (e.g., K^+) in the bulk of the oxide or the formation of specific compounds (e.g., K-aluminate). A further problem arises because the catalytic activities of these alkaline and alkaline-earth oxides depend strongly on the pretreatment temperature. Usually, below a pretreatment temperature of about 450 °C they do not show activity, but this is the critical temperature range in which they can start to react with the support.

For this reason, recent attention has been dedicated to micro- and meso-porous materials (MMM), and also for their possible use as shape-selectivity controlled reactions (space-restricted transition states, preferential diffusion and back-diffusion). Even if the latter aspects are better known regarding the performances of microporous materials, other aspects can be evidenced. In fact, also in mesoporous materials, where the dimensions of the channels are larger, as required for shape-selectivity effects, a change of reactivity of molecules inside the channels could be present due to confinement effects [281]. Therefore, the catalytic reactivity shown by basic sites located inside mesoporous channels could differ from that of the same sites located instead on the external surface of the mesoporous ordered material.

Basic MMM can be prepared mainly by two possible post-synthesis approaches: (i) ion exchange of protons by alkali metal or rare earth cations and (ii) generation of nanoparticles of alkali metal or alkaline earth metal oxides within the host substrate channels and cavities. The Lewis basicity is associated with the negatively charged framework oxygens and thus increases with the size of the counter cation, for example, in the order Li < Na < Ka < Cs. Alkali-exchanged zeolites contain a large number of relatively weak basic sites capable of abstracting a proton from molecules with a pK_a in the range 9–11 and a few more basic ones (up to $pK_a = 13.3$) [279]. These sites are able to catalyze Knoevenagel and Michael reactions, but, in general, reactivities are lower than those observed, for example, by hydrotalcites.

For bulky molecules, larger channel materials such as alkali-exchanged mesoporous silica should be used. Na- and Cs-exchanged MCM-41, for example, are active in Knoevenagel or aldol condensation, depending on the substrate [282]. Stronger basic sites could be created by loading alkali and alkaline earth metal oxides, instead of the ions. They can be generated within the pores and cavities of MMM by over-exchanging them with an appropriate metal salt, for example, an acetate, followed by thermal decomposition of the excess metal salt, to afford highly dispersed basic oxides occluded in the pores and cavities. For example, caesium oxide loaded faujasites, exhibiting super-basicities, have been prepared by impregnation of CsNa-X or CsNa-Y with caesium acetate and subsequent thermal decomposition [283].

A different strategy to generate superbasic sites has been reported recently by Zhu *et al.* [284]. They first coated SBA-15 with MgO to passivate the silanol groups on the surface of siliceous SBA-15 and stabilized it again by high-temperature treatment. Then, they dispersed KNO_3 and decomposed it at high temperature to generate the superbasic sites. They also prepared CaNS-loaded SBA-15 superbasic catalysts by decomposition of Ca-nitrate [285] and K-functionalized mesoporous γ-alumina solid superbase by a one-pot synthesis [286].

There has thus also been great interest recently in preparing novel solid base catalysts. One motivation is also given by the use of these basic catalysts in the production of biofuels. The most relevant example is the transesterification of vegetable oils (palm oil, soybean oil, jatropha oil, coconut oil, rapeseed oil, etc.). Figure 2.48 shows the scheme of the process. Transesterification reactions predominantly use homogeneous base catalysts, for example, sodium methoxide, sodium hydroxide and potassium hydroxide. The main differences between the commercial processes lie in the following:

- *Reactor design*: continuous stirred tank reactor (CSTR), loop reactor, tubular reactor.
- *Purification steps*: residual catalysts and soaps need to be removed from biodiesel and glycerol.

There are basically two options in the purification, that is, the water washing process and adsorbent treatment process (water-free process). In the water washing process the main drawbacks are the amount of wastewater produced and the energy costs to evaporate and recover water for re-use. In the adsorbent treatment process the problems are the high cost of adsorbent (e.g., Mg-silicate) and the disposal of the spent adsorbents. A potential cleaner process should thus eliminate the catalyst clean-up step and simplify biodiesel and glycerol purification. The options are: (i) the use of heterogeneous solid catalysts, (ii) the use of an enzymatic transesterification processes and (iii) a catalyst-free process, using, for example, supercritical methanol.

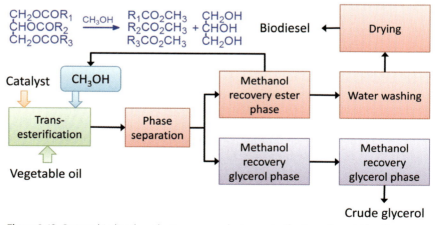

Figure 2.48 Current biodiesel production process by transesterification of vegetable oils.

Enzymatic transesterification is under investigation [287, 288], but the cost of lipase production is the main hurdle for commercialization. Intracellular lipase as a whole cell biocatalyst could lower the lipase production cost. Another problem is how to maintain lipase activity in the presence of a high concentration of methanol and glycerol. Industrialization is under investigation, but still not realized.

Transesterification using supercritical methanol is also possible [289–291]. No catalyst is needed and in 5 min over 90% conversion could be reached. However, high temperatures (350 °C) and pressures (45 MPa) are necessary, as well as a high methanol to oil ratio (>40). The advantage is that free fatty acids in crude oils and fats could also be converted into methyl esters. However, the main limitations are the high investment and energy costs, and the use of excessive methanol. There are good advances to overcome these limits, but the industrialization is still far.

The use of heterogeneous solid catalysts is thus the preferable option to develop a cleaner process. A commercial biodiesel process using a heterogeneous catalyst has been developed by IFP (ESTERFIP-H Process), and licensed by Axens. A plant using this technology has been operating in Séte (France) since 2006. It uses as catalyst ZnO and $ZnAl_2O_4$ on Al_2O_3. It operates in a fixed bed reactor at 210–250 °C and 30–50 bar. The yield of fatty acid methyl esters (FAME) is 91% at 240 °C, 50 bar and 160 min contact time. The process tolerates small amounts of H_2O and of free fatty acids (FFA). For higher amounts a preliminary step of esterification is necessary. The process shows high costs due to the high temperature and pressure requested, and the quality of the final product is also not high, requiring further processing steps to meet specifications. There are many other main technology providers for biodiesel processes (AT-Agrar-Technik, BDI – BioDiesel International, CD-Process Technology, Christof Group, Desmet Ballestra Oleo, Energea Umwelttechnologie GmbH, Lurgi AG, and so on), but most operate with homogeneous, alkaline catalysts [292, 293].

A large variety of solid base catalysts have been investigated, from alkali metal or alkaline earth oxides, carbonates, or hydroxides (alone or supported), to anion-exchange resins (Amberlyst A26, A27), silicates and layered clay minerals. Calcium carbonate is often used, because it is readily available at low-cost. However, it requires high reaction temperatures and pressures as well as high alcohol volumes. Similar drawbacks have to be attested to for alkali metal or alkaline earth metal salts of carboxylic acids. The use of strong alkaline ion-exchange resins, in contrast, is limited by their low stability at temperatures higher than 40 °C and by the fact that free fatty acids in the feedstock neutralize the catalysts even in low concentrations. Finally, glycerol released during the transesterification process has a strong affinity to polymeric resin material, which can result in complete impermeability of the catalysts.

Some of the characteristics required by the solid base catalyst are the following:
- high activity (conversion higher than 99% to meet specifications on mono-, di- and triglycerides, at temperatures below 100 °C to reduce costs of operations and of reactors);
- high catalyst stability (no or minimum leaching of soluble species, which also increase the costs of purification to meet biodiesel specification);
- need of low methanol to oil molar ratios;
- reusability and easy handling of the catalyst.

In the transesterification of palm oil a catalyst based on 1.5% K loaded-calcined (500 °C) Mg-Al hydrotalcite gave 96.9% methyl ester content and 86.6% yield at 100 °C in 6 h (methanol to oil ratio = 30) [294]. In the transesterification of blended vegetable oil with ethanol using different mesoporous silicas (MCM-41, KIT-6 and SBA-15) loaded with MgO a temperature of 220 °C was necessary for 96% conversion [295]. CaO (14% wt) supported on SBA-15 [296] gave a conversion of 95% with sunflower oil and 65% with castor oil at 60 °C after 5 and 1 h of reaction time, respectively, with a 1:12 molar ratio of the oil to MeOH. It is claimed that no leaching of active phase occurred due to the strong interaction between CaO and silica, but this aspect has to be further verified in longer-term experiments. The performances depend considerably on the type of vegetable oil. For example, in the transesterification of soybean oil using calcium ethoxide and a 12:1 molar ratio of methanol to oil, at 65 °C, a 95.0% biodiesel yield was observed after 1.5 h. The catalytic performances of calcium ethoxide are better than those of CaO. In addition, a 91.8% biodiesel yield was obtained when it catalyzes soybean oil to biodiesel with ethanol [297].

There are thus already quite interesting results using solid heterogeneous catalysts in the transesterification of vegetable oil, although performances and stability have still to be improved, as well as their applicability for use with mixtures of vegetable oils.

2.6
Redox Catalysis

Although in catalytic reactions, in particular on the surface of solid catalysts, it is not formally correct to distinguish between acid–base and redox catalysis, because usually they are both involved, this distinction is often common. The two main classes of reactions are selective hydrogenation and selective oxidation.

2.6.1
Hydrogenation

Catalytic hydrogenation is a common operation in both industrial and laboratory catalytic syntheses. As an example, in the synthesis of vitamins 10–20% of all reaction steps are catalytic hydrogenations.

Many different functions can be hydrogenated with a few catalytically active metals. However, when dealing with multifunctional molecules, catalytic activity for the desired transformation is not enough, because the catalyst should also be chemoselective. In other words, it should not affect other reducible functional groups.

Knowledge on selective reduction is growing and today several commercial catalysts can be selected from a catalogue. In pharmaceutical syntheses the time to develop the industrial production of a new product or to scale-up laboratory preparations are the critical factors. Therefore, the availability of various catalysts and knowledge would reduce considerably the time, even if the synthesis is not optimized (which could be a less serious problem in the production of high value products). This

Table 2.13 Important catalytic hydrogenation reactions, and
preferred metal and solvent types. Source: adapted from
Blaser *et al.* [298].

Substrate	Reaction	Metal	Solvent
Azides	$RN_3 \rightarrow RNH_2$	Pd, Pt, Ni	Polar
Aromatic nitro groups	$ArNO_2 \rightarrow ArNH_2$	Ni, Pd, Pt	Various
Benzyl derivatives (debenzylation)	$ArCH_2X \rightarrow ArCH_3 + HX$		
X = OR, NR$_2$		Pd	Protic, acidic or basic
Alkenes	$R_2C=CR_2 \rightarrow R_2HC\text{-}CHR_2$	Pd, Pt, Rh, Ni	Various
Alkynes	$RC\equiv CR \rightarrow RHC=CHR$	Pd/Pb	Low polarity
Aliphatic C=O groups	$R_2CO \rightarrow R_2CHOH$	Ni, Ru, Pt, Rh	Polar
Aromatic C=O groups	$ArCOR \rightarrow ArCH(OH)R$	Pd, Pt, Cu	Polar
Aryl halides	$ArX \rightarrow ArH \; X = Cl, Br, I$	Pd	Basic
Nitriles	$RCN \rightarrow RCH_2NH_2$	Ni, Ph/Pd, Pt	Basic/acidic
Imines	$R_2C=NR \rightarrow R_2CHNHR$	Pd, Pt	Various
Oximes	$R_2C=NOR \; R_2CHNH_2$	Ni/Pt, Pd	Basic/acidic
(Hetero)aromatic rings		Rh, Ru, Pt	Various

is why catalytic hydrogenations are by far the most utilized reactions catalyzed by
solids that are used in fine and specialty chemical production.

Table 2.13 reports a list of important catalytic hydrogenation reactions, together
with preferred metal and solvent types. The chemoselectivity is the most critical issue
when using multifunctional molecules, as is usually the case in the pharmaceutical
area. There are many combinations of reducible functions; Figure 2.49 gives an
overview of general rules for chemoselective hydrogenations of various substrate
types [298]. Of course, depending on the particular combination of functional groups,
there are exceptions to these rules.

Most heterogeneous hydrogenation catalysts are safe and easy to handle and can
be readily separated from the reaction mixture by simple filtration, allowing
convenient work-up and isolation of the desired product. The classical and most-
used hydrogenation catalysts are the noble metals Pt, Pd, Rh and Ru supported on
active carbon, along with Raney nickel and a few supported Ni and Cu catalysts.
Various manufacturers supply a full range of hydrogenation catalysts, including
Degussa, Engelhard, Grace, Heraeus, and Johnson Matthey. Many manufacturers
have developed specialized catalysts for the most important and widely used
transformations.

Among the numerous parameters affecting the performances of heterogeneous
hydrogenation catalysts, the following are the most important [298, 299]:

- *Type of metal*: as already mentioned, Pd, Pt, Rh, Ru, Ni and Cu are most often used.
  Each metal has its own activity and selectivity profile (Table 2.13).

- *Type of catalyst*: noble metals are usually supported on a carrier; sometimes they
  are used as fine powders (Pd black and Pt black, PtO$_2$), Ni is most often applied as
  skeletal Raney nickel or supported on silica; Cu is used as Cu-chromite.

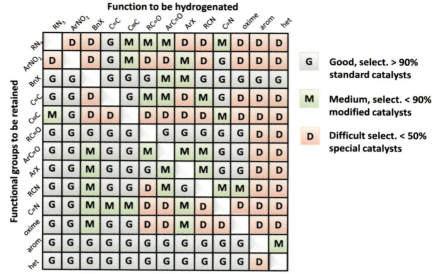

Figure 2.49 Rules for chemoselective hydrogenation of important functions. Source: adapted from Blaser *et al.* [298, 299].

- *Metal loading*: For noble metal catalysts 5% loading is standard. For Ni/SiO_2 the loading is usually 20–50%.

- *Type of support*: charcoal (also called active carbon) is most common; charcoals can adsorb large amounts of water; for safety reasons, many catalysts are sold with a water content of 50%. Aluminium and silicon oxides as well as $CaCO_3$ and $BaSO_4$ are also used as supports, but usually for special applications.

A major trend in fine chemicals and pharmaceuticals is towards increasingly complex molecules, which translates to a need for high degrees of chemo-, regio- and stereoselectivity. An illustrative example is the synthesis of Saquinavir, an intermediate for the Roche HIV protease inhibitor (Figure 2.50) [300]. It involves chemo- and diastereoselective hydrogenation of an aromatic ring, while avoiding racemization at the stereogenic center present in the substrate.

An interesting example of chemoselective hydrogenation of one functional group in the presence of other reactive groups is the selective conversion of an aromatic nitro group in the presence of other functional groups. Whereas the hydrogenation of simple nitroarenes poses few selectivity problems and is routinely carried out on a very large scale, the situation is different when other reducible functional groups are present in the molecule. Solvias [298, 299] has developed two modified Pt catalysts (Figure 2.51) able to hydrogenate an aromatic nitro group with high selectivities in the presence of various other functional group. Figure 2.51 shows several examples together with the reaction conditions and yields obtained for a specific example: the chemoselective hydrogenation of an aromatic nitro group in the presence of both

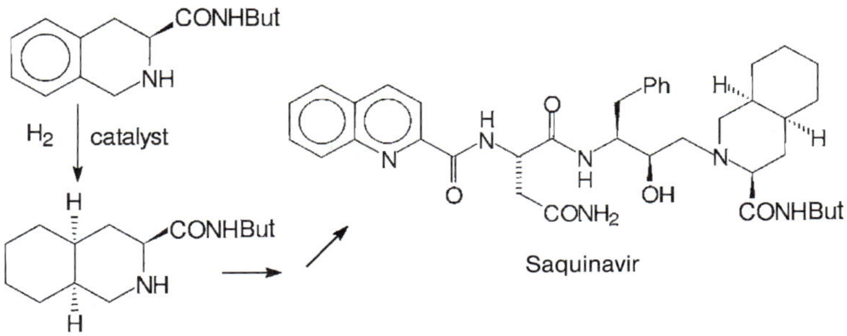

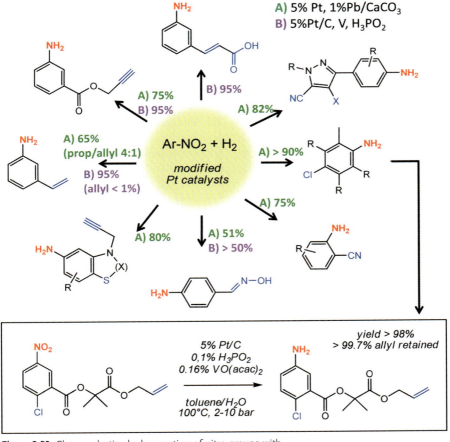

Figure 2.50 Synthesis of a Saquinavir intermediate. Source: adapted from Roessler [300].

Figure 2.51 Chemoselective hydrogenation of nitro groups with modified Pt catalysts developed by Solvias. Functional groups not converted are in blue. The boxed reaction is an example of chemoselective hydrogenation of a nitro group in the presence of an allyl ester. Source: adapted from Blaser *et al.* [298, 299].

an olefinic double bond and a chlorine substituent in the aromatic ring. The product is used for the synthesis of an herbicide.

There are still many developments in selective hydrogenation, both in terms of new catalysts and process operations. An example of the first is the discovery that Sn-substituted zeolite beta is the most active heterogeneous catalyst for the Meerwein–Pondorff–Verley reduction of aldehydes and ketones to the corresponding alcohols, with high cis-selectivity (99–100%) in the reduction of 4-alkylcyclohexanones [301]. An example of process development is in the heterogeneous catalytic hydrogenation of organic compounds in supercritical fluids (SCFs) [302].

2.6.2
Asymmetric Hydrogenation

A quite important area of hydrogenation, in particular for the fine and pharmaceutical area, is that of asymmetric or enantioselective hydrogenation. The stereoconfiguration in several biologically active molecules determines the functionality (Figure 2.52). Catalysis plays an important role in the industrial production of enantiomerically enriched chiral fine chemicals [303–305]. A survey of industrial asymmetric processes (commercial, under development or bench scale) found that 59 of a total of 77 processes are hydrogenations, followed by oxidation reactions and other types of reactions [303]. Bio- and homogeneous catalysts are mainly used, while more limited uses for heterogeneous enantioselective catalysts were found. However, knowledge on the latter has rapidly advanced from the initial use of chiral biopolymers, such as silk, as a support for metal catalysts to modern research areas. Mesoporous supports, noncovalent immobilization, metal–organic catalysts, chiral modifiers: many areas are rapidly evolving [306].

Surface science results have thrown new light on modified (chiral) metal surfaces, highlighting new phenomena such as complex adsorption phases, two-dimensional organization and the creation of extended chiral surfaces [307]. New strategies to immobilize an asymmetric catalyst onto a support (adsorption, encapsulation, tethering using a covalent bond and electrostatic interaction) also result in an

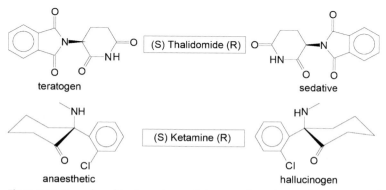

Figure 2.52 Examples of the importance of stereo-configuration on the properties of molecules.

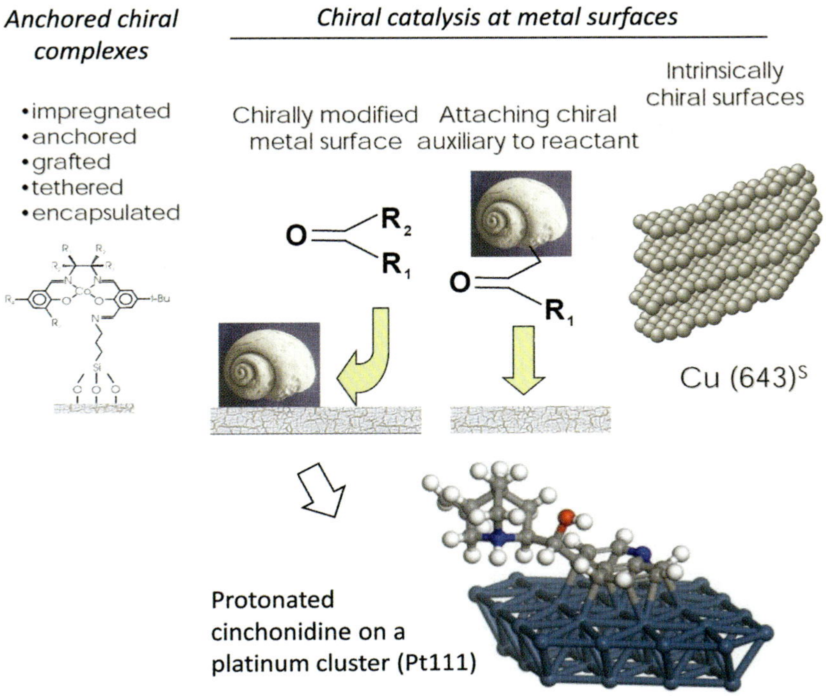

Anchored chiral complexes

Chiral catalysis at metal surfaces

Intrinsically chiral surfaces

•impregnated
•anchored
•grafted
•tethered
•encapsulated

Chirally modified metal surface

Attaching chiral auxiliary to reactant

Cu (643)^{S}

Protonated cinchonidine on a platinum cluster (Pt111)

Figure 2.53 Overview of some of the possibilities for developing enantioselective heterogeneous catalysts, including a model of cinchonidine adsorbed on an Pt surface.

efficient heterogeneous asymmetric catalyst that can be re-used [308]. Figure 2.53 overviews some of the possibilities for developing enantioselective heterogeneous catalysts, including a model of cinchonidine adsorbed on a Pt surface. Enantiose-lective hydrogenations of α-ketoesters on the Pt/Al$_2$O$_3$ catalyst surface in the presence of alkaloid modifiers, such as cinchonine and cinchonidine, are among the first and best known examples of heterogeneous enantioselective hydrogena-tions [309]. The inclusion of cinchonidine in the otherwise unselective reaction mixture leads to an enantiomeric excess of up to 90%, with a high conversion rate. For enantioselective hydrogenations of ethyl pyruvate, an e.e. as high as 99% has been achieved on cinchonidine-modified Pt/γ-Al$_2$O$_3$. In the enantioselective hydro-genation of ethyl 2-oxo-4-phenylbutyrate (EOPB), the TOF can reach as high as 20 000 h^{-1} with 86% e.e. The hydrogenation product (R)-($+$)-EHPB is an important building block for the synthesis of several commercially important ACE inhibitors (angiotensin-converting enzyme, a group of pharmaceuticals that are used primarily in the treatment of hypertension and congestive heart failure).

In some cases immobilized catalysts can give higher enantioselection than their non-immobilized counterparts, although the most common case is the contrary. One of the most-interesting ways to achieve this objective is to exploit the nanospace in solids (ordered mesoporous materials, in particular), for example, that the spatial confinement of prochiral reactants (and transition states formed at the chiral active center) would provide an altogether new method of boosting the enantioselectivity of anchored chiral catalysts [310, 311]. However, homogeneous (unsupported) catalysts are still the most used [312].

The production of commercial stereochemically defined products using catalytic routes is increasing [313]. An example, is the synthesis of metolachlor developed by Solvias/Novartis [312, 314]. This compound is the active ingredient of Dual, an herbicide for maize and other crops; over 20 000 tons per year of the racemate form is produced. Only the *(S)*-enantiomer is active and the enriched form requires about 35% less product for an equivalent effect, with a consequent relevant positive effect for the environment. The original process was based on a Pt-catalyzed reductive alkylation of 2-methyl-5-ethylaniline (MEA) with aqueous methoxyacetone in the presence of traces of sulfuric acid followed by chloroacetylation (Figure 2.54). The process produces a racemic mixture. By using an iridium complex of a chiral ferrocenyldiphosphine, complete conversion is achieved within 4 h at a substrate/catalyst ratio of about one million and a very high TOF, giving a product with an e.e. of 80%.

Figure 2.55 reports the preparation and structure of ferrocenyl diphosphine ligands and the dependence of the performances on the substituents in the ligand.

Another well-known example of catalytic asymmetric hydrogenation is the synthesis of L-Dopa (an anti-Parkinson drug) developed by Monsanto [315], which is schematically reported in Figure 2.56, where the role of the chiral phosphine ligand is also highlighted. Various attempts have been also reported to use heterogeneous asymmetric catalysis (cinchonine modified Pd/C) to produce L-Dopa, although the results are still unsatisfactory [316].

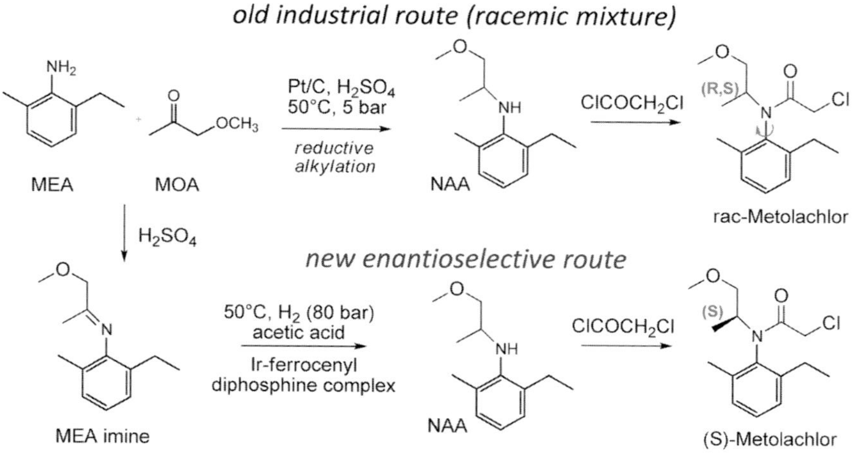

Figure 2.54 Old industrial racemic process for the synthesis of metolachlor versus the new enantioselective process developed by Solvias/Novartis. Source: adapted from Blaser *et al.* [312].

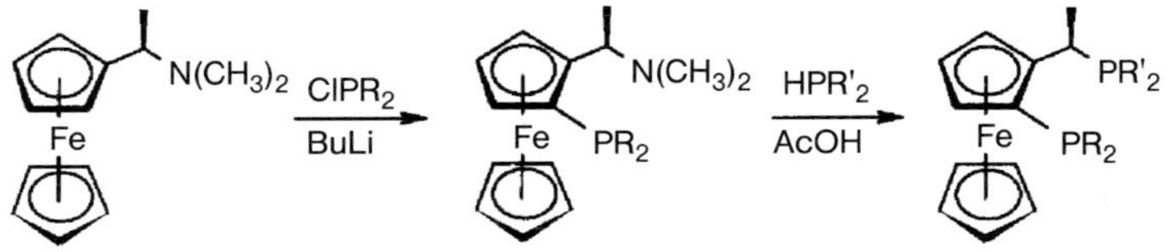

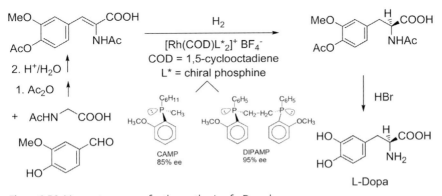

R	R'	Ton	tof (per h)	ee
Ph	3,5-xylyl	1 000 000	> 200 000	79
p-CF$_3$C$_6$H$_4$	3,5-xylyl	800	400	82
Ph	4-^{t}Bu-C$_6$H$_4$	5 000	80	87
Ph	4-(^{n}Pr)$_2$N-3,5-xyl	100 000	28 000	83

Figure 2.55 Preparation and structure of ferrocenyl diphosphine ligands for enantioselective Metolachlor synthesis and dependence of performance on the substituents in the ligand. Source: adapted from Blaser et al. [312].

Table 2.14 gives an overview of industrial processes using asymmetric catalysis, including examples both of asymmetric hydrogenation and other types of reactions, as well as of the use of biocatalysts for the reaction. Selected chiral ligands used in asymmetric catalytic reactions are also shown in the table.

One of the most important examples reported in Table 2.14 is the Takasago process for the manufacture of L-menthol, an important flavor and fragrance product. The key step is an enantioselective catalytic isomerization of a prochiral enamine to a chiral imine. The catalyst is a Rh-binap complex (developed by the winner of the 2001 Nobel Prize in Chemistry R. Noyori) and the product is obtained in 99% e.e. using a substrate/catalyst ratio of 8000; recycling of the catalyst affords total turnover numbers of up to 300 000. The Takasago process is used to produce several thousand tons of L-menthol on an annual basis.

Figure 2.56 Monsanto process for the synthesis of L-Dopa by asymmetric hydrogenation. Source: adapted from Knowles [315].

Table 2.14 Example of industrial processes or processes under development based on asymmetric catalysis; selected chiral ligands are shown at the end of the table. Source: adapted from Blaser and Scmidt [312].

Product	Reaction and catalyst	Scale	Company
L-Menthol	Isomerization of allyl amine with Rh-binap	$>1000 \, t \, y^{-1}$	Takasago
Vitamin E	Hydrogenation of allyl alcohol with Ru-binap	$300 \, t \, y^{-1}$	Takasago
Carbapenem	Hydrogenation of α-substituted β-keto ester, Ru-tolbinap	$50–120 \, t \, y^{-1}$	Takasago
(S)-Oxafloxazin	Hydrogenation of a-hydroxy ketone, Ru-tolbinap	$50 \, t \, y^{-1}$	Takasago
Aspartame	Dipeptide formation from D,L-phenylalanine with thermolysin	kilo/tons	Holland Sweetener Co.
(+)-*cis*-Methyl dihydrojasmonate	Hydrogenation of α,β-unsaturated ester with Ru-josiphos	Multi-$t \, y^{-1}$	Firmenich
L-Malic acid	C−O bond formation, from fumaric acid with fumarase (enzyme)	$2000 \, t \, y^{-1}$	Amino GmbH, Tanabe Seiyaku
1-Phenylethyl-amines	Chiral transesterification, with lipases 8enzyme)	$>100 \, t \, y^{-1}$	BASF
Tipranavir	Hydrogenation of C=C bond with Rh-duphos	Multi-tons	Chirotech
Orlistat	Hydrogenation of β-keto ester with Ru-biphep or Raney-Ni-tartrate (modified heterogeneous catalyst)	ton	Roche
HMR 2906	Hydrogenation of enamide with Rh-bpm	Multi-kg	Aventis
SK&F 107 647	Hydrogenation of di-enamide with Rh-duphos	Multi-kg	Chirotech

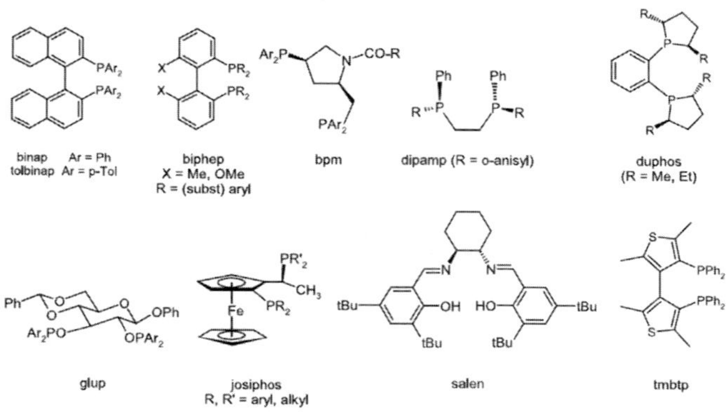

binap   Ar = Ph
tolbinap Ar = p-Tol

biphep
X = Me, OMe
R = (subst) aryl

bpm

dipamp (R = o-anisyl)

duphos
(R = Me, Et)

glup

josiphos
R, R' = aryl, alkyl

salen

tmbtp

2.6.3
Selective Oxidation

Catalytic selective oxidation processes play a central role in the chemicals industry as well as in the development of new catalysts as the key contributions to modern manufacturing, because a large proportion of materials and commodities in daily use undergo a selective oxidation process as a critical step in their production cycle. More than half of the products obtained using catalytic processes are obtained by selective oxidation, and nearly all the monomers used in the production of fibers and plastics are obtained in this way [241–243].

Selective oxidation processes are therefore one of the central building blocks of the chemicals industry, even if a great need for innovation is still necessary both in catalyst development and in the engineering of the processes, in particular in reactor design. Technological challenges include increasing product selectivity, stereoselectivity, decreasing undesirable by-products, minimizing energy consumption, utilizing and controlling exothermicities, designing catalysts for aqueous environments, and reducing process steps with multifunctional catalysts. These goals require not only a better understanding of the chemistry of the catalytic phenomena associated with selective oxidation catalysts but also a closer relationship between catalysts and reactor design. Recent major developments in this area (e.g., a new process for the oxidation of n-butane to maleic anhydride in a riser reactor, selective oxidations at very short contact times, a new process to formaldehyde synthesis in a combined fixed bed-monolith reactor, and a new process for the oxidation of o-xylene to phthalic anhydride in a composite bed reactor) derive from an integrated view of the catalyst, the reactor and their relationship.

The conversion of organic substrates is the largest class of selective oxidation processes, but it should be remembered that the conversion of inorganic molecules (e.g., ammonia and H_2S oxidation) and the synthesis of base chemicals by selective oxidation processes (e.g., methane ammoxidation to HCN) are also important industrial sectors. The conversion of organic substrates can be divided into processes occurring in the liquid phase (using homogeneous or heterogeneous catalysts) and in the vapor phase on solid catalysts.

Table 2.15 summarizes the main reactions of industrial interest in the liquid and vapor phases, the type of catalysts used, conversions and selectivities in the industrial processes. While O_2 is the only oxidizing agent in the second sub-class (apart from the recent case of benzene hydroxylation to phenol using N_2O as the oxidant), O_2 and oxygen transfer agents (alkyl hydroperoxide or H_2O_2) are used in the liquid phase.

The environmental factor has been also the driving force behind innovation over the last two decades in the area of selective oxidation of organic substrates, after the development of the basic processes that were commercialized in the 1960s and 1970s. The main characteristic of the technological progress in oxidation processes during these years was the increasing complexity of the type of catalysts used and the parallel change in the type of reactor technology used to improve performances. An example of this is the propene ammoxidation process, which in the early 1960s displaced the older process based on the acetylene–HCN route. After the introduction of this new, more economical and safer process, a marked surge in the use of acrylonitrile

Table 2.15 Main industrial catalytic selective oxidation processes and typical catalytic performances obtained.
Source: adapted from Centi et al. [241, 242].

Reactants	Main products/co-products	VP/LP[a]	Type of catalyst	Conversion[b] (%)	Selectivity[b] (%)
Ammonia/air	NO	VP	Pt/Rh gauze	100	94–98
Ammonia/air	N_2	VP	Pd-V_2O_5-WO_3/TiO_2-SiO_2	>99	>99
H_2S/air	S_n ($n=6$–8)	VP	Fe_2O_3/SiO_2	90–95	95–98
Methane/O_2/NH_3	HCN	VP	Pt/Rh gauze	100	60–70
CH_4 or $(CH_2)_x$/O_2	Syn gas (CO/H_2)	VP	Supported Rh or Ni	>99	90–95
Methanol/air	Formaldehyde	VP	Ag, Fe-Mo-oxide	95–99	91–98
Ethene/O_2	Ethene oxide	VP	Ag(K,Cl)/α-Al_2O_3	13–18[c]	72–76
Ethene/air or O_2/HCl	1,2-Dichloroethane	VP	Cu/Cl/O/K-γAl_2O_3	>95	93–96
Ethanol/O_2	Acetaldehyde	VP	Ag, Cu	45–50[c]	94–96
Ethene/acetic acid/air	Vinyl acetate	LP	Pd/Au/K-αAl_2O_3	8–12[c]	92
Ethene/O_2	Acetaldehyde	LP	Pd/Cl/acetate	90	>95
Propene/air	Acrolein	VP	Supported Bi/Mo/Fe/Co/K-oxide	92–97	80–88
Propene/air/NH_3	Acrylonitrile	VP	Supported Bi/Mo/Fe/Co/K-oxide	98–100	75–83
Propene/hydroperoxide	Epoxide/alcohol	LP	Mo-complexes or silica supported Ti-oxide	10	>90
Acrolein/air	Acrylic acid	VP	V/Mo/W-oxide	>95	90–95
n-Butane/air	Maleic anhydride	VP	V-P-oxide	75–80	67–72
n-Butane/O_2 or air	Acetic acid	LP	Co, Mn salts	>90	50–65
n-Butane/air	Butenes/butadiene	VP	Bi-Mo-P-oxide	55–65	93–95
t-Butyl alcohol	Methacrolein	VP	Bi-Mo-Fe-Co-K-oxide	99	85–90

Table 2.15 (Continued)

Reactants	Main products/co-products	VP/LP[a]	Type of catalyst	Conversion[b] (%)	Selectivity[b] (%)
Isobutene/air	Methacrolein	VP	Bi-Mo-Fe-Co-K-oxide	>97	85–90
Methacrolein/air	Methacrylic acid	VP	V/Mo/W-oxide	97–99	95–98
Cyclohexane/air	Cyclohexanone	LP	Co salts	5–15	70–90
Cyclohexanone/HNO₃	Adipic acid	LP	Co-salts	>95	92–96
Cyclohexanone/H₂O₂/NH₃	Cyclohexanone oxime	LP	Ti-silicalite	100	>98
Benzene/air	Maleic anhydride	VP	V-Mo-oxide	98	75
o-Xylene/air	Phthalic anhydride	VP	V-W-P-Cs-TiO₂	93–98	81–87
p-Xylene/air	Terephthalic acid	LP	Co/Mn/Br	95	90–95
Cumene/O₂ or air	Phenol/acetone	LP	Co-Mn salts	25–35	90–97
Naphthalene/air	Phthalic anhydride	VP	V-K-oxide/SiO₂	100	84

[a] VP: vapor phase on a solid catalyst. LP: in liquid phase with a catalyst in the same phase or present solution.
[b] Conversion of reactant and selectivity of product with respect to oxidized substrate.
[c] Conversion per pass, in those processes that operate with recycling of unconverted reactant.

occurred, since it is a cheap and a highly versatile raw material for the synthesis of fibers, resins, rubbers and specialty products.

The first generation of catalysts based on supported $Bi_9PMo_{12}O_{52}$ gave a yield under commercial conditions of around 55%, which increased to around 65% with the second generation of catalysts containing redox elements such as iron and to around 75% with the third generation of multiphase, multicomponent catalysts containing more than ten elements. These catalysts were further improved and present yields of over 80%. At the same time, reactor technology also improved, passing from the fixed bed to the bubbling fluid-bed and finally to the "braked" fluid bed reactor, which produces plug-flow type conditions and smaller bubbles.

Recent trends and developments in industrial selective oxidation technology can be summarized as follows:

- *Use of new raw materials and alternative oxidizing agents.* Alkanes are increasingly replacing aromatics and alkenes as raw materials. Phenol can be synthesized directly from benzene using N_2O and zeolite-type catalysts. In liquid-phase oxidations, hydrogen peroxide is being used more frequently as an oxidizing agent in place of traditional oxygen transfer agents. In some cases, the hydrogen peroxide is generated *in situ*.

- *Development of new catalytic systems and processes.* Heterogeneous rather than homogeneous catalysts are used, and oxidative dehydrogenation is increasingly used in place of simple dehydrogenation. New processes are under development that generate fewer or no undesired co-products.

- *Conversion of air-based processes into oxygen-based processes in vapor-phase oxidation to reduce polluting emissions.* Examples are: (i) synthesis of formaldehyde from methanol, (ii) ethene epoxidation to ethene oxide and (iii) oxychlorination of ethene to 1,2-dichloroethane.

- *Existing processes are being fine-tuned to improve each stage of the overall process.* Profit margins are being increased through changes in process engineering rather than through economies of scale.

2.6.3.1 Selective Oxidation: Liquid Phase

Vapor phase oxidation processes prevail over liquid phase processes, although the latter are sometimes used in large-scale chemical production when the products (i) can be easily recovered from the reaction medium, as in terephthalic acid production, for example; (ii) are thermally unstable (i.e., in the production of hydroperoxides and carboxylic acids, except for β-unsaturated compounds); and (iii) are very reactive at high temperature (i.e., epoxides, aldehydes and ketoses, with the exception of ethene oxide and formaldehyde). Liquid-phase oxidation is also preferred in fine chemicals production, although most processes are still non-catalytic.

The main industrial processes in the liquid phase are:

- *Production of hydroperoxides*: (i) isobutene to *t*-butyl hydroperoxide and ethylbenzene to ethylbenzene hydroperoxide, both subsequently used as the oxidizing agent for alkene epoxidation; (ii) cumene to cumyl hydroperoxide (subsequently converted into phenol and acetone); and (iii) *p*-diisopropylbenzene to the corresponding dihydroperoxide to form, subsequently, hydroquinone.

- *Production of acids*: (i) terephthalic acid from *p*-xylene, (ii) acetic acid from *n*-butane, (iii) benzoic acid from toluene (as an intermediate step in an alternative process of phenol synthesis), (iv) cyclohexane to cyclohexanone and cyclohexanol, intermediate to adipic acid, (v) acetaldehyde to acetic acid or acetic anhydride and *n*-butyraldehyde to *n*-butyric acid, (vi) pseudocumene to trimellitic acid or *m*-xylene to isophthalic acid and (vii) production of fatty acids from alkanes.

- *Partial oxidation of heavier substrates to promote their functionalization*: (i) fluorene to fluorenone, naphthalene to naphthoquinone or anthracene to anthraquinone, (ii) naphthalene to naphthene-2,6-dicarboxylic acid, (iii) *p*-methylisopropylbenzene to cresol and (iv) long-chain *n*-alkanes to secondary alcohols.

The most important industrial processes are: (i) the oxidation of cumene to cumyl hydroperoxide, (ii) the oxidation of cyclohexane to cyclohexanol and cyclohexanone, (iii) the oxidation of *n*-butane to acetic acid, (iv) the oxidation of *p*-xylene to terephthalic acid or dimethyl terephthalate, (v) the epoxidation of propene to propene oxide and (vi) the oxidation of ethene to acetaldehyde.

Liquid-phase catalytic oxidations can be classified according to the predominant reaction mechanism: (i) free-radicals with O_2 usually being the oxidizing agent, (ii) redox mechanism with Pd or Cu complexes and O_2 as the oxidizing agent and (iii) mono-oxygen transfer, with either alkyl hydroperoxide or H_2O_2 as the oxidizing agent, using either homogeneous or heterogeneous catalysts. Other types of processes (e.g., photo- or electro-catalytic) are also possible, but usually are utilized only in specific production areas. The oxidizing agents used are O_2 or air, or oxygen transfer agents such as alkyl hydroperoxide or H_2O_2.

The main technological issues are: (i) selectivity, (ii) removal of reaction heat, (iii) efficient gas–liquid and solid (when a solid catalyst is used) contact and (iv) safety. The choice and engineering of the reactor are thus critical aspects of the process development.

2.6.3.2 Selective Oxidation: Vapor Phase

Unlike liquid phase processes, air or O_2 are the only oxidizing agents used in commercial processes, although recently phenol synthesis from benzene using N_2O as the oxidizing agent has been developed, although it has not reached commercial viability. The process is only economical when a cheap source of N_2O is available, that is, when N_2O is recovered from waste streams such as in adipic acid production.

The different classes of industrial catalytic selective oxidation reactions in the vapor phase (over solid catalysts) are:

- *Allylic oxidation*: (i) propene to acrolein or acrylic acid and (ii) isobutene to methacrolein or methacrylic acid; the synthesis of the acids may be effected in a single step, but commercially a two-step process is used due to its better selectivity.

- *Oxidative dehydrogenation*: (i) butanes to butadiene and isopentene to isoprene, (ii) methanol to formaldehyde and (iii) isobutyric acid to methacrylic acid.
- *Electrophilic mono-oxygen insertion*: (i) ethene epoxidation to ethene oxide using O_2 and (ii) direct phenol synthesis from benzene using N_2O.
- *Acetoxylation*: vinyl acetate synthesis from ethene and acetic acid.
- *Oxychlorination*: 1,2-dichloroethane synthesis from ethene and HCl.
- *Ammoxidation*: (i) propene to acrylonitrile, (ii) isobutene to methacrylonitrile and (iii) α-methylstyrene to atroponitrile.
- *Synthesis of anhydrides*: (i) *n*-butane to maleic anhydride and (ii) *o*-xylene to phthalic anhydride.

Important classes of reactions not included in the above list, because they are not yet used on a commercial scale, are: (i) the oxidative dehydrogenation of C_2–C_5 alkanes, (ii) the selective oxidation of alkanes, such as the synthesis of maleic and phthalic anhydride from *n*-pentane and methacrolein or methacrylic acid from isobutene, and (iii) propane ammoxidation to acrylonitrile [317–319].

2.6.3.3 Selective Oxidation: Examples of Directions to Improve Sustainability

Some main directions towards improving sustainability that should be highlighted are: (i) the use of new and clear oxidants, which is exemplified by the use of H_2O_2, (ii) the use of new feedstocks, which can be illustrated by the substitution of alkenes with alkanes, and (iii) the process development, which can be discussed by showing the developments arising from using new reactor opportunities and an integrated reactor/catalyst design, as well as by substituting air with oxygen. Together with the development of new catalysts, which open up new opportunities to develop innovative routes of transformation, these directions allow a reduction of the environmental impact and use of resources (raw materials, energy) and often also allow a significant process simplification. Some examples are discussed below.

H_2O_2 as a Clean Oxidant In the area of liquid phase oxidation, perhaps the more relevant development is the substitution of either stoichiometric reagents (e.g., Cr^{VI}, but also several other reagents such as permanganate, MnO_2, and so on) or of organic peroxides with a clean reagent such as H_2O_2 in combination with a catalyst able to give selective oxidation such as titanium-silicalite (TS-1). This aspect is discussed in detail in the case of the new processes for the synthesis of propene oxide and phenol, and thus will not be discussed here. A further important development regards the new synthesis of caprolactam, which is discussed above (Figure 2.32). However, the same reaction of ammoxidation catalyzed by TS-1 in the presence of ammonia and H_2O_2 could be usefully applied in the synthesis of a large variety of substrates of interest for fine chemistry, as exemplified in Figure 2.57.

H_2O_2/TS-1 also finds application in the hydroxylation of aromatics. A relevant example is the Rhodia process for the manufacture of the flavor ingredient vanillin [229]. The process involves four steps, all performed with a heterogeneous catalyst, starting from phenol (Figure 2.58). Overall, one equivalent of phenol, H_2O_2, CH_3OH,

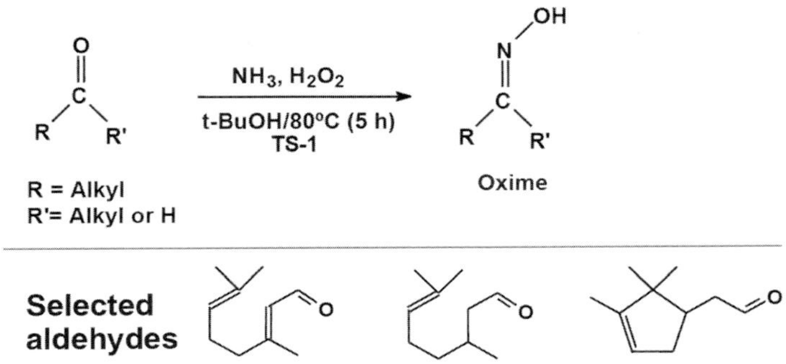

Selected aldehydes

	Citral	Citronellal	Campholenal
Conversion (%Mol.)	97	99	90
Selectivity to Oxime (%Mol.)	98	98	95

Figure 2.57 Ammoximation of carbonyl groups with TS-1. Source: adapted from Corma [320].

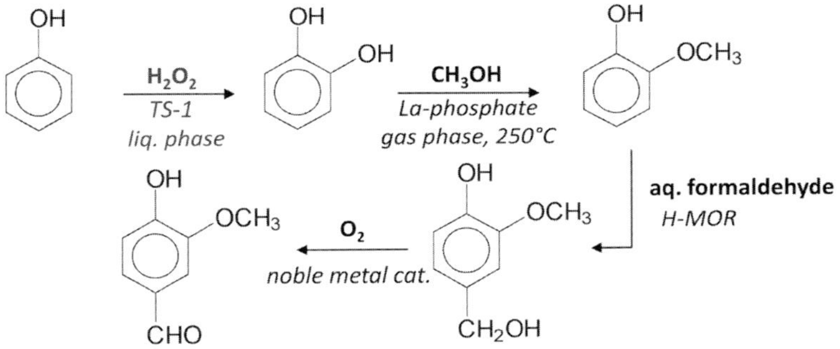

Figure 2.58 Rhodia vanillin process. Source: adapted from Ratton [229].

formaldehyde and O_2 are converted into one equivalent of vanillin and three equivalents of water.

Titanium-silicalite was the first example of a wider class of materials based on zeolites and micro- or mesoporous materials in which Si and/or Al atoms have been substituted by transition metals. A limitation of these materials was often the easy leaching of the transition metal, making them inapplicable on an industrial scale. A very successful example was, instead, Sn-Beta, which is characterized by a bifunctional active site that involves a Lewis acid tin center and an adjacent oxygen atom capable of accepting hydrogen bonding from water or hydrogen peroxide [321]. In the area of selective oxidation it can be applied to the Baeyer–Villiger oxidation of ketones and aldehydes [322] with aqueous H_2O_2 (Figure 2.59).

Another very interesting example from the same group is the use of Sn-Beta for the synthesis of melonal (2,6-dimethyl-5-hepten-1-al), a fragrance that is produced industrially by a Darzens reaction from 6-methyl-5-hepten-2-one, with ethyl chloroacetate as reagent. A novel halogen-free synthesis involves the chemoselective

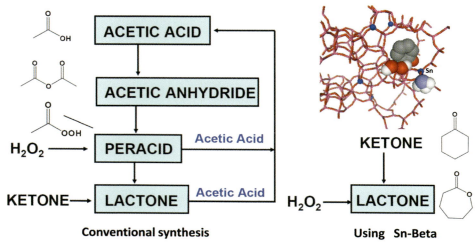

Conventional synthesis   **Using Sn-Beta**

Figure 2.59 Baeyer–Villiger oxidation of ketones using Sb-Beta. Source: adapted from Corma [320].

oxidation of citral (3,7-dimethyl-6-octen-1-al), a common compound in the fragrance industry, with H_2O_2 and Sn-Beta or Sn-MCM-41 as catalysts [323] (Figure 2.60). Aluminium Brønsted acid sites and zirconium or titanium Lewis acid sites are less efficient and selective than Sn in this Lewis acid site catalyzed reaction.

Notably, there are other types of catalysts active in clean organic reactions using H_2O_2 [324]. Among the homogeneous catalysts, the use of tungsten-based catalysts (phosphotungstate of the Keggin type [325], silicadecatungstate [326] and

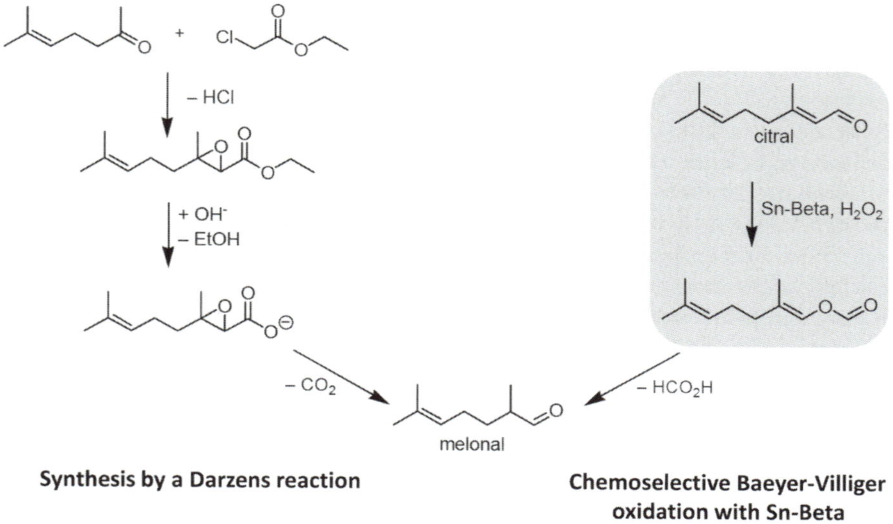

Synthesis by a Darzens reaction   **Chemoselective Baeyer-Villiger oxidation with Sn-Beta**

Figure 2.60 Baeyer–Villiger chemoselective oxidation of melonal using Sb-Beta. Source: adapted from Corma [320].

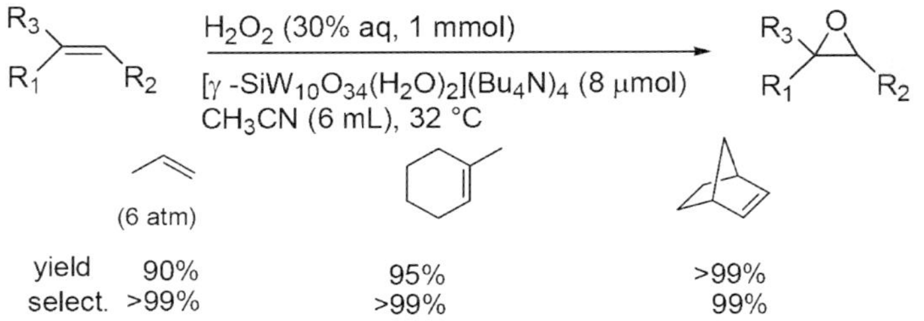

Figure 2.61 Epoxidation of alkenes promoted by silicadecatungstate and H_2O_2. Source: adapted from Goti and Cardona [324].

Na_2WO_4 [327]) for the selective epoxidation of a large variety of alkenes is notable. Figure 2.61 gives an example with the silicadecatungstate catalyst developed by Mizuno et al. [326].

Some of these W-based catalysts could be transformed into insoluble salts; for example, the phosphotungstate-pyridinium salt in a toluene–tributyl phosphate (4:3) solvent mixture [328]. In this solvent the pre-catalyst is insoluble, but upon reaction with H_2O_2 it gives the catalytically active W-peroxo complex, which is soluble. The catalytic action is then performed under homogeneous conditions and, at the end of the reaction, H_2O_2 being completely consumed, the precatalyst precipitates and can be easily filtered off and recovered. The method is smart, but from an industrial point of view the separation and recovery is costly.

Another interesting alternative in the epoxidation of alkenes is the use of methyltrioxorhenium (MTO), originally developed by Herrmann et al. [329]. MTO activates hydrogen peroxide by forming a mono-peroxo complex that undergoes further reaction to yield a bis-peroxorhenium complex. Both complexes are active as oxygen transfer species. A problem is the formation of 1,2-diol via ring opening of the epoxide. The addition of urea limits this problem.

Novel Pathways and Reactants This is a very broad area. We will thus restrict discussion to few examples. The first regards the important reaction of phenol synthesis and the possibility to realize it in one step directly from benzene using molecular oxygen as the oxidant. Various aspects of direct phenol synthesis from benzene are discussed in Chapter 13. We highlight here only recent results that exemplify how starting from the previously cited activity of Re complexes in the epoxidation in homogeneous phase could lead to investigation of the behavior of Re complexes when inserted into the channels of zeolites (ZSM-5) and in gas-phase selective oxidations. This has opened a new unexpected direction.

Bai et al. [330] found a remarkable selectivity (88% in the steady-state reaction and 94% in the pulse reaction) in the direct synthesis of phenol from benzene with molecular oxygen over a Re/zeolite catalyst prepared by chemical vapor deposition (Figure 3.62). However, stable performances could be obtained only by continuous feeding of relatively high concentrations of NH_3 (around 30%), which is necessary to stabilize the active complex containing interstitial N atoms (see the model of the complex in

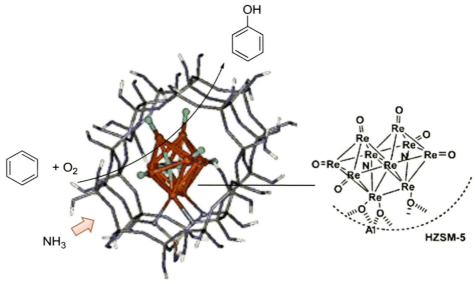

Figure 2.62 Direct synthesis of phenol from benzene using molecular oxygen on rhenium complexes in ZSM-5. Source: adapted from Iwasawa *et al.* [330].

Figure 2.62). This is the main drawback for industrial development, because part of the ammonia is also side converted. Nevertheless, this reaction and catalyst exemplify the novel possibilities opened to develop sustainable processes of selective oxidation.

A second example concerns the use of alkanes instead of alkenes to both use alternative feedstocks and reduce the environmental impact. An interesting example is the oxidation of isobutane to methacrylic acid [331].

The methyl ester of methacrylic acid (CH_2=C(CH_3)–COOH) is used to produce vinyl polymers used as cast sheets, molding and extrusion powders and coatings, besides being used in various copolymers. Poly(methyl methacrylate) production currently stands at over 2.8 million tons per year on a worldwide scale.

The traditional acetone cyanohydrin (ACH) process is the most widely used in Europe and North America, while other processes are more often used in Asia. In the ACH process (Figure 2.63), acetone and hydrogen cyanide react to yield acetone cyanohydrin; the latter is then reacted with an excess of concentrated sulfuric acid to form methacrylamide sulfate. In a later stage, methacrylamide is treated with excess aqueous methanol; the amide is hydrolyzed and esterified, with formation of a mixture of methyl methacrylate and methacrylic acid. The ACH process offers economical advantages, especially in Europe, where large plants are in use – most of them have been in operation for decades. The process also suffers from drawbacks that have been the driving forces for the development of alternative technologies.

Specifically, the process makes use of HCN, a very toxic reactant. Difficulties in its acquisition, though, can be met; in fact, HCN is a by-product of propylene ammoxidation. Integration of acrylonitrile and MMA products requires the balance

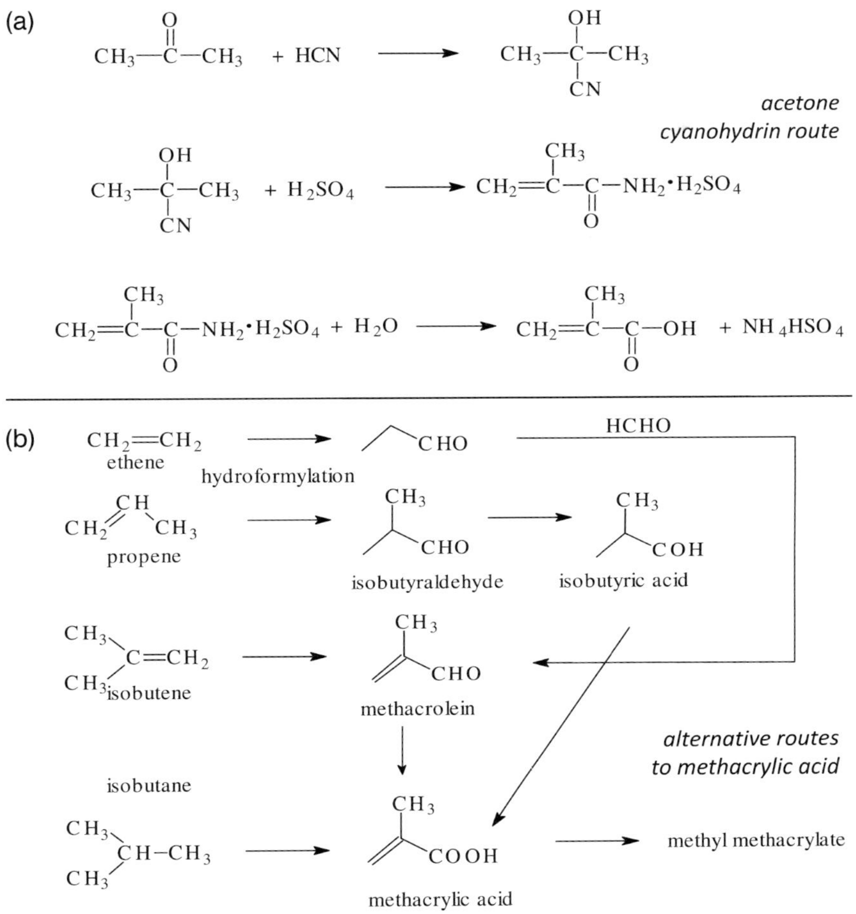

(a)

$$CH_3-\overset{O}{\overset{||}{C}}-CH_3 + HCN \longrightarrow CH_3-\overset{OH}{\underset{CN}{\overset{|}{C}}}-CH_3$$

acetone cyanohydrin route

$$CH_3-\overset{OH}{\underset{CN}{\overset{|}{C}}}-CH_3 + H_2SO_4 \longrightarrow CH_2{=}\overset{CH_3}{\underset{}{\overset{|}{C}}}-\overset{}{\underset{O}{\overset{||}{C}}}-NH_2{\cdot}H_2SO_4$$

$$CH_2{=}\overset{CH_3}{\overset{|}{C}}-\overset{}{\underset{O}{\overset{||}{C}}}-NH_2{\cdot}H_2SO_4 + H_2O \longrightarrow CH_2{=}\overset{CH_3}{\overset{|}{C}}-\overset{}{\underset{O}{\overset{||}{C}}}-OH + NH_4HSO_4$$

(b)

$$CH_2{=}CH_2 \longrightarrow {}^{\wedge}CHO$$
ethene
hydroformylation

HCHO

$$CH_2{\overset{CH}{{=}}}CH_3 \longrightarrow \overset{CH_3}{\underset{}{}}CHO \longrightarrow \overset{CH_3}{\underset{}{}}COH$$
propene
isobutyraldehyde   isobutyric acid

$$\overset{CH_3}{\underset{CH_3}{}}C{=}CH_2 \longrightarrow \overset{CH_3}{\underset{}{}}CHO$$
isobutene
methacrolein

isobutane

alternative routes to methacrylic acid

$$\overset{CH_3}{\underset{CH_3}{}}CH{-}CH_3 \longrightarrow \overset{CH_3}{\underset{}{}}COOH$$

methyl methacrylate

methacrylic acid

Figure 2.63 Acetone cyanohydrin (a) and alternative routes (b) in the synthesis of methacrylic acid.

of the two processes. Alternatively, HCN can be produced on purpose, but this is feasible only for large production capacities. The second major drawback of the process is the disposal of ammonium bisulfate, the co-product of the process. Additional costs are necessary for its recovery or pyrolysis.

The ACH process has been improved by Mitsubishi Gas [332]. Acetone cyanohydrin is first hydrolyzed to 2-hydroxyisobutylamide with a MnO_2 catalyst; the amide is then reacted with methyl formate to produce the methyl ester of 2-hydroxyisobutyric acid, with co-production of formamide (this reaction is catalyzed by sodium methoxide). The ester is finally dehydrated with an Na-Y zeolite to methyl methacrylate. Formamide is converted into cyanhydric acid, which is used to produce acetone cyanohydrin by reaction with acetone. The process is elegant, since it avoids the co-production of ammonium bisulfate, and no net income of HCN is present. However, there are many synthesis steps, and a high energy consumption.

Other technologies, already commercially applied or under development, are summarized in Figure 2.63b. Alternative routes of synthesis include (i) ethene hydroformylation to propionaldehyde, which then forms methacrolein by condensation with formaldehyde; methacrolein is then oxidized to methacrylic acid (BASF process); (ii) isobuthyraldehyde conversion into isobutyric acid and then oxidative dehydrogenation to methacrylic acid (Mitsubishi Kasei/Asahi process); and (iii) oxidation of *tert*-butyl alcohol to methacrolein followed by oxidation to methacrylic acid and esterification.

An attractive new route under development is direct (one-step) gas-phase isobutane conversion into methacrylic acid, because of the (i) low cost of the raw material, (ii) simplicity of the one-step process, (iii) very low environmental impact and (iv) absence of inorganic co-products. Several patents claiming the use of polyoxometalates (POMs, Figure 2.64) as heterogeneous catalysts for this reaction started to appear in the 1980s and 1990s [331]. Rohm and Haas was the first (1981) to claim the use of P/Mo/(Sb) mixed oxides for isobutane oxidation [333]. Later, several patents (issued to Asahi Kasei, Sumitomo Chem, Mitsubishi Rayon, and others) described the use of modified Keggin-type POMs as catalysts. Specifically, most attention has been focused on the possibility of improving the conversion of isobutane and the selectivity to methacrylic acid by developing POMs that contain specific transition metal cations. The major problem is the stability of this catalyst, because the reaction temperatures of operations are close to those necessary for the activation of the alkane. This problem is exacerbated by the fact that the control of the reaction temperature is difficult, due the very high heat of reaction that develops.

A peculiarity of the processes described in the patents is that all of them use isobutane-rich conditions, with isobutane-to-dioxygen molar ratios between 2 (for processes that include a relatively low concentration of inert components) and 0.8, and so closer to the stoichiometric value 0.5 (for those processes where a large amount of inert components is present). Low isobutane conversions are achieved in all cases, and recirculation of unconverted isobutane becomes compulsory.

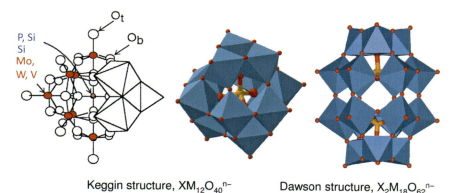

Keggin structure, $XM_{12}O_{40}{}^{n-}$ Dawson structure, $X_2M_{18}O_{62}{}^{n-}$

Figure 2.64 Structure of polyoxometalate catalysts used for the conversion of isobutane into methacrylic acid.

In all cases steam is present as the main ballast. The role of steam is to decrease the concentration of isobutane and oxygen in the recycle loop and thus keep the reactant mixture outside the flammability region. Water can be easily separated from the other components of the effluent stream, playing also a positive role in the catalytic performance of POMs. It is also possible that the presence of water favors the surface reconstruction of the Keggin structure, which decomposes during the reaction at high temperature, and also promotes desorption of methacrylic acid, saving it from unselective consecutive reactions.

Under the reaction conditions described in the patents, methacrolein is always present in non-negligible amounts, and therefore a commercial process necessitates an economical method for recycling it. Figure 2.65 shows a simplified flow-sheet of the Sumitomo process. CO_2 is maintained in the recycle loop to act as a ballast component; the desired concentration of CO_2 is obtained by combustion of CO, while excess CO_2 is separated. Methacrolein is separated and recycled to the oxidation reactor. An overall recycle yield of 52% to methacrylic acid is reported, with a recycle conversion of 96% and a per-pass isobutane conversion of 10%. The heat of reaction produced, mainly deriving from the combustion reaction, is recovered as steam. However, commercialization of this process is still hindered by the actual productivity (about $0.7\ \mathrm{mmol\ h^{-1}\ g_{cat}^{-1}}$), which is still too low [334]. Note that the productivity is limited by the oxygen conversion, the maximum concentration of which is dictated by the flammability limits, and by temperature, since the POM decomposes above 380 °C. Therefore, a possible development is to use microreactor technology, where, due to the high wall-to-volume ratio, operations inside the explosion limits are possible and also where the heat of reaction can be removed efficiently, thus improving catalyst stability.

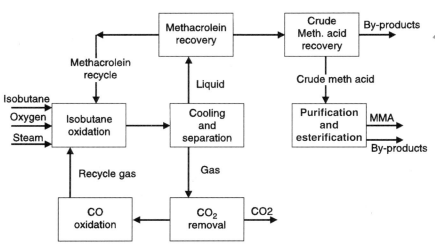

Figure 2.65 Simplified flow-sheet for isobutane oxidation to methacrylic acid proposed by Sumitomo. Source: adapted from Cavani *et al.* [331].

A recent interesting example of the use of alkane feedstocks to develop more sustainable processes is the direct conversion of ethane into acetic acid developed by Sabic. Acetic acid is the raw material for many key petrochemical intermediates and products, including vinyl acetate monomer (VAM), purified terephthalic acid (PTA), acetate esters, cellulose acetate, acetic anhydride, monochloroacetic acid (MCA), and so on. Acetic acid is produced commercially from several feedstocks and by several different technologies.

Methanol carbonylation technology using syngas accounts for over 60% of global capacity. This share is growing because it is the preferred technology for most new plants. Direct ethane oxidation to acetic acid has been an area of great interest to many chemical companies. Among the many patent holders, the most active players in this area are Hoechst Research and Technology Deutschland GmbH & Company, Saudi Basic Industries Corporation, Mitsubishi Chemical Corp., and BP Chemicals Ltd. In 2006 Sabic began operations for the direct conversion of ethane into acetic acid with an initial 34 000 metric ton per year capacity plant.

The Sabic acetic acid technology is characterized by a novel catalyst (a Mo-V-Nb mixed oxide [335]) and a novel oxidation reactor design, which is different from the conventional methanol-based technology. In the process ethane is mixed with oxygen and compressed, then passed over the catalyst to produce acetic acid and some ethylene, which is then separated and purified for use as a feedstock in other associated plants.

As cited in Chapter 1, the first example of commercial process using an alkane as feedstock, in substitution of the older process starting from benzene, was the synthesis of maleic anhydride from *n*-butane. Figure 2.66 briefly recalls the reaction scheme on the model surface of the catalyst (vanadyl pyrophosphate) to evidence the

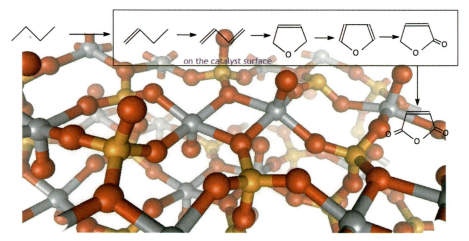

$(VO)_2P_2O_7$: *n-butane oxidation to maleic anhydride*

Figure 2.66 Multifunctionality of solid catalysts: synthesis of maleic anhydride from *n*-butane over a $(VO)_2P_2O_7$ catalyst.

concept that this complex reaction (a 19-electron oxidation that involves the abstraction of eight hydrogen atoms and the insertion of three oxygen molecules in the hydrocarbon substrate) can occur with high selectivity (over 80%) on the catalyst surface without desorption of any intermediate [336]. This is a good example of a sustainable process, not only because it has substituted the older and less sustainable process from benzene but also because it is a prototype of the possibility of realizing very complex reactions in a single stage, which is one of the main objectives of sustainable chemical processes.

Role of Reactor and Process Design   Optimization of catalytic performances, in terms of reactant conversion, yield, productivity and selectivity to the desired product, is not only related to a thorough knowledge of the nature of the catalyst and the interactions between reacting components and surface active phases, the reaction mechanism, thermodynamics and kinetics but also to the development and use of a suitable reactor configuration, where all the above-mentioned features can be successfully exploited.

Industrial reactors used in the petrochemical industry for exothermic reactions, with a few exceptions, are either fixed-beds (adiabatic or non-adiabatic) or fluidized-beds when the heat developed is too high to be removed in a fixed-bed reactor. In the last few decades, interest has been mainly directed towards the control of these reactors, which is strictly related to an understanding of the complex phenomena that occur at the interface between the different phases present in the reaction environment, and of the heat and mass transfer influence on the reaction kinetics.

Substantial improvements in the performance of several processes of hydrocarbon selective oxidation can be achieved solely by developing new reactor configurations. An important step in this direction is exemplified by the circulating fluidized bed reactor, which over the years has been proposed for use in several selective oxidation reactions and has, finally, found application in n-butane selective oxidation to maleic anhydride. Although production at the plant (built in Spain) was later stopped, because it was uneconomic, it remains an interesting example that may find application in other reactions.

The principle exploited in this kind of configuration is the decoupling of the classical redox mechanism (which operates in the selective oxidation of most hydrocarbons) into two separate steps, each of which can be optimized, thus improving the overall performance. Besides better control of overall reaction exothermicity, further advantages of this operation are: (i) higher selectivity, which is usually achieved, because the hydrocarbon never comes into contact with the molecular oxygen, and (ii) elimination of hazards associated with possible formation of flammable gas mixtures. This reactor configuration also offers the opportunity to work with catalysts in a partially reduced state, which is not possible when oxygen is co-fed, and thus opens a new area of investigation.

Another example is monolithic-type reactors, which have found their main application in the field of combustion. A monolith bed allows better autothermic operations with a minimal pressure-drop. This concept was used to improve performances in commercial methanol into formaldehyde conversion by adding a final monolithic reactor stage. Cross-flow monoliths have been applied to improve performances in highly exothermic oxidation reactions. Examples are the oxidation of ammonia to nitrogen with a $Co/\alpha\text{-}Al_2O_3$ catalyst, the oxidation of SO_2 with a Pt catalyst, the oxidation of polychlorinated biphenyls, the oxidative dehydrogenation of light alkanes and the partial oxidation of methane [241–243]. Layers of gauze, stacked one over the other to form a bed several millimeters deep are used in important industrial applications of oxidation reactions such as (i) the oxidation of ammonia to NO and (ii) methane ammoxidation to HCN. Using these reactors, operations at extremely short contact times are possible, which allow significant improvements in selectivity in several oxidation reactions [337] such as (i) methane selective catalytic oxidation to syngas and (ii) alkane oxidative dehydrogenation to alkenes. In all these cases, the new reactor option implies the design of new oxidation catalysts that can operate in the new reactor configuration as well as being able to take advantage of the opportunities offered by the new reactor design.

Membrane technology offers interesting potential advantages in allowing better control of the reaction kinetics [338]. Further aspects of this are discussed in Chapter. The catalytically active component can be either deposited on the membrane or, simply, the catalyst bed is contained in a reactor having membrane walls. Membrane reactors can be used with the aim of distributing oxygen along the catalyst bed. A gradual feeding of oxygen can (i) maintain the optimal O_2/hydrocarbon ratio for selectivity, (ii) limit the formation of hot spots and (iii) avoid the occurrence of runaway phenomena. Moreover, a controlled distribution of oxygen may keep the catalyst at the desired average oxidation level.

Conventional fixed-bed reactor operations can also be improved through better integration of catalyst and reactor design. An important commercial example is the new *o*-xylene oxidation to phthalic anhydride process introduced by Lonza-Alusuisse, which uses a dual-bed configuration. The catalytic beds are arranged in two parts, each containing a catalyst, the formulation of which has been optimized, that is, less active for the first part of the reactor (close to the reactor inlet), where the reaction rate is the highest, and more active for the final part of the reactor, that is, for finishing the reaction. In each section the catalyst is essentially made of α-alumina or steatite pellets, coated with a thin film of V_2O_5/TiO_2-based catalysts. The activity of the catalyst in each section is optimized by controlling the vanadia content, as well as by the addition of dopants (Cs and P, principally). In this way the hot spot temperature is considerably lowered, and the hottest region becomes spread over a longer reactor length, with a considerable improvement in selectivity to the partial oxidation product, as well as longer catalyst life.

Alternative reactor options are also offered by:

- *Periodic flow reversal inducing forced unsteady-state conditions* [339]. The flow to the reactor is continuously reversed before the steady state is attained. A dual hot-spot temperature profile, characterized by a considerably lower temperature than in the single hot spot that would develop in the traditional flow configuration, forms in exothermic oxidation reactions. An increase in selectivity and better reactor control (lower risk of runaway) is possible over fixed-bed reactor operations, but compared to the analogous advantages possible with multi-bed reactors (see case of *o*-xylene), which have lower reactor and operation costs as well as fewer safety problems, the periodic flow reversal option will be applied to specific cases only.

- *Decoupling of the exothermal reaction into two steps.* In this reactor configuration, the overall heat of reaction is subdivided into two less exothermic steps. In the first the hydrocarbon is brought into contact with the catalyst, while in the second step the reduced catalyst is reoxidized by contact with gaseous oxygen. Besides better control of the overall reaction exothermicity, further advantages of this operation are: (i) higher selectivity, which is usually achieved, because the hydrocarbon is never brought into contact with molecular oxygen and (ii) elimination of the hazards associated with the possible formation of flammable gas mixtures. This configuration can be carried out either (i) in two (or more) parallel reactors, where one (or more) reactor is at the reaction stage, and the other is at the catalyst reoxidation stage or (ii) in a circulating-bed reactor, where the catalyst is continuously transported from the reaction vessel to the regeneration vessel and vice versa.

All these reactor options allow not only the development of new processes, and the holding of proprietary technologies, but also an improvement in productivity (thus saving energy per ton of product) and often a reduction in the emissions of waste, greenhouse gases and improved safety of operations, for example, improved process sustainability.

A relevant issue in industrial selective oxidation is the substitution of air-based to O_2-based processes. Air has been the preferred oxidant for years, but several oxidation processes, both in the liquid and in the gas phase, have been modified over the years to allow the use of pure oxygen [241–243, 340]. These changes have been driven by improvements in productivity and yield, while more recently revamping or modifications aimed at pure oxygen use have been undertaken due to environmental constraints.

The use of oxygen instead of air implies that the same partial pressure can be used with a much lower total pressure than with air, thus making it feasible to possibly reduce the total pressure, with obvious energy advantages. Moreover, an increase in the reaction rate makes it possible to reach the same productivity while lowering the reaction temperature, with possible benefits from the selectivity point of view when several products are formed in the reaction. Lower nitrogen contents, a ballast with very poor thermal conduction properties, also allow better control of the temperature profile in the presence of strongly exothermic reactions.

An even more important benefit originates from the considerable decrease in polluting emissions released into the atmosphere, as a consequence of the fact that spent gases can be recycled when oxygen is used in place of air. Less waste gas is produced, with energy savings during incineration. In addition, the heating value of the stream is much higher than for the air-based process (the concentration of nitrogen is lower, while that of hydrocarbons and carbon oxides is expected to be higher). Therefore, the purge stream, instead of being treated, can be used as a fuel to incinerate other wastes.

The following chemical processes make use of oxygen-enriched air or of oxygen, as an alternative to air [243]:

- partial oxidation of oil fractions and coke to synthesis gas;
- oxidation of methanol to formaldehyde (either air or oxygen-enriched air);
- oxidation of ethene to ethene oxide (either air or oxygen; new plants oxygen);
- oxychlorination of ethene to 1,2-dichloroethane (either air or oxygen; new plants mostly oxygen);
- oxidative acetoxylation of ethene to vinyl acetate (oxygen);
- oxidation of *n*-butane to acetic acid (either air or oxygen);
- oxidation of ethene to acetaldehyde (either air or oxygen);
- oxidation of acetaldehyde to acetic anhydride (either air or oxygen);
- ammoxidation of propene to acrylonitrile (oxygen-enriched air);
- oxidation of cyclohexane to cyclohexanone (either air or oxygen);
- oxidation of isobutane to *t*-butyl hydroperoxide (the latter is used for propene epoxidation) (oxygen).

Some additional aspects are also discussed in Chapter 5, which is dedicated to accounting for the sustainability of chemical processes.

2.7
Cascade and Domino Catalytic Reactions

This chapter has reported various examples to evidence how the new advances in catalyst design allow us to perform selectively in one-pot multistep reactions. These, which apply in particular for fine chemicals production, are often called "cascade" reactions, because they effectively involve the desorption from one site to react on another site physically distant from the first. It is thus a different concept from that shown in Figure 2.66: *n*-butane to maleic anhydride on $(VO)_2P_2O_7$ catalysts, where all the reaction pathway proceeds only on the catalyst surface without desorption of reaction intermediates. However, in cascade reactions there is no separation of the products and thus they all occur in the same reactor. This is the reason for the indication "one-pot" reactions.

A further interesting example is the synthesis of nabumetone, an anti-inflammatory agent, with a multifunctional base/acid/Pd catalyst through a cascade reaction [341]. The commercial synthesis (Hoechst/Celanese) is based on a two-step process involving either a Heck reaction between 6-bromo-2-methoxy-naphthalene and methyl vinyl ketone or a condensation between 6-methoxy-2-naphthaldehyde and acetone to give an intermediate that is separated, purified and hydrogenated in a second, separate process to give the final product, while producing a large amount of waste products. Corma *et al.* [341] have reported a residue-free catalytic process for the production of nabumetone in 98% yield and 100% selectivity, achieved through a cascade reaction system involving a multifunctional base/acid/hydrogenation catalyst based on nanocrystalline (3 nm) MgO (Figure 2.67).

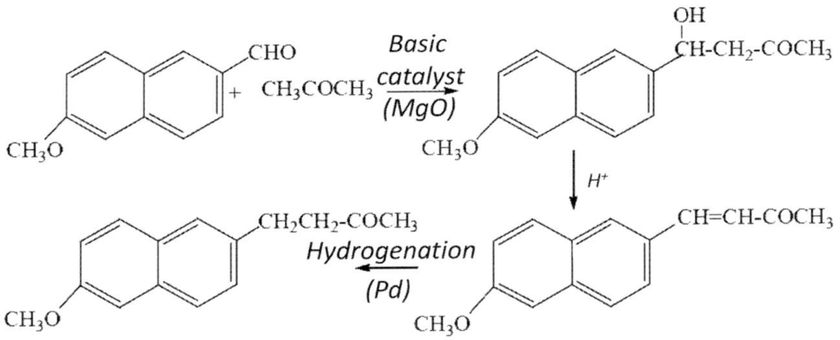

Figure 2.67 Synthesis of nabumetone with a multifunctional base/acid/Pd catalyst through a cascade reaction. Source: adapted from Corma *et al.* [341].

Another recent example from the Corma group concerns the one-pot synthesis of a topoisomerase-I inhibitor, an agent designed to interfere with the action of topo-isomerase enzymes. The commercial synthesis (Hoechst AG, US Pat 3 538 097) involves a seven-step procedure with a final cumulative yield of about 13%, due to the loss of selectivity in each step. In the one-pot synthesis reported by Corma's group [320], a combination of homogeneous and heterogeneous steps is used to obtain a global yield of 72% (Figure 2.68). This catalytic process has a clear advantage over the commercial organic synthesis procedure (which also produces more waste).

Another concept is that of domino reactions. This concept is known in organic chemistry [342], but less so in catalysis. The usual procedure for the synthesis of

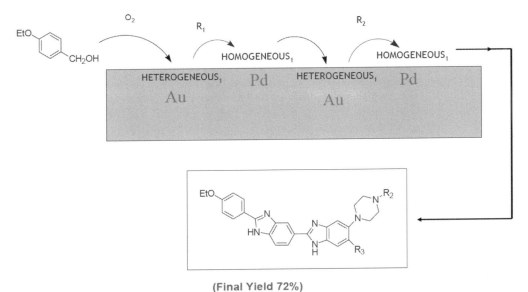

(Final Yield 72%)

Figure 2.68 One-pot synthesis of a topoisomerase-I inhibitor. Source: adapted from Corma *et al.* [320].

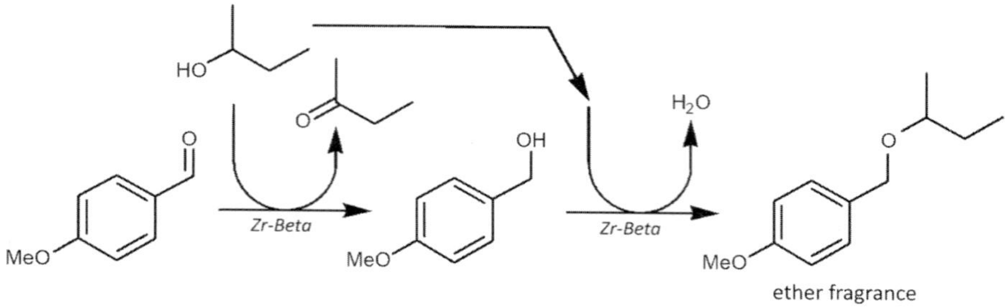

Figure 2.69 Domino reaction for the synthesis of an ether fragrance. Source: adapted from Corma *et al.* [320, 343].

organic compounds is the stepwise formation of the individual bonds in the target molecule. However, it would be more efficient if one could form several bonds in one sequence without isolating the intermediates, changing the reaction conditions, or adding reagents. This "domino" type procedure allows the minimization of waste, and the reduced use of solvents, reagents, adsorbents and energy with respect to stepwise reactions. The concept of a domino reaction is thus analogous to that of cascade reactions. It is used when a single catalyst, instead of different catalytic components, can perform the entire sequence of steps necessary, and the sequence of transformations is induced by a modification of the reaction conditions, or stepwise addition of the reagents. This is a unique opportunity for solid heterogeneous catalysts, because homogeneous catalysts are often characterized by a defined single functionality.

An interesting example of catalytic domino reaction is the synthesis of an ether fragrance reported by Corma and Renz [343] using either Sn-beta or Zr-beta, although the latter is preferable, giving at 100 °C (8 h) complete conversion with virtually 100% overall selectivity, while the Sn-beta is equally selective but less active. Figure 2.69 reports the scheme of this reaction. In a first step, the alcohol is produced by a Meerwein–Ponndorf–Verley reduction at 100 °C.

A one-pot reaction with cascade catalysis is not only particularly relevant in the case of fine chemistry but also for the new area of bioresource use. Several interesting examples have been discussed recently by Gallezot [344].

2.8
Multicomponent Catalytic Reactions

A further concept that is important for the development of more sustainable industrial processes is the ability of a catalyst to handle at the same time multiple reactants, to avoid the need of separation. This is a relevant area for the transformation of bioresources and has been discussed with various examples by Gallezot [344]. In particular, it is evidenced how a mixture of products suitable for a particular application, for example, in paper, paint, polymer and cosmetic industries, can be prepared in a one-pot process starting from raw materials such as starch, cellulose and

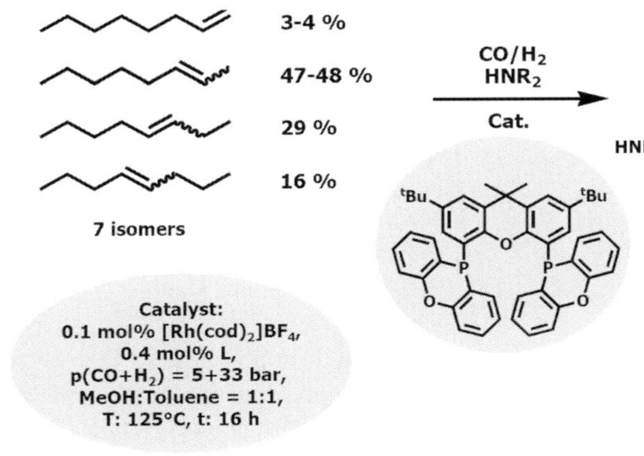

Figure 2.70 Selective hydroaminomethylation of mixtures using Rh complexes. Source: adapted from Beller *et al.* [345].

triglycerides. Examples of the direct transformation of starch into a mixture of products have been given that can be used as such to manufacture end-products.

Another interesting example has been reported recently by Beller *et al.* [345] regarding the selective hydroaminomethylation of mixtures instead of single compounds (Figure 2.70). The application of phenoxaphosphino-modified Xantphos-type ligands in the rhodium-catalyzed hydroaminomethylation of internal olefins to give linear amines shows excellent chemo- and regioselectivities, providing a practical and environmentally attractive synthetic route for the preparation of amines from internal alkenes. For the first time, both functionalized internal olefins and mixtures of internal and terminal olefins have been converted highly selectively into linear amines.

2.9
Organocatalysis

A new promising area that should be cited for its relevance to developing novel sustainable chemical processes is that of organocatalysis, that is, a purely organic and metal-free small molecule catalysis for a chemical reaction [346–348]. In addition to being another useful strategy for catalysis, this approach has some important advantages. Small organic molecule catalysts are generally stable and fairly easy to design and synthesize. They are often based on nontoxic compounds, such as sugars, peptides or even amino acids, and can be easily linked to a solid support, making them useful for industrial applications.

Organocatalysts were used sporadically throughout the last century, but recently this area has undergone exponential growth. A particularly appealing discovery of great potential is the use of chiral Brønsted acids as organocatalysts. Organic Brønsted acids function by donating a proton to the substrate. They have been used as catalysts for various reactions since the beginnings of modern chemistry, but applications in asymmetric catalysis have been extremely rare. A breakthrough in this area came when researchers developed highly active Brønsted acid organocatalysts that incorporate a urea motif as the active principle.

Other examples of asymmetric reactions involving organocatalysts are: (i) Diels–Alder reactions, (ii) Michael reactions, (iii) Mannich reactions and (iv) Shi epoxidation and organocatalytic transfer hydrogenation.

2.10
Conclusions

Catalysis is a main enabling factor to achieving sustainability through chemistry. We have discussed in this chapter several examples of homogeneous, heterogeneous and biocatalysis, which, although clearly not exhaustive of all the possibilities, give an overview of the tools offered for chemistry and chemical engineering researchers to develop novel and improved processes.

Progress in research on catalysis science has been impressive, and several new concepts and catalysts are now available to open up new directions. Progress in understanding catalysis, through the use of advanced characterization methods, which have been not discussed here for conciseness, has created a large platform of knowledge that provides not only an a-posteriori interpretation of the results, after catalytic screening, but also a-priori predictive methods. At the same time, progresses in experimental techniques for high-throughput screening have made available faster methods for testing and optimization of the catalysts.

We thus conclude this chapter by indicating that the intense research on catalysis in the last decade, although apparently of limited success in developing new industrial applications, has instead provided the background for breakthrough new applications in all areas of sustainable chemical industrial processes.

References

1 Jenck, J.F., Agterberg, F. and Droescherc, M.J. (2004) *Green Chem.*, **6**, 544.
2 Sheldon, R.A., Arends, I. and Hanefeld, U. (2007) *Green Chemistry and Catalysis*, Wiley-VCH Verlag GmbH, Weinheim.
3 Clark, J. and Macquarrie, D. (eds) (2002) *Handbook of Green Chemistry and Technology*, Blackwell Science Ltd, Oxford (UK).
4 Anastas, P.T. and Warner, J.C. (1998) *Green Chemistry, Theory and Practice*, Oxford University Press, Oxford.
5 Rothenberg, G. (2008) *Catalysis – Concepts and Green Applications*, Wiley-VCH Verlag, Weinheim, Germany.
6 Tundo, P. and Esposito, V. (eds) (2008) *Green Chemical Reactions*, Springer, Dordrecht, The Netherlands, (Proceedings of the NATO Advanced Study Institute on New Organic Chemistry Reactions and Methodologies for Green Production, Lecce, Italy, 29 Oct – 10 Nov 2006).
7 Doble, M. and Kruthiventi, A.K. (2007) *Green Chemistry and Engineering*, Elsevier Science & Technology Books, Amsterdam.
8 Afonso, C.A.M. and Crespo, J.G. (2005) *Green Separation Processes*, Wiley-VCH Verlag, Weinheim, Germany.

9 Nelson, W.M. (2003) *Green Solvents for Chemistry: Perspectives and Practice,* Oxford University Press, New York.

10 Tundo, P., Perosa, A. and Zecchini, F. (2007) *Methods and Reagents for Green Chemistry: an Introduction,* John Wiley & Sons, Inc., Hoboken, New Jersey, USA.

11 Growl, D.A. (ed.) (1996) *Inherently Safer Chemical Processes. A Life Cycle Approach,* American Institute of Chemical Engineers, New York.

12 Seeberger, P.H. and Blume, T. (eds) (2007) *New Avenues to Efficient Chemical Synthesis,* Springer-Verlag, Berlin.

13 Stankiewivz, A. and Moulijn, J.A. (2004) *Re-Engineering the Chemical Processing Plant,* Marcel Dekker, New York.

14 Steinbach, J. (1999) *Safety Assessment for Chemical Processes,* Wiley-VCH Verlag, Weinheim.

15 Azapagic, A., Perdan, S. and Clift, R. (2004) *Sustainable Development in Practice. Case Studies for Engineers and Scientists,* John Wiley & Sons, Ltd, Chichester, UK.

16 Boeschen, S., Lenoir, D. and Scheringer, M. (2003) *Naturwissenschaften,* **90**, 93.

17 Gennari, F., Pizzio, F. and Miertus, S. (2000) *Chim. Ind. (Milan Italy),* **82**, 514.

18 Clark, J.H. (2005) Green chemistry and environmentally friendly technologies, in *Green Separation Processes* (eds C.A.M. Afonso and J.G. Crespo), Wiley-VCH Verlag, Weinheim, p. 3.

19 Constable, D.J.C., Dunn, P.J., Hayler, J.D., Humphrey, G.R., Leazer, J.L. Jr, Linderman, R.J., Lorenz, K., Manley, J., Pearlman, B.A., Wells, A., Zaks, A. and Zhang, T.Y. (2007) *Green Chem.,* **9**, 411.

20 Anastas, P.T. and Zimmerman, J.B. (2003) *Environ. Sci. Technol.,* **37**, 95.

21 Anastas, P.T. and Kirchhoff, M.M. (2002) *Acc. Chem. Res.,* **35**, 686.

22 Lancaster, M. (2002) *Green Chemistry: An Introductory Text,* Royal Society of Chemistry, Cambridge, p. 242.

23 Garcia-Serna, J., Perez-Barrigon, L. and Cocero, M.J. (2007) *Chem. Eng. J. (Amsterdam),* **133**, 7.

24 Albini, A. and Fagnoni, M. (2008) *ChemSusChem,* **1**, 63.

25 Brett, C.M.A. (2007) *Pure. Appl. Chem.,* **79**, 1969.

26 Strauss, C.R. and Varma, R.S. (2006) *Top. Curr. Chem.,* **266**, 199 (Microwave Methods in Organic Synthesis).

27 Fukuoka, S., Tojo, M., Hachiya, H., Aminaka, M. and Hasegawa, K. (2007) *Polym. J. (Tokyo),* **39**, 91.

28 Kaneda, K., Ebitani, K., Mizugaki, T. and Mori, K. (2006) *Bull. Chem. Soc. Jpn.,* **79**, 981.

29 Misono, M. (2002) *Top. Catal.,* **21**, 89.

30 Brennan, D. (2007) *Environmentally Conscious Materials and Chemicals Processing* (ed Kutz, M.), Wiley Series in Environmentally Conscious Engineering, **3**, p. 59.

31 Dahl, J.A., Maddux, B.L.S. and Hutchison, J.E. (2007) *Chem. Rev.,* **107**, 2228.

32 Ishida, T. and Haruta, M. (2007) *Angew. Chem. Int. Ed.,* **46**, 7154.

33 Ranke, J., Stolte, S., Stoermann, R., Arning, J. and Jastorff, B. (2007) *Chem. Rev.,* **107**, 2183.

34 Walsh, P.J., Li, H. and de Parrodi, C.A. (2007) *Chem. Rev.,* **107**, 2503.

35 Clark, J.H. (2002) *Acc. Chem. Res.,* **35**, 79.

36 Horváth, I.T. and Anastas, P.T. (2007) (guest editors) *Chem. Rev.,* **107**, 2169.

37 Horváth, I.T. (guest editor) (2002) *Acc. Chem. Res.,* **35**, 685.

38 Sheldon, R.A. (2008) *Green Chem.,* **10**, 359.

39 Warner, J. (2007) *Green Chem. Lett. Rev.,* **1**, 1.

40 Centi, G. (2008) *ChemSusChem,* **1**, 7.

41 Anastas, P.T., Kirchhoff, M.M. and Williamson, T.C. (2001) *Appl. Catal. A-Gen.,* **221**, 3.

42 Jones, A.J. (2005) *Chem. Aust., June,* 4.

43 Centi, G. and Perathoner, S. (2003) *Catal. Today,* **77**, 287.

44 Thomas, J.M. and Raja, R. (2005) *Annu. Rev. Mater. Sci.,* **35**, 315.

45 Poliakoff, M. and Licence, P. (2007) *Nature,* **450**, 810.

46 Tundo, P.(preface) (2007) *Pure. Appl. Chem. (Special Issue)*, **79**, 1831.

47 Freedonia Group (2007) World Catalysts to 2010 - Demand and Sales Forecasts, Market Share, Market Size, Market Leaders, Study nr. 2125.

48 Frost & Sullivan (2007) European Market for Catalysts, Sep.

49 Smith, W.T. (2007) Catalysts Moving on the Fast Track, BASF CC Investor Relation.

50 Beller, M. (2007) Catalysis – a key to sustainability. *Leibniz Perspect. – Res. Sustainable*, 33.

51 Centi, G. and Perathoner, S. (2008) Catalysis, a driver for sustainability and societal challenges. *Catal. Today*, **138**, 69. (Opening lecture at 5th European Conference on Catalysis, Ottrott, France, September 12–16th, 2007).

52 Centi, G. and Perathoner, S. (2008) Catalysis: role and challenges for a sustainability energy. *Top. Catal.*, **52**, 948 (Invited lecture at the conference "Catalysis for Society", Cracow, Poland, May12–15th, 2008).

53 Bell, A.T., Gates, B.C., Ray, D. and Thompson, M.R. (2008) *Basic Research Needs: Catalysis for Energy*, PNNL-17214, US Department of Energy. http://www.pnl.gov/main/publications/external/technical_reports/PNNL-17214.pdf.

54 Centi, G. and van Santen, R.A. (eds) (2007) *Catalysis for Renewables*, Wiley-VCH Verlag GmbH, Weinheim.

55 Kamm, B., Gruber, P.R. and Kamm, M. (eds) (2006) *Biorefineries – Industrial Processes and Products*, Wiley-VCH Verlag, Weinheim.

56 Zhou, B., Han, S., Raja, R. and Somorjai, G.A. (eds) (2006) *Nanotechnology in Catalysis*, vol. 3, Springer, Dordrecht, The Netherlands.

57 Heiz, U. and Landman, U. (eds) (2007) *Nanocatalysis*, Springer, Dordrecht, The Netherlands.

58 van Leeuwen, P.W. (2004) *Homogeneous Catalysis: Understanding the Art*, Springer, Dordrecht, The Netherlands.

59 Chorkendorff, I. and Niemantsverdriet, J.W. (2003) *Concepts of Modern Catalysis and Kinetics*, Wiley-VCH Verlag, Weinheim.

60 Hagen, J. (2006) *Industrial Catalysis: A Practical Approach*, 2nd edn, Wiley-VCH Verlag, Weinheim.

61 van Santen, R.A. and Neurock, M. (2006) *Molecular Heterogeneous Catalysis, A Conceptual and Computational Approach*, Wiley-VCH Verlag, Weinheim.

62 Nilsson, A., Pettersson, L.G.M. and Nørskov, J.K. (eds) (2008) *Chemical Bonding at Surfaces and Interfaces*, Elsevier B.V., Amsterdam.

63 Cornils, B., Herrmann, W.A., Horváth, I.T., Leitner, W., Mecking, S., Olivier-Bourbigou, H. and Vogt, D.(eds) (2005) *Multiphase Homogeneous Catalysis*, Wiley-VCH Verlag, Weinheim.

64 Cole-Hamilton, D.J. and Tooze, R.P. (2006) *Catalyst Separation, Recovery and Recycling*, Springer, Dordrecht, The Netherlands.

65 Cornils, B. and Herrmann, W.A. (2004) *Aqueous-Phase Organometallic Catalysis*, 2nd edn, Wiley-VCH Verlag, Weinheim.

66 Kohlpaintner, C.W., Fischer, Ri.W. and Cornils, B. (2001) *Appl. Catal. A-Gen.*, **221**, 219.

67 Cole-Hamilton, D.J. (2003) *Science*, **299**, 1702.

68 Carrea, G. and Riva, S. (eds) (2008) *Organic Synthesis with Enzymes in Non-Aqueous Media*, Wiley-VCH Verlag, Weinheim.

69 Hobbs, H.R., Kirke, H.M., Poliakoff, M. and Thomas, N.R. (2007) *Angew. Chem. Int. Ed.*, **46**, 7860.

70 van Rantwijk, F., Madeira Lau, R. and Sheldon, R.A. (2003) *Trends Biotechnol.*, **21**, 131.

71 Sardessai, Y.N. and Bhosle, S. (2004) *Biotechnol. Prog.*, **20**, 655.

72 Schmid, A., Dordick, J.S., Hauer, B., Kiener, A., Wubbolts, M. and Witholt, B. (2001) *Nature*, **409**, 258.

73 Weissermel, K. and Arpe, H.-J. (2003) *Industrial Organic Chemistry*, Wiley-VCH Verlag, Weinheim.

74 Haumann, M., Koch, H. and Schomäcker, R. (2003) *Catal. Today*, **79**, 43.

75 Lee, J.K., Yoon, T.J. and Chung, Y.K. (2001) *Chem. Commun.*, 1164.

76 Wiese, K.D., Möller, O., Protzmann, G. and Trocha, M. (2003) *Catal. Today*, **79**, 97.

77 Zheng, X.L., Jiang, J.Y., Liu, X.Z. and Jin, Z.L. (1998) *Catal. Today*, **44**, 175.

78 Yasuo, T. and Noriaki, Y. (2005) *Catalysts Catal.*, **47**, 625.

79 Kohlpaintner, C.W. and Beller, M. (1997) *J. Mol. Catal.*, **116**, 259.

80 Cornils, B. (1999) *J. Mol. Catal. A: Chem.*, **143**, 1.

81 Bertleff, W., Roeper, M. and Sava, X. (2007) Carbonylation, in *Ullmann's Encyclopedia of Industrial Chemistry*, Wiley-VCH Verlag, Weinheim.

82 Asenjo, J.A. (2004) *J. Chem. Technol. Biotechnol.*, **59**, 109.

83 Kuhn, P., Sémeril, D., Matt, D., Chetcuti, M.J. and Lutz, P. (2007) *Dalton Trans.*, 515.

84 Sinou, D. (2004) Two-phase catalysis, in *Transition Metals for Organic Synthesis*, 2nd edn (eds M. Beller and C. Bolm), Wiley-VCH Verlag, Weinheim, p. 510.

85 Vogt, D. (2005) Economical applications (SHOP process), in *Multiphase Homogeneous Catalysis* (eds B. Cornils et al.), Wiley-VCH Verlag, Weinheim, p. 309.

86 Keim, W. (2003) *Green Chem.*, **5**, 105.

87 Baker, R.T. and Tumas, W. (1999) *Science*, **284**, 1477.

88 Sheldon, R.A. (2008) *Chem. Commun.*, 3352.

89 Sheldon, R.A., Arends, I.W.C.E. and Hanefeld, U. (2007) Catalysis in novel reaction media, in *Green Chemistry and Catalysis* (eds R.A. Sheldon, I. Arends and U. Hanefeld), Wiley-VCH Verlag, Weinheim.

90 Horváth, I.T. and Rábai, J. (1994) *Science*, **266**, 72.

91 Wende, M. and Gladysz, J.A. (2003) *J. Am. Chem. Soc.*, **125**, 5861.

92 Wasserscheid, P. and Welton, T. (eds) (2007) *Ionic Liquids in Synthesis*, 2nd edn, Wiley-VCH Verlag, Weinheim.

93 Swatloski, R.P., Holbrey, J.D. and Rogers, R.D. (2003) *Green Chem.*, **5**, 361.

94 Forestière, A. and Favre, F. (2005) Commercial applications and aspects, in *Multiphase Homogeneous Catalysis* (eds B. Cornils et al.), Wiley-VCH Verlag, Weinheim.

95 Chauvin, Y., Commereuc, D., Hirschauer, A., Hugues, F. and Saussine, L. (1988) Process and catalyst for the dimerization or codimerization of olefins, French Patent, FR, 2,611,700.

96 Chauvin, Y., Gilbert, B. and Guibard, I. (1990) *Chem Commun.*, 1715.

97 Chauvin, Y. and Olivier-Bourbigou, H. (1995) *CHEMTECH*, **25**, 26.

98 Seddon, K.R. (2003) *Nat. Mater.*, **2**, 363.

99 Plechkova, N.V. and Seddon, K.R. (2008) *Chem. Soc. Rev.*, **37**, 123.

100 Jastorff, B., Stormann, R., Ranke, J., Molter, K., Stock, F., Oberheitmann, B., Hoffmann, W., Hoffmann, J., Nuchter, M., Ondruschka, B. and Filser, J. (2003) *Green Chem.*, **5**, 136.

101 Webb, P.B., Sellin, M.F., Kunene, T.E., Williamson, S., Slawin, A.M.Z. and Cole-Hamilton, D.J. (2003) *J. Am. Chem. Soc.*, **125**, 15577.

102 Jastorff, B., Stormann, R., Ranke, J., Molter, K., Stock, F., Oberheitmann, B., Hoffmann, W., Hoffmann, J., Nuchter, M., Ondruschka, B. and Filser, J. (2003) *Green Chem.*, **5**, 136.

103 Swatloski, R.P., Holbrey, J.D., Memon, S.B., Caldwell, G.A., Caldwell, K.A. and Rogers, R.D. (2004) *Chem. Commun.*, 668.

104 Uerdingen, M., Treber, C., Balser, M., Schmitt, G. and Werner, C. (2005) *Green Chem.*, **7**, 321.

105 Ranke, J., Stock, F., Störmann, R., Mölter, K., Hoffmann, J., Ondruschka, B. and Jastorff, B. (2005) Preliminary (eco-) toxicological risk profiles of ionic liquids, in *Multiphase Homogeneous Catalysis* (eds B. Cornils et al.), Wiley-VCH Verlag, Weinheim, p. 588.

106 Cho, C.W., Jeon, Y.C., Pham, T.P., Vijayaraghavan, K. and Yun, Y.S. (2008) *Ecotoxicol. Environ. Saf.*, **71**, 166.

107 Wells, A.S. and Coombe, V.T. (2006) *Org. Process Res. Dev.*, **10**, 794.

108 Matzke, M., Stolte, S., Arning, J., Uebers, U. and Filser, J. (2008) *Green Chem.*, **10**, 584.

109 Meindersma, G.W., Maase, M. and De Haan, A.B. (2007) Ionic liquids, *Ullmann's Encyclopedia of Industrial Chemistry*, Wiley-VCH Verlag, Weinheim, doi: 10.1002/14356007.114_101.

110 Scurto, A.M. (2005) Introduction to catalysis using supercritical solvents, in *Multiphase Homogeneous Catalysis* (eds B. Cornils *et al.*), Wiley-VCH Verlag, Weinheim, p. 607.

111 Beckman, E.J. (2004) *J. Supercrit. Fluids*, **28**, 121.

112 Gladysz, J.A. (1994) *Science*, **266**, 55.

113 Zhao, F., Fujita, S.-i. and Arai, M. (2006) *Curr. Org. Chem.*, **10**, 1681.

114 Reilly, C.R. and Lerou, J.J. (1998) *Catal. Today*, **41**, 433.

115 Bhanage, B.M., Fujita, S.-i. and Arai, M. (2003) *J. Organomet. Chem.*, **687**, 211.

116 Villadsen, J. and Livbjerg, H. (1978) *Catal. Rev., Sci. Eng.*, **17**, 203.

117 Bennardi, D.O., Romanelli, G.P., Autino, J.C. and Pizzio, L.R. (2007) *Appl. Catal. A-Gen.*, **324**, 62.

118 de Angelis, A., Flego, C., Ingallina, P., Montanari, L., Clerici, M.G., Carati, C. and Perego, C. (2001) *Catal. Today*, **65**, 363.

119 Harmer, M.A., Junk, C., Rostovtsev, V., Carcani, L.G., Vickery, J. and Schnepp, Z. (2007) *Green Chem.*, **9**, 30.

120 Mehnert, C.P. (2004) *Chem. Eur. J.*, **11**, 50.

121 Mehnert, C.P., Cook, R.A., Dispenziere, N.C. and Afeworki, M. (2002) *J. Am. Chem. Soc.*, **124**, 12932.

122 Riisager, A., Jørgensen, B., Wasserscheid, P. and Fehrmann, R. (2006) *Chem. Commun.*, 994.

123 Dioosa, B.M.L. and Jacobs, P.A. (2006) *J. Catal.*, **243**, 217.

124 Gruttadauria, M., Riela, S., Lo Meo, P., D'Anna, F. and Noto, R. (2004) *Tetrahedron. Lett.*, **45**, 6113.

125 Lapis, A.A.M., de Oliveira, L.F., Neto, B.A.D. and Dupont, J. (2008) *ChemSusChem*, **1**, 759.

126 Gamba, M., Lapis, A.A.M. and Dupont, J. (2007) *Adv. Synth. Catal.*, **350**, 160.

127 Hintermair, U., Zhao, G., Santini, C.C., Muldoon, M.J. and Cole-Hamilton, D.J. (2007) *Chem. Commun.*, 1462.

128 Riisager, A., Fehrmann, R., Haumann, M. and Wasserscheid, P. (2006) *Top. Catal.*, **40**, 91.

129 Matsumoto, M., Inomoto, Y. and Kondo, K. (2005) *J. Membr. Sci.*, **246**, 77.

130 Tangab, J., Tanga, H., Sunb, W., Planchera, H., Radosza, M. and Shen, Y. (2005) *Chem. Commun.*, 3325.

131 Ilconich, J., Myers, C., Pennline, H. and Luebke, D. (2007) *J. Membrane. Sci.*, **298**, 41.

132 Choplin, A. and Quignard, F. (1998) *Coord. Chem. Rev.*, **178–180**, 1679.

133 Lindner, E., Schneller, T., Auer, F. and Mayer, H.A. (1999) *Angew. Chem. Int. Ed.*, **38**, 2154.

134 Arhancet, J.P., Davis, M.E., Merola, J.S. and Hanson, B.E. (1989) *Nature*, **339**, 454.

135 Wan, K.T. and Davis, M.E. (1994) *Nature*, **370**, 449.

136 Fache, E., Mercier, C., Pagnier, N., Despeyroux, B. and Panster, P. (1993) *J. Mol. Catal.*, **79**, 117.

137 Wan, K.T. and Davis, M.E. (1994) *J. Catal.*, **148**, 1.

138 Davis, M.E. (1992) *CHEMTECH*, **22**, 498.

139 Topea, B., Zhua, Y. and Lercher, J.A. (2007) *Catal. Today*, **123**, 113.

140 Centi, G. and Perathoner, S. Nano-reactor engineering of catalytic surfaces. Invited lecture at 14th ICC (Seoul, Korea, July 13–18, 2008).

141 Centi, G. and Perathoner, S. (2008) *Catalysis* (ed. J. Spivey), Royal Chemical Society Pub, Cambridge, UK, **20**, 367.

142 Kieboom, T. (2007) Integration of biocatalysis with chemocatalysis: cascade catalysis and multi-step conversions in concert, in *Catalysis for Renewables* (eds G. Centi and R.A. van Santen), Wiley-VCH Verlag, Weinheim, Ch. 13, p. 273.

143 Buchholz, K., Kasche, V. and Bornscheuer, U.T. (2005) *Biocatalysts & Enzyme Technology*, Wiley-VCH Verlag, Weinheim.

144 Straathof, A.J.J., Panke, S. and Schmid, A. (2002) *Curr. Opin. Biotechnol.*, **13**, 548.

145 Faber, K. (1997) *Pure Appl. Chem.*, **69**, 1613.

146 Adamczak, M. and Krishna, S.H. (2004) *Food Technol. Biotechnol.*, **42**, 251.

147 Matsuda, T. (ed.) (2007) *Future Directions in Biocatalysis*, Elsevier, Amsterdam.

148 Illanes, A. (ed.) (2008) *Enzyme Biocatalysis*, Springer Science + Business Media, Heidelberg.

149 Hou, C.T. (ed.) (2005) *Handbook of Industrial Biocatalysis*, Taylor & Francis (CRC), Boca Raton, FL, USA.

150 Thayer, A.M. (2006) *C&EN*, **84** (33), 29.

151 Paetzold, J. and Bäckvall, J.E. (2005) *J. Am. Chem. Soc.*, **127**, 17620.

152 Carey, J.S., Laffan, D., Thomson, C. and Williams, M.T. (2006) *Org. Biomol. Chem.*, 4, 2337.

153 EuropaBio (2003) *White Biotechnology: Gateway to a More Sustainable Future*, EuropaBio Pub, Brussels.

154 Wubbolts, M.(Jan. 15th, 2007) *The Future is White! Prospects of White Biotechnology*, DSM Innovation Center.

155 Hagen, J. (2006) Biocatalysis, in *Industrial, Catalysis*, 2nd edn, Wiley-VCH Verlag, Weinheim, Ch. 4, p. 83.

156 Thomas, S.M., DiCosimo, R. and Nagarajan, V. (2002) *Trends Biotechnol.*, **20**, 238.

157 Petersen, M. and Kiener, A. (1999) *Green Chem.*, **4**, 99.

158 Chuck, R.(10–15 Sept, 2006) Green sustainable chemistry in the production of nicotinates. Presented at the 1st International IUPAC Conference on Green-Sustainable Chemistry, Dresden, Germany.

159 Cheng, H.N. and Gross, R.A. (2005) *Polymer Biocatalysis and Biomaterials*, American Chemical Society Pub, Washington DC.

160 Uyama, H. (2007) Enzymatic polymerization, in *Future Directions in Biocatalysis* (ed. T. Matsuda), Elsevier B.V., Amsterdam, Ch. 10, p. 205.

161 Gross, R.A. and Kalra, B. (2002) *Science*, **202**, 803.

162 Mecking, S. (2004) *Angew. Chem. Int. Ed.*, **43**, 1078.

163 Matsumura, S. (2006) *Adv. Polym. Sci.*, **194**, 95.

164 Uyama, H. and Kobayashi, S. (2006) *Adv. Polym. Sci.*, **194**, 133.

165 Uyama, H. and Kobayashi, S. (2003) *Curr. Org. Chem.*, **7**, 1397.

166 Reihmann, M. and Ritter, H. (2006) *Adv. Polym. Sci.*, **194**, 1.

167 Ogawa, J. and Shimizu, S. (1999) *Trends Biotechnol.*, **17**, 13.

168 Drepper, T., Eggert, T., Hummel, W., Leggewie, C., Pohl, M., Rosenau, F., Wilhelm, S. and Jaeger, K.-E. (2006) *Biotechnol. J.*, **1** (7–8), 777.

169 Burton, S.G., Cowan, D.A. and Woodley, J.M. (2002) *Nat. Biotechnol.*, **20**, 37.

170 Pollard, D.J. and Woodley, J.M. (2007) *Trends Biotechnol.*, **25**, 66.

171 Woodley, J.M. (2008) *Trends Biotechnol.*, **26**, 321.

172 Duetz, W.A. (2007) *Trends Microbiol.*, **15**, 469.

173 Betts, J.I. and Baganz, F. (2006) *Microb. Cell. Fact.*, **5**, 21.

174 Micheletti, M. and Lye, G.J. (2006) *Curr. Opin. Biotechnol.*, **17**, 611.

175 Bornscheuer, U.T. and Buchholz, K. (2005) *Eng. Life Sci.*, **5**, 309.

176 Alcalde, M., Ferrer, M., Plou, F.J. and Ballesteros, A. (2006) *Trends Biotechnol.*, **24**, 281.

177 Hou C.T. and Shaw J-.F. (eds) (2008) *Biocatalysis and Bioenergy*, Wiley-VCH Verlag, Weinheim.

178 Kim, J., Jia, H. and Wang, P. (2006) *Biotechnol. Adv.*, **24**, 296.

179 Tischer, W. and Wedekind, F. (1999) *Top. Curr. Chem.*, **200**, 95 (Biocatalysis: From Discovery to Application).

180 Mukhopadhyay, K., Phadtare, S., Vinod, V.P., Kumar, A., Rao, M., Chaudhari, R.V. and Sastry, M. (2003) *Langmuir*, **19**, 3858.

181 Lei, C., Shin, Y., Liu, J. and Ackerman, E.J. (2002) *J. Am. Chem. Soc.*, **124**, 11242.

182 Fadnavis, N.W., Bhaskar, V., Kantam, M.L. and Choudary, B.M. (2003) *Biotechnol. Prog.*, **19**, 346.

183 Kim, J. and Grate, J.W. (2003) *Nano Lett.*, **3**, 1219.

184 Rege, K., Raravikar, N.R., Kim, D.-Y., Schadler, L.S., Ajayan, P.M. and Dordick, J.S. (2003) *Nano Lett.*, **3**, 829.

185 Kim, J., Grate, J.W. and Wang, P. (2006) *Chem. Eng. Sci.*, **61**, 1017.

186 Ran, N., Zhao, L., Chen, Z. and Tao, J. (2008) *Green Chem.*, **10**, 361.

187 Breslow, R. (ed.) (2005) *Artificial Enzymes*, Wiley-VCH Verlag, Weinheim.

188 Mart, R.J. and Webb, S.J. (2008) *Annu. Rep. Prog. Chem., Sect. B: Org. Chem.*, **104**, 370.

189 Bar-Cohen, Y. (ed.) (2006) *Biomimetics. Biologically Inspired Technologies*, Taylor & Francis (CRC Press), Boca Raton, FL, US.

190 van Leeuwen, W.N.M. (ed.) (2008) *Supramolecular Catalysis*, Wiley-VCH Verlag, Weinheim.

191 Likhtenshtein, G.I. (2003) *New Trends in Enzyme Catalysis and Biomimetic Chemical Reactions*, Kluwer Academic Publishers, Springer.

192 Schwartz, S., Masciangioli, T. and Boonyaratanakornkit, B.(rapporteurs) (2008) Bioinspired Chemistry for Energy: A Workshop Summary to the Chemical Sciences Roundtable, US National Research Council (NRC), Washington DC.

193 Gamez, P., Aubel, P.G., Driessen, W.L. and Reedijk, J. (2001) *Chem. Soc. Rev.*, **30**, 376.

194 Mahadevan, V., Klein Gebbink, R.J.M., Daniel, T. and Stack, P. (2000) *Curr. Opin. Chem. Biol.*, **4**, 228.

195 Koval, I.A., Gamez, P. and Reedijk, J. (2008) Mimicking nature: bio-inspired models of copper proteins, in *Tomorrow's Chemistry Today* (ed. B. Pignataro),

Wiley-VCH Verlag, Weinheim, Ch. 5, p. 101.

196 Thomas, F. (2007) *Eur. J. Inorg. Chem.*, **17**, 2379.

197 Mansuy, D. (2007) *C. R. Chim.*, **10**, 392.

198 Spasojevic, I., Colvin, O.M., Warshany, K.R. and Batinic-Haberle, I. (2006) *J. Inorg. Biochem.*, **100**, 1897.

199 Breslow, R. (1995) *Acc. Chem. Res.*, **28**, 146.

200 Breslow, R. and Schmuck, C. (1996) *J. Am. Chem. Soc.*, **118**, 6601.

201 Reetz, M.T. and Waldvogel, S.R. (1997) *Angew. Chem., Int. Ed. Engl.*, **36**, 865.

202 Vriezema, D.M., Aragones, M.C., Elemans, J.A.A.W., Cornelissen, J.J.L.M., Rowan, A.E. and Nolte, R.J.M. (2005) *Chem. Rev.*, **105**, 1445.

203 Nishioka, Y., Yamaguchi, T., Yoshizawa, M. and Fujita, M. (2007) *J. Am. Chem. Soc.*, **129**, 7000.

204 Maurizot, V., Yoshizawa, M., Kawano, M. and Fujita, M. (2006) *Dalton Trans.*, 2750.

205 Yoshizawa, M. and Fujita, M. (2005) *Pure Appl. Chem.*, **77**, 1107.

206 Lu, Y. and Mosier, N.S. (2006) *Biotechnol. Prog.*, **23**, 116.

207 Corma, A. (1995) *Chem. Rev.*, **95**, 559.

208 Pastore, H.O., Coluccia, S. and Marchese, L. (2005) *Ann. Rev. Mater. Res.*, **35**, 351.

209 Fryxell, G.E. (2006) *Inorg. Chem. Commun.*, **9**, 1141.

210 Taguchi, A. and Schüth, F. (2005) *Microporous Mesoporous Mater.*, **77**, 1.

211 Centi, G. and Perathoner, S. (2008) *Microporous Mesoporous Mater.*, **107**, 3.

212 Cavani, F., Trifiro, F. and Vaccari, A. (1991) *Catal. Today*, **11**, 173.

213 Fierro, J.L.G. (ed.) (2005) *Metal Oxides: Chemistry and Applications*, Taylor & Francis (CRC Press), Boca Raton, FL.

214 Carreon, M.A. and Guliants, V.V. (2005) *Eur. J. Inorg. Chem.*, **1**, 27.

215 Misono, M. (2000) *C. R. Acad. Sci. Paris, Ser. IIc, Chem.*, **3**, 471.

216 Hüsing, N. and Schubert, U. (2004) Porous inorganic-organic hybrid materials, in *Functional Hybrid Materials* (eds P. Gómez-Romero and C. Sanchez),

Wiley-VCH Verlag, Weinheim, Ch. 4, p. 86.

217 Stein, A., Melde, B.J. and Schroden, R.C. (2000) *Adv. Mater.*, **12**, 1403.

218 Harmer, M.A. and Sun, Q. (2001) *Appl. Catal. A-Gen.*, **221**, 45.

219 Motokura, K., Fujita, N., Mori, K., Mizugaki, T., Ebitani, K. and Kaneda, K. (2005) *J. Am. Chem. Soc.*, **127**, 9674.

220 Selva, M., Perosa, A. and Fabris, M. (2008) *Green Chem.*, **10**, 457.

221 Perego, C. and Ingallina, P. (2002) *Catal. Today*, **73**, 3.

222 Perego, C. and Ingallina, P. (2004) *Green Chem.*, **6**, 274.

223 Nowak, F.-M., Himes, J.F. and Mehlberg, R.L.(March 23–25, 2003) Advances in hydrofluoric (HF) acid catalyzed alkylation. Presented at NPRA Annual Meeting, San Antonio, TX, USA.

224 Hommeltoft, S.I. (2001) *Appl. Catal. A-Gen.*, **221**, 421.

225 Cavani, F., Corazzari, M., Bencini, E. and Goffredi, G. (2002) *Appl. Catal.*, **226**, 31.

226 Woodle, G.B. (2006) Ethylbenzene, in *Encyclopedia of Chemical Processing* (ed. S. Lee), Taylor & Francis (CRC Press), New York, p. 903.

227 Bellussi, G. and Perego, C. (2000) *CATTECH*, **7**, 4.

228 Schmidt, R.J. (2006) Cumene production, in *Encyclopedia of Chemical Processing* (ed. S. Lee), Taylor & Francis (CRC Press), New York, p. 603.

229 Ratton, S. (1997) *Chem. Today*, **3–4**, 33.

230 Derouane, E.G. (ed.) (2006) *Catalysts for Fine Chemical Synthesis*, Vol. 4, Wiley-Interscience.

231 Hitzler, M.G., Smail, F.R., Ross, S.K. and Poliakoff, M. (1998) *Chem. Commun.*, 359.

232 Zayed, F., Greiner, L., Schulz, P.S., Lapkin, A. and Leitner, W. (2008) *Chem. Commun.*, 79.

233 Corma, A. (1997) *Chem. Rev.*, **97**, 2373.

234 Mohd, A.Y., Raja Saadiah, R.S., Siti, R.O., Juan, J.C. and Haron, R. (2006) *Mater. Sci. Forum*, **517**, 117.

235 Harmer, M.A. and Sun, Q. (2001) *Appl. Catal. A-Gen.*, **221**, 45.

236 Sheldon, R.A. and van Bekkum, H. (eds) (2001) *Fine Chemicals through Heterogeneous Catalysis*, Wiley-VCH Verlag, Weinheim.

237 de Angelis, A., Bellussi, G., Bosetti, A., Flego, C., Paludetto, R. and Perego, C.(July, 2004) New green and "shape selective" process for methylendianyline production. Presented at 13th International Catalysis Conference, Paris, France.

238 de Angelis, A., Ingallina, P. and Perego, C. (2004) *Ind. Eng. Chem. Res.*, **43**, 1169.

239 de Angelis, A., Perego, C., Farias, O. and Bosetti, A. (2006) US Patent 7105700. assigned to Enitecnologie (Italy).

240 Perego, C., de Angelis, A., Carati, A., Flego, C., Millini, R., Rizzo, C. and Bellussi, G. (2006) *Appl. Catal. A-Gen*, **307**, 128.

241 Centi, G., Cavani, F. and Trifirò, F. (2001) *Selective Oxidation by Heterogeneous Catalysis* (eds M.V. Twigg and M.S. Spencer), Series: Fundamental and Applied Catalysis, Springer (Kluwer Acad.-Plenum Pub.), New York.

242 Centi, G. and Perathoner, S. (2003) Selective Oxida-tion. Industrial, in *Encyclopedia of Catalysis*, vol. 6 (ed. I.T. Horváth), John Wiley & Sons, Ltd, Hoboken, NJ, p. 239.

243 Arpentinier, P., Cavani, F. and Trifirò, F. (1999) *The Technology of Catalytic Oxidations*, Editions Technip, Paris, France.

244 Ichihashi, H. and Sato, H. (2001) *Appl. Catal. A-Gen.*, **221**, 359.

245 Ichihashi, H. and Kitamura, M. (2002) *Catal. Today*, **73**, 23.

246 Ishida, M., Suzuki, T., Ichihashi, H. and Shiga, A. (2003) *Catal. Today*, **87**, 187.

247 Nagahara, H., Ono, M., Konishi, M. and Fukuoka, Y. (1997) *Appl. Surf. Sci.*, **121/122**, 448.

248 Ono, Y. (2003) *J. Catal.*, **216**, 406.

249 Hattori, H. (2001) *Appl. Catal. A-Gen*, **222**, 247.

250 Figueras, F., Kantam, M.L.i and Choudary, B.M. (2006) *Curr. Org. Chem.*, **10** (13), 1627.

251 Hattori, H. (2004) *J. Jpn. Petroleum Institute*, **47**, 67.

252 Tanabe, K. and Hölderich, W.F. (1999) *Appl. Catal. A-Gen.*, **181**, 399.

253 Hattori, H. (1995) *Chem. Rev.*, **95**, 537.

254 Davis, R.J. (2003) *J. Catal.*, **216**, 396.

255 Weitkamp, J., Hunger, M. and Rymsa, U. (2001) *Microporous Mesoporous Mater.*, **48**, 255.

256 Sels, B.F., De Vos, D.E. and Jacobs, P.A. (2001) *Catal. Rev. Sci. Eng.*, **43**, 443.

257 Kannan, S. (2006) *Catal. Surveys Asia*, **10**, 117.

258 Rives, V. (ed.) (2001) *Layered Double Hydroxides: Present and Future*, Nova Publishers, Hauppauge, NY.

259 Shiraga, M., Kawabata, T., Li, D., Shishido, T., Komaguchi, K., Sano, T. and Takehira, K. (2006) *Appl. Clay Sci.*, **33**, 247.

260 Kawabata, T., Fujisaki, N., Shishido, T., Nomura, K., Sano, T. and Takehira, K. (2006) *J. Mol. Catal. A: Chem.*, **253**, 279.

261 Abello, S., Medina, F., Tichit, D., Perez-Ramirez, J., Groen, J.C., Sueiras, J.E., Salagre, P. and Cesteros, Y. (2005) *Chem.-Eur. J.*, **11**, 728.

262 Rao, K.K., Gravelle, M., Sanchez-Valente, J. and Figueras, F. (1998) *J. Catal.*, **173**, 115.

263 Chen, Y.Z., Hwang, C.M. and Liaw, C.W. (1998) *Appl. Catal. A-Gen.*, **169**, 207.

264 Nikolopoulos, A.A., Jang, B.W.-L. and Spivey, J.J. (2005) *Appl. Catal. A-Gen.*, **296**, 128.

265 Winter, F., van Dillen, A.J. and de Jong, K.P. (2005) *Chem. Commun.*, 3977.

266 Winter, F., Koot, V., van Dillen, A.J., Geus, J.W. and de Jong, K.P. (2005) *J. Catal.*, **236**, 91.

267 Abelló, S., Medina, F., Tichit, D., Pérez-Ramírez, J., Cesteros, Y., Salagre, P. and Sueiras, J.E. (2005) *Chem. Commun.*, 1453.

268 Bastiani, R. I, Zonno, I.V. I, Santos, I.A.V. II, Henriques, C.A. II, and Monteiro, J.L.F. I, (2004) *Braz. J. Chem. Eng.*, **21**, 193.

269 Roelofs, J.C.A.A., van Dillen, A.J. and de Jong, K.P. (2000) *Catal. Today*, **60**, 297.

270 Climent, M.J., Corma, A., Iborra, S. and Primo, J. (1995) *J. Catal.*, **151**, 60.

271 Corma, A. and Martin-Aranda, R.M. (1993) *Appl. Catal. A-Gen.*, **105**, 271.

272 Corma, A., Iborra, S., Primo, J. and Rey, F. (1994) *Appl. Catal. A-Gen.*, **114**, 215.

273 Kishore, D. and Kannan, S. (2002) *Green Chem.*, **4**, 607.

274 Kishore, D. and Kannan, S. (2004) *J. Mol. Catal. A: Chem.*, **223**, 225.

275 Ueno, S., Ebitani, K., Ookubo, A. and Kaneda, K. (1997) *Appl. Surf. Sci.*, **121/122**, 366.

276 Kaneda, K., Ueno, S. and Imanaka, T. (1995) *J. Mol. Catal. A: Chem.*, **102**, 135.

277 Yamaguchi, K., Ebitani, K., Yoshida, T., Yoshida, H. and Kaneda, K. (1999) *J. Am. Chem. Soc.*, **121**, 4526.

278 Motokura, K., Fujita, N., Mori, K., Mizugaki, T., Ebitani, K. and Kaneda, K. (2005) *J. Am. Chem. Soc.*, **127**, 9674.

279 Corma, A. and Iborra, S. (2001) Solid base catalysis, in *Fine Chemicals through Heterogeneous Catalysis* (eds R.A. Sheldon and H. van Bekkum), Wiley-VCH Verlag, Weinheim, p. 309.

280 Coluccia, S., Marchese, L. and Martra, G. (1999) *Microporous Mesoporous Mater.*, **30**, 43.

281 Goettmann, F. and Sanchez, C. (2007) *J. Mater. Chem.*, **17**, 24.

282 Ono, Y. (1997) *CATTECH*, **3**, 31.

283 Kloetstra, K.R. and van Bekkum, H. (1995) *J. Chem. Soc. Chem. Commun.*, 1005.

284 Wu, Z.Y., Jiang, Q., Wang, Y.M., Wang, H.J., Sun, L.B., Shi, L.Y., Xu, J.H., Wang, Y., Chun, Y. and Zhu, J.H. (2006) *Chem. Mater.*, **18**, 4600.

285 Sun, L.B., Kou, J.H., Chun, Y., Yang, J., Gu, F.N., Wang, Y., Zhu, J.H. and Zou, Z.G. (2008) *Inorg Chem.*, **47** (10), 4199.

286 Sun, L.B., Yang, J., Kou, J.H., Gu, F.N., Chun, Y., Wang, Y., Zhu, J.H. and Zou, Z.G. (2008) *Angew. Chem. Int. Ed.*, **47**, 3418.

287 Li, L., Du, W., Liu, D., Wang, L. and Li, Z. (2006) *J. Mol. Catal., B Enzym.*, **43**, 58.

288 Moreira, A.B.R., Perez, V.H., Zanin, G.M. and de Castro, H.F. (2007) *Energy Fuels*, **21**, 3689.

289 Han, H., Cao, W. and Zhang, J. (2005) *Process. Biochem.*, **40**, 3148.

290 Bunyakiat, K., Makmee, S., Sawangkeaw, R. and Ngamprasertsith, S. (2006) *Energy Fuels*, **20**, 812.

291 Ha, H., Wang, T. and Zhu, S. (2007) *Fuel*, **86**, 442.

292 Bacovsky, D., Körbitz, W., Mittelbach, M. and Wörgetter, M. (July, 2007) Biodiesel production: technologies and European providers. A Report to IEA Bioenergy Task 39, Report T39-B6.

293 Mittelbach, M. and Remschmidt, C. (2006) *Biodiesel, the Comprehensive Handbook*, M. Mittelbach Pub, Graz, Austria.

294 Trakarnpruk, W. and Porntangjitlikit, S. (2008) *Renew. Energy*, **33**, 1558.

295 Li, E. and Rudolph, V. (2008) *Energy Fuels*, **22**, 145.

296 Albuquerque, M.C.G., Jiménez-Urbistondo, I., Santamaría-González, J., Mérida-Robles, J.M., Moreno-Tost, R., Rodríguez-Castellón, E., Jiménez-López, A., Azevedo, D.C.S., Cavalcante, C.L. and Maireles-Torres, P. (2008) *Appl. Catal. A – Gen.*, **334**, 35.

297 Liu, X., Piao, X., Wang, Y. and Zhu, S. (2008) *Energy Fuels*, **22**, 1313.

298 Blaser, H.-U., Steiner, H. and Studer, M. (2004) Selective heterogeneous hydrogenation: a valuable tool for the synthetic chemist, in *Transition Metals for Organic Synthesis*, 2nd Edn (eds M. Beller and C. Bolm), Wiley-VCH Verlag, Weinheim, Part 1: Reductions, Ch. 12, p. 125.

299 Blaser, H.U., Malan, C., Pugin, B., Spindler, F., Steiner, H. and Studer, M. (2003) *Adv. Synth. Catal.*, **345**, 103.

300 Roessler, F. (1996) *Chimia*, **50**, 106.

301 Corma, A., Domine, M.E., Nemeth, L. and Valencia, S. (2002) *J. Am. Chem. Soc.*, **124**, 3194.

302 Seki, T., Grunwaldt, J.-D. and Baiker, A. (2008) *Ind. Eng. Chem. Res.*, **47**, 4561.

303 Blaser, H.U., Spindler, F. and Studer, M. (2001) *Appl. Catal. A-Gen.*, **221**, 119.

304 Roberts, S.M. and Whittall, J. (eds) (2007) *Regio- and Stereo- Controlled Oxidations and Reductions, Series: Catalysts for Fine Chemical Synthesis*, Vol. 5, Wiley-Interscience.

305 Trost, B.M. (2004) *Proc. Natl. Acad. Sci. U.S.A.*, **101**, 5348.

306 Heitbaum, M., Glorius, F. and Escher, I. (2006) *Angew. Chem. Int. Ed.*, **45**, 4732.

307 Raval, R. (2001) *CaTTech*, **5**, 12.

308 Mcmorn, P. and Hutchings, G.J. (2004) *Chem. Soc. Rev.*, **33**, 108.

309 Baiker, A. (1997) *J. Mol. Catal. A*, **115**, 473.

310 Thomas, J.M. and Raja, R. (2008) *Acc. Chem. Res.*, **41**, 708.

311 Li, C., Zhang, H., Jiang, D. and Yang, Q. (2007) *Chem. Commun.*, 547.

312 Blaser, H.U. and Schmidt, E. (2007) *Asymmetric Catalysis on Industrial Scale: Challenges, Approaches and Solutions*, Wiley-VCH Verlag, Weinheim.

313 Federsel, H.-J. (2005) *Nat. Rev. Drug. Discov.*, **4**, 685.

314 Blaser, H.U. (2002) *Adv. Synth. Catal.*, **344**, 17.

315 Knowles, W.S. (2007) Asymmetric hydrogenations – the Monsanto L-Dopa process, in *Asymmetric Catalysis on Industrial Scale: Challenges, Approaches and Solutions* (eds H.U. Blaser and E. Schmidt), Wiley-VCH Verlag, Weinheim, Ch. 1, p. 23.

316 Valdés, R.H., Puzer, L., Gomes, M., Marques, C.E.S.J., Aranda, D.A.G., Bastos, M.L., Gemal, A.L. and Antunes, O.A.C. (2004) *Catal. Commun.*, **5**, 631.

317 Centi, G., Perathoner, S. and Trifirò, F. (1997) *Appl. Catal. A*, **157**, 143.

318 Centi, G. and Perathoner, S. (1998) *CHEMTECH*, **28**, 13.

319 Sokolovskii, V.D., Davydov, A.A. and Ovsitser, O.Yu. (1995) *Catal. Rev.-Sci. Eng.*, **37**, 425.

320 Corma, A.(10–11 Dec, 2007) Selective solid catalysts by design. Presented at Workshop Applied Catalysis: The Key to

European Prosperity and Sustainability, Lisbon, Portugal.

321 Boronat, M., Concepción, P., Corma, A. and Renz, M. (2007) *Catal. Today*, **121**, 39.

322 Corma, A., Fornes, V., Iborra, S., Mifsud, M. and Renz, M. (2004) *J. Catal.*, **221**, 67.

323 Corma, A., Iborra, S., Mifsud, M. and Renz, M. (2005) *J. Catal.*, **234**, 96.

324 Goti, A. and Cardona, F. (2008) Hydrogen peroxide in green oxidation reactions: recent catalytic processes, in *Green Chemical Reactions* (eds P. Tundo and V. Esposito), Springer, Dordrecht, The Netherlands, p. 191.

325 Venturello, C., Alneri, E. and Ricci, M. (1983) *J. Org. Chem.*, **48**, 3831.

326 Kamata, K., Yonehara, K., Sumida, Y., Yamaguchi, K., Hikichi, S. and Mizuno, N. (2003) *Science*, **300**, 964.

327 Sato, K., Aoki, M., Ogawa, M., Hashimoto, T. and Noyori, R. (1996) *J. Org. Chem.*, **61**, 8310.

328 Zuwei, X., Ning, Z., Yu, S. and Kunlan, L. (2001) *Science*, **292**, 1139.

329 Herrmann, W.A., Fischer, R.W. and Marz, D.W. (1991) *Angew. Chem. Int. Ed.*, **30**, 1638.

330 Bal, R., Tada, M., Sasaki, T. and Iwasawa, Y. (2006) *Angew. Chem. Int. Ed.*, **45**, 448.

331 Ballarini, N., Cavani, F., Degrand, H., Etienne, E., Pigamo, A., Trifirò, F. and Dubois, J.L. (2007) *Methods and Reagents for Green Chemistry: an Introduction* (eds P. Tundo *et al.*), John Wiley & Sons, Hoboken, NJ, Ch. 14, p. 265.

332 Abe, T. (1999) *Science and Technology in Catalysis 1998*, Kodansha Pub, Tokyo, Japan, p. 461.

333 Krieger, H. and Kirch, L.S. (1981) U.S. Patent 4,260,822. assigned to Rohm and Haas Company.

334 Schindler, G.-P., Knapp, C., Ui, T. and Nagai, K. (2003) *Top. Catal.*, **22**, 117.

335 Roussel, M., Bouchard, M., Bordes-Richard, E., Karim, K. and Al-Sayari, S. (2005) *Catal. Today*, **99**, 77.

336 Contractor, R.M. and Sleight, A.W. (1987) *Catal. Today*, **1**, 587.

337 Schmidt, L.D. and Goralski, C.T. (1997) *Stud. Surf. Sci. Catal.*, **110**, 491.

338 Armor, J.N. (1989) *Appl. Catal.*, **49**, 1.

339 Matros, Y.Sh. and Bunimovich, G.A. (1996) *Catal. Rev.-Sci. Eng.*, **38**, 1.

340 Centi, G. and Perathoner, S. (1999) *Curr. Opin. Solid. St. M.*, **4**, 74.

341 Climent, M.J., Corma, A., Iborra, S. and Mifsud, M. (2007) *J. Catal.*, **247**, 223.

342 Tietze, L.F. (1996) *Chem. Rev.*, **96**, 115.

343 Corma, A. and Renz, M. (2007) *Angew. Chem. Int. Ed.*, **46**, 298.

344 Gallezot, P. (2007) Process options for the catalytic conversion of renewables into bioproducts, in *Catalysis for Renewables* (eds G. Centi and R.A. van Santen), Wiley-VCH Verlag, Weinheim, Ch. 3, p. 53.

345 Ahmed, M., Bronger, R.P.J., Jackstell, R., Kamer, P.C.J., van Leenwen, P.W.N.M. and Beller, M. (2006) *Chem.*, **12**, 8979.

346 Jaroch, S., Weinmann, H. and Zeitler, K. (2007) *ChemMedChem*, **2** (9), 1261.

347 Berkessel, A. and Gröger, H. (2005) *Asymmetric Organocatalysis*, Wiley-VCH Verlag, Weinheim.

348 List, B. and Yang, J.W. (2006) *Science*, **313**, 1584.

3
Methods and Tools of Sustainable Industrial Chemistry: Process Intensification

Gabriele Centi and Siglinda Perathoner

3.1
Introduction

Process intensification (PI) indicates the ensemble of technologies that lead to a substantially smaller, cleaner, safer and more energy efficient process technology [1]. For example, the scale reduction made possible by using high gravity fields to separate liquids can reduce a distillation column from 75 to 4–5 m in height. The term "process intensification" was introduced by Colin Ramshaw (ICI) in the 1980s to describe the approach to making smaller (i.e., intensified) chemical plants that would be significantly cheaper and safer than existing ones [2]. The initial focus was on process unit operations, in particular gas/liquid mass transfer. Later the concept extended to cover reaction intensification (microreactor, elimination of solvents, use of alternative energy sources, and so on) and the whole process.

Ramshaw and his coworkers at ICI defined PI as a "reduction in plant size by at least a factor 100." The choice of this factor is arbitrarily, but indicates the aim of a significant reduction in plant size. This reduction would significantly improve the process economics, by reducing fixed and operating capital costs. However, it will also allow implementation of the concept of modular design of chemical processes (see Chapter 1).

A reduction in size means also a better controlled process and the managing of lower amounts of chemicals per reactor. In addition, microreactors often offer intrinsically safer operations [3]. In fact, with respect to conventional reactors, they offer a much better control of heat transfer (e.g., reduced possibility of runaway reactions) and a higher wall to volume ratio, which minimize the possibility of side radical reactions (e.g., of explosion). PI concepts are thus essential to develop inherently safe processes as well as energy efficient operations.

PI encompasses not only the development of novel, more compact equipment but also the development of intensified methods of processing, such as the use of alternative energy sources [4]. In addition, one of the objectives of process intensification is to move away from batch processing to small continuous reactors. The better control of the reaction conditions often allows us to improve not only the

Sustainable Industrial Processes. Edited by F. Cavani, G. Centi, S. Perathoner, and F. Trifiró
Copyright © 2009 WILEY-VCH Verlag GmbH & Co. KGaA, Weinheim
ISBN: 978-3-527-31552-9

productivity but also the selectivity. PI is thus essential to achieve a lower consumption of raw materials (chemicals and energy). In recent times PI has been seen as providing processing flexibility, just-in-time (JIT) manufacturing capabilities and the opportunity for distributed manufacturing. We may thus conclude that PI concepts are a enabling tool to achieve the sustainability of industrial chemical processes.

3.1.1
Opportunities and Perspectives for a Sustainable Process Design

PI is a hot topic in chemical and process engineering, and is now reaching a maturity that is seeing PI concepts applied to a wide range of processes and technologies. Already, over 30 industrial examples are available, but this number is rapidly increasing. Originally developed for the bulk chemical industry, PI developments have more recently been focused on the higher added-value chemicals and pharmaceutical active ingredient sectors. Examples reported by the Process Intensification Network (www.pinetwork.org) and the BHR Group (www.bhrgroup.co.uk) indicate:

- a capital cost reduced by 60%;
- a 99% reduction in impurity levels, resulting in significantly more valuable product;
- a 70% plus reduction in energy usage and hence a substantial reduction in operating cost;
- a 93% yield first time out – better than fully optimized batch process;
- a 99.8% reduction in reactor volume for a potentially hazardous process, leading to inherently safe operation.

Although these values cannot always be obtained, they give an indication of the potential that may be reached by using the PI concepts. The general approach of PI is to reconsider from scratch the process or unit operation design, and determine the key issues in a process based on recent advances in chemical engineering knowledge. Distillation for example is all about gas–liquid mass transfer, and the keys to mass transfer for a given system are the following:

- well mixed liquid and gas phases,
- high interfacial surface area,
- thin liquid film,
- counter-current operation.

Well mixing of gas and liquid phases is a clear issue. Smaller, finer packing allows us to increase the surface area between the gas and liquid contact, for example, the mass transfer could be increased in a column with very fine packing with counter-current gas flow. However, a liquid film running through a bed of fine material is problematic when the liquid film thickness is around the same as the clearance between the bits of packing. Liquid flow essentially stops and the column floods. The key is the thickness of the liquid film and the factors that control that. The equations describing the problem indicate that most factors relate to the physical properties of

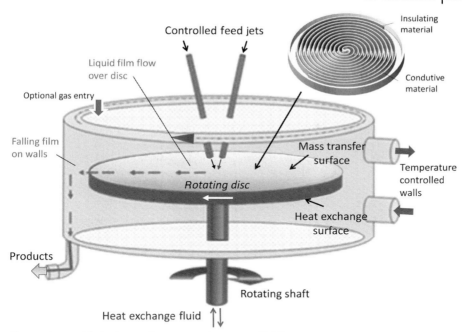

Controlled feed jets

Liquid film flow
over disc

Insulating
material

Condutive
material

Optional gas entry

Falling film
on walls

Mass transfer
surface

Rotating disc

Temperature
controlled
walls

Heat exchange
surface

Products

Rotating shaft

Heat exchange fluid

Figure 3.1 Simplified scheme of a spinning disc reactor (SDR).

the fluid, but one is independent, namely, gravity. The higher the applied gravity the thinner the film and the smaller the packing can be. One way to enhance throughputs and the interphase transfer rate is to replace the gravitational field with centrifugal fields that are higher by a few orders of magnitude. Process intensification research has therefore naturally focused on the use of rotating packed beds for the miniaturization of reactors and separators.

One example of "high-g" equipment is the spinning disc reactor (SDR), developed by Ramshaw's group at Newcastle University (Newcastle, UK) and commercialized, for example, by Protensive (www.protensive.co.uk). Figure 3.1 reports a simplified scheme of the equipment. An SDR can operate as a multifunction device. The disc can be horizontally or vertically mounted on an axle. Liquid fed near or at the center flows across the surface of a spinning disc under the influence of centrifugal force. This force stretches and contorts the film. The thin liquid film allows high rates of mass transfer and it favors absorption, stripping, mixing and reaction processes. Residence times on the disc are low, typically in the range of 3 s down to tenths of a second. Both film thickness and residence time depend on fluid physical properties, rotational speed and radial location of the fluid. On exiting the periphery of the disc, the liquid is thrown onto an enclosing wall whereupon it drains away. Heating or cooling can be applied to the disc surface and the enclosing wall to control fluid temperature.

Centrifugal forces are used not only in SDRs. High gravity (HIGEE) technology [5] is another example (Figure 3.2). The main difference with respect to SDR is the

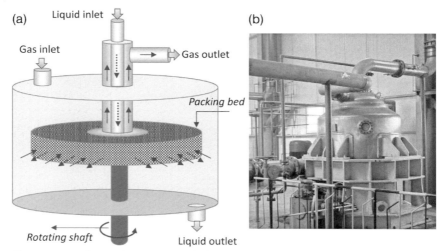

Figure 3.2 (a) Scheme of a HIGEE unit; (b) high-gravity rotating packed bed reactor for $CaCO_3$ nanoparticles production with a capacity of 10 000 tons a^{-1}. Source: courtesy of Research Center of the Ministry of Education for High Gravity Engineering & Technology, Beijing.

presence of a packing bed. The liquid flows as a thin film over the packing due to the high centrifugal acceleration (100–1000g) and therefore raises the upper limit of flooding and permits the use of packing of high surface area, in the range of 1000–5000 $m^2 m^{-3}$, which is 3–10-times that used in conventional packed columns. The rotating-bed equipment, originally dedicated to separation processes (such as absorption, extraction and distillation), can also be utilized for reacting systems. It potentially can be applied not only to gas/liquid systems but also to other phase combinations, including three-phase gas/liquid/solid systems. The HIGRAVITEC Center (Beijing, China) has successfully applied rotating (500–2000 rpm) packed beds on a commercial scale for deaeration of flooding water in oil fields. The equipment, about 1-m diameter, replaced conventional vacuum towers over 30 m high [5].

Heat exchangers are another example where PI concepts largely enhance performance. The heat transfer area is the critical factor, but conventional heat exchangers are based on pipes that have a minimum surface area, because the design reflects more mechanical engineering considerations than process ones. Core to the heat transfer performance of an SDR or analogous equipment is the characteristic of the film as it moves across the disc. Waves tend to form in the film that significantly enhance heat transfer (as well as mass transfer), for example, with high film coefficients. However, the film coefficient is only part of the problem. The film coefficient for the heating/cooling fluid and the thermal resistance of the disc itself are also critical components. By incorporating special channels for the service fluids in the rotating disc it is possible to further enhance the heat transfer thanks to the fin effect of the channels. When used with suitable heat transfer fluids, it can significantly enhance the effective transfer coefficient. Therefore, an SDR has an overall heat transfer coefficient of approximately $10\,kW\,m^{-2}\,K^{-1}$ even for organic liquids. This is typically 5–10-times that achieved by most heat transfer devices and enables small discs with low process fluid inventory to handle significant thermal duties.

Often, slow reactions (in a batch-type reactor) are not due to intrinsic slow chemical kinetics but, instead, heat transfer, mass transfer or mixing limitations of the reactor are the factors that control the rate. An exothermic reaction might need an hour to be carried out in a batch reactor, not because of kinetic constraints but because it takes an hour to remove the heat of reaction. A reaction that involves a gas to liquid interface might well be controlled by the mass transfer between phases. A liquid–liquid reaction might be controlled by mixing rather than chemical kinetics. By using a SDR (or similar equipment like the HIGEE) to carry out a chemical reaction it is possible to greatly reduce these limitations.

Not only the reaction rate increases, but often there are improvements in the selectivity of the reaction as the short residence time, coupled with rapid quenching, will reduce by-product formation caused by further reaction of the desired product. Other benefits in the selectivity are likely to come from the lack of back-mixing (plug flow) and the intense mixing of the SDR, which will minimize concentration gradients of the same reactants.

Further examples and concepts are discussed in the following sections, but already this introduction evidences the possibilities and opportunities for a sustainable industrial chemistry that derive from the application of process intensification equipment and methodology.

3.1.2
Process Intensification and Inherently Safer Processes

When dealing with highly exothermic reactions, process industries have long-established techniques for safe processing, often by slowing down such reactions, to a rate where heat evolution can be matched by the limited mass and heat transfers available in conventional reactors. Nevertheless, many serious accidents with exothermic reactions can be counted. A larger reactor requires careful calculation of potential heat flows, due to the high potential risks related to run-away reactions, loss of control and containment.

Polymerization and nitration are two particular examples, which are well understood but regularly misjudged at considerable cost, even by experienced companies. PI equipments (e.g., SDR) potentially offer a scale-independent way of eliminating such hazards. They offer the possibility of faster process development cycles, tiny inventories at risk and negligible risk of safety or quality pit-falls as production is scaled-up.

Fast exothermic reactions can be conducted in the thin turbulent film on a SDR (or analogous equipment), using higher temperatures than those possible in stirred tanks, thanks to the higher heat transfer performance. Inventories in the disc chamber are relatively tiny, making trivial the consequences of any localized run-away. Material exiting the disc chamber can be cooled as it descends the disc chamber walls, and transferred to storage or further processing under safe conditions.

An example is a two-component pre-polymer being manufactured as a 30% solution in water (Protensive Company, www.protensive.co.uk). Batch reactors demanded an eight-hour feed of one component to the other, in solution, and a maximum temperature of 35 °C. It is not possible to operate at higher temperature (faster reaction rate) due to the significant risk of runaway reaction in which, for example, that the temperature cannot be controlled and a major accident occurs. Spinning disc tests have shown that the two components can be reacted as a 70% solution in water at 100 °C in a 1.5-second single pass over the disc reactor. The product was cooled, diluted and quenched as it exited the reactor to yield the required product quality. Scale-up of the reactor to a 400 kg day^{-1} production (70 cm diameter disc reactor) was possible, with good reproducibility of the results.

3.1.3
A Critical Toolbox for a Sustainable Industrial Chemistry

PI is more than just a series of novel reactors or equipments. It offers the full potential of redesigning industrial chemical process in line with sustainability concepts. PI presents a range of exciting processing tools/opportunities for chemists. The smaller footprints of processes offer new opportunities to insert them in a factory, as visually illustrated in Figure 1.2 (Chapter 1), allowing thus a significant reduction in terms of an eyesore for the general public. In some cases the plant may be mobile, thereby offering opportunity for distributed manufacturing of chemicals. This will reduce the quantities of chemicals currently being transported by road and rail, thereby improving safety. The improved energy efficiency foreseeable in intensified unit operations constitutes yet another highly attractive benefit of PI in a world where energy cost is dominant. Better use of energy also means a reduction of emissions of greenhouse gases and in general of pollutants. PI is thus a path for the future of chemical and process engineering demands, and a critical toolbox for sustainable industrial chemistry (Figure 3.3) [7, 8].

The core of PI refers to the development of more or less complex technologies that replace large, expensive, energy-intensive equipment or processes with ones that are smaller, less costly, more efficient plants, minimizing environmental impact, increasing safety and improving remote control and automation, or that combine multiple operations into a single apparatus or into fewer devices [7]. However, we prefer to consider PI as an holistic approach starting with an analysis of economic constraints followed by the selection or development of a production process. PI aims at drastic improvements in performance of a process, by rethinking the process as a whole. In particular it can lead to the manufacture of new products that could not be produced by conventional process technology. PI is thus strictly related to innovation and sustainability in the chemical industry, and together with catalysis is the enabling factor to proceed in this direction.

Although initially PI had a conservative reception from industries due to their unwillingness to take the risks with a new technology, subsequently many companies have started to understood the potential. Many companies, for example, ICI [9, 10], Sulzer [11], SmithKline Beecham [12], Eastman Chemical [13], Dow [14],

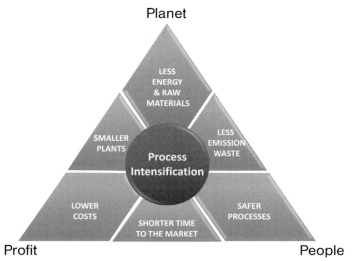

Planet

Profit   People

Figure 3.3 PI provides radically innovative principles in process and equipment design that can benefit process and chain efficiency, capital and operating expenses, quality, wastes, process safety and more, and align perfectly with the "Triple-P" philosophy of sustainable industrial chemistry. Source: adapted from "EU Roadmap for Process Intensification" (www.creative-energy.org).

Degussa [15], and so on, have embraced the PI philosophy and adopted it in several of their recent processes with commercial success.

The "European Roadmap for Process Intensification" (www.creative-energy.org) promoted by the Dutch Ministry of Economic Affairs and prepared by a group of experts of Netherlands (from both academy and industry) in 2008, demonstrates how PI offers important opportunities to modernize the process industry (oil refinery, petrochemicals, bulk chemicals, specialty chemicals, pharmaceuticals and food). Substantial savings are possible over time (energy, CO_2 emissions, waste production, etc.). It is estimated that process industry can achieve a 20% reduction in energy consumption by 2050 through PI implementation alone. However, the PI Roadmap is at the same time considered as the key part of the Business Plan for Innovation of the Chemical Industry. Four main areas of application of PI have been considered in the roadmap: (i) petrochemicals and bulk chemicals, (ii) specialty chemicals and pharmaceuticals, (iii) food ingredients and (iv) consumer food. By analyzing many process intensification technologies, the following potential benefits of PI have been identified:

- Petrochemicals, bulk chemicals:
  higher overall energy efficiency – 5% (10–20 years), 20% (30–40 years).

- Specialty chemicals, pharmaceuticals:
  overall cost reduction (and related energy savings due to higher raw material yield) – 20% (5–10 years), 50% (10–15 years).

- Food ingredients:
  higher energy efficiency in water removal – 25% (5–10 years), 75% (10–15 years). Lower costs through intensified processes throughout the value chain – 30% (10 years), 60% (30–40 years).

- Consumer food:
  higher energy efficiency in preservation processes – 10–15% (10 years), 30–40% (40 years). Through capacity increase – 60% (40 years). Through move from batch to continuous processes – 30% (40 years).

In particular, two main streams of PI applications have been identified: (i) PI innovations for reactors (e.g., microreactors, monolith reactors, spinning disc reactors, reactive separations) and (ii) PI technologies for more efficient energy transfer (e.g., ultrasound, pulse, plasma, microwave). Several PI technologies offer important potential, but require important fundamental/strategic research to reach proof-of-concept on the laboratory scale. These PI technologies are:

- foam reactors,
- monolith reactors,
- microreactors,
- membrane reactors,
- membrane absorption/stripping,
- membrane adsorption,
- HEX reactors,
- reactive extraction,
- reactive extrusion,
- rotating packed beds,
- rotor-stator mixers,
- spinning disc reactors.

Several other novel PI technologies have already been implemented for a limited number of applications, but further applied research is necessary for a broader implementation. The skill of designing PI equipment on an industrial scale (materials, robustness, economics) is still lacking, but for some technologies industrialization is in progress:

- plate, plate-fin, plate-and-shell, flat tube-and-fin heat exchangers,
- static mixer reactors,
- membrane extraction,
- reactive absorption,
- reactive distillation,
- centrifugal extractors.

However, there are already examples of (semi)-commercially available equipment:

- Sulzer SMR static mixer, which has mixing elements made of heat-transfer tubes, and Sulzer's open-crossflow structure catalysts, so-called Katapaks; Sulzer (www.sulzer.com) is a Swiss company active in the field of machinery and equipments;

- HIGEE (high gravity rotating contactor with a compact design, Figure 3.2); Protensive, Newcastle upon Tyne, UK, and GasTran Systems, Cleveland, Ohio, provide this technology, which also is known as the rotating packed bed;
- HIGRAVITECs rotating packed beds (Higravitec Center of the Beijing University of Chemical Technology); a 50 and a 300 ton h^{-1} Higrav machine for oil field flooding water deaeration at Shengli Oilfield, Sandong, China have been installed [6];
- BHRs improved mixing equipment, HEX reactors (integrated reactor-heat exchangers), FlexReactor (flexible and re-configurable); (www.bhrgroup.co.uk);
- high-pressure homogenizers for emulsifications;
- the spinning disc reactor (SDR); (www.protensive.co.uk);
- supersonic gas/liquid reactor.

Various ultrasonic transducers and reactors are now commercially available. The cited roadmap identified 46 key PI technologies, even if some of them still not available commercially, and reported also their potential for saving energy and CO_2 emissions, and for improving cost competiveness. Table 3.1 lists these key PI technologies and their potential, but the roadmap advises that, sometimes, the identified potential reflects in part personal opinions or expectations.

For the successful industrial implementation of PI technologies, the following enabling technologies need to be developed:

- Process analytical technology:
  (*in situ*) measurement and analysis methods for better understanding of kinetic and thermodynamic characteristics of chemical processes at the molecular level.

- Numerical process modeling:
  faster, more robust, often nonlinear numerical modeling of chemical reactions;

- Process control systems.
  to cope with the incorporation of (often continuous) PI modules in (often batch) processes.

A critical element for success is to link PI to chain optimization, but this often requires a socio/economic paradigm shift, and calls for optimization studies along the value chain as well as the development of longer-term transition paths. For example, milk separation into water, proteins and fats can be conducted on-site (i.e., at the farm) with low energy-consuming micro-separators. Product transportation to, and handling at, the factory can be limited to the relevant components, proteins and fat, saving energy and reducing CO_2 emissions by avoiding the unnecessary transportation of water and its removal at the factory.

The cited roadmap also reports some examples that demonstrate the role of PI towards achieving sustainability in industrial chemical processes:

- Petrochemicals, bulk chemicals:
  a common traditional technology is absorption-stripping; in the application to HClO synthesis, the use of HIGEE rotating packed beds allows (i) a reduction in equipment size by a factor about 40, (ii) an increase in product yield of about

Table 3.1 General overview of the PI key technologies. Source: adapted from "EU Roadmap for Process Intensification" (www.creative-energy.org).

PI technology	Class		Potential for		
			Energy saving	Eco impact CO_2	Cost effectiveness
Advanced plate-type heat exchangers	Structured devices	Non-reactive	M	M	H
Advanced shell and tube type heat exchangers			M	M	M
Static mixers			M	M	M
Heterog. catalyzed solid foam reactors		Reactive	L	L	L
Monolithic reactors			M	M	M
Millisecond (gauze) reactors			L	L	M
Structured reactors			M	M	M
Micro-channel reactors			L	L	L
Membrane reactors (nonselective)			L	L	L
Static mixer reactors for cont. reactions			H	H	M
Adsorptive distillation	Hybrid	Non-reactive	M	M	L
Extractive distillation			M	L	L
Heat-integrated distillation			H	H	H
Membrane crystallization technology			M	M	M
Membrane distillation technology			M	M	M
Distillation–pervaporization			M	M	M
HEX reactors		Reactive	L	L	H
Simulated moving bed reactors			L	L	L
Rotating annular chromatograp. reactors			L	L	H
Gas–solid–solid trickle flow reactors			L	L	H
Reactive extraction columns, HT and HS			M	M	L
Reactive absorption			H	H	L
Reactive distillation			H	H	H

Membrane-assisted reactive distillation		H	H	H
Centrifugal liquid–liquid contractors	Energy transfer	Rotating	H	M	H
Rotating packed beds		M	M	H
Rotor stator devices		H	M	M
Hydrodynamic cavitation reactors	Impulse	M	L	M
Impinging streams reactor		M	L	M
Pulsed compression reactor		H	H	M
Sonochemical reactors (ultrasound and low frequency sonics)		M	L	M
Ultrasound enhanced crystallization		M	L	M
Ultrasound reactors for enhanced disintegrat./phase dispersion/mass transfer		M	M	M
Supersonic gas–liquid reactors		L	L	L
Electric field-enhanced extraction	Electro-magnetic	H	L	L
Induction and ohmic heating		L	L	H
Microwave heating/microwave drying		H	L	H
Microwave reactors for non-catal. and homog. catalyzed liquid phase process		L	L	L
Microwave reactors for heterogeneously catalyzed chemical processes		L	M	H
Microwave reactors for polymerization reactors and polymer processing		L	L	L
Photochemical		H	M	M
Plasma (GlidArc) reactors	Dynamic	L	L	M
Oscillatory		M	L	H
Reverse flow reactor operation		M	H	M
Pulse combustion drying		M	L	M
Supercritical separations	Other	M	H	H

L: low; M: medium; H: high.

15%, (iii) a reduction to half the stripping gas and (iv) a reduction of one-third waste water.

- Specialty chemicals, pharmaceuticals:
  by substituting a stirred tank reactor with microreactor the equipment size can be reduced to one-third and selectivity increased by 20%.

3.1.4
Fundaments of PI

Process intensification is driven by four generic principles:

1. Maximize the effectiveness of intra and intermolecular events.
2. Give each molecule the same processing experience.
3. Optimize the driving forces on every scale and maximize the specific areas to which those driving forces apply.
4. Maximize the synergistic effects from events and partial processes.

Four main domains, which constitute the pillars of the PI scientific approach, may be identified:

1. Structure (spatial domain).
2. Energy (thermodynamic domain).
3. Synergy (functional domain).
4. Time (temporal domain).

Figure 3.4 presents the most relevant issues addressed by PI in each of these areas. Within PI, two basic categories of approaches can be distinguished, depending on the main focus, even though, often, the two aspects are highly integrated:

- equipment, for example, focus on new devices for PI;
- methodology, for example, focus on new processing methods).

Figure 3.5 overviews this classification.

Several recent reviews have discussed the fundaments and applications of PI concepts. Doble [16] has discussed the concept of a green reactor, for example, how process intensification could be achieved by microreactor technology using very high forces, ultra-high pressures, electrical fields, ultrasonics, surfactant-based separations, shorter diffusion and conduction pathways, flow field and fluid microstructure interactions, and/or size-dependent phenomena.

The concept of nanofluids to further intensify microreactors has been discussed by Fan *et al.* [17]. The nano-fluids are suspensions of solid nano-particles with sizes typically of 1–100 nm in traditional liquids such as water, glycol and oils. These solid–liquid composites are very stable and show higher thermal conductivity and higher convective heat transfer performance than traditional liquids. They can thus be used to enhance the heat transfer in nanofluids in compact multifunctional reactors. A nanofluid based on TiO_2 material dispersed in ethylene glycol showed an up to 35% increase in the overall heat transfer coefficient and a

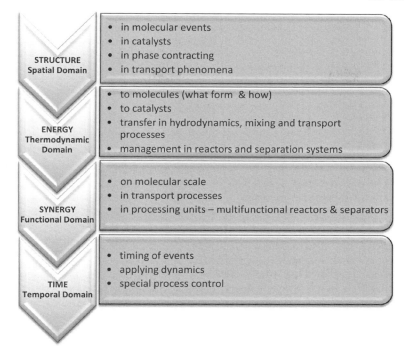

Figure 3.4 Relevant issues in the main four domains that constitute the pillars of the PI approach. Source: adapted from "EU Roadmap for Process Intensification" (www.creative-energy.org).

closer temperature control in the selective reduction of an aromatic aldehyde by hydrogen.

Gogate [18] has reviewed the use of cavitational reactors for PI, for example, how ultrasound irradiation can be used to enhance the rate in chemical processes. These reactors are a novel and promising form of multiphase reactors, based on the principle of the release of a large magnitude of energy due to the violent collapse of cavities. Overall it appears that considerable economic savings are possible by means of harnessing the spectacular effects of cavitation in chemical processing applications.

Wu et al. [19] have analyzed the intensification of industrial mixing. An important element for process intensification is to increase mixing rate, heat transfer rate and mass transfer rate. A range of methods have been reviewed by Wu et al. [19], for both conventional stirred reactors and alternative forms of reactors, all with the aim of achieving process intensification through enhanced mixing, heat and mass transfer.

PI by microreactor technologies applied to combined methanol/power production and to decentralized H_2 production have been discussed by VandenBussche [20]. The use of alternative sources and forms of energy for PI of chemical and biochemical

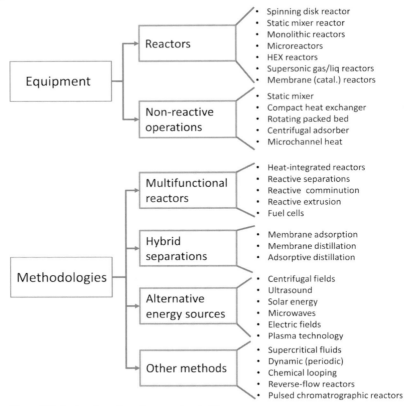

Figure 3.5 Basic categories of approaches in PI.

processes has been analyzed by Stankiewicz [21]. Although the PI potential of many of alternative sources and forms of energy has already been proven in the laboratory, their application on an industrial scale still presents a challenge.

A general review of continuous processing/process intensification in the pharmaceutical industry has been made by Rubin *et al.* [22], while the use of PI novel technologies to reshape the petrochemical and biotechnology industries has been analyzed by Hahn [23] and Akay [24], respectively. Recent advances in biotechnology process intensification have also been reviewed by Choe *et al.* [25] and Akay *et al.* [26]. The use of monoliths as biocatalytic reactors to achieve PI by smart gas–liquid contact has been reviewed by Kreutzer *et al.* [27].

The interaction between reactor miniaturization and catalyst nanotailoring to achieve an effective PI has been emphasized by Charpentier [28] and Dautzenberg [29]. The role of membranes in PI has been discussed by Drioli *et al.* [30]. This short overview of the recent state-of-the-art, although limited to reviews published in the last few years, evidences the intense research effort and broad-range type of applications for PI, from refinery and petrochemistry to biotechnology, fine and specialty chemical production.

Various books have also discussed recent advances in PI. In addition to the recent book of Reay *et al.* [1], the book edited by Stankiewicz and Moulijn [31] is one of the first to collect a series of contributions to give a comprehensive approach on how to reengineer the chemical processing plant by means of PI. The book edited by Keil [32] instead focuses on modeling in process intensification. New equipment like micro-reactors, membrane reactors and ultrasound reactors, and those in simulated moving-bed chromatography, magnetic fields in multiphase processes or reactive distillation, requires new modeling approaches. The same applies to nonstationary process operation or the use of supercritical media.

Hessel *et al.* [33] centered their book on the analysis of a series of specific examples, from gas- and liquid-phase, to gas/liquid-phase and liquid/liquid-phase reactions, where the use of a microreactor (or more generally microprocess technology) allows significantly enhance in performance. It is a very valuable source of examples taken from over 1500 publications analyzed. The recent book of Wirth [34] focuses instead on the analysis of the opportunities for organic synthesis and catalysis in the use of microreactor technology.

The intertwined roles of micro-instrumentation, high-throughput experimentation and process intensification in both academic and industrial investigations has been discussed by Kock *et al.* [35]. The potential of PI to improve processing competitiveness in the polymer, fine chemicals, pharmaceutical, food and household products indus-tries has been discussed by Jachuck [36], while Wang and Holladay [37] have collected a series of contributions given during an ACS meeting. The role of catalytic membranes and membrane reactors for PI is discussed in the book of Marcano and Tsotsis [38].

Finally, the role of PI from an industrial perspective has been discussed in two papers by Costello [39, 40], dealing with the use of PI for reactors, and in distillation, extraction and heat transfer, respectively.

3.1.5
Methodologies

Following Charpentier's [7] classification, we may distinguish as follows the meth-odologies used to obtain process intensification:

- unit operation hybridization,
- new operating modes of production,
- microengineering and microtechnology.

3.1.5.1 Hybrid Unit Operations

The first case deals with "multifunctional" equipment that couples or uncouples elementary processes (transfer–reaction–separation) to increase productivity and/or selectivity with respect to the desired product and to facilitate the separation of undesired by-products. Numerous reactive separation processes involving unit operation hybridization exist.

The concept of reactive or catalytic distillation has been commercialized success-fully, both in petroleum processing, where packed bed catalytic distillation columns are used, and in the manufacture of chemicals where reactive distillation is often employed [41–43]. Catalytic distillation combines reaction and distillation in one vessel, using structured catalysts as the enabling element [44]. The combination results in a constant-pressure boiling system, ensuring precise temperature control in the catalyst zone. The heat of reaction directly vaporizes the reaction products for efficient energy utilization. By distilling the products from the reactants in the reactor, catalytic distillation breaks the reaction equilibrium barrier. It eliminates the need for additional fractionation and reaction stages, while increasing conversion and improving product quality. Both investment and operating costs are far lower than with a conventional reaction followed by distillation [45].

We may distinguish between homogeneous and heterogeneous reactive distillation. Homogeneous reactive distillations (Figure 3.6a) make use of the good mixing and heat transfer inside a distillation column. The residence time can be adjusted and the rate of reaction can be enhanced by constant removal, via distillation of the resulting products. Typically, tray columns are used for homogeneous reactive distillation. Two unit operations, reaction and distillation, are performed in the same piece of equipment, which can give considerable cost savings.

Heterogeneous reactive distillations (Figure 3.6b) are performed in distillation columns that have three sections. The reactor section is the middle section, where a special packing (like Katamax$^{\times}$, developed by Koch Engineering Co., or Katapak-S$^{\times}$, developed by Sulzer Chemtech) is used that contains a solid catalyst. The liquid phase reaction takes place as the liquid film passes across the surface of the packing/catalyst. Above and below the reaction section is a rectifying and stripping section of traditional packing. Once again, two unit operations, reaction and distillation, are performed in the same piece of equipment, which can lead to considerable cost savings.

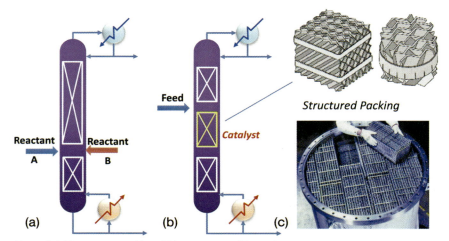

Figure 3.6 Homogeneous (a) and heterogeneous (b) reactive distillation; (c) structured packing (Katapak-S, from Sulzer) for reactive distillation, consisting of layers of metal wire gauze in which a catalyst is embedded and immobilized in "pockets."

Kaibel *et al.* [46] have specifically analyzed the role of catalytic reactive distillation for process integration. The main advantages of catalytic distillation were a decrease in equipment size (lower capital costs), lower energy consumption, higher conversion and lower recycle costs, improved selectivity, breaking of azeotropes, isothermal operation, effective cooling, use of reaction heat and exergetic advantages. Catalytic distillation could be applied on systems with homogeneous catalysis, autocatalytic processes, and heterogeneous catalysis. Schoenmakers and Bessling have reviewed reactive and catalytic distillation from an industrial perspective [47]. They showed how reactive and catalytic distillation has attracted growing industrial interest. Several examples for various reactions (etherification, hydration of olefins, dehydration of alcohols, alkylation of benzene, esterification and hydrolysis, olefin oligomerization, hydrogenation, addition of alcohols to aldehydes, addition of amines to ketones) have also been discussed by Podrebarac *et al.* [48].

The benefits of using reactive distillation are clearly proven in the production of fuel components (ethers) such as *tert*-amyl methyl ether (TAME), methyl *tert*-butyl ether (MTBE) and methyl acetate. The latter is synthesized from acetic acid and methanol with a reversible liquid-phase reaction:

$$CH_3COOH + CH_3OH \rightleftharpoons CH_3COOCH_3 + H_2O \tag{3.1}$$

The equilibrium point lies far to the left and little methyl acetate (CH_3COOCH_3) is formed if water in not removed. By reactive distillation it is possible to continuously remove water and considerably intensify the reaction. Eastman Chemical pioneered one of the first major applications of reactive distillation, to significantly simplify the production of methyl acetate (Figure 3.7). This unit first went into operation in 1983. Among typical reactions where a by-product prevents the reaction from going to the right are esterification, trans-esterification, hydrolysis, acetalization and amination. Other types of reactions that could benefit from reactive distillation include alkylation/transalkylation/dealkylation, isomerization and chlorination.

Other industrial processes that have taken advantage of the process intensification deriving from the introduction of reactive (catalytic) distillation are: (i) production of high purity isobutene, for aromatic alkylation; (ii) production of isopropyl alcohol by hydration of propylene; (iii) selective production of ethylene glycol, which involves a great number of competitive reactions and (iv) selective desulfurization of fluid catalytic cracker gasoline fractions; as well as various selective hydrogenations. Extraction distillation is also used for the production of anhydrous ethanol.

The next generation of commercial processes using catalytic distillation technology will be in the manufacture of oxygenates and fuel additives or in the synthesis of a range of fatty acid esters used in the manufacture of cosmetics, detergents and surfactants [49]. In general, many reactions in the field of use of bioresources can gain significant advantage from using reactive distillation for PI.

Another approach intensifies distillation by combining two columns into one, a so-called dividing wall column (Figure 3.8). This arrangement obviates a second separate column and its evaporator and condenser. The unit features a vertical wall in the middle part of the column, creating a feed and draw-off section in this part of the column. The dividing wall, which is designed to be gas- and liquid-sealed, permits

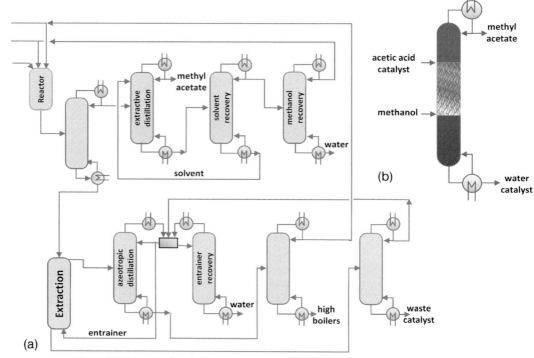

Figure 3.7 Process of methyl acetate synthesis without (a) and with (b) reactive distillation.

low energy separation of low- and high-boiling fractions in the feed section. The medium-boiling fraction is concentrated in the draw-off part of the column. Dividing wall columns can be used wherever multicomponent mixtures must be split into high purity individual components. They are well suited to obtaining pure medium-boiling fractions (sometimes called heart cuts). For instance, separating a three-component mixture into its pure components in conventional systems requires at least two main columns and a side column. In contrast, a single dividing wall column can handle this task – and cut installation costs by 20–30% and operating costs by around 25%. This approach also significantly simplifies process control and reduces maintenance work. A major supplier of dividing wall columns is Julius Montz GmbH (Hilden, Germany; www.montz.de), which has more than 60 installations.

Distillation is the most energy intensive unit of operation. There are many other options to intensify this process and reduce energy cost. One of them is the Higee, for example, the high gravity rotating contactor previously discussed. Currently there are approximately 30 units in operation worldwide, although most are still at pilot plant stage. Typically, Higee reduces the height of a theoretical stage to 0.05–0.12 m from the 0.15–0.60 m in a standard column.

An alternative reaction–separation unit is the chromatographic reactor. It uses differences in the adsorptivity of the different components involved rather than

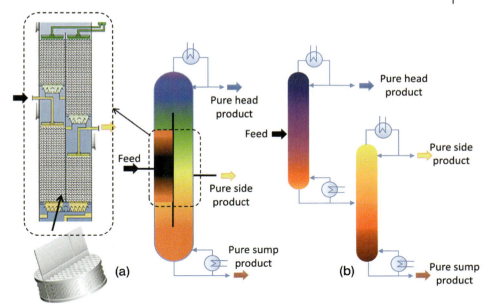

Figure 3.8 Dividing wall distillation column (a) and conventional column system (b). Source: adapted from Montz, 2008 (www.montz.de).

differences in their volatility. It is especially interesting as an alternative to reactive distillation when the species involved exhibit small volatility differences, are non-volatile or are sensitive to temperature, as in the case of small fine chemical or pharmaceutical applications. Typical adsorbents used are activated carbons, zeolites, alumina, ion-exchange resins and even immobilized enzymes. An example application is the esterification reaction catalyzed by acidic ion-exchange resins, as the polarity difference between the two products (ester and water) makes their separation easy on many different adsorbents. Other applications include trans-esterifications, alkylation, etherification, (de)hydrogenations and reactions involving sugars.

Another interesting example is the coupling of reaction and crystallization. Many basic chemicals, pharmaceuticals, agricultural products, ceramic powders and pigments could be produced by reactive crystallization-based processes. Crystallization and distillation could also be combined to overcome barriers in the use of the single unit operations, such as the presence of azeotropes in distillation, or of eutectics in crystallization. Extractive crystallization uses a solvent to change the relative solubility of the solutes to affect separations. The distillation column is used to create solvent swings and to recycle the solvent. Commercial examples include solvent dewaxing, solvent deoiling and separation of sterols. Another advantage of such crystallization–distillation hybrid separation processes is that they do not require the addition of solvents, which may increase the process flows, create waste streams, propagate throughout a chemical plant and require costly separation and recycle equipment.

Realizing hybrid approaches by coupling either a reactor or separation unit with a membrane is another very interesting possibility [50, 51]. There are two options in which membranes could be used coupled to reactors. The first is to distribute the feed of one of the reactants to a packed bed of catalyst, and thus realize a better profile of concentration along the reactor to minimize hot spots and consecutive reactions. This approach has resulted in increased selectivity in partial oxidation reactions. The second approach is to either remove a product that inhibits a reaction (e.g., in enzymatic reactions) or to remove a product to shift the equilibrium. Examples are in the continuous removal of water in dehydration reactions, or of H_2 in dehydrogenation reactions.

In addition, increasing use of membranes is being made in bioreactors to provide continuous operations with products that diffuse through the membrane but a biocatalyst that cannot pass through it. Bioreactors based on the hollow-fiber design are used to produce monoclonal antibodies for diagnostic tests, to mimic biological processes or to produce pure enantiomers, when a membrane separation is combined with an enantiospecific reaction. This technology can respond to the increasing demand for food additives, feeds, flavors, fragrances, pharmaceuticals and agrochemicals. Furthermore, membrane bioreactors find increasing application in water purification treatments, and will be a key technology in biofuel production.

Phase-transfer catalysis can also be performed in membrane reactor configurations by immobilizing the appropriate catalysts in the microporous structure of the hydrophobic membrane. Catalytic membrane reactors are also proposed for selective product removal to by-pass equilibrium limitations, that is, catalytic permselective or non-permselective membrane reactors, packed bed (catalytic) permselective membrane reactors an fluidized bed (catalytic) permselective membrane reactors. An example is given by membrane reactors for dehydrogenation reactions where hydrogen is withdrawn from the reaction mixture using permselective Pd-membranes, thereby shifting the reaction equilibrium to the desired products. In H_2 synthesis by steam reforming, the use of Pd-based permeoselective membranes allows operation of the steam reforming at lower temperatures (from about 900–950 °C down to below 600 °C) with an estimated saving of energy of about 15%. Similar advantage is also possible in syngas production, allowing a lower reaction temperature and operation with heterogeneous catalysts that would deactivate too quickly under conventional reaction conditions.

In the catalytic partial oxidation of methane to produce syngas the use of permselective dense perovskite membranes avoids (or minimizes) the need of air separation, the most costly step in the process. Although both these O_2- and H_2-permeoselective membranes (based on perovskites or thin supported Pd-based dense films, respectively) have still to be further developed for commercial applications the outlook appears quite interesting for intensifying various large chemical processes.

3.1.5.2 New Operating Modes of Production

Several new operating modes of production allow process intensification, although they have been mainly investigated at a laboratory and/or pilot stage.

The more relevant for practical applications were the following according to Charpentier [7]:

- reversed flow for reaction–regeneration,
- unsteady operations, cyclic processes,
- extreme conditions,
- pultrusion (a variation of the extrusion process),
- low-frequency vibrations to improve gas–liquid contact in bubble columns,
- high-temperature and high-pressure technologies,
- supercritical media and ionic liquids.

In reverse flow reactor operations, one or more process variables are intentionally and permanently (cyclic) perturbed according to design schedule [52, 53]. The technology was first introduced for removal of pollutants. However, it is well suited to optimize the use of the heat of reaction (for feed preheating) in the case of exothermic processes. The periodic flow reversal in such units allows the reaction heat to be retained within the reactor bed. After reversing the direction of the feed, this heat of reaction is used to pre-heat the cold reactant gases. Figure 3.9 shows the operating principle of the reverse flow reactor (RFR), exemplified for catalytic partial oxidation of methane. This dynamic operation creates process improvements that cannot be achieved by steady state operation. Expected benefits are energy savings, increased conversion, selectivity and productivity. Barriers are reaction kinetics (exothermic reactions, endothermic reactions as well as equilibrium reactions), energy storage, as well as reactor design operation and control. All these barriers stem from the dynamics of high flow reversal frequency.

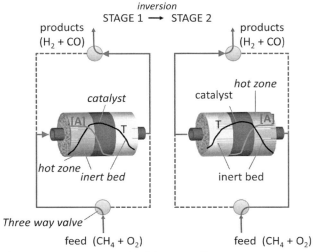

Figure 3.9 Operating principle of the reverse flow reactor (RFR). The reactor consists of three zones – a catalyst bed between two beds of packing of heat-accumulating material.

Reverse-flow reactors have been used in three industrial processes: SO_2 oxidation, total oxidation of hydrocarbons in off-gases and NOx reduction. The reverse-flow principle has also been applied in rotating monolith reactors used industrially for removal of undesired components from the gas streams and continuous heat regeneration. Studies have also been carried out on the use of reversed flow reactors for endothermic processes. Low-level contaminants or waste products such as volatile organic compounds can be efficiently removed in adiabatic fixed beds with periodic reversal by taking advantage of higher outlet temperatures generated in earlier cycles to accelerate exothermic reactions. Energy and cost savings are affected by this substitution of internal heat transfer for external exchange [54].

The RFR is a case of heat-integrated reactor concepts that have found fairly wide-spread application in PI, typically for reactions where the adiabatic temperature rise does not allow auto-thermal reactor operation. The RFR concept is also suited for extreme conditions of auto-thermal high-temperature millisecond contact-time catalysis. Liu *et al.* [55] have demonstrated that thermodynamic limitations in catalytic partial oxidation of methane to synthesis gas as well as kinetic limitations in the oxidative dehydrogenation of ethane to ethylene at high-temperature conditions can be overcome via regenerative heat-integration in a reverse-flow reactor. Strong improvements in product yields in comparison to conventionally operated fixed bed reactors are obtained while maintaining the compactness of the short contact-time reactor and keeping the process independent of external heat sources. Overall, the application of heat-integrated reactors to high-temperature catalysis is open to strongly intensified processes. The same group [56] has also demonstrated that RFR operation leads to strong improvements in synthesis gas yields over steady state (SS) operations for various catalysts, with particularly strong improvements for poorly performing catalysts. Furthermore, while the increased catalyst temperatures result in an accelerated deactivation of the unstable catalysts (Pt,Ir), heat integration leads to a complete compensation of this acceleration. RFR operation thus has an intrinsic "equalizing" effect on catalyst performance and thus offers a widely applicable reactor engineering approach to compensate for poor or degrading catalysts in high temperature partial oxidations.

Under most operating conditions a RFR eventually converges to a symmetric single-period operation so that the concentrations and temperature profiles after one flow reversal are a mirror image of those after the previous flow reversal. However, a cooled RFR may attain, under certain conditions, states with more complex periodicity, that is, states with period $n > 1$, nonsymmetric states and even complex quasi-periodic and chaotic states.

There are other possible unsteady (periodic) operation modes for a packed bed reactor that could lead to process intensification [57]. Indeed, there are several unsteady state strategies available to run a process unit such as a reactor. Pulses of different magnitude can be imposed on an input, or the input could be either changed progressively or varied according to an analytical function. However, not all unsteady state strategies are feasible in a commercial situation. Table 3.2 gives examples of the possibilities.

The pressure swing reactor (PSR) was developed from pressure swing adsorption (PSA) by simply adding a catalyst to the adsorbent. Consequently, PSR designs follow

Table 3.2 Examples of periodical operations in reactors to improve performances. Source: adapted from Aida and Silveston [57].

System	Manipulating variable	Performance enhancement
Water gas shift reaction in a pressure swing reactor	Pressure, composition and flow	Exceeded equilibrium conversion
NO_x reduction by NH_3 in a packed bed, catalytic reactor	Flow and flow direction	Reduced ammonia consumption
SO_2 absorption and oxidation to sulfuric acid in a trickle bed reactor	Flow	40% increase in SO_2 removal
Synthesis of ammonia in a packed bed, catalytic reactor	Composition	100-fold increase in nitrogen conversion
Steam reforming of methane	Temperature	>95% H_2 and less than 30 ppm CO with a dolomite CO_2 acceptor

those used for adsorption systems. Perhaps the most important incentive for exploring PSRs is that they can operate at lower temperatures for equilibrium-limited endothermic reactions than those usually employed. Dehydrogenation is a good example of such a reaction. Lower temperatures could reduce the importance of secondary reactions and would certainly lower capital and/or operating costs. In addition, for equilibrium-limited exothermic reactions with large activation energies, a PSR might increase reaction yield without requiring a reduction in temperature.

Temperature swing reactors (TSR) are an alternative to reverse flow reactors for the combustion of low concentration volatile organic compounds (VOCs). The most common case is the use of a rotating packed catalyst bed, where the upper part of the rotating disk is exposed to the cold stream containing the diluted VOC and acts as absorber for them. The lower part of the rotating disk is exposed to hot air. The adsorbed VOCs desorb and are combusted on the catalyst. The TSR is advantageous over RFR for very dilute VOC streams, but very careful design of the properties of the catalyst bed is necessary.

The concept of trapping a contaminant in low concentration by adsorption with periodic regeneration of the adsorbent-catalyst has been applied commercially by Toyota for NOx-trap catalysts used in converting NOx in diesel or lean burn engine emissions, for example, for reduction of NOx in the presence of O_2 [63]. The catalyst acts as absorbent of NOx (in the form of surface nitrate-like species) in the presence of O_2 (lean conditions), but a periodic switch of the air to fuel ratio to rich conditions (deficit of O_2 with respect to stoichiometry for the complete oxidation of CO and hydrocarbons present in the car emissions to CO_2) leads to regeneration by reducing trapped NO_x to N_2.

Process intensification is also possible by induced pulsing a liquid flow in trickle beds to improve liquid–solid contacting at low liquid mass velocities in the cocurrent downflow mode [64]. In a trickle bed reactor the liquid and gas phases flow

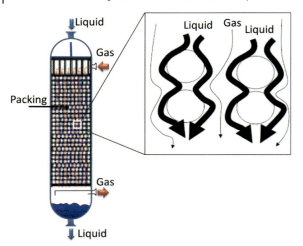

Figure 3.10 Schematic diagram of a trickle bed reactor. Inset: schematic of the trickle flow.

cocurrently downwards through a fixed bed of catalyst particles while the reaction takes place (Figure 3.10). The cocurrent upward flow operation provides better radial and axial mixing than the downward flow operation, thus facilitating better heat transfer between the liquid and solid phases. However, the downflow scheme is usually utilized due to better mechanical stability and less flooding, thus facilitating processing of higher flow rates and increased reactor capacity [65].

Trickle bed reactors operate in various flow regimes, ranging from gas-continuous to liquid-continuous patterns. They usually fall into two broad categories, referred to as low interaction regime (trickle flow regime) and high interaction regime (pulse, spray, bubble and dispersed bubble flow regimes). The low interaction regime is observed at low gas and liquid flow rates and is characterized by a weak gas–liquid interfacial activity and a gravity-driven liquid flow. The high interaction regime is characterized by a moderate to intense gas–liquid shear due to a moderate to high flow rate of one or both of the fluids.

In the trickle flow regime (inset in Figure 3.10) the liquid flows down the reactor on the surface of the packing in the form of rivulets and films while the gas phase travels in the remaining void space. The trickle flow regime can be further divided into two regions. At very low gas and liquid flow rates, the liquid flow is laminar and a fraction of the packing remains unwetted. This regime is called the partial wetting regime. If the liquid flow rate is increased, the partial wetting regime changes to the complete wetting trickling regime in which the packing is totally covered by a liquid film.

The pulse flow occurs at relatively high gas and liquid input flow rates. It refers to the formation of slugs that have a higher liquid content than the remainder of the bed. The pulsing behavior refers to gas and liquid slugs traversing the reactor alternately. It begins when the flow channels between packing are plugged by a slug of liquid, followed by blowing off the slug by the gas flow (Figure 3.11). Pulses always begin at the bottom of the bed, where the gas velocity is higher due to the lower pressure. As

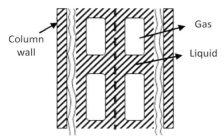

Figure 3.11 Schematic diagram of pulse flow.

the gas flow rate is increased, the incipient point of pulsing moves to the upper part of the reactor.

At present, trickle flow is the most common flow regime encountered in industrial applications. For process intensification, pulsing flow allows an increase in mass and heat transfer rates, complete catalyst wetting and a decrease in axial dispersion compared to trickle flow [66, 67]. The operation of a trickle bed reactor in the pulsing flow regime is favorable in terms of a capacity increase and the elimination of hot spots. Axial dispersion is less than with trickle flow due to increased radial mixing and disappearance of stagnant liquid holdup. Wu *et al.* [68] have demonstrated that pulsing flow has a positive effect, particularly on selectivity, with respect to trickle flow. Also, such periodic operation with respect to liquid flow may help in getting process intensification for gas-limiting reactions or for petroleum applications where filtration and bed plugging are serious threats [69].

Forced dynamic (periodic, pulsing) operation of chemical reactors as a means for improving the reactor performance has been investigated since late 1960s. Through dynamic operation one can advantageously influence the kinetics of the adsorption–reaction–desorption processes on the catalyst surface (solid-catalyzed gas-phase reactions), increase interfacial mass transfer rates (e.g., pulsing operation of trickle-bed reactors), shift the process beyond the equilibrium limitation or improve heat transfer (e.g., reversed-flow operation of fixed-bed catalytic reactors), or improve mixing characteristics of the system (e.g., variable-volume operation of the stirred tank reactors). The process rates have been improved by 50% or more in bench-scale experiments.

Another example of dynamic operations is the oscillatory baffled reactor (OBR). This technology consists of a cylindrical column containing equally spaced orifice baffles. Vortices are generated when fluid flow pass through the baffles, enabling significant radial motions, where events at the wall are of the same magnitude as those at the center. The generation and cessation of eddies creates uniform mixing in each baffled cell, collectively along the column. The degree of mixing is independent of the net flow, which makes it possible to realize a nearly plug-flow character (many CSTRs in series) in a flow system at long residence times. OBR offers enhanced mass and heat transfers over stirred tank reactors, the workhorse in fine chemical and pharmaceutical production. Using OBR it is possible to realize plug flow conditions even at low (laminar) flow rates, and thus to change batch to continuous production.

Table 3.3 Intensification effects of alternative energy forms.
Source: adapted from Stankiewicz [21].

Energy source	Intensified element	Degree of possible intensification	Sustainability effect
Electric field	Interfacial area	500×	Energy
	Heat transfer	10×	
Microwaves	Reaction time	1250×	Energy, material efficiency
	Distillation time	20×	
Light	Product yield/selectivity	Improved selectivity	Material efficiency, waste reduction, safety
Ultrasound	Reaction time	25×	Energy, material efficiency
	Gas–liquid mass transfer	5×	
	Liquid–solid mass transfer	20×	
Supersonic shockwave	Gas–liquid mass transfer coefficient	10×	Energy, material efficiency

Major benefits are significant energy/utility savings, higher yields and less side products/high product consistency. In addition, capital cost savings are achieved through much more compact designs.

Various other non-conventional reactor operational modes to improve performances have been proposed, based in particular on the use of alternative ways to either supply energy (microwave, light) or to induce high-energy microenvironments (e.g., cavitation effects by sonochemistry). The use of light, ultrasonic and microwave technologies to enhance the rates and improve the selectivities of catalytic reactions [21, 70, 71] is discussed in a more detail later. Table 3.3 summarizes the intensification effects possible by using alternative energy forms [21].

Three alternative energy forms have particularly attracted research interest: microwaves, light and ultrasound. Microwave frequencies ranges from 300 to about 300 000 MHz. Polar molecules subjected to microwave irradiation exhibit dipole rotation, trying to align with the rapidly changing electric field of the microwave. The rotational motion of the molecule results in a transfer of energy. Additionally, in substances where free ions or ionic species are present, the energy is also transferred by the ionic motion in an oscillating microwave field. As a result of both these mechanisms the substance is heated directly and almost evenly. Heating with microwaves is therefore fundamentally different from conventional heating by conduction.

Microwaves accelerate chemical reactions, often by factors of hundreds, and in many cases significantly better product yields are reported. A relevant challenge to further amplify the effect of microwaves is to couple with microreactor technology. This coupling offers a great opportunity to increase process selectivity by an instantaneous heating of the reactants and a fast quenching of the reaction products. Fundamental challenges here include equipment materials and their interaction with microwave radiation, application of microwave energy to micro-volumes, modeling and optimization of microwave-driven process in microequipment. The influence of the molecular effects induced by the microwave field on transport and interfacial phenomena in multiphase separation systems, such as distillation, extraction or crystallization, presents another exciting and largely unexplored research area.

The use of light, either artificial or solar, in catalysis can lead in principle to high product selectivities, although examples are still limited. Another problem with the present types of photocatalytic reactors is that a significant portion of emitted light is absorbed or dissipated before it reaches the catalytic site. This results in high energy demands and often makes photocatalytic reactors economically unattractive. An ideal photocatalytic reactor should be able to emit photons exactly where and when they are needed, that is, in the direct vicinity of the catalytic site and upon contact of the reacting molecules with that site.

Ultrasonics uses sound in the 0.1–100 nm wavelength range to enhance mixing and chemical reactions. Two levels of sonics are in common use: a low level with sufficient energy to enhance mixing but not sufficient to perform or directly assist the chemical reaction and a high level that imparts significantly more energy. The high level creates voids, acoustic cavitation, and this cavitation results in bubbles being formed. When the bubbles collapse they generate a micro-jet of fluid and a high-energy environment that can enhance the reaction rate.

Another interesting opportunity is magnetic-driven process intensification [72]. External inhomogeneous magnetic fields exert a magnetization body force. This force may be used in the case of electrically non-conducting and magnetically permeable fluids for hydrodynamic intensification of the chemical processes. Hence, in acting on paramagnetic or diamagnetic gases and liquids, this force can modify the direction as well as the magnitude of the gravitational force. An example of application is the possibility to influence two-phase flows through packed bed reactors by application of external inhomogeneous magnetic fields. The effect could be used to intensify performances of mini-trickle-bed reactors for applications in fine and pharmaceuticals chemical processes. A positive-gradient in homogeneous magnetic fields promotes larger values of liquid holdup (and thus wetting efficiency in the trickle flow regime) and two-phase pressure drop [73].

3.1.5.3 Microengineering and Microtechnology

Substantial progress in microtechnologies has largely pushed the development and availability of novel micro-sized chemical apparatus (reactors, mixers, separators, heat-exchangers and analyzers) in recent years, making possible the accurate control of reaction conditions with respect to mixing, quenching and temperature profile [33–35, 74, 75]. Microfabricated chemical systems are now expected to have several advantages for chemical kinetic studies, chemical synthesis and, more generally, for process development. Indeed, the reduction in size, integration of multiple functions, the continuous on-line monitoring with embedded sensors and quality control micro-devices has the potential to produce structures with capabilities that exceed those of the conventional macroscopic systems and to add new functionality, while potentially making possible mass production at low cost, improved quality with reduced waste production, and improved safety.

The high heat and mass transfer rates possible in microfluidic systems allow reactions to be performed under more aggressive conditions with higher yields than in conventional reactors [75–78]. The flow and heat transfer in heated microchannels are accompanied by several thermohydrodynamic processes, such as liquid heating and vaporization, boiling, formation of two-phase mixtures with a very complicated inner structure, and so on, which affect significantly the hydrodynamic and thermal characteristics of the cooling systems. The multiplicity of phenomena characteristic of flow in heated microchannels makes, on the one hand, quite complex the understanding and simulation of behavior of pure liquid and two-phase mixture in microchannels, but, on the other hand, greatly increases the possibilities of improving the performances. Therefore, microdevices (reactors, mixer, heat exchangers, etc.) have advantages over conventional devices in terms of (i) compactness and low capital cost, (ii) low energy consumption and other operating expenses, (iii) short mixing time and well-defined mixing behavior, (iv) narrow residence-time distribution and (v) performance independent of pressure and temperature. The particular characteristics of the flow in microchannels, the behavior of fluids in confined environments (micro spaces) and the high wall to volume ratio determine new properties that are the bases for nonlinear effects and their use in process intensification.

If the characteristic linear dimension of the flow field is small enough, then the measured hydrodynamic data differ from those predicted by the Navier–Stokes equations [79]. With respect to the value in macrochannels, in microchannels (around 50 microns of section) (i) the friction factor is about 20–30% lower, (ii) the critical Reynolds number below which the flow remains laminar is lower (e.g., the change to turbulent flow occurs at lower linear velocities) and (iii) the Nusselt number, for example, heat transfer characteristics, is quite different [80]. The Nusselt number for the microchannel is lower than the conventional value when the flow rate is small. As the flow rate through the microchannel is increased, the Nusselt number significantly increases and exceeds the value for the fully developed flow in the conventional channel. These effects have been investigated extensively in relation to the development of more efficient cooling devices for electronic applications, but have clear implications also for chemical applications.

In microdevices, more possibilities for controlling their performances are also offered by using externally induced effects. Micromixers may not only be passive devices, for example, which use part of the flow energy for mixing (e.g., lamellae to induce a mixing by diffusion) but can be also based on active effects that rely on applied forcing functions such as a pressure or an electric field. Possible external energy sources for active mixing are ultrasound, acoustic, bubble-induced vibrations, electrokinetic instabilities, periodic variation of flow rate, electrowetting induced merging of droplets, piezoelectric vibrating membranes, magneto-hydrodynamic action, small impellers, integrated micro valves/pumps and many others. Clearly, these possibilities are much less in conventional devices.

New reaction pathways considered too difficult for application in conventional equipments could be pursued because, if the microreactor fails, the small amount of chemicals released accidently could be easily contained. The high wall to volume ratio allows safe operation inside the explosion limits. By using multiple parallel micro-reactors, due to their inherent safety characteristics, a distributed point-of-use synthesis of chemicals with storage and shipping limitations, such as highly reactive and toxic intermediates like cyanides, peroxides and azides, is possible.

Therefore, microengineering and microtechnology is a new tool for process intensification, opening up new possibilities. For these reasons, the number of patent publications in the field of microreactor engineering has increased steadily during the last five years [81]. Most patents are initiated by German, US and Japanese inventors or applicants. Major company players were Merck, Degussa, Bayer, and some commercial providers (e.g., Velocys, CPC). Active Institutes were Forschungs-zentrum Karlsruhe, Battelle Memorial Institute, and the Institut für Mikrotechnik Mainz GmbH. Seven German entities appear in the top-ten positions. [81].

Initially, the focus was on fabrication, device and total system related inventions. Meanwhile, an increasing number of process patent publications have been released, showing the further stage of implementation of the technique, which is moving to microprocess engineering. Microreactor technology is concerned with microstruc-tured reactors, capillary and tube reactors of small inner dimensions and, to a lesser extent, mini-fixed bed or small-sized foam reactors. Microprocess technology is the integration of microreaction technology into a plant architecture, the combination with other operations, preferably also with separation units, and the way of designing the process. The focus of patents and published patent applications is on fine-chemical processes. This goes hand in hand with the number of pilot plants reported in the field of fine chemistry, where commercial implementation is much more advanced. There is increasing interest in combining microreactors and microprocess devices to produce an integrated system [82].

Microreactor-based tools have already been well demonstrated as stand-alone devices for high-throughput catalyst screening and combinatorial materials science. Owing to increased process intensification offered by microreactor technology, more traditional pilot-scale reactor systems and pilot plant systems may be replaced, at least in some cases, by smaller, faster responding, more flexible mini-plants with reduced capital and operating costs [74, 83–85]. In addition, scale-up to production by replication of microreactors units used in the laboratory would eliminate costly redesign and pilot plant experiments, thereby shortening the development time from laboratory to commercial-scale production. This aspect is particularly relevant for pharmaceutical and fine chemical industries where production amounts are often less than a few metric tonnes per year. Others, more recently, have begun to apply microchannel technology to larger scale applications such as methane steam reforming, gas to liquid, propylene oxide, hydrogen peroxide production, and so on, some with industrial partners such as Degussa, UOP, and so on [75, 86].

One likely application is the use of microreactors in distributed energy systems. Significant efforts are being placed on the development of highly efficient fuel cells for the generation of portable power, and potentially even supplementing electric utilities. Fuel cells need a source of hydrogen or methanol, which could be provided by a microreactor system. Various microstructured components for hydrogen production from hydrocarbons have been presented by Pfeifer *et al.* [87].

In conclusion, microengineered reactors and devices have unique characteristics for the implementation of process intensification. They can also provide significant advantages for high-throughput experimentation and process development. In terms of chemical manufacture, they allow distributed, mobile and intensified processing [88]. The role of microreaction technology in the discovery, development and commercialization phases of the catalyst invention cycle is expected to become increasingly important, due to improved safety characteristics, enhanced rates of heat and mass transfer, reduced hardware footprint, lower reagent costs and ease of creating parallel systems for higher data throughput and improved workflow efficiency. Problems in terms of fabrication still exist, but they cannot be considered a real barrier.

In the case of multiphase reactions, such as those involving gas–liquid, gas–liquid–solid and gas–liquid–liquid systems, microreaction technology is still in an early stage of development with respect to single-phase applications. However, this is also a rapidly developing area, but the fluidodynamics in microchannels have to be better understood.

Regarding the role of micro-device technology for process intensification, it is essential to focus on systems where microfabrication can provide unique process advantages resulting from the small dimension, that is, not only the high transport rates but the forces associated with high surface area-to-volume ratio. Microreactor technology provides more flexibility than traditional plants and thus the integration of microreactors within existing plants is facilitated. Microstructured mixers can already be considered an industrial reality, with throughputs barrier lifted to several tens of cubic meters per hour flow rate domain. These micromixers are applied to carry out advanced syntheses of fine chemicals or pharmaceuticals, as well as for the generation of dispersions, foams, creams and emulsions [89]. Also, the use of micro heat exchangers in industrial processes is increasing [90, 91]. The combination of various functionalities, such as on-board sensors and control systems, with microreactors and other microprocess units (fluid transport devices, valves, mixers, heat exchangers, separation systems, etc.) into an integrated package, for example, integrating microreactors into functioning microprocess systems and mini-plants, is still at an earlier stage of development. However, progress in this area will be driven by the possibility of extending the use of microreactors from niche to large-scale applications.

3.1.6
Role for the Reduction of Emissions of Greenhouse Gases

Process intensification (PI) offers several opportunities to improve energy efficiency and reduce environmental impact [92]. Many chemical reactions currently carried out as batch processes in stirred tanks could be carried out in continuously operated, intensified reactors such as spinning disc or oscillatory baffle types. The plant used for separations can be made highly compact, and even for large-scale plants (nitric acid production) the concept of "pocket-sized plant" has been introduced to reduce energy needs in the process [93]. PI is thus a key element in reducing the "carbon footprint" of the chemical industry. The opportunities arise from the following [92]:

- less unwanted by-products – so reduced downstream processing;
- moving from batch to continuous processing will reduce the energy need for cleaning the plant;
- more scope for process heat recovery – and higher grade heat may be available;
- reaction rates may be increased, and hence there will be reduced energy losses due to shorter processing times;
- reduced system losses.

Table 3.4 summarizes the main results (as reported by Reay [92, 94]) of a study made by Arthur D. Little for SenterNovem (The Netherlands) as a contribution to the European Roadmap of Process Intensification (EUROPIN). The overall energy

Table 3.4 Energy saving estimates of PI technologies as identified by the A.D. Little study for SenterNovem. Energy saving is in peta-joules (PJ) per year (1 PJ $= 1 \times 10^{15}$ J). Source: adapted from Reay [92, 94].

	Bulk chemicals	Fine chemicals	Food
Multifunctional equipment (advanced distillation)	50–80% energy savings in 15% of processes; 9–18 PJ	Limited to separation processes, that is, 10% of sector. Increase efficiency by 50%, saving <1 PJ	Drying and crystalli-zation. 10% total energy saving, worth 3–5 PJ
Micro/milli-reactors	A study by ECN in Holland suggests 20 PJ savings using heat exchanger-reactors. Micro-reactors extend this to 25 PJ	Applications in 20% of processes in the sector saving 20% of energy – 1 PJ Reduce feedstock and additives by 30% in 10% of processes saving 5–7 PJ	Spill-over from fine chemicals: <1 PJ
Microwaves (electri-cal enhancement)	[a]	Reduce feedstock and additives by 20–40% in 5% of processes: 2–3 PJ	20–50% saving in 10% of drying market: 1–1.5 PJ 10% energy reduction in product processing: 1–1.5 PJ
High gravity fields (e.g., spinning disc reactor. HiGee)	[a]	Reduce feedstock, solvents, and so on by 50% in 5% of processes: 1–3 PJ	Assuming 20% of electricity in food production goes to emulsification, mixing and so on, 10–20% saving worth 0.5 PJ

[a] Information of the effects of microwaves and HiGee in the bulk chemicals sector are still to limited to make estimations.

savings across the three sectors were estimated to be of the order of 50–100 PJ per year by 2050. The energy savings were largely due to better selectivity and reduced energy use in separation processes, as well as improved control.

The US Department of Energy has also initiated substantial programs supporting PI in the process industries [95]. Relevant to PI are the activities on hybrid distillation and novel reactors, together with advanced water removal (drying/evaporation) methods. The potential estimated savings of >70 PJ yr^{-1} by 2020 is greater than the European target.

Process intensification thus offers significant opportunities for carbon reductions in sectors ranging from chemicals to food and glass manufacture, even if it is actually still not established in the industrial sector [96, 97].

3.2
Alternative Sources and Forms of Energy for Process Intensification

Different alternative sources and forms of energy can be applied to intensify a chemical or biochemical process [98, 99]. The most important are the following:

- high-gravity fields;
- electric fields – stationary and dynamic;
- electromagnetic radiation – microwaves and light;
- acoustic fields (sonochemistry);
- flow.

3.2.1
High-Gravity Fields

As commented in Section 3.1.1, the use of high-gravity fields generated by the centrifugal operation for process intensification started with the pioneering work of Ramshaw and Mallison from the ICI New Science Group (around 1980) on the application of high-gravity fields (so-called "HiGee") in distillation processes [100]. Various other authors have later confirmed such a large increase in the mass transfer rates in the high-gravity field of rotating packed beds (RPBs). For example, Lin *et al.* [101], studying the methanol–ethanol distillation process in a rotating packed bed, have reported a HETP (height equivalent of theoretical plate) of 3–9 cm, compared to 30–40 cm for the conventional structured packings. In their review on rotating packed beds (HIGEE), Rao *et al.* [102] indicate that an up to about 200-fold increase in $k_L a$ (k_L is the mass transfer coefficient in liquid phase and a the interfacial area between gas and liquid phase) is possible in the RPBs with respect to the conventional packed columns. Such excellent mass-transfer properties of RPBs have found several commercial applications.

Table 3.5 compares the conventional technology for water deaeration in oil fields (vacuum desorption tower) with the HiGee technology designed for off-shore application (see also Figure 3.2) [6, 103]. High gravity technology has been also used in Dow's hypochlorous acid process [104].

Table 3.5 Comparison between a Higrav deaerator (two sets, 6000 t d^{-1}) and a vacuum tower deaeration system (one set, 10 000 t d^{-1}) for off-shore application. Source: adapted from Stankiewicz [98] and Zheng *et al.* [103].

	Vacuum tower	Higrav deaerator
Platform area (m^2)	30	2×10
Height (m)	14	3
Weight in operation (t)	130	2×10.5
Residual oxygen in winter (ppm)	2–3	<0.05
Power (kW)	155	2×160

The HIGEE technology consists of intensifying the mass transfer processes by carrying them out in rotating packed beds in which high centrifugal forces occur. Not only the mass transfer but also heat and momentum transfer can be intensified. The rotating bed equipment, originally dedicated to separation processes (such as absorption, extraction, distillation), can also be applied to reacting systems (especially those mass transfer limited). It can potentially be applied not only to gas–liquid but also to other phase combinations, including gas–liquid–solid systems. In a rotating packed bed the liquid (or the heavy phase for liquid–liquid systems) enters at the eye of the rotor, being distributed on the rotor packing at the inside diameter (Figures 3.1 and 3.2). Gas (or light phase in the case of liquid–liquid systems) enters the stationary housing and passes through the rotor from outside to inside. Woven wire screens, randomly packed pellets, foam metal or structured packings are used as the RPB internals. The micromixing and mass transfer can be 1–3 orders of magnitude larger. Barriers are reliability of rotating equipment, and modeling reactors in multiphase systems (flooding, micro-mixing and solid/liquid systems).

The spinning disc reactor (SDR), developed by Ramshaw and coworkers at the University of Newcastle [105], is a variation of the RPB concept. Here, the rotating surface of the disc enables the generation of a highly sheared liquid film. The film flow over the surface is intrinsically unstable and an array of spiral ripples is formed. This provides an additional improvement in the mass and heat transfer performance of the device. High heat transfer rates present the most important feature of SDRs and heat transfer coefficients exceeding $20 \, \text{kW} \, \text{m}^{-2} \, \text{K}^{-1}$ are reported. Spinning disc reactors are particularly attractive for applications in fast, highly exothermic reactions, and also those involving highly viscous liquids.

Excellent heat and mass transfer characteristics of the SDR have been confirmed by the study of a phase-transfer-catalyzed Darzens reaction for preparing a drug intermediate. The SDR allowed for a 99.9% reduction in reaction time, 99% reduction of inventory and 93% reduction in the level of impurities [106]. Other possible applications of the SDR include polymerizations and polycondensations (in both cases considerable time savings and more uniform product) as well as precipitation/crystallization (smaller crystals with much narrower size distribution). Two large chemical companies have patented processes based on spinning-disc technology. SmithKline Beecham has claimed a method to epoxidize substituted cyclohexanones [107], while Procter and Gamble has patented a process for making esters and amides using the SDR [108].

3.2.2
Electric Fields

Even if the use of electric fields, either static or dynamic, to improve chemical processes has long been known, only more recently it has been applied for process intensification. The application of an electric field in liquid–liquid systems induces an increase in the mass transfer (up to a factor 10), due to a higher degree of turbulence within and around the dispersed phase, as a result of interaction between the field and the interface [109]. Four different mechanisms are noted:

- higher terminal drop velocities resulting from electrical forces of attraction exerted on the drops in the direction of motion;
- generation of the electrically driven circulating flow in the neighborhood of the interface;
- alteration of the velocity profiles within and around individual droplets due to the oscillations by pulsed electric fields;
- interfacial tension-induced surface flows (Marangoni effects) due to the presence of electric charges.

An electric field applied to a nozzle results in a jetting phenomenon and the formation of an electrically charged emulsion of micron-sized droplets. This method may lead to a 200–500-fold increase in the surface area per unit volume, compared to the millimeter-sized droplets obtained in the conventional process.

Heat transfer can also be enhanced by application of an electric field. In boiling heat transfer, electric fields have been successfully used to control nucleation rates and achieve a continuous rise in heat transfer coefficient. Up to a sevenfold heat transfer enhancement by the electric field in falling film evaporators has been reported. In the presence of an electric field, both AC and DC, the mixing length in microchannels is shortened considerably, by a factor 30 or more.

Electric fields can be used, for instance, to reduce fouling phenomena in systems involving electrically charged macromolecules (e.g., proteins). In microsystems used for capillary zone electrophoresis an external electric field applied across the capillary tube induces electrostatic repulsion between the macromolecules and the inner surface. The reduced adsorption of macromolecules enhances separation resolution and efficiency.

3.2.3
Microwaves

Microwave-assisted operations are a very promising type of electromagnetic field application to intensify chemical processes [110–116]. Microwave frequencies range from 0.3 to 300 GHz but, to avoid interference, industrial and domestic microwave appliances operate at standard allocated frequencies, most often at 2.45 GHz.

Molecules that have a permanent dipole moment (e.g., water) can rotate in a fast changing electric field of microwave radiation. Additionally, in substances where free ions or ionic species are present the energy is also transferred by the ionic motion in an oscillating microwave field. Owing to both these mechanisms the substance is heated directly and almost evenly. Heating with microwaves is therefore fundamentally different from conventional heating by conduction. The magnitude of this effect depends on dielectric properties of the substance to be heated.

Also, in solid materials, microwaves are used on an industrial scale for heating purposes. The ability of the solid material to absorb microwave heating depends on two properties: the dielectric constant and the loss tangent. Some materials absorb the microwave energy very easily, while others are transparent or impermeable to it. The difference in sensitivity of various substances to microwaves makes the latter an interesting technology for the selective heating of materials/products.

Microwave-enhanced drying is used in the food, wood, textile and pharmaceutical industries. Microwaves (MW) enable selective heating of the moisture contained in microwave-transparent materials. The whole process proceeds at lower bulk temperatures and allows considerable energy savings. Also, the speed of the MW drying avoids unwanted degradation of some less stable components of the dried materials.

Commercial equipment for heating and drying is available from several vendors. The technology is widely applied: fine chemicals, pharma, food, polymers. New applications are in the field of materials production (nanomaterials, zeolites). Energy savings and occasionally, because of the much shorter process times, space savings can be substantial. Barriers are investment costs, reactor design (increase of efficiency, irradiation depth on the scale of cms, safety as the radiation is dangerous), limited knowledge of the physical properties of the materials to be processed (dielectric properties).

Several authors have reported the ability of microwave heating to accelerate organic reactions by factors ranging from ten to more than a thousand. Table 3.6 gives some

Table 3.6 Effect of microwave heating on reaction time and product yield. Source: adapted from Stankiewicz [98].

Reaction	Reaction time (min)		Product yield (%)	
	Conventional	Microwave	Conventional	Microwave
Hydrolysis of benzamide to benzoic acid	60	10	90	99
Oxidation of toluene to benzoic acid	25	5	40	40
Esterification of benzoic acid with methanol	480	5	74	76
S_N2 Reaction of 4-cyanophenoxide ion with benzyl chloride	960	4	89	93
Heck arylation of olefins	1200	3	68	68

examples. Not only the rate of the process is affected but also the product yield, which in some cases increases as a result of microwave heating.

Notably, however, an enhanced rate and product yield is not always observed. The reasons for the effect are still under question. Microscopic hot-spots, molecular agitation and improved transport properties of molecules have been mentioned as potential mechanisms of activation [117]. Other effects, such as positioning of the transitions states or decrease of the activation energy in the Arrhenius law, have also been suggested [118]. Several authors consider that the effect of microwaves is purely thermal [119], while others consider this explanation too simplistic [120].

Dramatic effects of microwave irradiation have also been reported for solvent-free synthesis systems [121] and in heterogeneous catalyzed reactions [122] but, as noted above, this effect is not always observed.

Despite the often large increase in the reaction rate the use of microwave-assisted reactions has still not been implemented on an industrial scale. One of the main barriers for industrial applications is reliable scale-up of microwave reactors [116], but there are also other engineering problems that have to be solved. The use of microwaves to speed-up distillation processes has also been indicated [123].

The use of microwave reactors in polymer processing (bonding/welding, curing and forming) is reasonably established and applied on an industrial scale. It offers a much better control of temperature than conventional heating (e.g., no baking, no deterioration due to high temperatures). However, their use in polymerization reactions is in its early stages of development. There are many indications that the use of microwave irradiation in polymerization processes may lead to a considerable increase of the process rates and can also influence product properties. Benefits arise from a better control of the molecular weight distribution compared to conventional heating. Major barriers are the small irradiation depth of the microwave versus required scale of production, narrow applicability of media in the process as well as materials of construction.

3.2.4
Light

Photochemical reactions have been well known for several years, but industrial applications have been limited to a few cases (chain reactions such as photochlorinations, photopolymerizations, photonitrosylations of alkanes or photosynthesis of previtamin D). The main applications are in the production of protective and decorative coatings, inks, packaging and electronic materials. Commercial applications in bulk chemical, pharma and agro sectors are scarce. Major benefits are the use of low temperatures (room temperature), thus providing energy savings, cost savings and very high conversion/yield/selectivity [124]. Major barriers are light penetration depth, the light source itself, light efficiency and wavelength. In addition, modeling, scale-up and optimal reactor configuration form a high hurdle.

Examples of possible photocatalytic applications developed on an laboratory scale cover a very broad range [125, 126]: from direct H_2 production from water, selective organic synthesis, water treatment and air cleaning to disinfection and anti-tumoral applications. However, industrial applications remain limited to date. The main problems are concerned with scale up and process efficiency. Several reactor designs have been proposed to overcome these limitations [127]: slurry reactors, annular reactors, immersion reactors, optical tube reactors and optical fiber reactors are among the most cited. However, a critical parameter often underestimated is the transport of light to the catalyst, for example, a low efficiency of illumination (photon transfer limitations) [124]. Another issue, particularly in liquid phase reactions, is the limited contact between activated catalyst and reagents (mass transfer limitations). New answers have been proposed to overcome these limitations [124].

Several other problems are encountered in exploiting photochemical processes on an industrial scale [128]:

- Specialized reaction vessels are required in which a light source may be incorporated. The most commonly used type of photoreactor is an immersion well, in which the light source is placed in the center of the reaction mixture. However, during scale-up, it is very difficult to reproduce the same ratio of area irradiated to volume of reactant.

- Light sources pose a difficulty on an industrial scale. Lamps used in photochemistry include medium- and high-pressure mercury lamps, xenon lamps and halogen lamps, all of which are costly to run. These have a limited lifetime and additionally tend to generate a large amount of heat and therefore require additional cooling systems.

- Photochemical reactions are typically carried out under batch reaction conditions. This method tends to be relatively inefficient compared to a continuous-flow process. Photochemical reactions operated in continuous-flow have been investigated and have proven to be far more effective at large-scale photochemical synthesis than the corresponding batch approach [129]. However, batch reactions continue to be the most common approach to photochemistry.

An approach to overcoming these problems, and which is directly linked to process intensification issues, is to combine microstructured reactor technology to photochemistry and photocatalysis [124,130]. Microstructured reactors have several possible advantages for photochemical processes [131]:

- They provide a mean of ensuring uniform irradiation to the entire reaction solution. As the depth of a microreactor is small, maximum penetration of light and thus irradiation of the reaction mixture can be achieved readily, even for relatively concentrated solutions.

- The high surface-to-volume ratio. In addition to an efficient illumination, in the case of photocatalytic reactions there is an efficient catalytic exposure to radiation and the reagent/catalyst contact is maximized [132, 133]. The small size of the channel also provides better control over variables such as temperature and flow rates, due to the fast heat and mass transfer.

- Microphotochemistry is commonly performed under continuous-flow rather than batch conditions. Consequently, the irradiation time for photochemical processes in a microreactor is easily altered, as this is directly proportional to the flow rate of the system. This feature allows rapid optimization of micro-photochemical reactions.

- Microstructured reactors possess high heat transfer coefficients. As a result, microstructured reactors are cooled very efficiently.

- Miniaturized light sources may be employed, for example light-emitting diodes (LEDs) [134–136]. These provide a clear alternative to conventional light sources, as they are available in a range of wavelengths, small in size and energy efficient. In addition, they produce less or no heat, thus reducing the need for coolant. Furthermore, the concept may be extended to pass from *macroscale* illumination to *microscale* illumination [137, 138] using a LED array, to even *nanoscale* illumination, for example, the direct integration of the light source on the catalyst surface. Gole *et al.* [139, 140] have suggested the combination of nitrogen-doped titania nanostructures and porous silicon (PS) to develop a device to produce visible light by electroluminescence of PS, thus activating the photocatalyst particles. This device could then be incorporated in a microreactor. Porous silicon emits visible light, hence the necessity to use nitrogen-doped TiO_2 samples to shift the absorption spectrum from UV to the visible spectrum.

- Microchip designs allow on-line monitoring of the reaction, for example, by UV-spectroscopic analysis of the effluent [141].

The photodegradation of chlorophenol in the presence of a titanium dioxide catalyst was one of the first microstructured reactor based catalytic reactions to be investigated [137, 142]. A microstructured reactor was prepared with TiO_2 as a photocatalyst on the walls of the channels. This was irradiated using an array of eleven UV-A LEDs, with a peak emission of 385 nm. Nakamura *et al.* [143] have focused instead on the coating of capillary tubes for use as microchannel reactors. In this case the microchannel is not embedded in a solid substrate; instead, the capillary acts as a single microchannel. The photocatalytic degradation of methylene blue was used as a model reaction. Matsushita *et al.* [144] have applied the concept of photocatalytic microstructured reactors in synthesis reactions, in particular the reduction of benzaldehyde and nitrotoluene. More examples have been reported by Coyle and Oelgemöller [130], also regarding homogeneous and heterogeneous liquid–gas reactions.

Although these results are preliminary, they demonstrate the feasibility of the concept of photocatalytic microstructured reactors, and their potential to increase by various orders of magnitude the catalyst coated surface per reaction liquid volume [124]. A critical problem to solve is the maximization of the illumination efficiency. The availability of a multitude of low-intensity light emitting sources on the micro- or even nanoscale near the catalyst particles appears to be a novel and promising approach to achieve uniform and maximized illumination. Combination of the latter approach with equipment to overcome the mass transfer limitations may prove to be the significant improvement that photoreactors need for industrial implementation.

3.2.5
Acoustic Energy

The use of ultrasound (sonochemistry), for example, acoustic energy effects, has a long tradition, similar to photochemistry. Although commercially applied in several applications, mainly in cleaning and decontamination or in the textile industry (dye dispersion and fixation), its use to improve performances in chemical reactions has essentially remained on a laboratory scale. The main focus of research has been on liquid-phase systems, where exposure to ultrasound results in formation, growth and subsequent collapse of microbubbles (microcavities), occurring over an extremely short period of time (milliseconds). The microimplosions are accompanied by an energy release with very high energy densities (of up to $10^{18}\,kW\,m^{-3}$), which leads to local generation of extremely high temperatures and pressures, up to about 5000 K and about 50 000 bar, respectively, as well as release of free radicals due to pyrolysis of water [145–148].

Similarly to microwaves, the use of ultrasound can dramatically speed-up chemical reactions and increase the product yield. Table 3.7 shows some examples of such ultrasound effects, both in homogeneous and in heterogeneous reaction systems. More data can be found in the review by Thompson and Doraiswamy [149].

Acoustic irradiation appears to be able not only to boost chemical reactions but also to intensify mass transfer processes in multiphase systems. A twofold increase of $k_L a$ using ultrasound has been observed [150], but depends strongly on the reaction conditions. Other authors have reported instead higher intensification factors. The enhancement is probably related to a reduction of the boundary layer thickness due to the microscale turbulence and reduction of the viscosity in the boundary layer.

The essence of the process intensification effect associated to the use of sound energy is related to the associated cavitation phenomena. Table 3.8 summarizes the different ways in which cavitation based on the use of sound and flow energy can be effectively used for process intensification [151]. Cavitation can be in general defined as the generation, subsequent growth and collapse of the cavities releasing large magnitudes

Table 3.7 Examples of the effect of ultrasound on reaction time and product yield. Source: adapted from Stankiewicz [98] and Thompson and Doraiswamy [149].

Reaction	Reaction time, min		Product yield (%)	
	Conventional	Ultrasound	Conventional	Ultrasound
Diels–Alder cyclization	2100	210	77.9	97.3
Epoxidation of long-chain un-saturated fatty esters	120	15	48	98
Oxidation of arylalkanes	240	240	12	80
Synthesis of chalcones by Claisen–Schmidt condensation	60	10	5	76
Ullmann coupling of 2-iodonitrobenzene	120	120	<1.5	70.4

Table 3.8 Process intensification effects of cavitation based on sound and flow energy. Source: adapted from Gogate [151].

Energy source	Form of application	Intensified element	Approximated intensification effect
Acoustic field	Ultrasound irradiation	Reaction time	25
		Product yield	In some cases 100% yield of the product
		Gas–liquid mass transfer	5
	Low frequency acoustics	Liquid–solid mass transfer	20
		Gas–solid mass transfer	3
		Gas–liquid mass transfer	2
Flow	Hydrodynamic cavitation	Reaction time and product yield	Similar to ultrasound, but up to 10× higher cavitational yield for same energy
	Supersonic flow	Gas–liquid mass transfer coefficient	10×
		Fluid bed reactor capacity	2×

of energy over a very small location, resulting in very high energy densities. Cavitation occurs at millions of locations in the reactor simultaneously and generates conditions of very high temperatures and pressures locally with overall ambient conditions. Thus, chemical reactions requiring stringent conditions can be effectively carried out using cavitation at ambient conditions. Moreover, free radicals are generated in the process due to the dissociation of vapors trapped in the cavitating bubbles, which results in either intensification of the chemical reactions or may even result in the propagation of certain reactions under ambient conditions [152]. Cavitation also results in generation of local turbulence and liquid micro-circulation (acoustic streaming) in the reactor, enhancing the rates of the transport processes, as noted above. These mechanical effects of cavitation are mainly responsible for the intensification of physical processing applications and also chemical processing limited by mass transfer whereas the chemical effects, such as generation of hot spots and reactive free radicals, are responsible for intensification of chemical processing applications.

The method of energy efficiently producing cavities of a desired quality (type of the dynamic behavior) can be taken as the main criterion in distinguishing among different types of cavitation. The four main types of cavitation and their causes can be summarized as follows [151]:

- **Acoustic cavitation:** In this case, the pressure variations in the liquid are effected using the sound waves, usually ultrasound (16 kHz–100 MHz). The chemistry taking place due to the cavitation induced by the passage of sound waves is commonly known as sonochemistry.

- **Hydrodynamic cavitation:** Cavitation is produced by pressure variations, which are obtained by using the geometry of the system, creating velocity variation. For example, based on the geometry of the system, the pressure and kinetic energy can be interchanged, resulting in the generation of cavities, as in the case of flow through an orifice, venturi, and so on.

- **Optic cavitation:** This is produced by photons of high intensity light (laser) rupturing the liquid continuum.

- **Particle cavitation:** This is produced by a beam of elementary particles, for example, a neutron beam, rupturing a liquid, as in the case of a bubble chamber.

Only the first two types of cavitation are of suitable intensity for chemical or physical processing. In the case of cavitation reactors, two aspects of cavity dynamics are of main importance, the maximum size reached by the cavity before its violent collapse and the life of the cavity. The maximum size reached by the cavity determines the magnitude of the pressure pulse produced on the collapse and hence the cavitation intensity that can be obtained in the system. The life of the cavity determines the distance traveled by the cavity from the point where it is generated before the collapse and hence it is a measure of the active volume of the reactor in which the actual cavitational effects are observed. The aim of the designer of the equipment should be to maximize both these quantities by suitable adjustment of the different parameters, including the methodology used for the generation of cavities (the type of cavity generated is a crucial parameter in deciding the intensity of the cavitation phenomena).

When the cavity is formed it contains vapor from the liquid medium or dissolved volatile reagents or gases. During the collapse, these vapors will be subjected to extreme conditions of high temperatures and pressures, causing molecules to fragment and generate highly reactive radical species. These radicals may then react either within the collapsing bubble or after migration into the bulk liquid. The collapse of the bubble also results in an inrush of the liquid to fill the void, producing shear forces in the surrounding bulk liquid that can break the chemical bonds of any materials, which are dissolved in the fluid or disturb the boundary layer, facilitating the transport.

When considering the reaction conditions for a cavitational process, the choice of the solvent and bulk operating temperature are thus significant factors and are often interrelated. Any increase in the solvent vapor pressure decreases the maximum bubble collapse temperature and pressure. Thus, for a reaction where cavitational collapse is the primary cause of the activation a low operating temperature is recommended, particularly if a low boiling solvent is used. Conversely, for a reaction requiring elevated temperatures, a high-boiling solvent is recommended. It is very important to decipher the controlling mechanism in the overall intensification of the chemical processing applications and then select the appropriate operating parameters. Sometimes, when only mass transfer is the rate-limiting step in deciding the overall rate of chemical reaction, the use of cavitational reactors may not be needed and similar effects can be achieved using mechanical agitation at higher speeds of rotation or by using a proper geometry/type of stirring device. Cavitation phenomena need to be effectively used for chemical processing applications limited by intrinsic kinetics or where greener chemical synthesis routes need to be established.

The cavitational activation in heterogeneous systems is mainly a consequence of the mechanical effects of cavitation. In a heterogeneous solid/liquid system, the collapse of the cavitation bubble results in significant structural and mechanical defects. Collapse near the surface produces an asymmetrical inrush of the fluid to fill the void, thereby forming a liquid jet targeted at the surface. This effect is equivalent to high-pressure/high-velocity liquid jets and is the reason why ultrasound is used for cleaning solid surfaces. These jets activate the solid catalyst and increase the mass transfer to the surface by the disruption of the interfacial boundary layers as well as by dislodging the material occupying the inactive sites. Collapse on the surface, particularly of powders, produces enough energy to cause fragmentation (even for finely divided metals). Thus, in this situation, ultrasound can increase the surface area for a reaction and provide additional activation through efficient mixing and enhanced mass transport.

In heterogeneous liquid/liquid reactions, cavitational collapse at or near the interface will cause disruption and mixing, resulting in the formation of very fine emulsions. When such emulsions are formed the surface area available for the reaction between the two phases is significantly increased, thus increasing the rates of reaction. Emulsions formed using cavitation are usually smaller in size and more stable than those obtained using conventional techniques, and often require little or no surfactant to maintain the stability. This is very beneficial, particularly in the case of phase-transfer catalyzed reactions or biphasic systems.

Cavitation can be used in chemical processing applications to obtain the following [151]:

- reaction time reduction;
- increase in reaction yield;
- use of less forcing conditions (temperature and pressure) than with the conventional routes;
- reduction in the induction period of the desired reaction;
- possible switching of the reaction pathways resulting in increased selectivity;
- increase in the effectiveness of the catalyst used in the reaction;
- initiation of the chemical reaction due to generation of highly reactive free radicals.

It is important to cite the following critical issues while selecting the application of cavitating conditions for intensification of chemical processing applications [151]:

1. For homogeneous reactions, only those reactions that proceed via radical or radical-ion intermediates are usually sensitive to the cavitational effect. This means that cavitation can intensify reactions proceeding through radicals whereas ionic reactions are not likely to be modified.

2. For heterogeneous reactions, reactions proceeding through ionic intermediates can also be stimulated by the mechanical effects of cavitation. However, in this case an appropriate balance needs to be maintained between the positive effects of cavitation and the associated costs of operation as sometimes the same effects might be achieved by using improved agitation conditions.

3. For heterogeneous reactions where mixed mechanisms, that is, radical and ionic, exist, cavitation can improve both mechanisms. In this case, if the two mechanisms lead to the same product an overall increase in the rate of reaction will be obtained, but if they lead to different product distributions, cavitational switching might occur due to enhancement in the radical pathway and the nature of reaction products is actually changed due to cavitational effects.

Therefore, the magnitudes of collapse pressures and temperatures as well as the number of free radicals generated at the end of cavitation events are strongly dependent on the operating parameters of the equipment, namely, the intensity and frequency of irradiation along with the geometrical arrangement of the transducers in the case of sonochemical reactors and the liquid phase physicochemical properties, which affect the initial size of the nuclei and the nucleation process. Table 3.9 summarizes the optimum operating conditions for sonochemical reactors. Gogate [151] reports in his review similar considerations on the optimum operating conditions for hydrodynamic cavitation reactors.

Table 3.9 Optimum operating conditions for sonochemical reactors.
Source: adapted from Gogate [151].

Property	Influences	Favorable conditions
Intensity of irradiation (range: $1–300\,W\,cm^{-2}$)	Number of cavities, collapse pressure of single cavity	Use power dissipation till an optimum value and over a wider area of irradiation
Frequency of irradiation (range: $20–200\,kHz$)	Collapse time of the cavity as well as final pressure/temperature pulse	Use enhanced frequencies till an optimum value
Liquid vapor pressure (range: $5–13\,kPa$ at room temperature)	Cavitation threshold, intensity of cavitation, rate of chemical reaction	Liquids with low vapor pressures
Viscosity (range: $1–6\,cP$) Surface tension (range: $0.03–0.072\,N\,m^{-1}$)	Transient threshold Size of the nuclei (cavitation threshold)	Low viscosity Low surface tension
Bulk liquid temperature (range: $30–70\,°C$)	Intensity of collapse, rate of the reaction, threshold/nucleation, almost all physical properties	Optimum value exits, generally lower temperatures are preferable
Dissolved gas: solubility	Gas content, nucleation, collapse phase	Low solubility
Dissolved gas: polytropic constant and thermal conductivity	Intensity of cavitation events	Gases with higher polytropic constant and lower thermal conductivity (monoatomic gases)
Geometry of the reactor	Number of cavitational events and cavitational activity distribution	Higher number of transducers of optimum shape so as to achieve uniform cavitational activity in the reactor

Gogate [151] also compares the energy efficiency of acoustic and hydrodynamic techniques with that of conventional reactors in the case of trans-esterification of vegetable oils. Cavitation can be successfully applied to trans-esterification reactions to afford a more than 90% yield of the product as per stoichiometry in a reaction time as low as 15 min. The technique hence appears to be very effective compared to the conventional approach. Hydrodynamic cavitation is about 40 times more energy efficient than acoustic cavitation and 160–400 times more efficient than the conventional agitation/heating/refluxing method.

For further process intensification, it is also possible to combine the use of microwave irradiation and sonochemistry [153, 154].

The use of acoustic energy to induce cavitational effects thus appears to be very effective for intensification of chemical processing operations, although a better understanding of the phenomena and their scaling-up is necessary to pass from laboratory-scale to industrial size applications. The future of sonochemical reactors lies in the design of multiple frequency transducer based reactors, whereas for hydrodynamic cavitation reactors the orifice plate type configuration appears to be most suitable. Hydrodynamic cavitation reactors offer more immediate potential for industrial-scale applications than the sonochemical reactors and the scale up of these reactors is comparatively easier. The operating costs of cavitational reactors, especially the sonochemical reactors are higher side than for conventional reactors, but the advantages obtained for a specific application (e.g., better product distribution ratios or global ambient operating conditions, etc.) should be weighed against the higher processing costs.

3.2.6
Energy of Flow

Owing to the analogous mechanism, Section 3.3.4 has already introduced the concept that the energy of liquid flow can be utilized to create cavitation for intensifying reactions and other operations. There are two main possible mechanisms:

- The liquid passes through a throttling valve, orifice plate or any other mechanical constriction (Figure 3.12a). If the pressure in *vena contracta* falls below the cavitation pressure (usually the vapor pressure of the medium), millions of microcavities will be generated. Those cavities will subsequently collapse as the liquid jet expands and pressure recovers.

- Use the so-called liquid whistle (Figure 3.12b), which is already applied in the food industry for homogenization and emulsification. The liquid is accelerated in a jet and then flows across a steel blade, which vibrates as liquid passes over it at high velocity. The frequency of those vibrations can be adjusted in such a way that cavitation is created. Large liquid volumes could be processed, but severe shortcomings are a low intensity of cavitation, high pumping costs and erosion of the blade.

As also mentioned in the previous section, hydrodynamic cavitation creates on average an order of magnitude higher cavitation yields than the acoustic cavitation. In

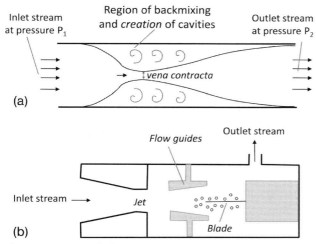

Inlet stream
at pressure P_1

Region of backmixing
and *creation* of cavities

Outlet stream
at pressure P_2

vena contracta

(a)

Outlet stream

Flow guides

Inlet stream →

Jet

Blade

(b)

Figure 3.12 Devices for hydrodynamic cavitation through an orifice (a) or in a liquid whistle (b). Source: adapted from Thompson and Doraiswamy [149].

addition, the processing volumes could be up to about 100-times larger than in conventional sonochemical reactors.

The energy of the supersonic shockwave presents another promising alternative method for intensification of the phase contacting and transport processes. In a supersonic shockwave reactor, a gas cooled to subpyrolysis temperature is expanded to supersonic speed by mixing with a supersonic flow of feedstock. The company Praxair has investigated a supersonic gas–liquid reactor for carrying out fast processes [155]. The energy of the supersonic shockwave is used to disperse gas into tiny micro-bubbles, thus enhancing the interfacial area for mass transfer by a factor of ten. The German company Messer Griesheim GmbH has patented and commercialized a supersonic nozzle for fluidized-bed applications [156]. The concept was subsequently applied on an industrial scale in a fluidized-bed reactor for iron sulfate decomposition at Bayer AG, increasing the capacity of the reactor by 124% [157]. The same technology has also been applied to sludge combustion reactors, increasing the throughput by approximately 40% [158]. DSM has also reported the use of a supersonic oxygen injection technology in large-scale fermentation processes.

3.3
Micro(structured)-Reactors

The role of micro(structured)-reactors for process intensification (PI) has been widely commented on in previous sections, as well as the strong interest on this sector of various innovative companies like DuPont, BASF, Merck, UOP, Shell, Schering, Degussa-Hüls, Bayer, and so on. However, due to the relevance of this area for PI some further comment is necessary; for a deeper understanding, several some books [33, 34, 75, 159] and specific reviews [37, 81, 82, 160, 161] are recommended.

Notably, two classes of microreactors exist, referring to applications in analysis, especially in the field of biochemistry and biology (e.g., continuous flow microfluidic devices, micro total analysis systems – μTAS, etc.) or chemical engineering and chemistry. Although these fields are distinct, there are clear overlaps and common areas of development.

3.3.1
Microreactor Materials and Fabrication Methods

Several materials and methods are available for fabrication of microreactors and microprocess components [162, 163]. Materials of construction [82] include:

- A special-purpose photo-structured glass called FOTURAN (available from Mikroglas Chemtech GmbH - www.mikroglas.com), which is based on lithium aluminum silicate and is especially useful for creating microchannels and related structures with high aspect ratios.

- Wafer-grade silicon, due to the significant micromachining experience developed in both the integrated circuit (IC) and MEMS industries. Silicon is an excellent material because of its large operating temperature range and its chemical inertness. A limit (besides cost), however, is that fabrication of integrated microprocess devices for safe operation at elevated pressures and temperatures is not feasible, or difficult, due to problems at the interface with the fluid inlet and exit ports.

- Plastics, which are at a low cost and are easily worked by micromachinery devices, but have a clear limit in operational temperature.

- Metals, which combine robust characteristics with relatively low costs, high thermal conductivity and good resistance to mechanical stresses. Microfabrication techniques for metals include mechanical micromachining, laser micromachining, wet chemical etching and selective laser melting [164]. The quality of the surface generated by mechanical micromachining is dependent on the metal or alloy type.

In terms of fabrication methods [82], the most commons are:

- LIGA (lithography, electroforming and molding), which is an electroplating technique for creating microstructures using a synchrotron radiation source.

- Micro-machining techniques (turning, sawing, embossing, punching and drilling processes based on precision engineering), which can be cost-effective for prototypes, and micro-milling machines [165].

- Bulk micromachining of monocrystalline materials, for example, silicon, using anisotropic wet chemical etching (including anisotropic etching of photosensitive glass), isotropic wet chemical etching, for example, of metal foils with a resist pattern, or dry etching processes using low pressure plasma or ion beams (reactive ion etching, reactive ion beam etching).

- Micromolding, for example, using mold inserts machined by precision techniques.

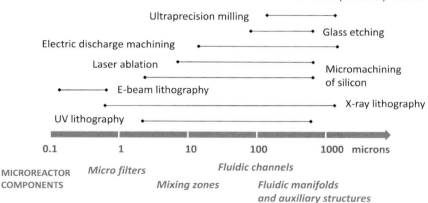

Figure 3.13 Ranking of fabrication technologies with respect to their accuracy and correlation to typical characteristic dimensions of microreactor components. Source: adapted from Ehrfeld *et al.* [159].

Usually, several technologies are combined in a process line for fabricating microreactors. In addition, auxiliary processes ranging from thin film technologies to mechanical surface modification are necessary. Figure 3.13 presents a summary of fabrication technologies with respect to their accuracy and correlation to typical characteristic dimensions of current microreactor components. In addition to applicability range, other factors to consider in the choice of the fabrication technology are: (i) process costs, (ii) process time, (iii) accuracy, (iv) reliability, (v) material choice and (vi) access.

3.3.2
Microreactors for Catalytic Gas-Phase Reactions

A major problem in using microstructured reactors for heterogeneously catalyzed gas-phase reactions is how to introduce the catalytic active phase. The possibilities are to (i) introduce the solid catalyst in the form of a micro-sized packed bed, (ii) use a catalytic wall reactor or (iii) to use novel designs. Kiwi-Minsker and Renken [160] have discussed in detail these alternatives.

The micro-packed bed reactor is the easiest way to incorporate the catalyst into a microstructured reactor (MSR), by filling the microchannels with catalyst powder. The advantage is the use of a conventional catalyst, although the catalyst particles should have diameters in the range 30–70 µm. The MSR operates with laminar flow. Problems arise from easy blocking of the microchannels, and blow out of the catalyst. The concept has been applied to methanol reforming over Pd/ZnO in a microscale fuel processor [166, 167] and to the on-demand production of phosgene in a MSR [168]. In the latter case, the microreactor was fabricated out of single silicon crystalline wafers (20 mm long, 625 µm wide and 300 µm deep reaction channel) that were then capped with Pyrex glass. The catalyst (based on activated carbon) was used as particles with diameter of 53–73 µm. Using a stoichiometric mixture of CO and Cl_2 and a temperature of 200 °C complete conversion was achieved, corresponding to a productivity of $0.4 \, g \, h^{-1}$ ($3.5 \, kg \, yr^{-1}$) phosgene from a single channel.

To avoid high-pressure drop and clogging problems in randomly packed micro-structured reactors, multichannel reactors with catalytically active walls were proposed. The main problem is how to deposit a uniform catalyst layer in the microchannels. The thickness and porosity of the catalyst layer should also be enough to guarantee an adequate surface area. It is also possible to use methods of *in situ* growth of an oxide layer (e.g., by anodic oxidation of a metal substrate [169]) to form a washcoat of sufficient thickness to deposit an active component (metal particles). Suzuki *et al.* [170] have used this method to prepare Pt supported on nanoporous alumina obtained by anodic oxidation and integrate it into a microcatalytic combustor. Zeolite-coated microchannel reactors could be also prepared and they demonstrate higher productivity per mass of catalyst than conventional packed beds [171]. Also, a MSR where the microchannels are coated by a carbon layer, could be prepared [172].

An alternative to filling or coating with a catalyst layer the microchannels, with the related problems of avoiding maldistribution, which leads to a broad residence time distribution (RTD), is to create the microchannels between the void space left from a close packing of parallel filaments or wires. This novel MSR concept has been applied for the oxidative steam reforming of methanol [173]. Thin linear metallic wires, with diameters in the millimeter range, were close packed and introduced into a "macro" tubular reactor. The catalyst layer was grown on the external surface of these wires by thermal treatment.

3.3.3
Microreactors for Catalytic Multiphase Systems

In multiphase reactions containing liquids, mixing of the reactants and the interfacial mass transfer is of primary importance. Laminar flow prevails in microreactors and the diffusion coefficients in liquids are four orders of magnitude smaller than for gases. An efficient micromixing and the modalities to introduce the different reactants are thus important issues for effective reactor performance. The design of micromixers has been discussed in detail by Hessel *et al.* [174], while microreactor designs for multiphase contacting has been reviewed by Doku *et al.* [161].

When the reaction involves two immiscible liquids (e.g., an organic-aqueous systems), a longitudinal interface could be established (Figure 3.14a), but the formation of a multiple-pulse (segmented) configuration (transverse contact interfaces; Taylor-slug-flow) is more likely (Figure 3.14b). The latter is also typically present in gas–liquid microreactor systems (Figure 3.14c). However, this view is a simplification of the real problem, because the surface tension and wall adhesion should also be considered. When the latter forces are relevant and flow velocities are low, a parallel flow with a wavy interface is present (Figure 3.14a). At higher flow velocities, the waves grow and finally they are as big as a slug. However, deformed interfaces could be present. The presence of surface roughness in the microchannel (e.g., due to the presence of the catalyst layer) further complicates the picture.

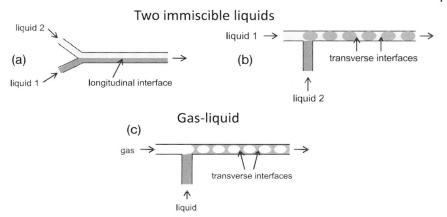

Figure 3.14 (a) Cocurrent mobilization of two immiscible liquids, and established longitudinal contact interface in a microchannel; (b) segmented-pulse injection of one liquid into the main flow of another immiscible liquid, and established transverse contact interfaces in a microchannel; (c) cocurrent continuous mobilization of gas and pulse injection of liquid stream, leading to single-line gas microbubble train in a microchannel. Source: adapted from Doku et al. [161].

In the slug flow, mass transfer takes place by two mechanisms: convection within the slug and diffusion between two slugs. The convection is due to the internal circulations within each slug because of the shearing action between slug axis and capillary wall, while the diffusion is because of concentration gradients between two consecutive slugs. The first phenomenon depends on the physical properties of fluids, slug geometry and flow velocity, while the latter depends on the interfacial area available for transfer and the concentration gradient between two slugs.

Mixing and mass/heat transfer are thus highly dependent on the specific characteristics of the flow, which in turn depends on the specific modalities of introduction of the different reactants, the characteristics of the fluids, their flow rates, and the features of the microchannels. Although significant progress has been made in understanding the fluidodynamic in multiphase MSR and their modeling, this is an area that deserves further study. A good overview of recent advances has been given by Hardt [175].

3.3.4
Industrial Microreactors for Fine and Functional Chemistry

Microreactor technology is a tool used by most large chemical and pharmaceutical companies and also by some SMEs. Various examples are discussed by Hessel et al. [176], based on the collaboration of the Institute for Micro-Technology at Mainz (Germany) with various companies. We recall here some relevant cases to further evidence how microreactors are not more only at a laboratory-scale stage of development, even though most of the applications are still on a small-size scale.

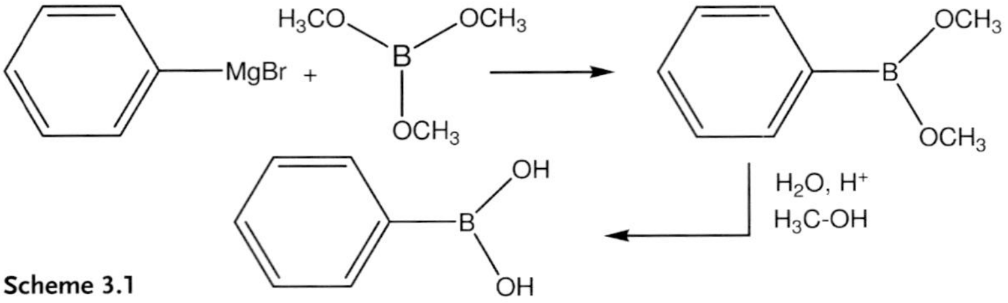

Scheme 3.1

3.3.4.1 Phenyl Boronic Acid Synthesis (Clariant)

Many organometallic reactions suffer from insufficient mixing, because often the mixing times in conventional mixers are longer than the conversion time. Therefore, the reactions are made under non-stoichiometric and/or in the presence of changing concentration profiles, which promote consecutive reactions. As a consequence, organometallic reactions are often carried out under cryogenic conditions to get acceptable selectivity, for example, the reaction is slowed down. This requires capital investment for cooling utilities and increases the process energy consumption and costs. It is instead possible to perform the reaction at room temperature using a microreactor coupled to an efficient micromixer.

This concept has been applied in the phenyl boronic acid synthesis (Scheme 3.1) [177].

Phenyl magnesium bromide and boronic acid trimethyl ester react to give phenyl-boronic acid with high selectivity (about 90%) even at room temperature, which saves energy costs and the respective CAPEX investment [177]. The yield was about 25% higher with respect to industrial batch production. Purity of the crude product could be enhanced by about 10%, thereby allowing purification by favorable crystallization only, avoiding thus the distillation steps needed in the conventional process.

3.3.4.2 Azo Pigment Yellow 12 (Trust Chem/Hangzhou)

Micromixers provide very fast mixing, down to milliseconds or even below [174]. This is an important aspect in the presence of very fast precipitations such as in the Yellow 12 manufacture. In fact, the seed formation and agglomeration processes are important aspects to obtain narrow-sized, morphologically uniform crystals. However, fouling of the mixer's tiny structures may plug the whole system and leads to plant shut down. By using a micromixer it is possible to decrease fouling problems, due to better geometry and shorter residence times. It is thus possible, by coupling the micromixer to a microreactor, to obtain finer particles with more uniform size distribution for the commercial azo pigment Yellow 12 (Scheme 3.2) [178].

Finer and more uniform sized particles improve the optical properties, such as the glossiness and transparency, improving thus the commercial value of the product.

3.3.4.3 Hydrogen Peroxide Synthesis (UOP)

Microreactors, due to the high wall-to-volume ratio and the absence of runaway, allow to safe operation within the explosion limits [179], thus enabling an increase in process productivity. A relevant case is the direct synthesis of hydrogen peroxide [180], for which

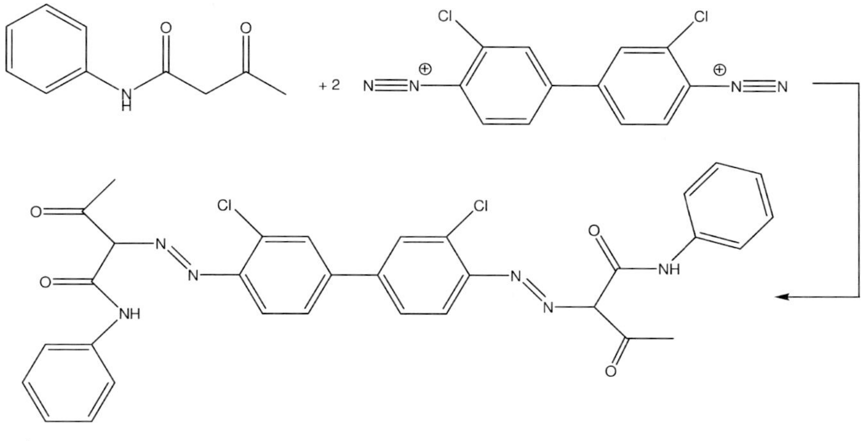

Scheme 3.2

virtually the whole concentration and operational range is in the explosive regime:

$$H_2 + O_2 \rightarrow H_2O_2$$

In addition to the inherently safe features of microchannel reactors, they allow rapid catalyst testing and have the option of the same catalyst type use, for example, washcoated, throughout the whole development cycle, up to pilot and production. Furthermore, mass transfer is improved through the increase in gas–liquid interfaces, and thermal control is also better due to larger exchange surfaces.

Using UOP process specifications, a space–time yield of 2 g hydrogen peroxide per $g_{catalyst}$-h was achieved. In addition, operation at only 20 bar were possible instead of the 100 bar required in batch-type reactors, and the use of smaller oxygen/hydrogen ratios allowed a decrease of material use. Selectivity as high as 85% at 90% conversion was achieved (at an oxygen/hydrogen ratio of 1.5–3). These laboratory experiments were followed by pilot tests by UOP, resulting in a basic engineering design for the production of about 150 000 tons H_2O_2 per year [83]. Moreover, this demonstrates that the performance of microstructured units within a plant and process may not necessarily be restricted to small-capacity production.

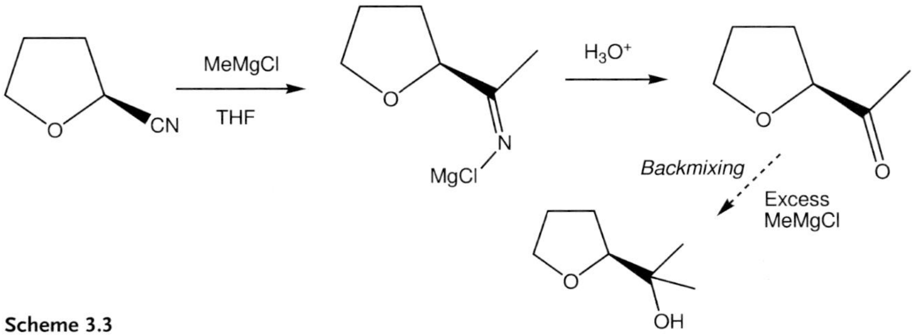

Scheme 3.3

3.3.4.4 *(S)*-2-Acetyltetrahydrofuran Synthesis (SK Corporation/Daejeon)

In the synthesis of *(S)*-2-acetyltetrahydrofuran, the Grignard reagent MeMgCl is very reactive and not easy to handle on a large scale. This results in safety and hazard issues at an industrial scale. In addition, selectivity issues result from over-alkylation to the tertiary alcohol, an undesired consecutive reaction. This impurity level must be kept <0.2% (Scheme 3.3).

Finally, chirality conservation must be maintained, because the α-hydrogen of the reactant is unstable under basic reaction conditions. Therefore, a small degree of racemization occurs, which needs to be minimized.

Using a microreactor the impurity by over-alkylation was 0.18%, while the batch impurity was 1.56% [181]. This was possible due the lower back-mixing in the microflow system. The optical purity of the microreactor product was 98.4% as compared with 97.9% at batch level.

References

1 Reay, D., Ramshaw, C. and Harvey, A. (2008) *Process Intensification - Engineering for Efficiency, Sustainability and Flexibility*, Elsevier (Butterworth Heinemann) Pub., Amsterdam.

2 Ramshaw, C. (1985) *Chem. Eng.*, **415**, 30.

3 Hendershot, D.C. (2000) *Chem. Eng. Prog.*, **96**, 35.

4 Stankiewicz, A.I. and Moulijn, J.A. (2000) *Chem. Eng. Prog.*, **96**, 22.

5 Ramshaw, C. (1983) *Chem. Eng.*, **389**, 13.

6 Zheng, C., Guo, K., Song, Y., Zhou, X., Al, D., Xin, Z. and Gardner, N.C. (1997) *Proceedings of the 2nd International Conference on Process Intensification in Practice* (ed. J. Semel), BHR Group Conference Series, publication no. **28**, BHR Group, London, p. 273.

7 Charpentier, J.-C. (2007) *Chem. Eng. J.*, **134**, 84; Charpentier, J.-C. (2007) *Ind. Eng. Chem. Res.*, **46**, 3465.

8 Tsouris, C. and Porcelli, J.V. (2003) *Chem. Eng. Prog.*, **99**, 50.

9 Ramshaw, C. (1983) *Chem. Eng.*, **389**, 13.

10 Ramshaw, C. and Mallinson, R.H. (1984) European Patent, 0,002,568.

11 Meili, A. (1997) *Proceedings of the 2nd International Conference on Process Intensification in Practice* (ed. J. Semel), BHR Group Conference Series, publication no. **28**, BHR Group, London.

12 Oxley, P., Brechtelsbauer, C., Ricard, F., Lewis, N. and Ramshaw, C. (2000) *Ind. Eng. Chem. Res.*, **39**, 2175.

13 Siirola, J.J. (1995) *AIChE Symp. Series*, **91**, 222.

14 Trent, D. and Tirtowidjojo, D. (2001) *Proceedings of the 4th International Conference on Process Intensification for the Chemical Industry* (ed. M. Gough), BHR Group Conference Series, BHR Group Ltd, Cranfield, UK.

15 Becht, S. and Hahn, H. (April 23–27, 2006) AIChE Spring National Meeting, Conference Proceedings, Orlando, FL, USA, p. P41655/1.

16 Doble, M. (2008) *Chem. Eng. Prog.*, **104**, 33.

17 Fan, X., Chen, H., Ding, Y., Plucinski, P.K. and Lapkin, A.A. (2008) *Green Chem.*, **10**, 670.

18 Gogate, P.R. (2008) *Chem. Eng. Proc.*, **47**, 515.

19 Wu, J., Graham, L.J. and Noui-Mehidi, N. (2007) *J. Chem. Eng. Jpn.*, **40**, 890.

20 VandenBussche, K.M. (2007) in *Micro Instrumentation* (eds M.V. Koch, K.

VandenBussche, and R.W. Chrisman),
Wiley-VCH Verlag, Weinheim, p. 43.

21 Stankiewicz, A. (2006) *Chem. Eng. Res. Design*, **84** (A7), 511.

22 Rubin, A.E., Tummala, S., Both, D.A., Wang, C. and Delaney, E.J. (2006) *Chem. Rev.*, **106**, 2794.

23 Hahn, A. (2000) *Proceedings of the 16th World Petroleum Congress, Calgary, Alberta Canada, June 11–15*, Institute of Petroleum, p. 141.

24 Akay, G.(July 10–14, 2005) *Proceedings of the 7th World Congress of Chemical Engineering, Glasgow, UK*, Institution of Chemical Engineers, Rugby, UK, p. 83240/1.

25 Choe, W.-S., Nian, R. and Lai, W.-B. (2005) *Chem. Eng. Sci.*, **61** (3), 886.

26 Akay, G., Bokhari, M.A., Byron, V.J. and Dogru, M. (2005) *Chemical Engineering* (eds M.A. Galan and E. Martin del Valle), John Wiley & Sons, Ltd, Chichester, UK, p. 171.

27 Kreutzer, M.T., Kapteijn, F., Moulijn, J.A., Ebrahimi, S., Kleerebezem, R. and van Loosdrecht, M.C.M. (2005) *Ind. Eng. Chem. Res.*, **44**, 9646.

28 Charpentier, J.-C. (2005) *Chem. Eng. J.*, **107**, 3.

29 Dautzenberg, F.M. (2004) *Catal. Rev. Sci. Eng.*, **46**, 335.

30 Drioli, E., Curcio, E. and Di Profio, G. (2005) *Chem. Eng. Res. Des.*, **83**, 223.

31 Stankiewicz, A. and Moulijn, J.A. (eds) (2004) *Re-Engineering the Chemical Processing Plant*, Marcel Dekker Inc., New York.

32 Keil, F.J. (ed.) (2007) *Modeling of Process Intensification*, Wiley-VCH Verlag, Weinheim.

33 Hessel, V., Hardt, S. and Löwe, H. (2004) *Chemical Micro Process Engineering, Fundamentals, Modelling and Reactions*, Wiley-VCH Verlag, Weinheim.

34 Wirth, T. (2008) *Microreactors in Organic Synthesis and Catalysis*, Wiley-VCH Verlag, Weinheim.

35 Koch, M.V., VandenBussche, K.M. and Chrisman, R.W. (eds) (2007) *Micro Instrumentation: for High Throughput Experimentation and Process Intensification*, Wiley-VCH Verlag, Weinheim.

36 Jachuck, R. (2006) *Process Intensification in the Chemical and Related Industries*, Wiley-Blackwell, Oxford, UK.

36 Jachuck, R. (2006) *Process Intensification in the Chemical and Related Industries*, Wiley (Blackwell), Oxford (UK).

37 Wang, Y. and Holladay, J.D. (2005) *Microreactor Technology and Process Intensifiation, ACS Symposium Series No. 914*, American Chemical Society, Washington DC.

38 Marcano, J.G.S. and Tsotsis, T.T. (2002) *Catalytic Membranes and Membrane Reactors*, Wiley-VCH Verlag, Weinheim.

39 Costello, R.C. (2006) *Chem. Process.* http://www.chemicalprocessing.com/articles/2006/176.html.

40 Costello, R.C. (2008) *Chem. Process.* http://www.chemicalprocessing.com/articles/2008/166.html.

41 Malone, M.F. and Doherty, M.F. (2000) *Ind. Eng. Chem. Res.*, **39**, 3953.

42 Taylor, R. and Krishna, R. (2000) *Chem. Eng. Sci.*, **55**, 5183.

43 Guo, Z. and Lee, J.W. (2004) *AICHE J.*, **50**, 1751.

44 Gorak, A. and Hoffmann, A. (2001) *AICHE J.*, **47**, 1067.

45 Xu, Y., Chuang, T. and Sanger, A.R. (2002) *Chem. Eng. Res. Des.*, **80**, 686.

46 Kaibel, G., Kons, G., Schoenmakers, H. and Schwab, E. (2002) DGMK Tagungsbericht. 2002-4 (Proceedings of the DGMK-Conference "Chances for Innovative Processes at the Interface between Refining and Petrochemistry", 2002), p. 175.

47 Schoenmakers, H.G. and Bessling, B. (2003) *Chem. Eng. Process.*, **42**, 145.

48 Podrebarac, G.G., Ng, F.T.T. and Rempel, G.L. (1997) *CHEMTECH*, **27**, 37.

49 Omota, F., Dimian, A. and Bliek, A. (2003) *Chem. Eng. Sci.*, **58**, 3159 and 3175.

50 Drioli, E., Curcio, E. and Di Profio, G. (2005) *Chem. Eng. Res. Des.*, **83**, 223.

51 Nyström, M. and Rios, G.M.(eds) (2005) *Chem. Eng. Res. Des.*, **83** (special issue), 221.

52 Matros, Y.Sh. and Bunimovich, G.A. (1996) *Catal. Rev.*, **38**, 1.

53 Kolios, G., Frauhammer, J. and Eigenberger, G. (2000) *Chem. Eng. Sci.*, **55**, 5945.

54 Dautzenberg, F.M. and Mukherjee, M. (2001) *Chem. Eng. Sci.*, **56**, 251.

55 Liu, T., Gepert, V. and Veser, G. (2005) *Chem. Eng. Res. Des.*, **83**, 611.

56 Mitri, A., Neumann, D., Liu, T. and Veser, G. (2004) *Chem. Eng. Sci.*, **59**, 5527.

57 Aida, T. and Silveston, P.L. (2005) *Cyclic Separating Reactors*, Blackwell Ltd Pub., Wiley Interscience.

58 Carvill, B.T., Hufton, J.R., Anand, M. and Sircar, S. (1996) *AIChE J.*, **42**, 2765.

59 Agar, D.W. and Ruppel, W. (1988) *Chem. Eng. Sci.*, **43**, 2073.

60 Haure, P.M., Hudgins, R.R. and Silveston, P.L. (1989) *AIChE J*, **35**, 1437.

61 Rambeau, G. and Amariglio, H. (1981) *Appl. Catal.*, **1**, 291.

62 Harrison, D.P. and aPeng, Z.-Y. (2003) *Int. J. Chem. React. Eng.*, **1**, A37.

63 Centi, G. and Perathoner, S. (2007) *Stud. Surf. Sci. Catal.*, **171**, 1.

64 Nigam, K.D.P. and Larachi, F. (2005) *Chem. Eng. Sci.*, **60**, 5880.

65 Dudukovic, M., Larachi, F. and Mills, P.L. (2002) *Catal. Rev. – Sci. Eng.*, **44**, 123.

66 Boelhouwer, J.G., Piepers, H.W. and Drinkenburg, A.A.H. (2001) *Chem. Eng. Sci.*, **56**, 1181.

67 Boelhouwer, J.G., Piepers, H.W. and Drinkenburg, A.A.H. (2002) *Chem. Eng. Sci.*, **57**, 4865.

68 Wu, R., McCready, M.J. and Varma, A. (1999) *Catal. Today*, **48**, 195.

69 Iliuta, I. and Larachi, F. (2005) *Chem. Eng. Sci.*, **60**, 6217.

70 Gogate, P.R. (2008) *Chem. Eng. Proc.: Process Intensification*, **47**, 515.

71 Toukoniitty, B., Kuusisto, J., Mikkola, J.-P., Salmi, T. and Murzin, D.Yu. (2005) *Ind. Eng. Chem. Res.*, **44**, 9370.

72 Iliuta, I. and Larachi, F. (2004) *Chem. Eng. Process.*, **43**, 141.

73 Iliuta, I. and Larachi, F. (2003) *AICHE J.*, **49**, 1525.

74 Charpentier, J.-C. (2005) *Chem. Eng. Technol.*, **28**, 255.

75 Hessel, V., Löwe, H., Müller, A. and Kolb, G. (2005) *Chemical Micro Process Engineering, Processing and Plants*, Wiley-VCH Verlag, Weinheim.

76 Yarin, L.P., Mosyak, A. and Hetsroni, G. (2009) *Fluid Flow, Heat Transfer and Boiling in Micro-Channels*, Springer-Verlag, Heidelberg.

77 Zhang, Z. (2007) *Nano/Micro-Scale Heat Transfer*, Nanoscience and Technology Series, McGraw-Hill, New York.

78 Kandlikar, S.G., Garimella, S., Li, D., Colin, S. and King, M.R. (2005) *Heat Transfer and Fluid Flow in Minichannels and Microchannels*, Elsevier Science & Technology, The Netherlands

79 Kariandakis, G.E. and Beskok, A. (2002) *Microflows Fundamental and Simulation*, Springer-Verlag, Berlin.

80 Jiang, J., Hao, Y. and Shi, M. (2008) *Heat Transfer - Asian Res.*, **37**, 197.

81 Hessel, V., Knobloch, C. and Löwe, H. (2008) *Recent Pat. Chem. Eng.*, **1**, 1.

82 Mills, P.L., Quiram, D.J. and Ryley, J.F. (2007) *Chem. Eng. Sci.*, **62**, 6992.

83 Pennemann, H., Hessel, V. and Lowe, H. (2004) *Chem. Eng. Sci.*, **59**, 4789.

84 Pennemann, H., Watts, P., Haswell, S.J., Hessel, V. and Lowe, H. (2004) *Org. Process Res. Dev.*, **8**, 422.

85 Mae, K. (2007) *Chem. Eng. Sci.*, **62**, 4842.

86 Tonkovich, A., Kuhlmann, D., Rogers, A., Mc Daniel, J., Fizgerald, S., Arora, R. and Yuschak, T. (2005) *Chem. Eng. Res. Des.*, **83**, 634.

87 Pfeifer, P., Bohn, L., Görke, O., Haas-Santo, K. and Schubert, K. (2005) *Chem. Eng. Technol.*, **28**, 474.

88 Gaviilidis, A., Angeli, P., Cao, E., Yeong, K.K. and Wan, Y.S.S. (2002) *Chem. Eng. Res. Des.*, **80**, 3.

89 Werner, B., Hessel, V. and Löb, P. (2005) *Chem. Eng. Technol.*, **28**, 401.

90 Luo, L., Fan, Y. and Tondeur, D. (2007) *Int. J. Energ. Res.*, **31**, 1266.

91 Anxionnaz, Z., Cabassud, M., Gourdon, C. and Tochon, P. (2008) *Chem. Eng. Process.: Process, Intensification*, **47**, 202.

92 Reay, D. (2008) *Appl. Thermal Eng.*, **28**, 201.

93 Perez-Ramirez, J. and Vigeland, B. (2005) *Catal. Today*, **105**, 436.

94 Reay, D.(21st November, 2007) PI & carbon reductions. Presented at the 15th PIN (Process Intensification Network) Meeting, Cranfield, UK.

95 US Department of Energy – Energy Efficiency and Renewable Energy (2007) See www.eree.energy.gov/industry/saveenergynow.

96 Dudukovic, M.P. (2009) *Chem. Eng. Commun.*, **196**, 252.

97 Reay, D., Ramshaw, C. and Harvey, A. (2008) Process intensification – an overview, in *Process Intensification (Engineering for Efficiency, Sustainability and Flexibility)* (eds D. Reay, C. Ramshaw, and A. Harvey), Elsevier Pub, Amsterdam, Ch. 2, p. 21.

98 Stankiewicz, A. (2006) *Chem. Eng. Res. Des.*, **84**, 511.

99 Stankiewicz, A. (2007) *Ind. Eng. Chem. Res.*, **46**, 4232.

100 Reay, D., Ramshaw, C. and Harvey, A. (2008) A brief history of process intensification, in *Process Intensification (Engineering for Efficiency, Sustainability and Flexibility)* (eds D. Reay, C. Ramshaw, and A. Harvey), Elsevier Pub, Amsterdam, Ch. 1, p. 1.

101 Lin, C.-C., Ho, T.-J. and Liu, W.-T. (2002) *J. Chem. Eng. Jpn.*, **35**, 1298.

102 Rao, D.P., Bhowal, A. and Goswami, P.S. (2004) *Ind. Eng. Chem. Res.*, **43**, 1150.

103 Zheng, C., Guo, K., Song, Y., Zhou, X. and Ai, D. (1997) Industrial practice of HIGRAVITEC in water deaeration, in *Process Intensification in Practice: Applications and Opportunities* (ed. J. Semel), Mechanical Engineering Pub. Ltd, Edmunds, UK, pp. 273.

104 Trent, D. (2004) Chemical processing in high-gravity fields, in *Re-Engineering the Chemical Processing Plant: Process Intensification* (eds A. Stankiewicz and J.A. Moulijn), Marcel Dekker Inc., New York, p. 33.

105 Ramshaw, C. (2004) The spinning disc reactor, in *Re-Engineering the Chemical Processing Plant: Process Intensification* (eds A. Stankiewicz and J.A. Moulijn), Marcel Dekker Inc., New York, pp. 69.

106 Oxley, P., Brechtelsbauer, C., Ricard, F., Lewis, N. and Ramshaw, C. (2000) *Ind. Eng. Chem. Res.*, **39**, 2175.

107 Brechtels-Bauer, C.M.H. and Oxley, P. (2001) Process for epoxidising substituted cyclohexanones, Patent WO 01/14357.

108 Burns, M.E., Gibson, M.S. and York, D.W. (2002) Process for reacting carboxylic acids and esters, Patent WO 02/18328.

109 Ptasinski, K.J. and Kerkhof, P.J.A.M. (1992) *Sep. Sci. Technol.*, **27**, 995.

110 Oliver Kappe, C. (2008) *Chem. Soc. Rev.*, **37**, 1127.

111 Baxendale, I.R., Hayward, J.J. and Ley, S.V. (2007) *Comb. Chem. High Throughput Screen.*, **10**, 802.

112 Man, A.K. and Shahidan, R. (2007) *J. Macromol. Sci., Part A: Pure Appl. Chem.*, **44**, 651.

113 Xu, J. (2007) *Prog. Chem.*, **19**, 700.

114 Lévêque, J.-M. and Cravotto, G. (2006) *Chimia*, **60**, 313.

115 Dallinger, D. and Kappe, C.O. (2007) *Chem. Rev.*, **107**, 2563.

116 Nüchter, M., Ondruschka, B., Bonrath, W. and Gum, A. (2004) *Green Chem.*, **6**, 128.

117 Jacob, J., Chia, L.H.L. and Boey, F.Y.C. (1995) *J. Mater. Sci.*, **30**, 5321.

118 Howarth, P. and Lockwood, M. (2004) *Chem. Eng.*, **756**, 291.

119 Kappe, C.O. (2004) *Angew. Chem. Int. Ed.*, **43**, 6250.

120 De la Hoz, A., Díaz-Ortiz, A. and Moreno, A. (2004) *Chem. Soc. Rev.*, **34**, 164.

121 Loupy, A., Petit, A., Hamelin, J., Texier-Boullet, F., Jacquault, P. and Mathé, D. (1998) *Synthesis*, **9**, 1213.

122 Will, H., Scholz, P. and Ondruschka, B. (2004) *Chem. Eng. Technol.*, **27**, 113.

123 Armstrong, B. (1995) Microwave distillation apparatus, US Patent 5,711,857.

124 Van Gerven, T., Mul, G., Moulijn, J. and Stankiewicz, A. (2007) *Chem. Eng. Process.*, **46**, 78.

125 Hoffmann, N. (2008) *Chem. Rev.*, **108**, 1052.

126 Balzani, V., Credi, A. and Venturi, M. (2008) *ChemSusChem*, **1**, 26.

127 de Lasa, H., Serrano, B. and Salaices, M. (2005) *Photocatalytic Reaction Engineering*, Springer, New York.

128 Ciana, C.L. and Bochet, C.G. (2007) *Chimia*, **61**, 650.

129 Hook, B.D.A., Dohle, W., Hirst, P.R., Pickworth, M., Berry, M.B. and Booker-Milburn, K.I. (2005) *J. Org. Chem.*, **70**, 7558.

130 Coyle, E.E. and Oelgemöller, M. (2008) *Photochem. Photobiol. Sci.*, **7**, 1313.

131 Jähnisch, K., Hessel, V., Löwe, H. and Baerns, M. (2004) *Angew. Chem. Int. Ed.*, **43**, 406.

132 Barthe, P.J., Letourneur, D.H., Themont, J.P. and Woehl, P. (2004) Method and microfluidic reactor for photocatalysis, Patent B01J19/00.

133 Takei, G., Kitamori, T. and Kim, H.B. (2005) *Catal. Commun.*, **6**, 357.

134 Lapkin, A.A., Boddu, V.M., Aliev, G.N., Goller, B., Polisski, S. and Kovalev, D. (2008) *Chem. Eng. J.*, **136**, 331.

135 Chen, H.W., Ku, Y. and Irawan, A. (2007) *Chemosphere*, **69**, 184.

136 Chen, D.H., Ye, X.J. and Li, K.Y. (2005) *Chem. Eng. Technol.*, **28**, 95.

137 Gorges, R., Meyer, S. and Kreisel, G. (2004) *J. Photochem. Photobiol. A: Chem.*, **167**, 95.

138 Lu, H., Schmidt, M.A. and Jensen, K.F. (2001) *Lab Chip*, **1**, 22.

139 Gole, J.L., Fedorov, A., Hesketh, P. and Burda, C. (2004) *Phys. Stat. Sol.*, **1**, 188.

140 Gole, J., Burda, C., Fedorov, A. and White, M. (2003) *Rev. Adv. Mater. Sci.*, **5**, 26.

141 Lu, H., Schmidt, M.A. and Jensen, K.F. (2001) *Lab. Chip.*, **1**, 22.

142 Kreisel, G., Meyer, S., Tietze, D., Fidler, T., Gorges, R., Kirsch, A., Schäfer, B. and Rau, S. (2007) *Chem.-Ing.-Tech.*, **79**, 153.

143 Nakamura, H., Li, X.Y., Wang, H.Z., Uehara, M., Miyazaki, M., Shimizu, H. and Maeda, H. (2004) *Chem. Eng. J.*, **101**, 261.

144 Matsushita, Y., Kumada, S., Wakabayashi, K., Sakeda, K. and Ichimura, T. (2006) *Chem. Lett.*, **35**, 410.

145 Gogate, P.R. and Pandit, A.B. (2001) *Rev. Chem. Eng.*, **17**, 1.

146 Gogate, P.R. and Pandit, A.B. (2004) *Ultrason. Sonochem.*, **11**, 105.

147 Gogate, P.R. and Pandit, A.B. (2005) *Ultrason. Sonochem.*, **12**, 21.

148 Gogate, P.R., Wilhelm, A.M. and Pandit, A.B. (2003) *Ultrason. Sonochem.*, **10**, 325.

149 Thompson, L.H. and Doraiswamy, L.K. (1999) *Ind. Eng. Chem. Res.*, **38**, 1215.

150 Kumar, A., Gogate, P.R., Pandit, A.B., Delmas, H. and Wilhelm, A.M. (2004) *Ind. Eng. Chem. Res.*, **43**, 1812.

151 Gogate, P.R. (2008) *Chem. Eng. Process: Process Intensification*, **47**, 515.

152 Lindley, J. and Mason, T.J. (1987) *Chem. Soc. Rev.*, **16**, 275.

153 Trotta, F., Martina, K., Robaldo, B., Barge, A. and Cravotto, G. (2007) *J. Inclusion Phenom. Macrocyclic Chem.*, **57**, 3.

154 Cravotto, G., Di Carlo, S., Curini, M., Tumiatti, V. and Rogerro, C. (2007) *J. Chem. Technol. Biotechnol.*, **82**, 205.

155 Cheng, A.T.Y. (1999) Process for accelerating fast reactions using high intensity plug flow tubular reactors, Patent EP 0995489.

156 Gross, G. (1998) Verfahren und Vorrichtung zur Umwandlung von Schwefelwasserstoff in elementaren Schwefel, Patent DE 197 18 261; Gross, G. (1998) Verfahren und Vorrichtung zur Umwandlung von Schwefelwasserstoff in elementaren Schwefel, Patent DE 197 22 382.

157 Gross, G. (2000) Supersonic oxygen injection doubles the capacity of fluidized-bed reactors, presented at ACHEMA

2000. Abstracts of the Lecture Groups Chemical Engineering and Reaction Technology, Dechema, Frankfurt am Main, Germany, pp. 161–162.

158 Gross, G. and Ludwig, P. (2003) *Chem.-Anlagen, Verfahren*, **36**, 84.

159 Ehrfeld, W., Hessel, V. and Löwe, H. (2000) *Microreactors (New Technology for Modern Chemistry)*, Wiley-VCH Verlag, Weinheim.

160 Kiwi-Minsker, L. and Renken, A. (2005) *Catal. Today*, **110**, 2.

161 Doku, G.N., Verboom, W., Reinhoudt, D.N. and van den Berg, A. (2005) *Tetrahedron*, **61**, 2733.

162 Madou, M.J. (2002) *Fundamentals of Microfabrication – The Science of Miniaturization*, CRC Press, Boca Raton, FL.

163 Allan, J.J. (2005) *Micro Electro Mechanical System Design*, CRC Press, Boca Raton, FL.

164 Brandner, J.J., Gietzelt, T., Henning, T., Kraut, M., Mortiz, H. and Pfleging, W. (2006) Microfabrication in metals and polymers, in *Micro Process Engineering: Fundamentals, Devices, Fabrication, and Applications* (eds N. Kockmann, O. Brand, G.K. Fedder, C. Hierold, J.G. Korvink, and O. Tabata), Wiley-VCH Verlag, Weinheim, p. 529.

165 Denkena, B., Hoffmeister, H.-W., Reichstein, M., Illenseer, S. and Hlavac, M. (2006) *Microsystem Technol.*, **12**, 659.

166 Cao, C., Xia, G., Holladay, J., Jones, E. and Wang, Y. (2004) *Appl. Catal. A: Gen.*, **262**, 19.

167 Cao, C.S., Wang, Y., Holladay, J.D., Jones, E.O. and Palo, D.R. (2005) *AIChE J.*, **51**, 982.

168 Ajmera, S.K., Losey, M.W., Jensen, K.F. and Schmidt, M.A. (2001) *AIChE J.*, **47**, 1639.

169 Ganley, J.C., Riechmann, K.L., Seebauer, E.G. and Masel, R.I. (2004) *J. Catal.*, **227**, 26.

170 Suzuki, Y., Saito, J. and Kasagi, N. (2004) *JSME Int. J. B-Fluid T.*, **47**, 522.

171 Coronas, J. and Santamaria, J. (2004) *Chem. Eng. Sci.*, **59**, 4879.

172 Schimpf, S., Bron, M. and Claus, P. (2004) *Chem. Eng. J.*, **101**, 11.

173 Horny, C., Kiwi-Minsker, L. and Renken, A. (2004) *Chem. Eng. J.*, **101**, 3.

174 Hessel, V., Lowe, H. and Schonfeld, F. (2005) *Chem. Eng. Sci.*, **60**, 2479.

175 Hardt, S. (2007) Modeling and simulation of microreactors, in *Modeling of Process Intensification* (ed. F.J. Keil), Wiley-VCH Verlag, Weinheim, Ch. 3, p. 25.

176 Hessel, V., Löb, P., Löwe, H. and Kolb, G. (2007) Microreactor concepts and processing, in *Micro Instrumentation for High Throughput Experimentation and Process Intensification – a Tool for PAT* (eds M.V. Koch, K.M. VandenBussche, and R.W. Chrisman), Wiley-VCH Verlag, Weinheim, Ch. 6, p. 85.

177 Hessel, V., Hofmann, C., Löwe, H., Meudt, A., Scherer, S., Schönfeld, F. and Werner, B. (2004) *Org. Proc. Res. Dev.*, **8**, 511.

178 Pennemann, H., Hessel, V., Löwe, H., Forster, S. and Kinkel, J. (2005) *Org. Proc. Res. Dev.*, **9**, 188.

179 Veser, G. (2001) *Chem. Eng. Sci.*, **56**, 1265.

180 Centi, G., Perathoner, S. and Abate, S. (2009) Direct synthesis of hydrogen peroxide: recent advances, in *Modern Heterogeneous Oxidation Catalysis: Design, Reactions and Characterization* (ed. N. Mizuno), Wiley-VCH Verlag, Weinheim, Ch. 8, p. 253.

181 Kim, J., Park, J.-K. and Kwak, B.-S.(12–15 June, 2005), Development of pharmaceutical fine chemicals using continuous microreactor technology (MRT). Proceedings of the 4th Asia-Pacific Chemical Reaction Engineering Symposium, APCRE05, Gyeongju, Korea, p. 441.

4
Membrane Technologies at the Service of Sustainable Development Through Process Intensification

Gilbert M. Rios, Marie-Pierre Belleville, Delphine Paolucci-Jeanjean, and José Sanchez

4.1
Introduction

In today's economy, engineers must respond to the very important changing needs of industrial processes that aim to satisfy, at the same time, stronger constraints on raw materials, energy saving and environmental impact imposed by a sustainable development and the ever-increasing demand of consumers for new products with specific end-use properties. To respond correctly these requests, which are a little contradictory, process intensification appears to be a privileged path.

Basically, process intensification aims at replacing large, expensive, energy-intensive equipments or processes with ones that are smaller, cheaper and more efficient. At the same time, they have to minimize environmental impact, increase safety, improve remote control and automation and ensure a better product quality.

Several routes to these numerous and complex goals may be envisaged [1, 2]: new multifunctional operations, new operating modes and new contacting devices. All these methods are based on strongly integrated multidisciplinary and multiscale approaches. Usually, they involve new functional materials designed at the micro- if not at the nano-, scale level. With regard to operations and processes, the new concept of "integration" replaces the old-fashioned idea of "addition." Chemical engineering provides the appropriate links between the different length scales, from the micro- or nano-scale up to the scale of multifunctional equipment and whole plants.

Undoubtedly, artificial membranes and related technologies meet these different requests. This is not entirely surprising if we consider that evolution has based all life mechanisms on the use of natural membranes. With cell membranes as an ultimate model, breakthrough biomimetic approaches offer a privileged direction to reach sustainable development and citizen well-being at the core of the social demand.

The term membrane technology has encompassed, in recent decades, various mass transfer operations using a thin layer at an interphase. They differ from classical operations of chemical engineering because they were built around materials with original properties acting under the effect of specific driving forces. Today the ideas have evolved, and a more total analysis, integrating in a far better way the process

Sustainable Industrial Processes. Edited by F. Cavani, G. Centi, S. Perathoner, and F. Trifiró
Copyright © 2009 WILEY-VCH Verlag GmbH & Co. KGaA, Weinheim
ISBN: 978-3-527-31552-9

aspects, has grown. This opens up broader and more promising prospects for applications, in terms of both the number of fields concerned and the quality of performance.

Today, advances towards intensification can be awaited through membrane technologies at three different scale levels:

1. by improving performance starting from local considerations (i.e., around the thin layer itself);
2. by creating new more efficient operations (by coupling functions and/or changing operating modes);
3. by rethinking the whole processes with their assistance. (How to integrate at best a membrane operation in an industrial plant without changing major requirements on inputs and outputs?).

After a short review of the basic principles of membrane technologies, and after having pointed out some important concepts for the correct positioning of these technologies as regards material sciences and chemical engineering, the three different scale levels mentioned above are analyzed and illustrated with relevant examples.

4.2
From Definitions to Function: A Few Fundamental Ideas

4.2.1
Membrane Operation

Membrane operation is a specific, but not exotic, operation. In fact it is a "hybrid" of classical heat and mass transfer processes (Figure 4.1). Direct contact mass transfer operations tend to reach equilibrium due to a difference of chemical potential between two phases that are put into contact. In the same way, temperature equilibrium is aimed at during heat transfer operations, for which driving force is a temperature gradient. In contrast, for membrane operations, by using the specific properties of separation of the thin layer material that constitutes the membrane, under the particular driving force that is applied, it is possible to deviate from the equilibrium that prevails at fluid-to-fluid interphase with classical direct contact mass exchange systems and to reorientate the mass transfer properties. In particular, this is the case with classical operations such as microfiltration (MF), ultrafiltration (UF), reverse osmosis (RO), gas separation (GS), pervaporation (PV), dialysis (DI) or electrodialysis (ED), for which a few characteristics are recalled in Table 4.1.

In Table 4.1 L and G refer to liquid and gas phases; ΔP, Δp, ΔC and ΔV are the differences in pressures, partial pressures, concentrations and voltages, respectively; porous and dense refer to the type of the material; and sieving, solubility-diffusion and Donnan are types of mass-transfer mechanism.

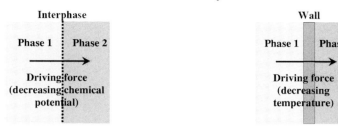

Direct contact
mass transfer operation

Heat transfer
operation

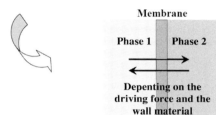

Membrane operation

Figure 4.1 Comparison between membrane operation and classical mass/heat transfer operations.

Table 4.1 Classical membrane operations (see text for details).

Process	Phase 1	Phase 2	Force	Material	Mechanism
MF	L	L	ΔP	Porous	Sieving
UF	L	L	ΔP	Porous	Sieving
RO	L	L	ΔP	Dense	Solubility-diffusion
GS	G	G	ΔP	Dense	Solubility-diffusion
PV	L	G	Δp	Dense	Solubility-diffusion
DI	L	L	ΔC	Dense	Solubility-diffusion
ED	L	L	ΔV	Dense	Donnan

4.2.2
Overall Performance: A Balance Between Material and Fluid Limitations

As in all unit operations in which a wall separates two fluids, detrimental effects of polarization and fouling may be expected on both sides of the wall following the build up of boundary layers and fluid-to-wall interactions. All these phenomena are classical with, for example, heat exchange surfaces.

With dense materials, control by the properties of the material acting as the main resistance in the system may be envisaged; as an example is "classical" membrane

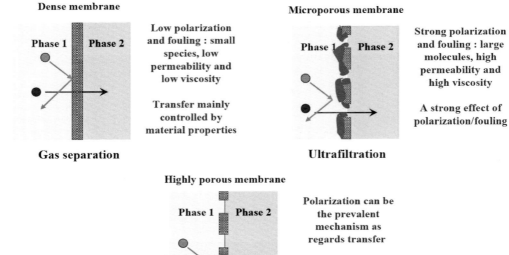

Dense membrane

Phase 1   Phase 2

Low polarization
and fouling : small
species, low
permeability and
low viscosity

Transfer mainly
controlled by
material properties

Gas separation

Microporous membrane

Phase 1   Phase 2

Strong polarization
and fouling : large
molecules, high
permeability and
high viscosity

A strong effect of
polarization/fouling

Ultrafiltration

Highly porous membrane

Phase 1   Phase 2

Polarization can be
the prevalent
mechanism as
regards transfer

The selectivity is a
function of phase
equilibrium at
interface

Membrane contactor

Figure 4.2 Comparison between the different effects as regards performance.

operations such as reverse osmosis or gas permeation. With porous materials (e.g., ultrafiltration or microfiltration), the influence of the material itself may become considerably lower, with performance also depending on polarization and fouling phenomena. Finally, at the other limit, we can imagine the sole use of the supporting material as creating and controlling interphases: this is the case with the newly developed concept of "membrane contactors." Figure 4.2 summarizes these trends. As a rule, the driving force and the related energy consumed will be more important the denser the material is, with the global flux/selectivity ordinarily varying in the reverse/same direction, respectively.

4.2.3
Membrane Material as a "High Tech Product" Contacting Device

When used mainly as a contacting device, the membrane material leads to considerably improved performance for transfer and/or reaction as compared to a classical packing. This is due to a better management of fluid dynamics and mixing conditions, to increased surface areas and to better controlled driving forces. In fact the membrane represents the assembly of a multitude of microsystems, which act in parallel and can be functionalized. With materials structured at the nanoscale level it would be possible to control phenomena at the molecular level. As compared to dispersed beds of microparticles, porous layers have the main advantage of keeping their shape until breaking. With ceramic layers, powders are sintered to avoid dispersion under process constraints. Through these media, mass transfer may be controlled easily using, for example, a pressure difference. To conclude on this point, and concerning the functionalities that may be reached, we can assert that the differences between classical packaging and membranes used for contacting purposes are similar to those between classical electric resistors and electronic chips!

4.2.4
A Clear Distinction Between the "Function" and the "Material"

The membrane is, firstly, a "function." As shown in Figure 4.3, the "membrane function" is a triptych involving a thin layer (the material), at least two fluid phases (the products) and some particular working conditions (the process). In this way, the performance of a membrane operation under real working conditions must be clearly distinguished from the dynamic characterization of the material as traditionally conducted with model fluids and under classical operating modes. Indeed, by changing only one of the "ingredients," the performance will be modified completely.

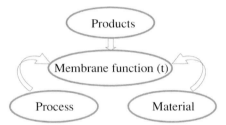

Figure 4.3 The membrane function.

This conceptual approach allows us to understand the respective roles of the specialists of material sciences on one side and chemical engineering on the other side. Undoubtedly, they must cooperate, but the development of new materials clearly cannot be the way to elucidate all the problems! This assessment shows the new directions in which the specific expertise of chemical engineers is strongly required: working with real and complex solutions, trying to involve new membrane operations in industrial processes, designing new hybrid operations, proposing improved models to make simulation and scale changes. Thus it should prevent them from limiting, as is too often the case, their investigations to the dynamic characterization of new materials. Not taking this into account has been responsible in the past for a strong bottleneck in the deployment of membrane technology in industry.

4.2.5
Enlarged Uses of Membrane Concepts

From the above, many functions may be imagined based on the employment of thin layers: selective mass transfer, chemical or biological reaction, fluid distribution or even energy transfer. Up to now, thin porous layers have been used mainly for mass transfer purposes. But there would be no impediment to take advantaging, for example, of the specific microporous structure and geometry of certain kinds of thin layers to use them for energy (heat) transfer. These new devices differ from classical dense metallic wall apparatuses to the same extent that classical dense membrane operations differ from membrane contactors.

Concurrently to these new unit operations based on the use of a local function defined on the thin layer itself, we can also recall many other achievements involving more complex functions attained by coupling membrane materials with other external unit operations of chemical engineering: this is the so-called concept of "hybridization."

Finally, the "integration" of membrane steps in an industrial process already running, to reduce its size, to simplify its flow sheet, to improve its global performance (energy consumption, waste rejects, final product quality, etc.) is certainly a more complex and enthralling challenge. It is probably also the most promising in terms of "intensification" and the more demanding in terms of specific expertise of chemical and process engineers.

While membrane technologies will not provide the solution to every problem in the near future, their specific properties and advantages (compactness, modularity, microstructures, improved functionalities, easy control and, most of the time, an isothermal use without requirement for the addition of chemical additives) will allow broader fields of application. The following examples testify to their huge potential, as expected from laboratory experiments, while showing also the root of their unexpected but practical use in present day industrial processes and plants.

4.3
The Need for More Integrated Views on Materials and Process Conditions

As already underlined, the balance between the limitations induced by the supporting material, on the one side, and the fluid phases flowing along it, on the other side, is the key factor for the overall performance of a membrane operation.

4.3.1
When Dense or Microporous Materials Control the Overall Process Performance

For membrane operations in which dense or microporous materials control the overall process performance there is no doubt that process intensification will follow directly from improvement of material properties. With the development of new materials having properties controlled at the nanoscale level, operations of this kind seem promised a really bright future.

An example is provided by "mixed matrix membranes" (MMM). Basically they are constituted of inorganic molecular nanoparticles (such as zeolites, carbon molecular sieves, etc.) imbedded in polymers. MMMs open up new perspectives in gas separation. A main application for sustainable development is the purification of

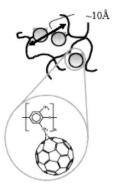

Figure 4.4 Mixed matrix membranes (MMM) with covalent
linking of rigid entities like buckyballs into polymers.

hydrogen cheaply and efficiently. Figure 4.4 pictures a membrane developed at
Twente University [3] by the chemical modification of polymers with covalent linking
of rigid entities like buckyballs or nanotubes.

Free volume changes are observed and it appears that covalent bonding enhances
the permeability without compromising the selectivity: a strong contrast to dispers-
ing the entities in the polymer matrix as most often classically done.

Another interesting example has been provided by Eric Hoek from the University
of California [4]. The membrane presented by the author promises to reduce the cost
of sea-water desalination and wastewater reclamation, with a driving pressure lower
than in conventional systems, and a considerably reduced fouling; thus, a reduction
of about 25% of the overall cost of desalination, including energy consumption and
environmental issues, follows. This new membrane is also of MMM type, with a
uniquely crosslinked matrix of polymers and engineered nanoparticles structured at
the nanoscale. Indeed, molecular tunnels are formed and water flows through them
much more easily than nearly all contaminants. The nanoparticles are designed to
attract water and are highly porous, soaking up water like a sponge, while repelling
dissolved salts and other impurities such as organics and bacteria, which tend to clog
up conventional membranes.

A recent publication of Silvestri *et al.* reports a synthesis on molecular imprinted
membranes designed for an improved recognition of biomolecules [5]. The unique
feature of molecularly imprinted membranes is the interplay of selective binding and
transport properties, making them potentially superior to state-of-the-art synthetic
separation membranes already applied in various biological and biomedical fields. By
using such materials based on the concept of "key–lock" (ideally, only molecules used
for the imprinting will be recognized and thus able to pass the selective layer) it is
possible to envisage the highly selective separation of highly toxic substances from
drinking water (hormones, pesticides, antibiotics, etc.).

Another interesting example was provided by Teplyakov *et al.* [6] at the last
EuroMembrane Congress in Taormina, in September 2006. The authors deal with
processes using porous ceramics with catalytic coating in microchannel walls. This

design is valuable in creating high speeds (residence time less than 10^{-3} s) and improve reactor compactness. Catalytic microporous inorganic membranes combining selective gas transport and catalytic activity can be considered as an "ensemble" of nanoreactors, and as then opening up new fields of applications of heterogeneous catalysis [7]. In such a configuration the classical counter-diffusion in catalyst particles is replaced by an unidirectional transport with the potential of intensified catalysis and increased selectivity.

This example can be applied to a broad class of catalytic reactions but it is much more obvious for partial oxidation reactions where secondary reactions (total combustion) result in a dramatic decrease of selectivity. This is the case with methanol decomposition and methane conversion, where the intensification of gas-phase catalytic operations in micro- or nanochannels clearly appears.

Other promising routes to preparing new kinds of efficient materials include molecular self-assembly, adaptive supramolecular and dynamic chemistry, and hydrothermal synthesis for zeolites [8–10].

4.3.2
Other Operations Using Meso- or Macroporous Membranes

In other operations using meso- or macroporous membranes – ultrafiltration, microfiltration, membrane contactors, and so on – the influence of the material itself is much less pronounced Considerations relating to properties of the material surface (roughness, hydrophobicity, and so on) or the core (porosity, pore size distribution, and so on) can remain significant, but do not always dominate. In fact polarization and fouling phenomena can also strongly influence the performance of the process. In those cases intensification will result from a clear distinction between the different resistances (analysis and comparison) and on the optimization of the "membrane function" by taking into account "three ingredients": material properties, process conditions and fluid characteristics.

Micro- or ultra-filtration provides our first example to illustrate these assessments. Here, polarization and fouling phenomena are important and, consequently, they may control process performance if the fluid to be treated has a high viscosity, the pore sizes of the thin layer are small and the dispersion of molecular size of species dissolved in the fluid is relatively large. With very complex fluids such as milk or suspensions, the main impediment to high flux is the building of a deposit at wall, which results from the accumulation in the boundary layer of various species (polarization with different species such as fat globules, proteins for milk, or microparticles, colloids for water and clogging). In the long run this partially "dynamic layer" will control the process performance based on its own properties. Surprisingly, good selectivity values may thus be obtained. In milk filtration, high retention values that are really characteristic of UF may be obtained with a nominal pore size of material in the MF range (e.g., 1 μm for ceramic membranes). In that case process intensification will follow from the optimization of the membrane function, through the choice of optimal working conditions [11].

Today there is growing academic and industrial interest in new membrane contactor processes [12], such as membrane distillation (MD) or membrane evaporation (ME). In ME the driving force is an activity gradient, while in MD it is a temperature difference between the two sides of the membrane. The membranes used in these processes are porous and they exhibit a hydrophobic character; as a consequence they prevent the penetration of aqueous solution into the pores. Consequently, a vapor–liquid interface is created at each pore entrance, in so far as the transmembrane pressure does not exceed a limit: the so-called "intrusion pressure." This value is a function of the material surface energy, liquid surface tension and mean pore diameter. In both cases water is transported in the vapor state through the pores under the effect of a partial pressure difference (Figure 4.5). In MD, a hot feed solution evaporates; the vapor then diffuses through the pores and undergoes condensation on the opposite side of the membrane.

A reduction of the number and dimensions of the pores on the membrane surface significantly affects the performance as evaluated in terms of gas (vapor) permeability: the flux decreases. In contrast, the presence of large pores on the membrane surface enables deposits into their core: as an example the deposition of $CaCO_3$ crystallites when ground water is treated. This can cause the wet ability of pores filled by deposit, or require frequent cleaning steps that in the mid–long run will themselves cause a flux decline.

This drawback can be avoided by applying specific treatments during preparation of the material: for example, by covering the grains or fibers that constitute this material with a thin layer of a hydrophobic polymer but keeping the porosity before and after the covering almost similar [13], or by covering the membrane material surface and then the large pores with a thin polymer layer having a small porosity. As a consequence, the wet ability of the membranes with a small surface porosity proceeds to a considerable lesser degree. This advantage largely counterbalances the effect of flux decrease due to pore size reduction [14].

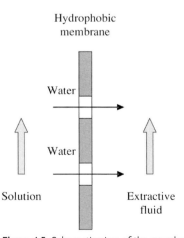

Figure 4.5 Schematic view of the membrane evaporation (ME) process.

One of the most interesting and promising extension of the MD concept is membrane crystallization (MC) [15]. This innovative technology uses the evaporative mass transfer of volatile solvents through macroporous hydrophobic membranes to concentrate feed solutions above their saturation limit, thus achieving a metastable state (supersaturation) in which crystals may nucleate and grow. The role of the membrane is not limited to providing a support for solvent evaporation. A crystallizing solution can be imagined as a certain number of solute molecules moving among the molecules of solvent and colliding with each other, so that a number of them converge to form clusters. There is, anyway, an energetic barrier that must be crossed to induce the formation of stable nuclei.

The presence of a foreign surface, specifically a polymeric membrane, decreases the work required to create critical nuclei and will increase locally the probability of nucleation with respect to other locations in the system: this phenomenon is known as heterogeneous nucleation. The reduction of ΔG due to heterogeneous nucleation (ΔG^{het}) compared to a homogeneous one (ΔG^{hom}) is related to the contact angle (θ) between solution and solid substrate (Equation 4.1):

$$\Delta G^{het} = \Delta G^{hom}\left(\tfrac{1}{2} - \tfrac{3}{4}\cos\theta + \tfrac{1}{4}\cos^3\theta\right) \tag{4.1}$$

Different polymeric membranes exhibit dissimilar interactions with a liquid phase: this concept might represent the starting point for a better engineering of the crystallization procedures, based on the use of nanoscale-tailored material surface properties. These new technologies open bright perspectives for desalination and the production of pure water. With MC it becomes possible to envisage leaving the concentration of salt until its ultimate stage, and to thus avoid the strongly detrimental effect for the environment of the rejection of very concentrated brines like those produced today at the exit of reverse osmosis units.

4.3.3
Two Important Remarks

4.3.3.1 Nano- and Micro-Engineering for New Porous Thin Layers
Beyond the traditional chemistry for material preparation, nano- and micro-engineering techniques today merge into new exciting technologies for tailoring new membrane supports, specifically dedicated to a well-defined task, in a very regular and reproducible way [16]. This can greatly help in optimizing the previously mentioned "membrane function."

Conventional micro-perforation methods with the use of laser drilling, precision etching, and so on allow us to obtain well-defined pore sizes larger than about 10 micron. With micro-engineering techniques, originating from the semiconductor industry (mask, lithography, etc.), it is relatively easy to downscale pore sizes between 10 and 0.1 micron. Silicon is therefore the primary substrate material. These membranes are durable filters, and especially suitable for applications where a long lifetime or easy cleaning conditions are required (e.g., for the crossflow filtration of industrial biological fluids, tissue engineering).

When disposable filters are preferred (which may be the case for medical or microbiological analysis), it can be better to elaborate lower cost materials. A new and efficient way to prepare them is by deposition of polymers on patterned substrate structures (such as with phase separation micro-molding).

Nanotechnology requires a different approach to fabrication from that of microtechnology. Whereas microscale structures are typically formed by top-down techniques (lithography, deposition, etc.), the practical formation of structures at nanoscale dimensions involves additional components, that is, bottom-up self-assembly.

Significant studies prove the interest in this kind of membrane. Among them we mention the work of Lambricht and Schubert [17] on emulsification with microchannels, and the study of Popat *et al.* [18] that proves that the bone cell response can be significantly improved using controlled nanoarchitecture (alumina membranes fabricated using a two-step anodization process and which present pore sizes of 30–80 nm).

4.3.3.2 Membrane Processes and Solid Bed Technologies: A Comparison

It is worth underlining the strong advantages of membrane pressure-driven processes or contacting systems, as compared to more classical unit operations based on the use of a dispersed solid phase in fixed, fluidized or suspended beds (ordinarily microparticles with specific properties for adsorption, ion exchange, heterogeneous catalysis). With solid beds, the performance is controlled only at the macroscopic level. Hydrodynamics and mixing are not easily foreseen. There is no possibility to act on diffusion phenomena at the interface, or on the way in which turbulence develops. Moreover, local driving forces are, most of the time, hazardous. Indeed, all of these phenomena harm the efficiency of the system. Conversely, with porous membrane systems the hydrodynamics may be managed at the very local level (microfluidics), with a laminar flow and easy-to-control, and a transmembrane pressure acting like a driving force.

4.4
Use of Hybrid Processes and New Operating Modes: The Key to Many Problems

Seeking synergistic effects is really a key principle for process intensification, as underlined recently by Huang *et al.* [19]. This section illustrates how to obtain synergistic effects either by bringing together separation–reaction unit operations into one single operation (examples 1–3) or by changing the mode in which operation is driven (example 4).

4.4.1
Nanofiltration-Coupled Catalysis

Even if not really new, the concept of a membrane reactor has not been applied so intensively in homogeneous catalysis as in heterogeneous catalysis [20]. Nevertheless, the application of this concept in this area seems to be an exciting new topic, which opens up enormous possibilities of research and development.

As an example the work of De Smet *et al.* in 2001 [21] on nanofiltration-coupled catalysis provides a good overview of the advantages that may be awaited. It can be applied in many different types of reactions and for a wide range of catalysts and substrates, especially in the field of fine chemical synthesis. The thought of these authors is initiated around the fact that, for industrial processes, heterogeneous catalysts are generally preferred as they facilitate removal of the catalyst after reaction and allow a continuous operating mode. However, preparing a heterogeneous catalyst can be tedious and might demand a high preparative effort. In certain cases, mass or heat transfer limitations in the solid state catalyst may lead to decreased activities, and homogeneous reactions generally show higher chemo- and enantios-electivity. To integrate as well as possible all these constraints, a hybrid process is proposed that combines nanofiltration with homogeneous catalysis (Figure 4.6).

The reaction takes place in a continuously stirred tank reactor, thus reaching the activity and selectivity found in homogeneous reactions. The liquid is contacted with a nanofiltration (NF)-membrane that allows products to permeate but rejects the dissolved catalyst. This set-up is made possible by the development of solvent resistant NF-membranes having a molecular weight cut-off (MWCO) in the range 200–700 Da and working conditions below 40 °C and 35 bar.

More specifically, chiral catalysts are among the preferred systems for this hybrid membrane/catalysis process, owing to their extremely high cost and their sensitivity towards traditional heterogenizing methods. Furthermore, most of these catalysts are transition metal complexes with a MW above 500 Da and high activity and selectivity at moderate reaction conditions. In the reported set-up, the hydrogen pressure needed for the hydrogenation of the substrates, is – without any additional cost or equipment – the driving force for the membrane permeation. The whole hybrid process is operated in such a way that a sufficient amount of product with high purity is yielded in catalysis, while a good rejection of the catalyst and reasonable fluxes are preserved by NF.

The continuous enantioselective hydrogenations of dimethyl itaconate (DMI) with Ru-BINAP (MW = 929 Da) and of methyl 2-acetamidoacrylate (MAA) with

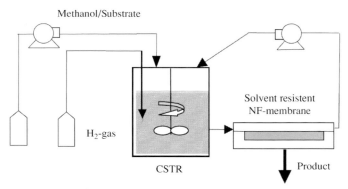

Figure 4.6 Nanofiltration coupled catalysis.

Rh-EtDUPHOS (MW = 723 Da) were selected as model systems because of the excellent performance of these catalysts and their industrial relevance.

4.4.2
Supercritical Fluid-Assisted Membrane Separation and/or Reaction

The viscosity of highly viscous compounds is a limiting factor for their filtration on a membrane in terms of fluxes and energy costs. In the past, a few attempts were made to decrease viscosity by either using high temperatures (a few hundred of degrees) or by adding chemicals (surfactants, solvents). Among the inconveniences of the first process, are difficulties in safe handling and the strongly detrimental degradation of temperature-sensitive compounds. The second method raises the problem of removing chemicals after filtration. Consequently, a new idea has been proposed that consists of lowering the viscosity of sticky fluids by injecting SC_{CO2} at moderate temperature. Working at room temperature allows the preservation of thermo-sensitive products, and the supercritical solvent can be removed easily by pressure release at the exit of the separation chamber.

The feasibility of this hybrid operation – fluidification by injection of pressurized CO_2 and cross-flow filtration – has been proved with model molecules [such as poly (ethylene glycol)s] or mineral oils using inorganic membranes. Particular interest in the new technology for recycling used motor oils was demonstrated [22]. It could provide a good alternative to the traditional way of used-oil burning, which has the main disadvantages of not being a very environmentally friendly process and of destroying valuable oils.

By preserving thermo-sensitive compounds, the new process is also particularly attractive for vegetable oils, which represent very important part of the food and health industry. In addition to reducing the viscosity of the oil for purification purposes, we can also foresee its composition being modified by specific enzymatic reactions at the pore level: this is the concept of the "active membrane," which follows the immobilization of enzymes at membrane walls. The goal would be to obtain high-added-value products from cheap and plentiful raw substrates.

Today, enzymatic reactions have become economically conceivable at an industrial scale. In comparison with chemical reactions they are less polluting (environmental interest) and more specific. Thus the enzymatic modification of fats and oils has emerged as a "hot topic," for which there is a huge demand in terms of products from end-users and in terms of new processes from producers. The presence of SC_{CO2} in the oil has no detrimental effect on enzyme activity. As previously established by various authors, SC_{CO2} is a good reaction medium for lipase activity and offers many advantages linked to its relatively mild critical temperature (31 °C) and pressure (74 bar). Its interesting properties such as low viscosity and high diffusivity often result in faster reaction kinetics compared to organic solvents.

This new concept has been validated with immobilized lipases, by performing interesterification between castor oil and methyl oleate to produce methyl ricinoleate (Equation 4.2) in the unit represented in Figure 4.7 [23, 24]:

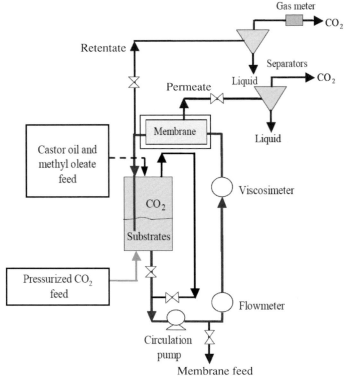

Figure 4.7 Experimental set-up for interesterification between castor oil and methyl oleate to produce methyl ricinoleate.

$$\text{Castor oil TG} + \text{Methyl oleate} \xrightarrow[\text{SCCO}_2]{\text{Lipase}} \text{Methyl ricinoleate} + \text{TG} \qquad (4.2)$$

On the whole, SC_{CO2} lowers appropriately the fluidity of the oil, thus allowing filtration under suitable enough conditions; it also seems to improve the activity of lipase. The enzymatic membrane, in turn, represents a very good microcontacting system (particularly pore mouth) that is well adapted to elevated reaction yield because of the high probability level of contact between substrate and biocatalyst.

4.4.3
Membrane-Assisted Fluidized Bed Reactors

The idea to limit polarization and fouling during tangential filtration of biological fluid by fluidizing small inert particles inside a tubular ceramic membrane had been presented at the end of the 1980s [25]. More recently, based on advances in the development of more stable membranes with increased permeance, the possibilities for integrating membranes into gas catalytic reactors to achieve a major increase in reactor performance by process integration and process intensification have also been examined [26].

Several reviews and even special issues of catalysis related journals illustrate the significant progress in the field of inorganic membrane reactors within the last two decades [7].

Chemical engineers and material scientists have joined forces and addressed this topic from various view-points. Despite their considerable efforts in these directions, the application of the membrane reactors, especially packed bed membrane reactors, in commercial processes has been very limited because of technical as well as economical drawbacks. The most recent trend in membrane reactor technology has been in the direction of incorporating inorganic membranes into fluidized beds to combine the permselective and controlled dosing capabilities of membranes with excellent gas–solid contact and heat transfer capabilities of fluidized beds, thereby overcoming the limitations often prevailing in packed bed membrane reactors. The work more specifically tackles ultrapure hydrogen production via methane steam reforming.

The proposed novel membrane-assisted fluidized bed reactor (MAFBR) consists of a partial oxidation bottom section and a steam reforming of methane/water-gas shift SRM/WGS top section (Figure 4.8). Because of the large difference in operating temperatures of the permselective H_2 membranes (below $700\,^\circ C$) and permselective O_2 membranes (above $900\,^\circ C$), two sections operate at these respective temperatures. The fraction of CH_4 feed is fed through the bottom section in such a way that the permeating O_2 is able to generate CPO (catalytic partial oxidation) equilibrium conditions in the bottom section, the temperature of which is in turn favorable for the permselective O_2 transport. The steam is also added to avoid coke formation in the bottom section. The top section is then fed with the remaining CH_4 and steam feed so that overall autothermal process is achieved when both sections are considered together. The endothermic heat demand of the top section is thus catered by the equilibrium mixture coming from the bottom section and the side feed of additional CH_4 and steam.

In the bottom section O_2 is introduced selectively via dense perovskite membranes to supply the required reaction energy via CPO for the steam reforming/WGS reactions in the top section, where H_2 is selectively extracted via dense Pd-based membranes, thereby surpassing the thermodynamic equilibrium limitations. Using thermodynamic equilibrium calculations and more detailed fluidized bed membrane reactor modeling, it has been demonstrated that autothermal operation and effective temperature control in both reaction sections can be achieved along with high CH_4 conversions and H_2 yields by tuning the overall CH_4 and steam feed ratios and the feed ratios to the bottom and top sections.

4.4.4
Electrodialysis with A Non-stationary Field

Electrodialysis as a method of demineralization of natural and wastewater is ecologically expedient and economically profitable. However, the use of ion-selective membranes is complicated by the phenomenon of concentration polarization, which

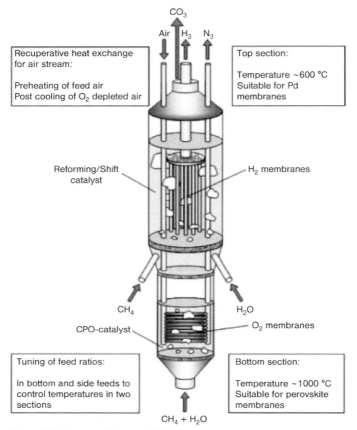

Figure 4.8 Scheme of the novel MAFBR.

is caused by a difference between the ion transfer numbers in the solution and in the membranes. Applying an electric field with special pulse characteristics can diminish the effect of the concentration polarization.

A theoretical analysis has been carried out for galvanostatic and potentiostatic pulse regimes [27]. The idea that developed is a bit the same as backflushing with pressure driven-membrane operation such as microfiltration or ultrafiltration. The time dependencies of the extent of the concentration polarization near the membrane surface during the pulse are described theoretically for both pulse regimes and a qualitative discussion of the pause duration is presented. The main characteristic of the non-stationary process is the transition time between the state without polarization and the state with stationary polarization.

In principle, the electrodialysis process can be intensified significantly when the applied pulse is sufficiently smaller than this transition time. Experimental results qualitatively support the model predictions. The desalination can be intensified several times, depending on the pulse–pause characteristics.

4.5
Safe Management of Membrane Integration in Industrial Processes: A Huge Challenge

Everybody agrees today that membrane technologies are very suitable downstream separation processes for chemical industrial plants, with a huge potential for improving their performance and reducing their size. We can imagine using membrane filtration for different purposes such as: (i) to avoid the addition of chemicals (such as salts for flocculation, solvent for extraction) ordinarily responsible for a negative impact on environment; (ii) to replace thermal unit operations (heat exchanges, distillation, evaporation) detrimental to the quality of heat sensitive molecules; (iii) to increase process flexibility by favoring less sensitive process output (e.g., following changes in the metabolic pathway of microorganisms in an upstream fermentation vessel); and (iv) to better put process by-products to good use. But in practice it appears that their use is somewhere limited.

The more classical explanation is really too "short" to be realistic and useful: membrane fouling, high investment costs. In fact a more thorough analysis shows that the reasons are to be sought elsewhere, and that the main bottlenecks with industrial processes lie within:

- how to take into account the complexity of real solutions;
- how to do scaling-up and down between laboratory and industrial plants;
- how to foresee performance with the new technology.

In other words, how to link in a reliable way, with limited information, the macroscale process performance to local phenomena at the meso- (membrane pore) if not at the nano- (solute–solute or solute–barrier interactions) scale levels. To answer these questions, there is a need for a new methodology, based on chemical engineering principles, with a holistic approach involving well-balanced experimental/simulation/modeling parts. A first attempt to tackle this question was carried out a few years ago with a particular example: the purification by ultrafiltration (UF) of a small neutral molecule from a fermentation broth, constituted of unknown peptides and proteins with sizes ranging on a very large scale [28–30].

Notably, at first, the scaling-up process as ordinarily carried out is very hazardous. There is no logic in the step-by-step progress and risks of error are significant:

- *First step:* work at laboratory level for the membrane choice (flux and selectivity).
- *Second step:* pilot plant trial – and error – assays to optimize performance.
- *Third step:* extrapolation towards industrial unit by simple rule-of-thumb.

In fact differences between laboratory, pilot and industrial plant implementations are numerous, including:

- membrane selected in steady-state (permeate recycling) at laboratory level, while working in continuous non-steady regime at industrial scale;
- strong differences in geometrical configurations and hydrodynamics, with membranes and modules involved in each case.

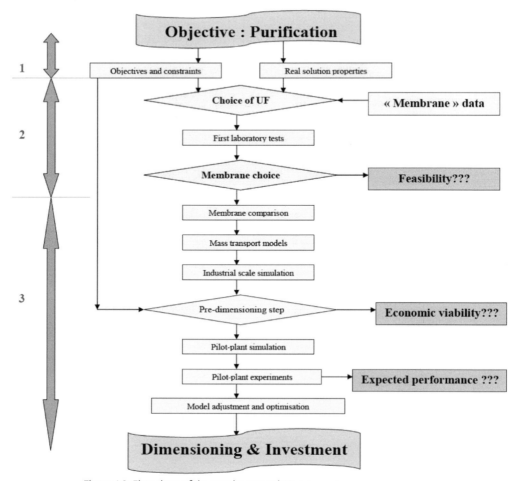

Figure 4.9 Flow-sheet of the step-by-step solution process.

Figures 4.9 and 4.10 summarize the new way that is proposed. It involves a more stable closed-loop progress, which replaces the inconsistent linear approach, with three different steps:

1. Process analysis, objectives to be reached, characterization of the real solution.

2. Membrane choice at laboratory level according to the objectives to be tackled; a few candidates are selected and then more thoroughly tested.

3. Modeling and simulation; making an appropriate hypothesis to get a simple representation of the real solution; identification of membrane transfer mechanisms; associated non-steady state mass balance equations, this information will serve as the ground for all further simulations.

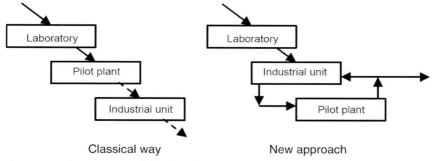

Figure 4.10 Schematic views of the classical and new approaches.

Concerning process analysis and membrane selection for purification, the purity level of the molecule to be extracted is the main criterion. Operation will be considered as "feasible at laboratory scale" for every membrane that meets this constraint. Such membranes, which must also meet other technological constraints (such as guaranteed availability on the market, correct cleaning, acceptable lifespan), are selected and compared. Notably, "feasibility at laboratory level" does not guarantee "industrial feasibility" due to large discrepancies between geometrical and working conditions, as already underlined! Ordinarily purity at industrial scale will be a mean value over time: this remark emphasizes the main part of further modeling/ simulation steps. We keep in mind also that, from a broader viewpoint, in the industrial process the key-criterion will be the compliance of the quality of the product obtained at the exit of the new process/plant with the standards (e.g., from the FDA with pharmaceutical substances)!

For modeling and simulation, mass-balances will play a central role in showing the evolution of concentrations versus time. For purification it appears that a key parameter is the time constant (Equation 4.3):

$$\tau = \frac{V/J_p S}{\alpha - 1 - \left(\alpha - \frac{1 + FD}{FP}\right)\left(\frac{\alpha - Tro}{\alpha}\right)} \tag{4.3}$$

where *FD*, *FP* and α are dimensionless numbers related to flow rates, and *Jp* and *Tro* are the permeate flux and transmission rate, respectively. Equation 4.3 shows that to preserve τ when changing scale (intra- or extrapolation), it will be absolutely necessary to keep unchanged also *V/S* (the volume to surface ratio). This is rarely done in a classical operation!

Another important issue is how to represent complex biological solutions, considering that most of the time a thorough description of it is not possible. The idea dividing it into a limited number of specific groups, defined by a generic property as regards the operation to be realized, is thus introduced. For purification by UF, the simple model that was successfully used consisted in considering three distinct fractions – the active molecule to recover, the compounds with zero rejection and the remaining dry matter – that are easy to analyze. Obviously this model is not relevant for any other topic!

To represent mass transfer at wall, the idea is to use a set of data (flux and rejection) as realistic (i.e., issued from laboratory experiments) and as easy-to-obtain (i.e., with a reduced number of experiments) as possible, and then to select by trial-and-error the simplest model (from constant transmission rate and flux if possible to more complex variation of these indicators with time if necessary).

Finally, the modeling and simulation will be implemented as indicated on the previous schemes noted above:

- Coarse pre-dimensioning of production facility (surface and flow) before pilot plant tests, to check the possibilities of extrapolating the technology at an industrial level (evaluation of the economic viability through a rough estimate of investment and running costs).

- Pilot-plant experiments to obtain complementary information – continuous tests, new geometries, new operating conditions, mid-long term fouling, cleaning – and simulation with a strict compliance with basic rules for scaling-up (all the dimensionless numbers are kept constant).

- Comparison of experimental and simulation results; a too strong discrepancy can justify adapting the model for new simulation and confrontation with results. An iterative calculation!

- A "well-controlled" dimensioning of industrial unit!

Undoubtedly, this new kind of integrated approach is well representative of what should be "membrane engineering," with final objectives clearly defined, the right hypothesis and choice of simple equations for modeling, a realistic representation of real complex solutions and the set-up of efficient simulation tools involving successive intra- and extrapolation steps. It appears to be easily extended to other membrane operations, in other fields of applications. It should provide stakeholders with information needed to make their decision: costs, safety, product quality, environment impact, and so on of new process. Coupled with the need to check the robustness of the new plant and the quality of final output, it should constitute the right way to develop the use of membranes as essential instruments for process intensification with industrial units at work.

4.6
Conclusions

In this chapter we have endeavored to show that membrane technologies are promising tools for sustainable development based on process intensification. On the way that leads from the structure (microscopic if not nanoscopic level) of the material supporting the membrane function to the level of the production plant facility, while passing by the intermediate stages of the module and of the process that hosts it, we have highlighted the different length scales at which intensification can be sought. Thus we met the different bottlenecks that block the applications of these technologies and prevent them from reaching the level of development that accords to their well-known intrinsic advantages. Our work underlines the strong need for a holistic and fully integrated approach of problems, involving all the scientific and technological expertise that is necessary to correctly address them, with everybody working at the level corresponding to his own field of expertise and collaborating with other disciplines within the framework of projects. That is undoubtedly the sound way for future action and faster development of membrane technologies at the industrial level.

References

1 Charpentier, J.C. (2007) *Chem. Eng. J.*, **134**, 84.

2 Charpentier, J.C. (2007) *Ind. Eng. Chem. Res.*, **46**, 3465.

3 Sterescu Sana, M., Bolhuis-Versteeg, L., van Der Vegt, N.F.A., Stamatialis, D.F. and Wessling, M. (2004) *Macromol. Rapid Commun.*, **25**, 1674.

4 Jeong, B.H., Hoek, E.M.V., Yan, Y., Subramani, A., Hurwitz, H.G., Ghosh, A.K. and Jawor, A. (2007) *J. Membr. Sci.*, **294**, 1.

5 Silvestri, D., Barbani, N., Cristallini, C., Giusti, P. and Ciardelli, G. (2006) *J. Membr. Sci.*, **282**, 284.

6 Teplyakov, V., Tsodikov, M., Magsumov, M., Moiseev, I. and Kapteijn, F. (2006) *Desalination*, **199**, 161.

7 Sanchez Marcano, J. and Tsotsis, T.T. (2002) *Catalytic Membranes and Membrane Reactors*, Wiley VCH Verlag, Weinheim.

8 Peinemann, K.V., Konrad, Ma. and Abetz, V. (2006) *Desalination*, **199**, 124.

9 Barboiu, M., Guizard, C., Hovnanian, N. and Cot, L. (2001) *Sep. Purif. Technol.*, **25**, 211.

10 Motuzas, J., Heng, S., Ze Lau, P.P.S., Yeung, K.L., Beresnevicius, Z.J. and Julbe, A. (2007) *Microporous Mesoporous Mat.*, **99**, 197.

11 Brans, G., Schroën, C.G.P.H., van der Sman, R.G.M. and Boom, R.M. (2004) *J. Membr. Sci.*, **243**, 263.

12 Rios, G.M. and Nyström, M. (2005) Editorial of the Special Issue on "Process developments in membrane contactors and reactors". *Chem. Eng. Res. Des.*, **83**, 221.

13 Hengl, N., Mourgues, A., Pomier, E., Belleville, M.P., Paolucci-Jeanjean, D., Sanchez, J. and Rios, G. (2007) *J. Membr. Sci.*, **289**, 169.

14 Gryta, M. (2007) *J. Membr. Sci.*, **287**, 67.

15 Curcio, E., Criscuoli, A. and Drioli, E. (2001) *Ind. Eng. Chem. Res.*, **40**, 2679.

16 Van Rijn, C.J.M. (2004) *Nano- and Micro-Engineered Membrane Technology*, Membrane Science and Technologies Series, Vol. 10, Elsevier, Netherlands.

17 Lambricht, U. and Schubert, H. (2005) *J. Membr. Sci.*, **257**, 76.

18 Popat, K.C., Swan, E.E.L., Mukhatyar, V., Chatvanichkul, K.I., Mor, G.K., Grimes, C.A. and Desai, T.A. (2005) *Biomaterials*, **26**, 4516.

19 Huang, K., Wang, S.J., Shan, L., Zhu, Q. and Qiang, J. (2007) *Sep. Purif. Technol.*, **57**, 111.

20 Sanchez Marcano, J. (2005) Membrane techniques, in *Multiphase Homogeneous Catalysis* (eds B. Cornils, W.A. Hermann, T.I. Horvath, W. Leitner, S. Mecking, H. Olivier-Bourbigou, and D. Vogt), Wiley VCH Verlag, Weinheim.

21 De Smet, K., Aerts, S., Ceulemans, E., Vankelecom, I.F.J. and Jacobs, P.A. (2001) *Chem. Commun.*, 597.

22 Sarrade, S., Schrive, L., Gourgouillon, G. and Rios, G.M. (2001) *Sep. Purif. Technol.*, **25**, 315.

23 Pomier, E., Galy, J., Paolucci-Jeanjean, D., Pina, M., Sarrade, S. and Rios, G.M. (2005) *J. Membr. Sci.*, **249**, 127.

24 Pomier, E., Delebecque, N., Paolucci-Jeanjean, D., Pina, M., Sarrade, S. and Rios, G.M. (2007) *J. Supercrit. Fluids*, **41**, 380–385.

25 Rios, G.M., Rakotoarisoa, H. and Tarodo De La Fuente, B. (1987) *J. Membr. Sci.*, **34**, 331.

26 Deshmukh, S.A.R.K. Heinrich, S., Mörl, L., van Sint Annaland, M. and Kuipers, J.A.M. (2007) *Chem. Eng. Sci.*, **62**, 416.

27 Mishchuk, N.A., Koopal, L.K. and Gonzalez-Caballero, F. (2001) *Colloid Surf. A*, **176**, 195.

28 Darnon, E., Belleville, M.P. and Rios, G.M. (2002) *AIChE J.*, **48**, 1727.

29 Rios, G.M., Belleville, M.P. and Darnon, E. (2006) Keys for a good membrane integration in industrial processes, ECI Conference on "Membrane technology for process intensification", Cetraro, Italy.

30 Darnon, E., Morin, E., Belleville, M.P. and Rios, G.M. (2003) *Chem. Eng. Proc.*, **42**, 299.

5
Accounting for Chemical Sustainability

Gabriele Centi and Siglinda Perathoner

5.1
Introduction

The initial chapters have evidenced the need to analyze in a quantitative way the impact on environment and society of chemical processes, technologies and products, in order to compare on more correct bases the alternative possibilities and their sustainability ("greenness"). This is a challenging task, but a necessary step forward, because metering the beneficial effect of a novel chemistry, process or product is the first step to account for chemical sustainability, for example, to measure the benefit for society and environment of the new developments in chemistry.

The problem of metering and assessing sustainability is an old and a new problem. The United Nations (UN) Conference on Environment and Development in 1992 successfully established the concept of sustainable development as an underlying principle for strategic policy and planning. But the translation of the principle of sustainable development into practice has presented new challenges in finding workable solutions to the complex trade-offs that can arise between the different, and often conflicting, dimensions of sustainable development [1]. At the Johannesburg World Summit on Sustainable Development (September 2002), the global chemical industry demonstrated what has been done to make sustainable development a reality and the plans for the future. But despite the guidance they provide, chemical companies and their associations worldwide are still grappling with what sustainable development means and how the industry's existing safety, health and environmental program (Responsible Care) and the principles of sustainable development mesh.

The concept of sustainability itself has evolved over recent years partly because sustainability is difficult to characterize and because environmental interactions are difficult to model quantitatively [2]. Consequently, viewpoints about sustainability are numerous and, unlike mature sciences, there are not yet established reliable modes of investigation or a uniform framework for dialogue. The key problem in the analysis of sustainability is to decouple the economic growth from the material and energy consumption and to integrate design in a broader manner in society [3]. We could also

Sustainable Industrial Processes. Edited by F. Cavani, G. Centi, S. Perathoner, and F. Trifiró
Copyright © 2009 WILEY-VCH Verlag GmbH & Co. KGaA, Weinheim
ISBN: 978-3-527-31552-9

note that the concept of infinite economical growth, which underlies the leading economical theories and the sustainability theory, is highly questionable. The validity of the concept of continuous growth is most certainly doubtful in the context of limitation of resources and of the environmental pump [4]. Therefore, the concept of sustainability requires profound revisions. However, progress in sustainability can be made only when practical steps are made in this direction.

It is thus necessary to look at the specific problem and use suitable quantitative methodologies of analysis, in order to pass from generic statements of "green" chemistry (we may argue that in several cases what is reported in the literature as greener chemistry do not appear as such upon deeper analysis) to more effective analysis methodologies that allow us to meter the sustainable ("green") content. This requires some simplification (of complex problems with many, sometimes unpredictable, relationships between the different aspects) to choose suitable parameters and methodologies for this quantification. However, the correctness of this procedure should be verified within the framework of the general problem of sustainability.

We may illustrate this concept with few examples. One of the principles of green chemistry is that raw materials should be renewable. Collins [5] in his article "Toward Sustainable Chemistry" indicated that to achieve sustainability one of the three key areas is that "the reagents used by the chemical industry, today mostly derived from oil, must increasingly be obtained from renewable sources to reduce our dependence on fossilized carbon." We know today that the alteration of the global market of food derived from the increasing use of biomass, although in significant part caused by speculations and other economic factors, has significantly changed the approach to the use of bioresources. For example, why use vegetable oils (e.g., palm oil) for biodiesel when these are very valuable raw materials for oleochemistry [6]? And why use these vegetable oils for chemicals when they are edible? A shift to use bioresources not in competition with food has been observed, but often their global impact (e.g., on water) is not really considered in depth. The push towards the use of biomasses for fuels and chemistry in general derives in large part from several external factors, but the sustainability of their use is still not fully demonstrated and in some cases it is instead shown that these are not sustainable. We may thus argue that more care should be made in considering as "greener" a chemistry that uses products derived from biomasses.

Another interesting example is the case of DDT [1,1'-(2,2,2-trichloroethylidene)bis(4-chlorobenzene)], a well-known synthetic pesticide. Between 1940s and 1950s, DDT was used in massive ways to control mosquitoes spreading malaria, typhus, and other insect-borne diseases among both military and civilian populations. In 1955, the World Health Organization commenced a program to eradicate malaria worldwide, relying largely on DDT. However, resistance soon emerged in many insect populations as a consequence of widespread agricultural use of DDT. The program was successful in eliminating malaria only in areas with "high socio-economic status." In 1962, the book *Silent Spring* by American biologist Rachel Carson was published. The book catalogued the environmental impacts of the indiscriminate spraying of DDT in the USA and questioned the logic of releasing large amounts of chemicals into the environment without fully understanding their effects on ecology or human health. *Silent Spring* resulted in a large public outcry that led to DDT being banned in the USA in 1972. DDT was subsequently banned for agricultural use worldwide under the Stockholm Convention. DDT is one of the so-called "dirty dozen," twelve persistent organic pollutants, banned by this convention. However, there are claims that restrictions on the use of DDT in vector control have resulted in substantial numbers of unnecessary deaths due to malaria. Estimates for the number of deaths that have been caused by an alleged lack of availability of DDT range from hundreds of thousands to much higher figures. Robert Gwadz of the National Institutes of Health said in 2007 that "The ban on DDT may have killed 20 million children" [7]. The problem is that in cost–benefit terms DDT could be still preferable, when applied under careful control and not in the indiscriminate manner used earlier. A more comprehensive approach to measure cost-effectiveness or efficacy of malarial control would not only measure the cost and the number of people saved, but would also take into account the negative aspects of insecticide use on human health and ecological damage. The use of DDT in malarial epidemic situations (e.g., several areas of Africa) is still the preferable solution. The situation is different in other countries. We conclude this example by remarking that a chemical that should be not used according to "green" chemistry principles may instead be used in specific cases and can be the most sustainable solution.

Transforming a sustainability strategy into action requires thus not only the development of a practical framework aimed to identify and assess options for improving the sustainability of chemical processes and products, address issues such as risk and security, green building design, globalization and corporate responsibility [8], but also to establish reliable and robust bases to characterize the various aspects of the complex phenomenon.

In this chapter, starting from an analysis of the first and perhaps best known parameter of "green" chemistry, the E-factor, the main features of green chemistry metrics [9] will be discussed to conclude with the more comprehensive approach based on life-cycle analysis (LCA) or related methods. However, to put the problems in a more general context, and develop the bases for accounting for chemical sustainability (as far as we know, there are still no reliable attempts in this direction), we will also discuss aspects related to metrics for environmental analysis in a sustainability perspective [2], environmental and resource accounting [10] for ecological footprint, impact assessment for sustainable development [11] and accounting for sustainability [12]. These aspects are not exhaustive for the complex problem of the sustainability assessment. However, we feel that these aspects provide the foundation on which to analyze how to account for chemical sustainability.

5.2
Ecological Footprint

The ecological footprint (EF) is one of the most widely used indicators of sustainability. It is a measure of human demand on the Earth's ecosystems and compares human demand with planet Earth's ecological capacity to regenerate it. It represents the amount of biologically productive land and sea area needed to regenerate the resources that the human population consumes and to absorb and render harmless the corresponding waste, given prevailing technology and resource management practice.

This approach can also be applied to an activity such as the manufacturing of a product, but has scarcely been used to assess chemical production. In contrast to other indicators of ecological wellbeing (biological diversity, soil erosion, over-grazing, forest loss, over-fishing, local air pollution in developing countries, over-abstraction and pollution of fresh water, and so on), the EF is well adapted to quantify the concept discussed in Chapter 1 regarding the need to evaluate chemical production not in abstract terms but with specific reference to the local impact on the eco-system. Owing to the dependence of EF on the land productivity (high-productivity agricultural mono-cultures give a higher biocapacity to such regions), EF should be used in conjunction with indicators of biodiversity. Indeed, WWFs Living Planet Report complements the biennial Footprint calculations with the Living Planet Index of biodiversity [13].

The EF clarifies the relationship of resource use to equity by explicitly tying activities to ecological demands and thus provides a systematic resource accounting tool to develop methodologies for accounting the chemical production. This resource accounting is similar to life cycle analysis wherein the consumption of energy, biomass (food, fiber), building material, water and other resources are converted into a normalized measure of land area called 'global hectares' (gha).

The value of EF as indicator also for chemical production lies in the explicit indication that the ecosystem has a value equivalent to its ecological yield valued as it would be on commodity markets: for the value of water, wood, fish or game that is purified or nurseried or generated or harbored in that ecosystem. Thus, a price can be put on the natural capital of an ecosystem based on the price of natural resources it yields each year. Therefore, EF provides the basis to estimate the ecological cost of production of a chemical that should be added to the industrial cost to develop an intrinsic cost that considers not only the product itself but also how it is produced. In Chapter 1 the need for a new model of global economy, which includes sustainability and ecosystem valuation in the value of goods, has already been discussed.

Related indicators to EF are the carbon and water footprints. A carbon footprint is a measure of the impact human activities have on the environment in terms of the amount of greenhouse gases produced, measured in units of carbon dioxide. The water footprint is an indicator of water use that includes both direct and indirect water use of a consumer or producer. Water use is measured in water volume consumed (evaporated) and/or polluted per unit of time.

The value of these indicators clearly emerges when the role of biomasses in fuel and chemical production is considered. Holzman [14], in discussing the carbon footprint of biofuels, introduced the concept of "carbon debt," for example, the time necessary before a net carbon saving from the use of biofuels is achieved. He introduced in the estimation the results of two recent studies published in *Science* [15, 16], evidencing that clearing new land and converting existing cropland to produce feedstocks incurs a "carbon debt." In fact, in analyzing how biofuels will reduce greenhouse gases it should not only be considered that biofuels sequester carbon through the growth of the feedstock but also the carbon emissions that occur as farmers worldwide respond to higher prices by converting forest and grassland into new cropland to replace the corn (or cropland) diverted to biofuels should also be counted. By using a worldwide agricultural model to estimate emissions from land-use change, it was found [15] that corn-based ethanol, instead of producing a 20% savings, nearly doubles greenhouse emissions over 30 years and greenhouse gases for 167 years increases.

Thus, within the critical time frame for avoiding a climate catastrophe, for example, probably a couple of decades, according to the February 2007 United Nations report "Confronting Climate Change: Avoiding the Unmanageable and Managing the Unavoidable," biofuel crops will only aggravate global climate disruption. The debt repayment time for biofuels varies from about 17 years for ethanol produced from highly productive sugarcane grown on the Brazilian Cerrado, to 420 years for biofuels from palm oil grown in tropical peatland [16]. Even switching existing croplands from food to biofuel feedstocks would result, albeit indirectly, in similar carbon debts [15].

Even if part of above results could be questionable, they evidence that the use of ecological footprint indicators are valuable tools to analyze the impact on the eco-system of chemical production.

5.3
Ecological Indicators

Agenda 21, for example, the plan of action to achieve sustainable development that was adopted at the United Nations Conference on Environment and Development held in Rio de Janeiro in June 1992, calls on countries and the international community to develop indicators of sustainable development. Such indicators are needed to increase focus on sustainable development and assist decision-makers at all levels to adopt sound national sustainable development policies.

The Commission on Sustainable Development (CSD) of the United Nations finalized a set of 96 indicators, including a subset of 50 core indicators, in 2006 [17]. CSD themes cover a range of topics, from poverty, to education, atmo-sphere, land, biodiversity, and so on. Most of them do not adapt to use for assessing sustainability of chemical production. The United Nations has also developed the Millennium Development Goals (MDGs) Indicators, a set of 48 indicators linked to the eight goals derived from the United Nations Millennium Declaration. Like the CSD indicators, the MDG Indicators are driven by policy relevance and an overlap between the indicators exists, even if they have different goals.

5.4
Metrics for Environmental Analysis and Eco-Efficiency

Various metrics and methods in engineering design are used to evaluate and measure the different aspects of the environmental impact of industrial activities and services. These metrics differ in terms of approach to manage this complex problem, involving various kinds of interactions. An in-depth analysis of this problems is beyond the scope of the present chapter, which will thus be limited to the classification used by Seager and Theis [18]. Sustainable metrics may be characterized in a classification that includes five broad categories:

1. **Financial metrics:** estimate environmental impacts or ecosystem services in terms of currency so that they may be compared with monetary transactions or industrial accounts. In practice, monetization may lead to the erroneous assumption that environmental exploitation can be reversible in a manner analogous to pecuniary transactions, even if in many cases ecological systems are damaged beyond recovery. Nevertheless, emission trading already exists. For example, the European Union Greenhouse Gas Emission Trading Scheme involves all the 25-member states of the European Union participating in the scheme. Notably, emissions trading contributes little to solve pollution problems, as groups that do not pollute sell their conservation to the highest polluters. However, as discussed in Section 5.2, monetization of the impact on the eco-system can be used in a different way as an additional cost to that of production so as to incentive companies to produce in a cleaner way, bypassing the excuse of global market competition.

2. **Thermodynamic metrics:** indicate the resource requirements of industrial activities or services, but usually do not include the specific environmental impacts associated with resource consumption and the export of exogenous material into the environment. Only a few, such as the concept of exergy (available work), attempt to indicate whether the resources consumed were used wisely and efficiently.

3. **Environmental (including health and safety) metrics:** estimate the potential for creating chemical changes or hazardous conditions in the environment. They may be simple measures of what is released to the environment, without chemical considerations such as pollutant degradation, catalysis or recombination to form new pollutants; or they may include potency factors, such as toxicity, reactivity or rarity, and the fate/transport of the pollutants. Most are directed at specific biological or ecological end points, such as death, cancer or mutation, while others may indicate a loss of environmental quality without suggesting any particular ecological manifestation. It is possible for environmental metrics to be expressed in chemical or thermodynamic units. However, environmental metrics are distinguished from thermodynamics by the fact that these are intended to measure environmental loads or changes rather than resource demands. They are generally measures of the waste created by industrial processes rather than by the use of raw materials.

4. **Ecological metrics:** attempt to estimate the effects of human intervention on natural systems in ways that are related to living things and ecosystem functions. The rates of species extinction and loss of biodiversity are examples, and may be incorporated in the concept of ecosystem health. Ecological metrics relate to biological processes, whereas environmental metrics relate to chemical or other hazardous conditions. For example, a pollution-free environment does not lead to

recovery of a depleted bear population if there is a total absence of quality sites due to of the human pressure.

5. **Socio-political metrics:** evaluate whether industrial activities are consistent with political or ethical goals.

In addition, there are aggregated metrics that combine features or metrics belonging to various categories, or they may group several metrics that belong to a single category.

An example of this composite index is the Environmental Sustainability Index (ESI), which tracks 21 elements of environmental sustainability, covering natural resource endowments, past and present pollution levels, environmental management efforts, contributions to protection of the global commons, and a society's capacity to improve its environmental performance over time [19]. The ESI was developed to evaluate environmental sustainability relative to the paths of other countries. Owing to a shift in focus by the teams developing the ESI, a new index was developed, the Environmental Performance Index (EPI), which uses outcome-oriented indicators, working as a benchmark index that can be more easily used by policy makers, environmental scientists, advocates and the general public [20]. The EPI focuses on two overarching environmental objectives: (i) reducing environmental stresses to human health and (ii) promoting ecosystem vitality and sound natural resource management. The two overarching objectives are gauged using 25 performance indicators tracked in six well-established policy categories, which are then combined to create a final score. Figure 5.1 gives the scheme of how the EPI is constructed.

For each indicator, a relevant long-term public health or ecosystem sustainability goal is identified. These targets are drawn from (i) treaties or other internationally agreements, (ii) standards set by international organizations, (iii) leading national regulatory requirements and (iv) prevailing scientific consensus. The indicators serve as a gauge of long-term environmental policy success. For each country and each indicator, a proximity-to-target value is calculated based on the distance from current results to the policy target. Then, giving a weight to each indicator, the overall Environmental Performance Index is calculated.

This method is designed for evaluation of performances of countries. The same approach, but adapting indicators, could be used also to assess the sustainability of chemical processes. Schwarz et al. [21] have proposed the use of sustainability metrics to guide decision-making managers in chemical companies. In fact, they suggested that a management strategy that incorporates eco-efficiency strives to create more value with less impact. They suggested the use of a few basic indicators of sustainability: (i) material intensity, (ii) energy intensity, (iii) water consumption, (iv) toxic emissions and (v) pollutant emissions and (CO_2) emissions. They suggested a limited core set of indicators, but the approach can be expanded using complementary metrics.

Each metric is constructed as a ratio, with impact, either resource consumption or pollutant emissions, in the numerator and a representation of output, in physical or financial terms, in the denominator. To calculate the metrics, all impact numerators and output denominators are normalized.

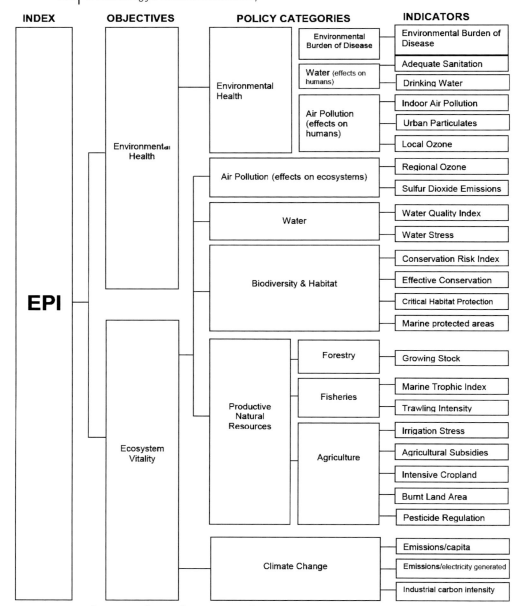

Figure 5.1 Scheme of construction of an Environmental Performance Index. Source: Yale Center for Environmental Law and Policy [20].

Metrics are useful decision-support tools for evaluating alternative processes for the manufacture of a given product. Table 5.1 reports an example, taken from Schwarz *et al.* [21], which illustrates the metrics for two hexamethylenediamine

Table 5.1 Comparing alternative production processes – metrics for hexamethylenediamine production (more favorable metrics are shown in bold). Source: Schwarz et al. [21].

Metric	Unit[a]	Hydrocyanation of butadiene	Electrohydrodimerization of acrylonitrile
Material	lb per $VA	1.44	**0.17**
Energy	kBtu per $VA	**59.4**	92.1
Water	lb per $VA	16.2	**15.4**
Toxics	lb per $VA	0.0023	**0.0000**
Pollutants	lb per $VA	0.81	**0.008**
CO_2	lb per $VA	**8.85**	13.2

[a]$VA = dollar value-added.

(HMDA) manufacturing processes. The two sets of metrics clearly show the trade-offs for these processes. The hydrocyanation process has lower energy-intensity and greenhouse-gases metrics, but its material-intensity, water consumption and pollutants metrics are higher than those for the electrohydrodimerization process.

Similarly, metrics provide a means for comparing resource consumption and pollutant emissions for the manufacture of various products. The comparison of the metrics for various processes serves to highlight those areas, such as high energy intensity or toxics emissions, that pose potential business risks. An important characteristic of the metrics is that they are stackable – that is, they can be combined (or stacked) to calculate environmental impact per pound of product over the series of processes that comprise a supply chain.

Figure 5.2, also taken from Schwarz et al. [21], illustrates how metrics for ethylene, chlorine, vinyl chloride and poly(vinyl chloride) (PVC) can be stacked to obtain metrics for the production of PVC, beginning with naphtha and brine. The metrics calculated with the mass denominator can be readily combined. Impact per dollar can also be calculated for a supply chain by combining the single values along the chain in

Metric	Material (lb/lb PVC)	Energy (kBtu/lb PVC)	Water (gal/lb PVC)	Toxics (lb/lb PVC)	Pollutants (lb/lb PVC)	CO_2 (lb/lb PVC)
Ethylene × 0.474	0.039	1.473	0.433	0.00027	0.00006	0.207
Chlorine × 0.601	0.007	5.050	0.380	0.00000	0.00014	0.748
Vinyl Chloride × 1.013	0.206	4.966	2.129	0.00000	0.00356	0.756
Polyvinyl Chloride	0.049	3.935	0.619	0.00203	0.00000	0.546
PVC Supply Chain	**0.300**	**15.424**	**3.561**	**0.00230**	**0.00376**	**2.257**

Figure 5.2 Stacking metrics for the PVC supply chain. Source: Schwarz et al. [21].

the same way that impact per pound of product is stacked. Note, however, that the values were not updated, and may thus actually be different.

The metrics can be integrated with other analysis tools, such as practical minimum-energy requirements, lifecycle inventories and total cost assessment (TCA), to develop integrated decision-support tools that can provide guidance for decision-makers [8].

BASF has also developed the tool of eco-efficiency analysis to address not only strategic issues but also issues posed by the marketplace, politics and research [22, 23]. The major elements of the environmental assessment include primary energy use, raw materials utilization, emissions to all media, toxicity, safety risk and land use. The basic preconditions in eco-efficiency analysis are [22]:

- the concrete (final) customer benefit is at the heart of the analysis;
- all products or processes studied have to meet the same customer benefit;
- the entire life cycle is considered;
- both an ecological and an economic assessment are carried out;
- impact on health and the danger to people is assessed.

Eco-efficiency assessment focuses in principle on the entire life cycle, but then concentrates on specific events in a life cycle where the alternatives under consideration differ. Eco-efficiency analysis includes the cost data as well as the straight life cycle data. Figure 5.3 shows that life cycle assessment is based on the environmental profile, which can be obtained, for example, from data provided by the plants and which includes the path from the cradle to the work-gate. On extending this approach to the entire life cycle, a life cycle assessment is obtained. Adding to these additional assessment criteria again, followed by an economic assessment, then leads to an eco-efficiency analysis (Figure 5.4).

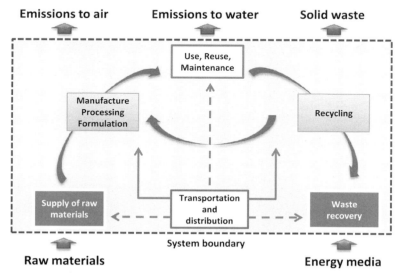

Figure 5.3 Life cycle assessment. Source: adapted from Saling *et al.* [22].

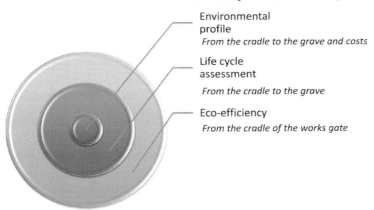

Environmental
profile
From the cradle to the grave and costs

Life cycle
assessment
From the cradle to the grave

Eco-efficiency
From the cradle of the works gate

Figure 5.4 Eco-efficiency analysis includes life cycle assessment and environmental profile. Source: adapted from Saling *et al.* [22].

First, the environmental impact is described based on six categories:

1. Raw material consumption.
2. Energy consumption.
3. Land use.
4. Air and water emissions and disposal methods.
5. Potential toxicity.
6. Potential risks.

Energy consumption is determined over the entire life cycle and describes the consumption of primary energy. Fossil energy media are included before production and renewable energy media before harvest or use. This captures conversion losses from electricity and steam generation.

Emission values are initially calculated separately as air, water and soil emissions (waste). The calculation includes not only values, for example, from electricity and steam production and transport but also values directly from the processes. The individual values are subsequently aggregated via a weighting scheme to form the overall value for the emissions. Air emissions considered were CO_2, SO_2, NO_x, CH_4, hydrocarbons (HC), halogen HC, NH_3, N_2O, HCl and HF. These are lumped into four impact categories:

1. Global warming potential (GWP).
2. Ozone depletion potential (ODP).
3. Photochemical ozone creation potential (POCP).
4. Acidification potential (AP).

For the inventory of emissions to water, the following aspects are considered: COD (chemical oxygen demand), BOD (biological oxygen demand), N-tot (total nitrogen), NH_4^+ (ammonium), PO_4^{3-} (phosphate), AOX (adsorbable organic halogen), heavy metals (HMs), hydrocarbons (HCs), SO_4^{2-} (sulfate) and Cl^- (chloride).

The results of the inventory on solid wastes are combined to form three waste categories: special wastes, wastes resembling domestic refuse and building rubble/gangue material.

Under raw material consumption, the mass of raw materials needed by the corresponding process is determined first. The individual materials are weighted according to their reserves. The toxicity potential is calculated using the classifications for hazardous materials under EU law. The abuse and risk potential reflects the dangers of accidents in the manufacture, use and recycling of the product. The approach adopted is similar to a risk assessment in the case of plant safety in which the probability of occurrence and the level of damage are estimated. Values used for the individual products are only comparative and not absolute.

The different categories, after normalization, are then aggregated using weighting factors. For more details on the estimations of the different categories and methods of their aggregation, reference could be made to the work of Saling *et al.* [22] and to the BASF web site of eco-efficient analysis (http://www.corporate.basf.com/en/sustainability).

The results can also be viewed using a special plot called the environmental fingerprint. Figure 5.5 reports an example taken from the work of Saling *et al.* [22] of the evaluation of different alternatives in the use of indigo, the dye that is used for coloring in blue the jeans. Clearly, the electrochemical variant is the most advantageous alternative in all categories. With the exception of risk potential, the least favorable variant on all criteria is indigo powder from plants.

Estimating the total costs it is then possible to plot normalized cost versus environmental impact and thus select the most favorable alternatives that are located top right in the plot (Figure 5.6). The distance of the individual alternatives to the plot diagonal is a measure of the respective eco-efficiency.

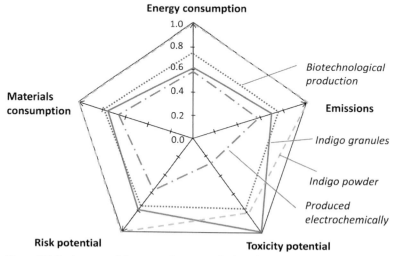

Figure 5.5 Environmental fingerprint by BASF of indigo dye (dyeing process only). Source: adapted from Saling *et al.* [22].

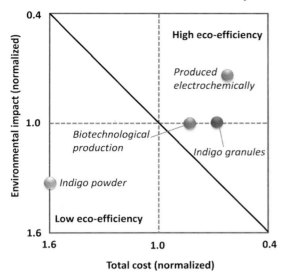

Figure 5.6 Eco-efficiency versus total cost (portfolio plot) for the indigo study (dyeing process only). Source: adapted from Saling *et al.* [22].

The value of the eco-efficiency analysis tool lies also in the recognition of dominant influences and in the illustration of "what if . . .?" scenarios. It is also possible to perform sensitivity analyses in every project. Not only the assumptions made, but also the system boundaries and the societal weighting factors, are varied and checked within realistic ranges.

Eco-efficiency analysis results make it also possible to identify weaknesses in products, processes and overall systems over the entire life cycle. This makes it possible to identify factors whose optimization would result in distinct improvements in the overall position of an alternative under consideration. From another point of view, it is possible to use this tool to develop marketing strategies with a joined-up focus and identify synergistic impacts in the overall process.

Eco-efficiency analysis can be expanded to include an assessment of the social dimension of the sustainability-analysis (SEEbalance methodology) [24]. In this way, it is possible to obtain an integrated assessment of economic, ecological and social aspects of products and processes, and introduce the ecotoxicity evaluation model as a standard tool to the environmental dimension of the analysis.

Although providing only comparative and not absolute values, the methodology of eco-efficiency is very useful for an evaluation of process alternatives. In Chapter 1 (Section 1.6.1), Table 1.10 summarizes the results of a study by EuropaBio to evaluate the contribution of biotechnologies to sustainability. One of the cited examples is the synthesis of vitamin B2, an essential nutrient found in meat, dairy foods, plant foods and corn products and which is required by the body to break down food components, maintain tissue and absorb other nutrients. Using the eco-efficiency and, particularly, the portfolio plot it is possible to further demonstrate the benefits

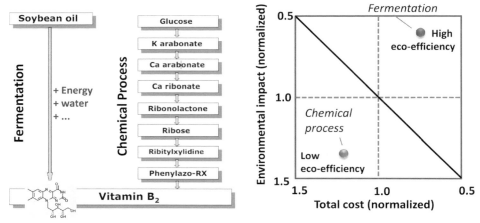

Figure 5.7 Comparison of process alternative pathways of the synthesis of vitamin B2. Source: adapted from Saling [25].

of the new biotechnology route with respect to the traditional chemical route (Figure 5.7) [25].

Another example regards an aspect discussed before on the sustainability of land use for biofuels. We have already remarked the various problems in this analysis that can be summarized, in a simplified form, in reply to the question: which use is the most sustainable for one ha of land? There are different possible options to produce: (i) bioliquids (by fast pyrolysis), bioethanol or biodiesel, (ii) food, (iii) chemicals, (iv) electricity or (v) to feed animals. There are more options (e.g., biogas) but the present already allow a good comparison, although theoretical (land productivity depends on several factors and, for example, palms to produce the vegetable oil raw material for biodiesel is not locally in competition with the production of bioethanol from biomasses for which different climate and land characteristics are required).

Despite these limitations, the eco-efficiency analysis provides a good bases for discussion. Figure 5.8 compares different alternatives, showing that, for example, the use of biomass for electricity production via the conventional combustion route is not different, in terms of eco-efficiency, with respect to biodiesel, while bioethanol is slightly worse. The best situation is for vitamin B2 by fermentation, but clearly the dimension of the market between vitamin B2 and biofuels is completely different.

5.5
Sustainability Accounting

Sustainability accounting [26–28] is a general term for the variety of measurements used as a quantitative basis for the informed management of environmental, social and economic sustainability. However, notably, there are no commonly accepted

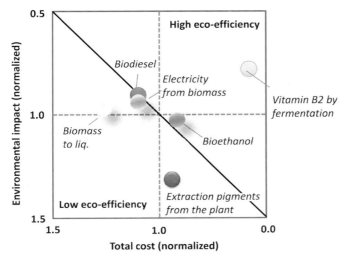

Figure 5.8 Eco-efficiency analysis of different options in using land. Source: adapted from Saling [25].

definitions of sustainability accounting and/or sustainability reporting, the latter being the practice of measuring, disclosing and being accountable to internal and external stakeholders for organizational performance towards the goal of sustainable development. Accounting and reporting are thus used with the same meaning in the context of sustainability.

In this respect, the Global Reporting Initiative's (GRI) (www.globalreporting.org) is a worldwide multi-stakeholder network that aims to provide guidance for organizations to use as the basis for disclosure about their sustainability performance, and also provides stakeholders a universally-applicable, comparable framework in which to understand disclosed information. A main output is the Sustainability Reporting Guidelines, known as the G3 Guidelines, which was published in the third version in 2006.

An important aspect regards the concept of sustainability context. G3 Guidelines indicate that reporting only on trends in individual performance (or the efficiency of the organization) will fail to respond to the critical question of how an organization contributes, or aims to contribute in the future, to the improvement or deterioration of economic, environmental, and social conditions, developments, and trends at the local, regional, or global level. Reports should therefore seek to present performance in relation to broader concepts of sustainability. This could mean that, in addition to reporting on trends in eco-efficiency, an organization might also present its absolute pollution loading in relation to the capacity of the regional ecosystem to absorb the pollutant. We thus see that the concept of sustainability accounting and/ or reporting, although apparently similar to that of environmental analysis and eco-efficiency discussed in the previous section, presents distinct differences. In fact, while in the latter case the focus is on the process and product evaluation, in the case of sustainability accounting/reporting the organization presents its performance with reference to broader sustainable development conditions and goals in appropriate geographical contexts. In addition, the focus is on how sustainability topics relate to long-term organizational strategy, risks and opportunities, including supply-chains. Sustainability accounting is considered a part of the sustainability science [29] that has emerged as a new academic discipline, although a clear distinction between sustainability, sustainable development and sustainability science is not clear.

Measure and monitor are the two central elements of accounting. The cited G3 Guidelines organize the sustainability performance indicators in economic, environmental and social categories. Social indicators are further categorized by labor, human rights, society, and product responsibility. Core indicators have been also identified, and are intended to identify generally applicable indicators.

The economic indicators illustrate the organization's impacts on the economic conditions of its stakeholders and on economic systems at local, national, and global levels. The objective is to evidence the organization's contribution to the sustainability of a larger economic system. If we refer to the example discussed in the previous section on eco-efficiency, the difference between the two approaches is now clearly evidenced. In eco-efficiency the objective is to develop a tool for correct management of the company based on sustainability criteria. In sustainability accounting/reporting the focus is on the organization as a dynamic part of a sustainable systems. Therefore, in eco-efficiency the focus is internal to the company, while sustainability accounting/reporting is external, for example, on the system.

The environmental indicators cover performances related to inputs (e.g., material, energy and water) and outputs (e.g., emissions, effluents and waste). In addition, they cover performances related to biodiversity, environmental compliance and other relevant information such as environmental expenditure and the impacts of products and services. These indicators are close to those discussed in the previous sections. The core indicators in this category are:

- **Materials:**
  - materials used by weight or volume;
  - percentage of materials used that are recycled.

- **Energy:**
  - direct energy consumption by primary energy sources;
  - indirect energy consumption by primary sources.

- **Water:**
  - total water withdrawal by source.

- **Biodiversity:**
  - location and size of land owned, leased, managed in, or adjacent to, protected areas and areas of high biodiversity value outside protected areas;
  - description of significant impacts of activities, products and services on biodiversity in protected areas and areas of high biodiversity value outside protected areas.

- **Emissions, Effluents and Waste:**
  - total direct and indirect greenhouse gas emissions by weight;
  - other relevant indirect greenhouse gas emissions by weight;
  - emissions of ozone-depleting substances by weight;
  - NOx, SOx and other significant air emissions by type and weight;
  - total water discharge by quality and destination;
  - total weight of waste by type and disposal method;
  - total number and volume of significant spills.

- **Products and Services:**
  - initiatives to mitigate environmental impacts of products and services, and extent of impact mitigation;
  - percentage of products sold and their packaging materials that are reclaimed by category.

- **Compliance:**
  - monetary value of significant fines and total number of non-monetary sanctions for noncompliance with environmental laws and regulations.

In this case, together with indicators analogous to those discussed in the previous sections, there are also different indicators (in particular, those of products and services and compliance sub-categories).

The Social Performance Indicators identify key performance aspects surrounding labor practices, human rights, society, and product responsibility.

The result of the sustainability accounting/reporting is to provide all the elements necessary for sustainability auditing (e.g., for ISO 14000 environmental management standards) and to evaluate the sustainability performance of a company, organization or other entity using various performance indicators.

5.5.1
System Boundary

Any evaluation of sustainability should first clearly define the boundary conditions of the analysis, as well as the definition of sustainable development (SD). Owing to several different definitions and interpretations of the SD concept and boundary limits, there are often contrasting indications in the literature. Boundary conditions must address two fundamental issues: time scale and spatial scale. Defining the temporal and spatial scales defines the limits to sustainability [30].

To date, most definitions deal with spatial boundaries set within the Earth system (generally as a closed system, with two forms of openness: absorption of energy – predominantly solar – and dissipation of heat into space). Within the spatial scale of Earth, all variations are permitted, from the most local (from rural, village to urban) to national, regional, sub- and full continental to global.

While spatial boundaries can be set with a useful degree of concreteness, the temporal boundaries are more elusive and more arbitrary (though from a systems perspective, political-administrative spatial boundaries are also arbitrary), and very often they are not set explicitly.

Even the best-known temporal definition of SD, that of the Brundtland Report (Brundtland 1987), speaks vaguely of "future generations;" in most other cases the time frame of SD is indefinite. The most common implicit time frame within which SD is discussed is tied to political processes and cycles. However, it is often hidden and can lead to different interpretations.

In conclusion, there is the need for a common interpretation of sustainable development (SD) in the context of evaluation [30], because the evaluation depends in turn on the definition of SD.

5.6
E-Factor and Atom Economy

Previous sections have outlined the general aspects of evaluation of sustainability, and the value and limits of the different methodologies. Even though not exhaustive of the state of the art in this broad topic, which is beyond the scope of this book, the discussion can provide the elements for an evaluation of the specific indexes and methodologies proposed for "green" (sustainable) chemistry. We also remark that some of the indicators and methodologies discussed in the previous sections do not adapt well to the specific case of assessing sustainability of chemical production, while others, in particular the eco-efficiency method, are well suited for the purpose. However, this method, and most of the indicators/methodologies we discuss in this and the following section, is based mainly on the estimation of the input/output process of a chemical production and has limited considerations on how the process and product is put in relation with the whole eco- and societal-system. This is why it would be necessary to move to evaluation models that also consider these aspects, for example, to develop accounting models for chemical sustainability.

Two of the most used indicators to assess "green" chemistry and evaluate the potential environmental acceptability of chemical processes are the following [31–34]:

- *Atom economy*, originally proposed by Barry Trost of Stanford University [35], which is a measure of how much of the reactants remain in the final product in a chemical reaction; mathematically, the atom economy (AE) is the molecular weight (MW) of the desired product divided by the sum of the molecular weights of all substances produced:

$$AE\,(\%) = 100 \cdot \frac{\text{MW desired product}}{\sum(\text{MW all reactants})}$$

It is a simple calculation based on the stoichiometry of the reaction, but does not account for solvents, reagents, reaction yield and reactant molar excess. Atom economy is one of the 12 principles of green chemistry [36]. The larger the number, the higher the percent of all reactants appearing in the product.

- The *E factor*, introduced by Sheldon [31, 32], is defined as the mass ratio of waste to desired product:

$$\text{E-factor} = \frac{\text{Total waste (tons)}}{\text{Product (tons)}}$$

Notably, the E-factor depends on the definition of waste, and may include process use only, or chemicals needed for scrubbing, for example. The E-factor can be split into different sub-categories: (i) organic waste and (ii) aqueous waste.

The E-factor takes the chemical yield into account and includes reagents, solvents losses, all process aids and, in principle, even fuel (although this is often difficult to quantify and it is usually not included). An exception is water as solvent, which is generally not included in the E factor.

The smaller the number the closer to zero waste is being produced.

A review of the effect that the E Factor concept has had over the last 15 years on developments in the (fine) chemical industry and pharmaceutical industry with regard to waste minimization and to assessing its current status in the broader context of green chemistry and sustainability has been presented recently by Sheldon [37]. It was concluded that the E Factor concept played a major role in focusing the attention of the chemical industry worldwide, and particularly the pharmaceutical industry, on the problem of waste generation in chemicals manufacture.

One of the most cited examples regarding this topic and the magnitude of the waste problem in chemicals manufacture regards the typical E factors in various segments of the chemical industry (Table 5.2).

There are various other related metrics: (i) atom efficiency, (ii) effective mass yield, (iii) carbon efficiency and (iv) reaction mass efficiency.

- **Atom efficiency:** the percentage yield (molar flow of the desired product divided by the molar flow of the limiting reactant, taking into account the stoichiometry of the reaction) multiplied by the atom economy. It could be used to replace yield and AE. For example, AE could be 100% and yield 5%, making this a not very green reaction.

- **Effective mass yield (EMY):** [38]: the percentage of the mass of the desired product relative to the mass of all non-benign materials used in its synthesis. However, what is "benign" it is not defined.

Table 5.2 The E factor (kg-waste[a] per kg-product). Source: Sheldon *et al.* [34].

Industry segment	Product tonnage[b]	E factor
Oil refining	10^6–10^8	<0.1
Bulk chemicals	10^4–10^6	<1–5
Fine chemicals	10^2–10^4	$5 \rightarrow 50$
Pharmaceuticals	10–10^3	$25 \rightarrow 100$

[a]Defined as everything produced except the desired product (including all inorganic salts, solvent losses, etc.).
[b]Typically represents annual production volume of a product at one site (lower end of range) or worldwide (upper end of range).

- **Carbon efficiency:** [39]: the percentage of carbon in the reactants that remains in the final products. It takes into account the yield and stoichiometry. The advantage is that it is directly related to greenhouse gases.

- **Reaction mass efficiency (RME):** the percentage of the mass of the reactants that remains in the product. It takes into account the atom economy, yield and reactant stoichiometry.

There are more indicators (e.g., mass intensity, MI, and mass productivity) that belong to this general class of resource intensity indicators, for example, which quantify "greenness" of chemical processes and products in terms of effectiveness of mass and energy intensity.

5.6.1
Limits to Their Use

All the indicators discussed above are based on mass intensity. These indicators have been developed specifically for their use in synthetic chemistry, either at the laboratory or industry level. They have in common the objective of showing to what extent a chemical reaction is wasteful in material terms; simplicity is their main merit, as most of them can be easily calculated as soon as the chemical reaction is defined and quantified.

However, they include neither information on energy issues nor on the toxicity or hazards associated with the inputs used and the waste produced. Another question is which indicator to use. Sometimes, different indicators can lead to different (conflicting) results, as shown by Constable *et al.* [39], who checked 28 chemical reactions against yield, atom economy, carbon efficiency, RMI, MI and mass productivity.

Scheme 5.1 shows the reaction taken as an example by Constable *et al.* [39] and the main elements needed to calculate the above indicators. The various mass intensity indicators are calculated considering a yield of 90%.

$$E-factor = \frac{[(10.81 + 21.9 + 500 + 15) - 23.6}{23.6} = \frac{22.2 \text{ kg waste}}{1 \text{ kg product}}$$

$$AE = 100 \cdot \frac{262.29}{(108.1 + 190.65 + 101)} = 65.8\%$$

$$\text{Atom Efficiency} = 90\% \cdot 65.8\% = 59.2\%$$

$$\text{Carbon Efficiency} = \left[\frac{(0.09 \cdot 14)}{(0.1 \cdot 7) + (0.115 \cdot 7)}\right] \cdot 100 = 83.7\%$$

$$\text{Mass Efficiency} = 100 \cdot \frac{23.6}{(10.81 + 21.9)} = 70.9\%$$

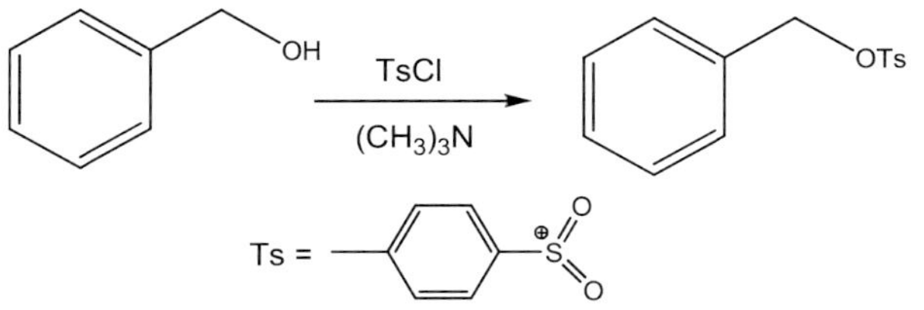

Reactant	Benzyl alcohol	10.81 g	0.10 mol	MW 108.1
Reactant	Tosyl alcohol	21.9 g	0.115 mol	MW 190.65
Solvent	Toluene	500 g		
Auxiliary	Triethylamine	15 g		MW 101
Product	Sulfonate ester	23.6 g	0.09 mol	MW 262.29

Scheme 5.1 Reaction used as an example to estimate the value of different mass intensity indicators. Source: adapted from Constable [39].

5.6.2
Applicability to Evaluating the Sustainability of Chemical Industrial Processes

Although the above indicators, due to their simple calculation and use, are an useful tool, and evidence the general problem of waste minimization in the chemical industry, their effective relevance in the sustainable industrial chemistry approach is questionable, at least in terms of effective driving forces (as qualifying element) for improving chemical sustainability. In fact, waste is always an additional cost. Avoiding waste and minimizing solvent use are thus primarily an economic motivation for a chemical process. Reducing the amount of by-products means, firstly, improving the yield to the desired product, but especially significantly reducing the costs of separation.

Co-production is still used in many relevant industrial processes. A typical example is the synthesis of phenol via the cumene process, which involves the formation of acetone as by-product. This example is discussed in a more detail later in Chapter 13. Direct oxidation of benzene to phenol using H_2O_2 is an attractive industrial alternative, but a main motivation is to avoid the formation of acetone, as there is actually excess acetone on the market. The process requires a single step versus three steps in the cumene process, one of which is related to the synthesis of a cumene hydroperoxide intermediate that shows significant problems of safety. There are thus various aspects that make more sustainable the direct synthesis of phenol over the cumene process, but atom economy (or related mass intensity indicators) is not the correct indicator to assess sustainability.

Scheme 5.2 reports the atom economy for two direct routes (using N_2O or H_2O_2 as the oxidant) and for the commercial route via cumene. This example shows some of the problems in using this indicator to evaluate the "greener" process. Are the direct routes better than the cumene route? The problem is related to whether to consider

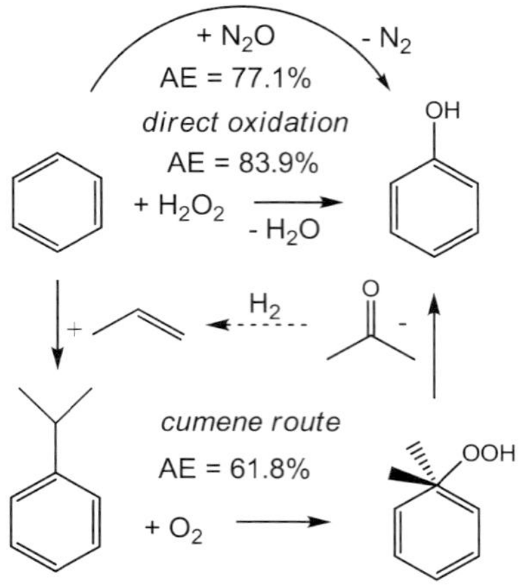

Scheme 5.2 Atom economy (AE) of different routes for phenol synthesis.

acetone as a by-product or as a co-product. If it is sold as a co-product the AE indicator is not correct. If it is not sold, acetone could be hydrogenated back to propene. In this case, the AE would be equal to that of the direct synthesis of phenol by oxidation of H_2O_2, but the sustainability of the two processes is different.

In terms of atom economy, direct synthesis of phenol using H_2O_2 would be preferable to the use of N_2O, but this is not in relation to a "greener" process, because in terms of by-products one gives H_2O and the other N_2, both equally ideal "green" by-products. However, the oxidation of benzene with H_2O_2 occurs in the liquid phase using organic solvents (sulfolane), while the oxidation with N_2O occurs in the gas phase (thus no solvents). Using H_2O_2 as an oxidant, conversion is low (around 10%) in order to limit further hydroxylation of phenol. For oxidation with N_2O the conversion could be higher, but carbonaceous deposits that deactivate the catalysts are formed. These carbonaceous deposits are not accounted for in the AE.

A further problem in the comparison is related to the source of the oxidant. N_2O could derive from side streams of chemical processes (adipic acid production). Being a greenhouse gas that should be eliminated, the use of N_2O from adipic acid production instead of its decomposition would thus be a further incentive for a "greener" process. However, the process never passed the stage of small-scale demonstration plant owing to several problems associated with the need of purification of N_2O, and catalyst stability and productivity. The use of H_2O_2 would be preferable from an industrial point of view, but the effective application of the process is mainly related to the industrial cost of H_2O_2. Chapter 13 gives a more in-depth analysis and comparison of the different processes, but the discussion reported here already evidences the great difficulty in using mass intensity indicators to evaluate whether a chemical industrial process would be more sustainable.

We may further add that, in the chemical industry, process improvement results in a significant reduction of waste, even if this aspect does not appear, at least in a transparent way, in the cited indicators.

The first example regards the use of pure O_2 versus air as the oxidizing agents in selective oxidation processes [40, 41]. Air has been the preferred oxidant for years, but several oxidation processes, both in the liquid and in the gas phase, have been modified over the years to allow the use of pure oxygen [41–43]. These changes have been driven by improvements in productivity and yield, while more recently revamping or modifications aimed towards pure oxygen use have been undertaken due to environmental constraints. It is likely that in the near future most industrial processes will be changed from air-based to oxygen-based, especially for debottle-necking the actual plants (increasing productivity) and improving energy savings. The use of oxygen instead of air implies that the same partial pressure can be used with a much lower total pressure than with air, thus making it feasible to possibly reduce the total pressure, with obvious energy advantages. Moreover, an increase in the reaction rate makes it possible to reach the same productivity while lowering the reaction temperature, with possible benefits from the selectivity point of view when several products are formed in the reaction. Lower nitrogen contents, a ballast with very poor thermal conduction properties, also allow better control of the temperature profile in the presence of strongly exothermal reactions.

An even more important benefit originates from the considerable decrease in polluting emissions released into the atmosphere because spent gases can be recycled when oxygen is used in place of air. Much less waste gas is produced, with energy savings during incineration. Other treatments are now possible, such as adsorption of some components or chemical treatments. In addition, the heating value of the stream is much higher than for air-based processes (the concentration of nitrogen is much lower, while that of hydrocarbons and carbon oxides is expected to be much higher). Therefore, the purge stream, instead of being treated, can be used as a fuel to incinerate other wastes.

An interesting example is ethene oxychlorination where the use of pure O_2 instead of air as the oxidizing agent improves productivity and reduces the environmental impact. The oxychlorination of ethene to 1,2-dichloroethane (DCE) reaction:

$$CH_2=CH_2 + 2HCl + 0.5O_2 \rightarrow CH_2Cl=CH_2Cl \qquad \Delta H = -238 \text{ kJ mol}^{-1}$$

is the basis for the manufacture of vinyl chloride monomer ($CH_2=CHCl$) through the so-called "balanced" oxychlorination process. Vinyl chloride monomer (VCM) is used for the production of major homo- and co-polymers, used for extrusions such as pipes, films, coatings and moldings.

Different options are available for the oxychlorination process, depending on the technology used and operation conditions [42]: (i) reactors may be either fixed bed or fluidized bed, (ii) air, or pure O_2 can be used as the oxidizing agents and (iii) excess ethene or a stoichiometric ratio can be used. In general, there are two different cases where the use of oxygen as the oxidizing agent instead of air is preferable: (i) when a large excess of ethene with respect to the stoichiometric amount is fed to the reactor, typically in a fixed-bed technology, and recycling is aimed to convert the remaining ethene, and (ii) when recycling is carried out to meet environmental constraints and avoid emission of pollutants into the atmosphere. In this case, the residual unconverted ethene is very low, since an almost stoichiometric composition is fed

Table 5.3 Vent composition for the air-based and O_2-based processes of ethene oxychlorination. Source: Centi et al. [40].

Component	Air-based process	Oxygen-based process
	Content (vol.%); flow-rate ($m^3 h^{-1}$)	
Oxygen + argon	4–8; 400–2400	0.1–2.5; <25
Ethane	0.1–0.8; 10–24	2–5; <50
COx (CO_2:CO = 3–4:1)	1–3; >100–900	15–30; <300
DCE and chlorinated comp.	0.02–0.2; 2–60	0.5–1; <10
Molecular nitrogen	rest	$\cong$600
Vent flow rate ($m^3 h^{-1}$)	10000–30000	<1000

to the reactor. In both cases the use of air would lead to concentrations of inert gas (nitrogen) that are too large, and the ratio of purge-to-recycle streams would be excessively high. When an operation is carried out with a large excess of ethene with respect to the stoichiometric requirement (fixed-bed technology), the purge stream also has a very high ethene concentration, and can be sent to the liquid-phase direct chlorination section, to further improve the yield to DCE. In this case, the gaseous vent from the direct chlorination reactor becomes the only source of pollution. Alternatively, the catalytic or thermal incineration of the purge stream can be carried out more economically than with the larger and less concentrated vent stream of the air-based process.

Table 5.3 shows typical vent compositions of air- and oxygen-based processes for fluid-bed processes (operation with almost stoichiometric feedstock).

When a large excess of ethene is used, the alkene is the major component in the gaseous stream entering the reactor. In this case, typically used in fixed-bed oxychlorination technology, the better heat transfer properties (higher heat capacity) of ethene as compared to nitrogen (which is the major gaseous component in air-based processes) allows the hot spot temperature and the average temperature in the reactor to be significantly lowered. This yields several advantages, including (i) improved heat transfer from the reaction medium to the heat-removal fluid, (ii) better selectivity due to the lower contribution of combustion reactions and the lower formation of chlorinated by-products such as trichloroethane (formed from vinyl chloride, itself formed by the cracking of DCE), (iii) improved HCl conversion (for temperatures above 300 °C in the hot spot the reaction becomes mass-transfer limited) and (iv) longer catalyst lifetime, owing to lower active phase volatilization and lower coke formation, which finally causes catalyst dusting. Thus, increased productivity is possible, because productivity is usually essentially limited by the heat transfer between the reaction medium and the cooling fluid.

The main problems with the use of pure O_2 are related to its higher cost and lower process safety. Several air-based processes have been revamped in the last few years to O_2-based process, mainly due to environmental constraints. In this case, the investment for fixed and operating costs must be compared with those necessary for the installation of a catalytic combustor for the treatment of air-based reactor effluents.

Safety is an issue that should be considered when an oxygen-based oxidation process is adopted as an alternative to the air-based process. The absence of large amounts of ballast (unlike the air-based process) makes the formation of flammable mixtures more likely. Besides the use of suitable instruments to continuously monitor the gas-phase composition, key requirements are a knowledge of flammability limits under reaction conditions (temperature and pressure), including knowledge of the auto-flammability temperature, careful reactor and apparatus design aimed to avoid the presence of ignition sources, and the presence of safety valves and membranes, to minimize the consequences of an explosion. The best way to mix components that may potentially form flammable mixtures is to use a fluidized bed, for example, a solid suspended in the gaseous stream. For fixed beds this is not possible; however, very often, fixed-bed reactors operate under flammable conditions, to reach higher productivities.

We have discussed in depth this example to evidence that the change of air to pure oxygen as oxidant could bring various benefits in terms of sustainability, but it also adds problems of safety. However, none of these aspects could be evidenced from using indicators such as atom economy, for which the two process options are perfectly equivalent. In fact, these mass intensity indicators are mainly suited for organic syntheses and related industries such as that of fine chemical production. In fact, we could note that the concepts of "green" chemistry have spread largely within this community, but less so in other sectors, which introduced alternative concepts such as those of green engineering, cleaner production, eco-efficiency, industrial ecology, ecodesign, and life cycle thinking – discussed in part in Chapter 1. All these concepts focus attention more on the process and product than on the chemistry and reaction, in opposition to "green" chemistry. They are thus complementary views, but in terms of a practical contribution to sustainability, process improvement has made, so far, the largest contribution.

A second example regards the need for advanced plant control (by using improved real-time monitoring and smart model systems) and supply chain management in chemical production. The European Technology Platform on Sustainable Chemistry (Chapter 1) has recognized these aspects between the priorities to achieve a flexible, inherently safe production plant with optimal market demand responsiveness, and minimal environmental impact. A quality real-time control in the production line with automatic optimization will allow maximization of productivity and minimization of by-products and waste, while improving at the same time safety. In the production of goods, malformed polymers, for example, should be either sold as scrap for much less than the properly formed parts or have to be reprocessed. Improved on-line control and optimization of the production would minimize production waste. An improved supply chain management would reduce the risks, minimizing storage of chemicals. This example also evidences that there are many factors outside the reaction chemistry, which belong to chemical production, that are important for reaching the objectives of sustainable industrial chemistry. These aspects are not described and considered in actual indicators for "green" chemistry.

The need for more advanced methodologies to evaluate process sustainability [43] and to perform an environmental assessment of chemical production [44] is thus evidenced.

5.7
Energy Intensity

With rapidly raising costs of fossil fuels, the reduction of energy intensity of chemical production is a primary objective, although more related to economy than to a push deriving from the consideration that less energy-intensive processes are more sustainable [45]. However, the increasing concerns regarding greenhouse gas emissions (which translate into legislative actions), certainly contribute to the need for companies to move to less-energy intensive processes.

Some of the indicators for energy intensity are [46, 47]:

- **Process energy:** the amount of energy consumed in a chemical reaction per mass unit of product.

- **Primary energy usage:** the amount of primary energy inputs consumed, either per mass unit of product or per unit value added; electricity and steam must be converted into primary energy equivalent values, taking into consideration the efficiency of the conversion.

- **Solvent recovery energy:** the total energy consumed to recover the solvents used during a chemical reaction, expressed per mass unit of product.

Process energy and solvent recovery energy are indicators focusing mainly on the level of chemical reaction, while primary energy usage is more industry oriented. The latter may be calculated with respect to the economic value of the product, in line with the Eco-efficiency concept, defined as a ratio of economic and environmental performance.

As discussed for mass intensity indicators, energy intensity indicators are simple to calculate but give only a partial overview of sustainability, as processes outside the chemical reaction or the chemical plant are not included. In addition, no discrimination is made between renewable and non-renewable sources, which have very different implications with regard to sustainability. Finally, the energy intensity related to material production (reactants, solvents, etc.) is also not included. As a consequence, energy intensity indicators must always be accompanied by material intensity indicators, which are measured in different units and cannot be directly aggregated.

An indicator related to energy that tries to solve the above problems is the analysis of *exergy*. In thermodynamics, the exergy of a system is the maximum work possible during a process that brings the system into equilibrium with a heat reservoir. When the surroundings are the reservoir, exergy is the potential of a system to cause a change as it achieves equilibrium with its environment. Exergy is then the energy that is available to be used. After the system and surroundings reach equilibrium, the exergy is zero. Because it can be measured for any material (fuels, minerals, etc.), exergy is a natural choice for a common measure of resource quantity for either mass or energy [48]. Several uses have been suggested in the literature for exergy in environmental management: as a measure of thermodynamic efficiency in conventional process balances [49], in calculations of life cycle impact [50, 51] and as the key parameter in measuring sustainability of processes [52, 53].

The exergy-based sustainability parameter developed by Dewulf *et al.* [52] can be defined as a mean of the renewability and efficiency parameters for a given process. Renewability is defined as the ratio of renewable exergy consumption to total exergy consumption during the process life cycle. The efficiency parameter is calculated from the contributions of exergy used for various purposes: manufacturing, emission abatement, product abatement at the end of life, exergy content of the product and losses due to irreversible processes. This sustainability parameter varies between 0 and 1. A sustainable process would achieve a value of 1 in the calculations, meaning that only renewable exergy sources are used, zero emissions are produced, product abatement is not required and there are no losses due to irreversibility [45].

5.8
Environmental Impact Indicators

Environmental impact indicators are used to characterize and aggregate the contribution of a given set of inputs and/or outputs to environmental impact, with the being latter understood as a general concept, or a particular problem such as climate change, toxicity, and so on. These indicators are more elaborate than those on resource intensity, since not only the inputs and outputs have to be identified and quantified, but also an impact function describing the unfriendliness of those inputs and outputs, and which allows their aggregation to be defined.

Two of the several environmental impact indicators are reported below:

- **Environmental quotient (EQ)**, which could be considered an extension of the E-factor; the environmental quotient [32, 54] assigns an environmental hazardous quotient Q to each kind of waste produced. The amount of waste produced is multiplied by its corresponding factor and the contribution of each waste is aggregated. Examples of application can be found in Heinzle *et al.* [55] and Koller *et al.* [56].
- **Environmental Assessment Tool for Organic Syntheses (EATOS)**, based on the environmental quotient, the computer program EATOS [57] can be used to compare and improve chemical reactions. It expands the EQ by considering the potential environmental impact (PEI) of both waste and reactants [58].

The EQ and EATOS both focus on the impact of synthetic processes, by assigning impact scores to the raw materials used (EQ case) or to both reagents and waste produced (EATOS case). These impact scores can be defined according to upstream and downstream effects of the material in question, meaning that a life cycle approach is addressed. These methods do not include energy use, an indicator that would have to be measured separately.

There are also a series of specific indicators regarding the impact on environment. Each of these indicators, developed by IChemE [47], focuses on a particular environmental problem. The amount of the substance emitted is multiplied by its potency factor (PF), which expresses the specific contribution of substance to the environmental problem in question, per kg emitted. Potency factors are expressed in kg of a reference substance. All indicators are expressed in environmental burden per unit value added. Similar indicators have been developed by the Center for Waste Reduction Technologies [59] and the European Environment Agency [60]:

- atmospheric acidification,
- global warming,
- human health (carcinogenic) effects,
- ozone depletion,
- photochemical ozone,
- aquatic acidification,
- aquatic oxygen demand,
- ecotoxicity for aquatic life by metals,
- ecotoxicity for aquatic life by nonmetals.

These indicators have been developed mainly to be applied at the industry level, for monitoring and benchmarking, using economic value to measure products or services. Another difference is that, while EATOS and the EQ summarize environmental impact as a general attribute, expressed by potential environmental impact/ environmental unfriendliness factors, IChemE indicators measure particular environmental problems, such as toxicity or acidification. This problem-oriented approach used by IChemE is based on the Life Cycle Impact Assessment framework that will be discussed later.

5.9
Sustainable Chemical Production Metrics

The assessment of a complex phenomenon such as that of the sustainability of the chemical production requires integration of several indicators into a useful metric. The usefulness of a metric depends strongly on the number of indicators: too few may not provide an adequate description of a phenomenon, whereas too many would make the cost of completing the metric prohibitively high.

Several organizations have been actively involved in developing sustainability metrics. The most prominent efforts are made by the Centre for Waste Reduction Technologies (CWRT) and the Institution of Chemical Engineers (IChemE). Interestingly, industry is playing a leading role in developing eco-efficiency indicators and various metrics. Not surprisingly, the main focus of industry-developed metrics is on economic profitability of products and processes, reflected by the wide acceptance of the value added as a normalizing factor.

CWRT metrics have been developed on the basis of the eco-efficiency concept, defined as a ratio of product (or service) value to its environmental influence, where value is understood to be capital creation, and the main environmental influences are due to consumed energy, materials and water, and released green house and ozone depleting gases. Important aspects of the CWRT metrics are (i) normalization to dollar sales or value added, (ii) consideration for relative environmental impacts of different pollutants and (iii) use of national databases.

The main categories included in CWRT metrics are energy, mass, water usage, pollutant, human health and eco-toxicity. Indicators of human health and eco-toxicity are based on the parameters already widely used in the assessment of chemical hazards, that is, permissible exposure limits and 50% of lethal concentration. The indicators also take into account the life-time of chemical pollutants in various media of the environment.

IChemE metrics of sustainability consist of 49 indicators classified into three main categories: economic, environmental and social. The environmental indicators within the IChemE metrics are similar to those in the CWRT metrics. However, there are some differences. The IChemE metrics include the area of land as an environmental indicator. The actual indicators are (i) the sum of directly occupied and affected land per value added and (ii) the rate of land restoration. Other differences relate to the assessment of the relative impacts of pollutants on the environment and human health. The IChemE indicators do not take into account the life-time of chemicals in various media of the environment. The human health indicator is limited to carcinogenic effects and is normalized to benzene.

Various attempts have been made also to define specific metrics for green chemistry [9, 39, 61, 62]. The ultimate metric can be considered to be life cycle assessment (LCA), but full LCA studies for any particular chemical product are difficult and time consuming [63, 64]. However, as discussed in the following section, this is the necessary direction and qualitative LCA could even be preferable in comparing routes or considering a significant change in any product supply chain. Various of the proposed green chemistry metrics, such as atom efficiency and attempts to measure overall process efficiency such as E factors, mass intensities, and mass efficiency, are instead an oversimplification, as discussed before. As with LCA, these metrics have to be applied with definite system boundaries, and, interestingly, for process metrics these boundaries generally do not include feedstock sources or product fate. Energy costs and water consumption are also normally not included, although specific indicators are possible, as discussed in the previous sections.

Constable *et al.* have compared the use of various proposed green chemistry metrics in the evaluation of four commercial pharmaceutical processes [39].

As indicated in Chapter 1, new REACH legislation will also completely redraw the scenario of chemical assessment, with increasing relevance of testing for human toxicity. More attention should be given to environmental impact, and measures of biodegradability, environmental persistence, ozone depletion and global-warming potential. It will be also necessary to add metrics for feedstock issues, and particularly on long-term availability and supply chain for a product.

It will be thus necessary to introduce a hierarchical approach. Lapkin *et al.* [65] have proposed four vertical hierarchy levels: (i) product and process, (ii) company, (iii) infrastructure and (iv) society. Each level should reflect different boundary limits and use an appropriate choice of indicators. It is proposed also that the choice of appropriate indicators depends on the specifics of the industry sector and even on the types of products. The indicators should reflect specific by-products, wastes and emissions that are characteristic of the process or the product. Of course, a limit of the approach is how to make uniform the comparison between different industrial sectors. On the other hand, we have already remarked that industrial chemical production is different from other manufacture industrial sectors, because (i) it includes very different types of productions, from several thousand tons per day in refinery to kg amounts per day in fine chemical production and (ii) it is characterized by a highly integrated structure in which a large part of the products are the input for other chemical processes.

In specialty chemicals and pharmaceuticals production many different types of synthetic organic reactions are used. The product volume may be small, but very high purities are expected. The syntheses are typically discontinuous (batch-type reactors) and dependent on solvents, as both reactants and products are often solids, and produced molecules may possess toxicity and have other effects on the environment. Therefore, clearly, "green" chemistry metrics, and related aspects such as atom economy, solvent recovery and reuse, use of benign solvents, product toxicity and product end-of-life, are conditioning factors. The same concepts, when applied to a refinery lose in large part their relevance, and other aspects related to process and energy use become relevant. LCA, which has more limited applicability for fine chemical production, even if some related indicator such as the eco-efficiency discussed in Section 5.4 has proved highly valuable also in this area [22–25], is instead a necessary methodology to compare biofuel alternatives [66], for example. The concept of sustainability has thus a different meaning depending on the specific sector considered in the frame of chemical production. This concept has often been forgotten, along with the definition of the correct boundary limits for the evaluation.

We could thus conclude that no single set of universally accepted indicators and metrics is acceptable [67]. Table 5.4 lists a proposed set of process indicators that are appropriate to use when comparing various technologies. These indicators not only include "within the boundary" factors but also consider factors that affect the environment as well as factors that affect the company.

These indicators can have different values if they are considered for the short, medium or long term and also will depend on the type of stakeholders. Notably, various of these indexes are substantially different from those discussed before, and can be collected into five categories:

1. Waste minimization.
2. Process indexes.
3. Efficiency of use of resources.
4. Eco-economics indexes.
5. Impact on local environment.

Table 5.4 Indicators for measuring the sustainability of a chemical process.

Waste minimization (amount produced per ton of product):

- greenhouse gases (GHG)
- ozone-depleting gases (ODG)
- gaseous pollutants (NOx, SOx, VOC, HC)
- fugitive lost
- waste (solid + liquid and gaseous, including catalysts and auxiliary)
- non-biodegradable material
- cyto-, eco- and phyto-toxic material

Process indexes:

- synthesis effectiveness: ratio between desired product and input materials (reactants, solvents, catalyst, auxiliary, etc.) flow rates (or weight, if discontinuous reactor)
- process intensification: product to reactor volume (cumulative volume for reactors, if multistep)
- process integration: number of steps, including separation, for the whole process
- recycle: ratio between waste and by-products recycled and produced
- energy efficiency: ratio between energy input (reactants, fuel and other energy sources, including utilities) and output (as valuable products, including energy streams – steam, for example – which may be used)
- intrinsic eco-efficiency: ratio between product amount and end-of-pipe waste amount (gas, liquid, solid) to be treated before being externally discharged
- safety control: number of process parameters under multiple automatic control with respect to parameters with single or human control, normalized to process degrees of freedom
- operators risk: number of operators exposed directly to hazardous chemicals with respect to those necessary for operations
- intrinsic safety: ratio of intrinsic safe operations to those requiring human control
- safety: time dedicated to training and safety operations (including maintenance) with respect to total working time
- hazard storage index: amount of hazard chemicals stored (as reactant, intermediate or end products) with respect to day production

Efficiency of use of resources (amount per ton of product):

- freshwater used
- solvent used and lost
- equivalent oil barrels of energy input (all forms, from heat to electrical energy, to sustain the process, including utilities and services)

Eco-economics indexes:

- ratio between cleanup costs and product value
- ratio between monetary compensation that must be paid due to toxic or pollutants release above legislation limits and total production value
- ratio between monetary compensation that must be paid due to accidents and total production value
- ratio between monetary compensation to local communities and total production value

Impact on local environment:

- change of biodiversity
- degree of increase of persistent pollutants
- degree of change of local use of land and water bodies for human activities

These represent a simplification over a true LCA methodology, and include also aspects related to safety and risk, and economics. They thus could be a basis, which can be adapted for specific cases, for a sustainable chemical production metric, which should be integrated with other assessment tools such as LCA, Sustainable Process Index, and Risk Analysis and evaluation.

5.10
Life Cycle Tools

Many assessment methodologies have been developed that include not only the product or process but also the entire supply chain and disposal, that is, the cradle-to-grave approach [68, 69]. Apart from life cycle assessment (LCA) and life-cycle costing (LCC) [70], the energy-based sustainability index (ESI) [52, 53], MIPS (materials input per unit service [71]) and MFA (materials flow accounting) [72] are also cradle-to-grave approaches. The main differences among all these indices lie in the way they normalize the environmental impacts. There is still no readily available simple, efficient and unambiguous methodology suitable for comparing alternative chemical technologies in such a way that they faithfully represent the benefits of implementation of sustainable chemical technologies. However, LCA is the best known and used methodology.

The original definition of LCA is that of a process aimed at evaluating the environmental burdens associated with a product, process or activity by identifying and quantifying energy and materials used and wastes released to the environment. The assessment includes the entire life cycle of the product, process or activity, encompassing (i) extracting and processing raw materials, (ii) manufacturing, transportation and distribution, (iii) use, (iv) re-use, maintenance and (v) recycling and final disposal. A general overview of LCA history and recent developments can be found in Sonneman *et al.* [73]. The ISO 14040 standard has provided very relevant input into the process of defining LCA.

The ISO 14040 standard determines four basic stages for LCA studies, schematically represented in Figure 5.9. "Goal and scope definition" is the first stage of the study and one of the most important, since the elements defined here, such as

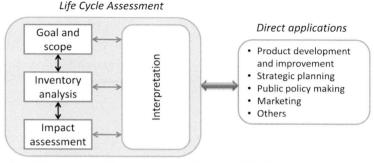

Figure 5.9 Stages in a life cycle assessment. Source: ISO [74].

purpose, scope and main hypothesis considered are the key of the study. In first place, the goal of the study is defined, as well as the reasons that have lead to its realization, the kind of decisions that will be made from the results obtained and if these will be of internal use (for a company, for instance) or external (to inform the general public or an institution). Secondly, the scope of the study is defined. This implies, among other elements, defining the system, its boundaries (conceptual, geographical and temporal), the quality of the data used, the main hypothesis, as well as the limitations of the study. A key issue is the definition of the functional unit. This is the unit of the product or service whose environmental impacts will be assessed or compared. It is often expressed in terms of amount of product, but should really be related to the amount of product needed to perform a given function.

The "inventory analysis" is a technical process of collecting data, in order to quantify the inputs and outputs of the system. Energy and raw materials consumed, emissions to air, water, soil and solid waste produced by the system are calculated for the entire life cycle of the product or service. To make this analysis easier, the system under study is split into several subsystems and unit processes, and the data obtained are grouped in different categories in a LCI table.

The "impact assessment" identifies and characterizes the potential effects produced in the environment by the system under study. The first step is "classification," in which the environmental interventions (resources consumed, emissions to the environment) identified in the inventory analysis are grouped in different impact categories or indicators, according to the environmental effects they are expected to produce. For example, CO_2 and CH_4 emissions are classified in the category global warming potential (GWP).

The second step, called "characterization," consists of weighting the different substances contributing to the same environmental impact. For each impact category included in the impact assessment, an aggregated result is produced, in a given unit of measure. For example, the GWP is calculated in kg eq. of CO_2, from the contribution of CO_2 and CH_4 emissions, among others. At this point, the so-called environmental profile of the system is obtained, consisting of a set of indicator scores.

The third step is "normalization," which involves the environmental profile of the system to a broader data set or situation, for example, relating the system's GWP to a country's yearly GWP.

The last step is "weighting," where the environmental profile is reduced from a set of indicators to a single impact score, by using weighting factors based on subjective value judgments. For instance, a panel of experts could be formed to weight the impact categories. The advantage of this stage is that different criteria (impact categories) are converted into a numerical score of environmental impact, thus making it easier to make decisions. However, a lot of information is lost, and reality is simplified.

"Interpretation" is the last stage of an LCA study, where the results obtained are presented in a synthetic way, presenting the critical sources of impacts and the options to reduce them. Interpretation involves a review of all the stages in the LCA process, to check the consistency of the assumptions and the data quality, in relation to the goal and scope of the study.

LCA is an holistic approach. All necessary inputs and emissions in many stages and operations of the life cycle are considered to be within the system boundaries. This includes not only inputs and emissions for production, distribution, use and disposal but also indirect inputs and emissions – such as from the initial production of the energy used – regardless of when or where they occur.

The LCA thus collects and evaluates the data on the emissions and their environmental impacts at every step in a process of production of a given product or provision of a service, starting from acquiring raw materials (including energy) and finishing with the end-of-life disposal and/or elimination of incurred emissions/wastes (so-called "cradle to grave" approach). This approach is the strength of LCA, but also its limitation, since the broad scope of analyzing the entire life cycle of products and processes can only be achieved at the expense of simplifying other aspects. Some of the limitations of the approach are:

- LCA addresses potential rather than actual impacts. This is because, in LCA, impacts are not specified in space and time.
- The LCA model focuses on physical characteristics of industrial activities and other economic processes. Market mechanisms or other secondary effects of technological development are not included.
- LCA generally regards all processes as linear, both in the economy and the environment, but this is not always true.
- LCA focuses on environmental issues associated with products and processes, excluding economic and social consequences.
- Availability of true and reliable data is often a limitation.

The suitability of LCA as a tool for environmental evaluation of chemical products and processes has been suggested by several authors to be involved in the development and promotion of green/sustainable chemistry [44, 76–79]. The tool is well known by the chemical industry, which uses it for product and process development, marketing and communication with public authorities and clients, among other purposes.

Development of suitable inventories and LCA databases are one of the necessary components for more extensive use of the methodology. One of these is the Ecoinvent (www.ecoinvent.ch) database, which contains over 200 datasets corresponding to the category of chemical products, divided into organic and inorganic chemicals. The database contains international industrial life cycle inventory data on energy supply, resource extraction, material supply, chemicals, metals, agriculture, waste management services and transport services.

In these databases, chemical products are typically inventoried applying "cradle to gate" boundaries, which means that only the first half of the life cycle is included: from raw material extraction (cradle) until the product is ready to be sold in the market (gate). It is thus necessary to include in the life cycle the distribution, use and end of life as waste.

It is important to include the full life cycle. A good recent example is the case of methyl *tert*-butyl ketone (MTBE). MTBE can be produced using an efficient catalytic distillation process and it performs well as a fuel additive. However, the combination of three facts [carcinogenic potency, significant solubility in water and difficulty in avoiding leakage of MTBE into the ground water table at the point-of-use (leakage of underground tanks)] make the use of this fuel additive undesirable. MTBE was banned for this reason in California. To identify such problems a broad system boundary is necessary. However, considering the overall environmental impact of a system within a broad system boundary often gives inconclusive results.

Another interesting example of inappropriate boundary conditions for environmental assessment is the use of water- or organic-type solvents for automotive-based coating [80]. Despite the lower level of emissions from water-based coatings, application of these coatings requires more energy (even including VOC abatement for solvent based processes) due to the slow evaporation of water. The two types of solvents have a similar global impact on environment from LCA analysis, an unexpected result based on green chemistry principles and metrics, for example.

Accessibility to data and accuracy of the data are of particular importance. Even if significant progress is made in data standardization and the development of new databases and search software, relevant issues are still present. The large disagreement in LCA data for biofuel assessment (e.g., the well-known case of ethanol) teaches us the need for more reliable data as well as of procedures in definition of the elements to consider in a LCA analysis.

The cost, time required and the difficulties in gathering all the necessary data are the main drawbacks for carrying out a complete LCA study. For this reason, some effort has been made to develop "simplified," but still reliable, methodologies.

A streamlined LCA is a simplified version of detailed LCA, in which the scope, cost and effort required is reduced [81, 82]. A complete LCA is considered to include all the relevant life cycle information in a quantitative manner. Streamlining can be based on increasing the amount of qualitative or semi-quantitative data used; this process progressively leads to applying *life cycle thinking* rather than LCA, that is, the concept instead of the tool. On the other hand, streamlining can be based on excluding processes or stages in the life cycle, but keeping the quantitative nature of the tool. In sustainable chemistry, streamlining approaches will be very often needed to assess products and processes by means of LCA.

Life cycle costing (LCC), also called life cycle management (LCM), is another toolbox based on life cycle concepts [83, 84]. It includes aspects related to the three pillars of sustainability: an environmental tool, an economic tool and a social tool. It is expected to become in the near future a standard addition to LCA, in order to evaluate the economic implications of a product's life cycle. Hunkeler and Rebitzer [85] have indicated that LCC is an assessment of all costs associated with the life cycle of a product that are directly covered by any one or more of the actors in the product life cycle (supplier, producer, user/consumer, end-of-life actor), with complementary inclusion of externalities that are anticipated to be internalized in the decision-relevant future. For example, a product manufacturer should include in an LCC study the costs incurred by the user of his product. On the other hand, it is important to note that only externalities expected to be internalized in the future by means of taxes or other regulatory measures must be included. The issue of externalities is one of the most controversial in environmental accounting.

There are two classes of costs to consider in LCC studies:

- **Internal costs**, also called private costs, which are those appearing in company's accounts, as well as those incurred by consumers or other stakeholders. These costs have a clear market value. In this group we find conventional costs (materials, fuels, labor, equipment, etc.) and potentially hidden and less tangible costs, which are usually assigned to overheads in company's accounts (permits, post-closure care, liability costs, etc.).

- **External costs**, also called social costs, societal costs or externalities, these are the monetized effects of environmental and social impacts caused by products and services, for which a company, consumer or another stakeholder is not obliged to pay, since neither the marketplace nor regulations assign such costs to a particular

Table 5.5 List of commonly used environmental accounting terms. Source: adapted from Muñoz Ortiz [86].

Environmental cost accounting	Addition of environmental cost information into existing cost accounting procedures and/or recognizing embedded environmental costs and allocating them to appropriate products or processes
Full cost (environmental) accounting	Allocation of all direct and indirect costs (highlighting environmental, safety and health costs) to a product or product line for the purposes of inventory valuation, profitability analysis and pricing decisions
Total cost assessment	Integrating environmental costs into a capital budgeting analysis. Similar to total cost accounting (TCA). Synonymous with true cost accounting
Cost–benefit analysis	Describes and quantifies the social advantages and disadvantages of a project in monetary units
Cost-effectiveness analysis	Determines the least cost option for a predetermined environmental target, or, conversely, the option involving the greatest environmental improvement for a given expenditure
Life cycle cost assessment	Evaluation of life cycle costs of a product, product line, process, system or facility by identifying environmental consequences and assigning a monetary value to these consequences
Life cycle accounting	Assignment and analysis of product-specific costs within a life cycle framework
Life cycle cost	Total of the direct, indirect, recurring, non-recurring and other related costs incurred by or estimated for the project in the design, development, production, operation, maintenance and support of a major system over its anticipated useful life span
Life cycle assessment, life cycle analysis	Identifying the environmental consequences of a product, process or activity through its entire life cycle and opportunities for achieving environmental improvements. Focuses on environmental impacts, not costs

person or activity. Examples of these costs are increased risk of asthma resulting from air pollutants or the expected impacts on global climate due to emissions of greenhouse gases.

Owing to the lack of standardization, it is usual to find in the literature different terms related to life cycle approaches and/or environmental accounting. Sometimes different terms are used as synonymous, while other times the same term is used for different approaches, causing some confusion. Table 5.5 summarizes an attempt to define commonly used terms.

5.11
Conclusions

The issues of sustainability and pollution clearly focus on a few important aspects: energy, raw materials, land, state of the environment, climate change and risks to human health that processes/products may cause. Long-term sustainability re-quires continuous availability of: (i) arable land for agriculture and for human habitat, (ii) land for preservation of biodiversity and as a climate regulator, (iii) fresh water and clean air and (iv) resources for manufacturing products and generation of energy [45].

Thus the key focus areas for sustainable chemical technologies are:

- energy efficiency and use of energy produced from renewable resources;
- use of renewable feedstocks;
- recycling, re-use or benign disposal of products, for example, biodegradation;
- zero pollution strategy;
- sustainable level of emissions of greenhouse gases;
- use of benign chemicals;
- use of intensive processes with a small footprint;
- site regeneration.

To proceed in this direction it is necessary to use indicators, metrics and tools of analysis/assessment that allow us to compare the different alternative and options, and quantify the benefits of adoption new solutions. Several of these indicators, metrics and tools have been developed, and applicability and limits have been discussed in this book. New aggregated indicators to assess sustainability of chemical production have also been proposed.

However, the need to use more extensively quantitative tools to support decision making has also been emphasized, and in particular LCA, even if complex and costly, is a very suitable tool for that purpose, since it attempts to capture the direct and indirect environmental consequences of a given product or process design, and benefits from an increasing international acceptance, endorsed by several ISO standards. The extension to methodologies that include also the costs, such as LCC methods, would also be necessary.

References

1 Reisch, M.S. (2001) *Chem. Eng. News*, **79**, 17.

2 Coatanéa, E., Kuuva, M., Makkonnen, P.E., Saarelainen, T. and Castillón-Solano, M.O. (2006) Analysis of the concept of sustainability: definition of conditions for using exergy as a uniform environmental metric. Presented at 13th CIRP International Conference on Life Cycle Engineering, Leuven, Belgium, May 31 – June 2.

3 Tomiyama, T. (2001) Service engineering to intensify service contents in product life cycles. Proceedings of EcoDesign 2001: 2nd International Symposium On Environmentally Conscious Design and Inverse Manufacturing, Tokyo, Japan.

4 Svirezhev, Y.M. (2000) *Ecol. Model.*, **132**, 11.

5 Collins, T. (2001) *Science*, **291**, 48.

6 Hill, K. (2007) Industrial development and application of biobased oleochemicals, in *Catalysis for Renewables* (eds G. Centi and R. van Santen), Wiley-VCH Verlag, Weinheim, Ch. 4, p. 75.

7 Finkel, M.(July 2007) Malaria, National Geographic.

8 Beloff, B., Lines, M. and Russell, W.G. (2005) *Transforming a Sustainability Strategy into Action: the Chemical Industry*, Wiley-Interscience, NJ.

9 Lapkin, A. and Constable, D.C. (eds) (2008) *Green Chemistry Metrics*, Wiley-VCH Verlag, Weinheim.

10 Lange, G.-M. (2007) Environmental and resource accounting, in *Handbook of Sustainable Development* (eds G. Atkinson, S. Dietz, and E. Neumayer), Edward Elgar Publishing Ltd, Cheltenham, UK.

11 George, C. and Kirkpatrick, C. (2007) *Impact Assessment and Sustainable Development*, Edward Elgar Publishing Ltd, Cheltenham, UK.

12 Accounting for Sustainability Group (www.sustainabilityatwork.org.uk) (2008) *Accounting for Sustainability*, Beacon Press, Lichfield, UK.

13 Loh, J., Green, R., Ricketts, T., Lamoreux, J., Jenkins, M., Kapos, V. and Randers, J. (2005) *Phil. Trans. R. Soc. B*, **360**, 289.

14 Holzman, D.C. (2008) *Environ. Health Perspect.*, **116**, A246.

15 Searchinger, T., Heimlich, R., Houghton, R.A., Dong, F., Elobeid, A., Fabiosa, J., Tokgoz, S., Hayes, D. and Yu, T.-H. (2008) *Science*, **319**, 1238.

16 Fargione, J., Hill, J., Tilman, D., Polasky, S. and Hawthorne, P. (2008) *Science*, **319**, 1235.

17 United Nations (2007) *Indicators of Sustainable Development: Guidelines and Methodologies*, 3rd edn, United Nations Pub., New York.

18 Seager, T.P. and Theis, T.L. (2004) *J. Cleaner Production*, **12**, 865.

19 Yale Center for Environmental Law and Policy (2005) *2005 Environmental Sustainability Index*, Yale University Pub., New Haven, CT.

20 Yale Center for Environmental Law and Policy (2008) *2008 Environmental Performance Index*, Yale University Pub., New Haven, CT.

21 Schwarz, J., Beloff, B. and Beaver, E. (2002) *Chem. Eng. Prog.*, **7**, 58.

22 Saling, P., Kicherer, A., Dittrich-Krämer, B., Wittlinger, R., Zombik, W., Schmidt, I., Schrott, W. and Schmidt, S. (2002) *Int. J. Life Cycle Assess.*, **7**, 203.

23 Shonnard, D.R., Kicherer, A. and Saling, P. (2003) *Environ. Sci. Technol.*, **37**, 5340.

24 Saling, P., Maisch, R., Silvani, M. and König, N. (2005) *Int. J. Life Cycle Assess.*, **10**, 364.

25 Saling, P. (2007) Using the eco-efficiency analysis and SEEBALANCE in the sustainability assessment of products and processes. Presented at Sustainable Neighbourhood – from Lisbon to Leipzig through Research, 4th BMBF-Forum for Sustainability, Leipzig, Germany, May 8–10.

26 Schaltegger, S., Bennett, M. and Burritt, R. (eds) (2006) *Sustainability Accounting and*

Reporting, Springer Pub., Dordrecht, The Netherlands.

27 Taplin, J.R.D. Bent, D. and Aeron-Thomas, D. (2006) *Bus. Strat. Environ.*, **15**, 347.

28 Lamberton, G. (2005) *Accounting Forum*, **29**, 7.

29 Komiyama, H. and Takeuchi, K. (2006) *Sustainability Sci.*, **1**, 1.

30 Hardi, P. (2007) The long and winding road of sustainable development evaluation, in *Impact Assessment and Sustainable Development* (eds C. George and C. Kirkpatrick), Edward Elgar Publishing Ltd, Cheltenham, UK, Ch. 2, p. 15.

31 Sheldon, R.A. (1992) *Chem. Ind. (London)*, 903;Sheldon, R.A. (1997) *Chem. Ind. (London)*, 12.

32 Sheldon, R.A. (1994) (March) *Chemtech.*, 38.

33 Sheldon, R.A. (2000) *Pure Appl. Chem.*, **72**, 1233–1246.

34 Sheldon, R.A., Arends, I. and Hanefeld, U. (2007) *Green Chemistry and Catalysis*, Wiley-VCH Verlag, Weinheim.

35 Trost, B.M. (1991) *Science*, **254**, 1471.

36 Anastas, P. and Warner, J.C. (eds) (1998) *Green Chemistry: Theory and Practice*, Oxford University Press, Oxford.

37 Sheldon, R.A. (2007) *Green Chem.*, **9**, 1273.

38 Hudlicky, T., Frey, D.A., Koroniak, L., Claeboe, C.C. and Brammer, L.E. Jr (1999) *Green Chem.*, 57.

39 Constable, D.J.C., Curzons, A.D. and Cunningham, V.L. (2002) *Green Chem.*, **4**, 521.

40 Centi, G., Cavani, F. and Trifirò, F. (2001) *Selective Oxidation by Heterogeneous Catalysis, in Series: Fundamental and Applied Catalysis* (eds M.V. Twigg and M.S. Spencer), Kluwer Acad. Plenum Pub., New York.

41 Centi, G. and Perathoner, S. (2003) Selective oxidation. Industrial, in *Encyclopedia of Catalysis*, Vol. 6, (Chief ed. I.T. Horváth), John Wiley & Sons, Inc., Hoboken, p. 239.

42 Arpentinier, P., Cavani, F. and Trifirò, F. (1999) *The Technology of Catalytic Oxidations*, Editions Technip, Paris.

43 Gonzalez, M.A. and Smith, R.L. (2003) *Environ. Prog.*, **22**, 269.

44 Hellweg, S., Fischer, U., Scheringer, M. and Hungerbühler, K. (2004) *Green Chem.*, **6**, 418.

45 Lapkin, A. (2002) Metrics of Green Chemical Technology, A Report for Crystal Faraday Partnership, University of Bath, UK.

46 Curzons, A.D., Constable, D.J.C., Mortimer, D.N. and Cunnigham, V.L. (2001) *Green Chem.*, **3**, 1.

47 IChemE (2002) *The Sustainability Metrics. Sustainable Development Progress Metrics*, Institution of Chemical Engineers.

48 Ayres, R.U., Ayres, L.W. and Martinás, K. (1998) *Energy*, **23**, 355.

49 Hinderink, P., van der Kooi, H.J. and de Swaan Arons, J. (1999) *Green Chem.*, **1** (6), G176.

50 Finnveden, G. and Östlund, P. (1997) *Energy*, **22** (9), 923.

51 Dewulf, J. and Van Langenhove, H. (2004) *Int. J. Energy Res.*, **28** (11), 969.

52 Dewulf, J., van Langenhove, H., Mulder, J., van den Berg, M.M.D., van der Kooi, H.J. and de Swaan Arons, J. (2000) *Green Chem.*, **2** (3), 108.

53 Dewulf, J., van Langenhove, H. and Dirckx, J. (2001) *Sci. Total Environ.*, **273** (1–3), 41.

54 Sheldon, R.A. (1997) *J. Chem. Technol. Biotechnol.*, **68**, 381.

55 Heinzle, E., Weirich, D., Brogli, F., Hoffmann, V.H., Koller, G., Verduyn, M.A. and Hungerbühler, K. (1998) *Ind. Eng. Chem. Res.*, **37**, 3395.

56 Koller, G., Weirich, D., Brogli, F., Heinzle, E., Hoffmann, V.H., Verduyn, M.A. and Hungerbühler, K. (1998) *Ind. Eng. Chem. Res.*, **37**, 3408.

57 Hungerbühler, K. (2000) *Ind. Eng. Chem. Res.*, **39**, 960.

58 Eissen, M. and Metzger, J.O. (2002) *Chem. Eur. J.*, **8** (16), 3580.

59 Center for Waste Reduction Technologies (2002) Collaborative projects: focus area sustainable development: development of baseline metrics.

60 European Environment Agency (1999) *A First Set of Eco-efficiency Indicators for Industry: Pilot Study. Final Report*, Anite Systems, Luxembourg.

61 Nüchter, M., Ondruschka, B., Bonrathard, W. and Gurn, A. (2004) *Green Chem.*, **6**, 128.

62 Eissen, M., Hungerbühler, K., Dirks, S. and Metzger, J. (2004) *Green Chem.*, **6**, G25.

63 Graedel, T.E. (1998) *Streamlined Life-Cycle Assessment*, Prentice Hall, New Jersey.

64 Azapagic, A., Perdan, S. and Clift, R. (2004) *Sustainable Development in Practice*, John Wiley & Sons, Ltd, Chichester.

65 Lapkin, A., Joyce, L. and Crittenden, B. (2004) *Environ. Sci. Technol.*, **38**, 5815.

66 Zah, R., Böni, H., Gauch, M., Hischier, R., Lehmann, M. and Wäger, P. (2007) *Life Cycle Assessment of Energy Products: Environmental Assessment of Biofuels*, EMPA, Switzerland.

67 Doble, M. and Kruthiventi, A.K. (2007) *Green Chemistry and Engineering*, Elsevier Science & Technology Books, Amsterdam.

68 Hoffmann, V.H., Hungerbühler, K. and McRae, G.J. (2001) *Ind. Eng. Chem. Res.*, **40**, 4513.

69 Sikdar, S.K. (2003) *AIChE*, **49**, 1928.

70 Goel, H.D., Herder, P.M. and Weijnen, M.P.C. (2001) *Chem-Ing-Tech.*, **73**, 622.

71 Schmidt-Bleek, F. (1993) *Fresenius Environ. Bull.*, **2**, 407.

72 Brunner, P.H. (2008) *J. Industrial Ecol.*, 8 (3), 4.

73 Sonneman, G., Castells, F. and Schuhmacher, M. (2004) *Integrated Life-Cycle and Risk Assessment for Industrial Processes*, Lewis Publishers, Florida.

74 ISO (1997) ISO 14040: *Environmental Management – Life cycle assessment –*

Principles and Framework, ISO, Geneva, Switzerland.

75 Lankey, R.L. and Anastas, P.T. (2002) *Ind. Eng. Chem. Res.*, **41**, 4498.

76 Anastas, P.T. and Lankey, R.L. (2000) *Green Chem.*, **2**, 289.

77 Domènech, X., Ayllón, J.A., Peral, J. and Rieradevall, J. (2002) *Environ. Sci. Technol.*, **36**, 5517.

78 Yasui, I. (2003) *Green Chem.*, **5**, G70.

79 Sikdar, S. (2003) *Environ. Prog.*, **22**, 227.

80 Hazel, N.J., Walker, P. and Bray, R. (1995) LCA based environmental evaluation of automotive OEM basecoat alternatives. Presented at International Body Engineering Conference, 2nd November, Detroit, USA.

81 Todd, J.A. (1995) Streamlining LCA concepts and thoughts, in *Life Cycle Assessment* (ed. M.A. Curran), McGraw-Hill.

82 Hochschorner, E. and Finnveden, G. (2003) *Int. J. Life Cycle Assess.*, **8**, 119.

83 Fabrycky, J.W. and Blanchard, B.S. (1991) *Life-Cycle Cost and Economic Analysis*, Prentice Hall, Englewood Cliffs, NJ.

84 Guinée, J.B., Gorree, M., Heijungs, R., Huppes, G., Kleijn, R., Udo de Haes, H.A., Van der Voet, E. and Wrisberg, M.N. (2002) *Life Cycle Assessment. An Operational Guide to ISO Standards*, Centre of Environmental Science, Leiden University, Leiden, The Netherlands.

85 Hunkeler, D. and Rebitzer, G. (2005) *Int. J. Life Cycle Assess.*, **10**, 305.

86 Muñoz Ortiz, I. (2006) Life cycle assessment as a tool for green chemistry: application to different advanced oxidation processes for wastewater treatment, PhD Thesis, Universitat Autònoma de Barcelona, Spain.

6
Synthesis of Propene Oxide: A Successful Example of Sustainable Industrial Chemistry

Fabrizio Cavani and Anne M. Gaffney

6.1
Introduction: Current Industrial Propene Oxide Production

Propene oxide (PO) is one of the largest propene derivatives in production, ranking second behind polypropene; it is used primarily as a chemical intermediate [1]. Worldwide, approximately 8 million tons are produced annually, the major suppliers are Dow (1.9 M-ton yr^{-1}), Lyondell (2.1 -ton yr^{-1}) and Shell/BASF (0.9 M-ton yr^{-1}); other key producers are Sumitomo, Repsol and Huntsman.

One of the fastest growing markets is China; in fact, nowadays there are more than 30 Chinese PO producers. From 1990 to 2000, PO consumption in China grew by more than 30% per year, and, with the rapid development in the fields of polyurethane, propylene glycol and surfactants, it is expected to increase even further over the next few years. By 2005, the output capacity of China reached almost 600 000 ton yr^{-1}, and by 2010, it is expected to surpass 1.1 M-ton yr^{-1}. For instance, in 2006, 250 000 ton of PO per year were put on stream in the CNOOC-Shell petrochemical complex in Nanhai. Sinopec Zhenhai Ref Chem and Lyondell Chemical have received approval for a joint PO/SM plant in Ningbo, China; the plant will be able to produce 270 000 ton of PO per year and start-up is expected in 2010. Outside China, one of the largest plants in the world, which use PO/SM technology, the Maasvlakte facility, The Netherlands, is a joint venture between Lyondell and Bayer and has an annual capacity of almost 300 000 ton of PO per year. The Dow facility in Stade, Germany, produces almost 630 000 tons of PO per year with CHPO technology.

Approximately 65% of the PO produced is used for the synthesis of polyether polyols (in a reaction with polyhydric alcohols), one of the main components used in the manufacture of polyurethanes, propene glycol (20%), glycol ethers (5%) and butanediol, amongst others. Figure 6.1 shows an overview of the PO industry.

Polyurethane products and formulated systems are used in rigid foams, flexible foams, adhesives, sealants, coatings and elastomers, as well as in many other applications. Propene glycols are used in a wide variety of end-use and industrial applications, from unsaturated polyester resins, cosmetics and household detergents, to paints and automotive brake fluids. Propene glycol ethers are commonly

Sustainable Industrial Processes. Edited by F. Cavani, G. Centi, S. Perathoner, and F. Trifiró
Copyright © 2009 WILEY-VCH Verlag GmbH & Co. KGaA, Weinheim
ISBN: 978-3-527-31552-9

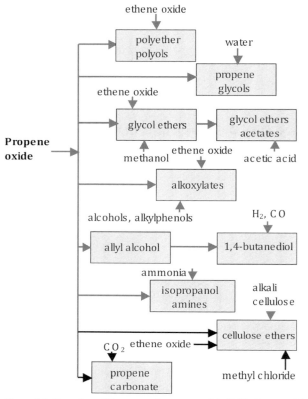

Figure 6.1 Overview of the uses of propene oxide (PO). Source: elaborated from [1e].

used as solvents and coupling agents in paints and in the production of coatings, inks, resins and cleaners. PO is also used to manufacture butanediol, propoxylated or specialty organic compounds, flame-retardants, modified carbohydrates (starches), synthetic lubricants, oil-field drilling chemicals and textile surfactants.

The overall annual growth rate for polyether polyols, propene glycol and propene glycol ethers is more than 3%, which corresponds to a demand of approximately 200 000–300 000 tons of PO. This demand is primarily driven by end-use applications, including unsaturated polyester resins, food, cosmetics, anti-freeze and aircraft de-icing fluids. The improved biodegradability and biocompatibility of propene glycol and polyalkylene glycols, with respect to the analogous products obtained from ethene oxide, allow them to be used in food product additives, pharmaceuticals and cosmetics. Demand for polyether polyols is expected to grow at a rate of 5–6% per year, driven by its considerable requirement in the automotive, construction and home furnishing industries. The global demand for PO is therefore forecast to grow at 4–5% per year, with the greatest growth observed in China and India. European demand is expected to grow at a slower rate of 3–3.5% per year. In the USA, demand grew at an average rate of 3.5% per year between 2001 and 2006.

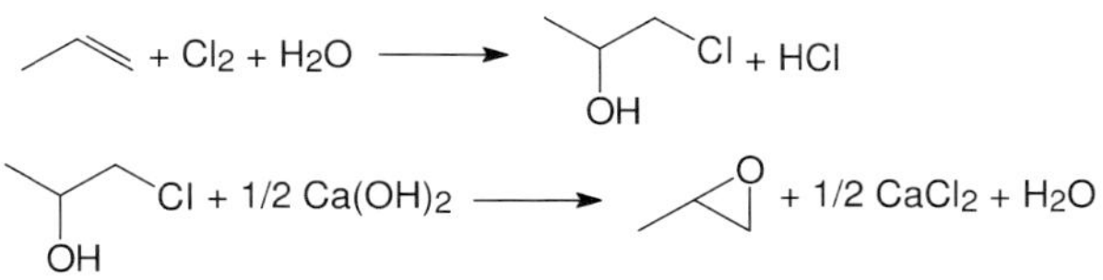

$$\text{propene} + Cl_2 + H_2O \longrightarrow \text{chlorohydrin} + HCl$$

$$\text{chlorohydrin} + 1/2\ Ca(OH)_2 \longrightarrow PO + 1/2\ CaCl_2 + H_2O$$

Scheme 6.1 CHPO technology for PO synthesis.

Currently, the industrial production of PO mainly comes from the oxidation of propene with other chemicals. The main technologies employed in its production are outlined below [2].

6.1.1
CHPO (Chlorohydrin) Technology

The overall process stoichiometry is:

$$\text{Propene} + Cl_2 + Ca(OH)_2 \rightarrow PO + H_2O + CaCl_2$$

Synthesis is carried out in two separate steps (Scheme 6.1). In the first reactor, propene reacts with Cl_2 to produce propene chlorohydrin via intermediate formation of the propene chloronium complex, then quenched by water. In the epoxidation reactor, the dehydrochlorination of propene chlorohydrin occurs using a base (usually calcium hydroxide).

6.1.2
PO/TBA Technology

The overall process stoichiometry is:

$$\text{Propene} + \text{isobutane} + O_2 \rightarrow PO + t - \text{butyl alcohol (TBA)}$$

Synthesis is carried out in two separate steps (Scheme 6.2). The oxidant for propene is t-butyl hydroperoxide (t-BuOOH). The coproduct TBA can be used as such, or it can be dehydrated to isobutene.

6.1.3
PO/SM Technology

The overall process stoichiometry is:

$$\text{Propene} + \text{ethylbenzene} + O_2 \rightarrow PO + \text{styrene (SM)} + H_2O$$

Synthesis is carried out in three separate steps (Scheme 6.3). The oxidant for propene is ethylbenzene hydroperoxide (EBHP).

While in the 1970s more than 95% of the PO world production capacity was achieved by means of CHPO technology, in the 1980s the TBA and SM processes

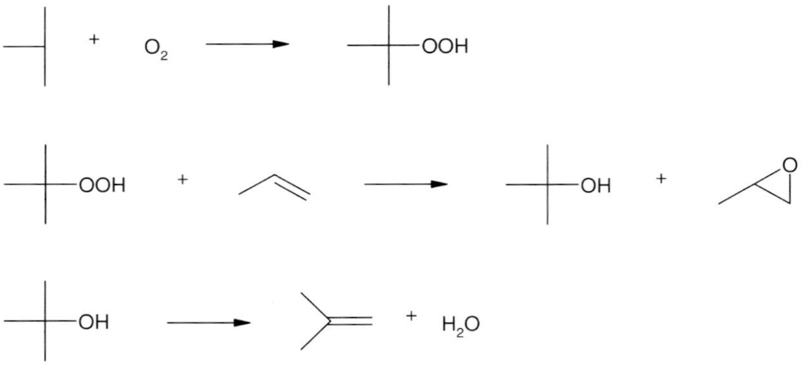

Scheme 6.2 TBA technology for PO synthesis.

gained increasing market shares. In 2005, CHPO technology accounted for approximately 45% of the capacity (Dow, Asahi Glass), SM technology for 35% (Lyondell, Shell/BASF, Repsol) and TBA technology for around 20% (Lyondell, Huntsman).

In the hydroperoxidation routes (PO/SM and PO/TBA), the conversion of the oxidant is greater than 95%, with a selectivity for PO greater than 95%; the main C_3 by-product is acetone. In the PO/TBA process, one disadvantage is that a portion of the t-BuOOH, once formed, immediately decomposes to TBA. In the PO/SM process, styrene may yield heavy by-products through oligomerization, causing a loss in catalyst activity.

The hydroperoxidation method was first developed by Halcon Corp and Atlantic Richfield Oil Corp (now Lyondell) in the 1970s, and was then also implemented commercially by Shell. The catalysts for this reaction are either homogeneous Mo^{6+} complexes (Halcon/ARCO) or heterogeneous silica-supported Ti^{4+} (Shell) [2, 3].

The main drawback of current technologies is that they generate additional coproducts. In the CHPO process, for each ton of PO, 1.4 tons of chlorine and 1.0 ton of calcium hydroxide (or sodium hydroxide) are needed, and approximately

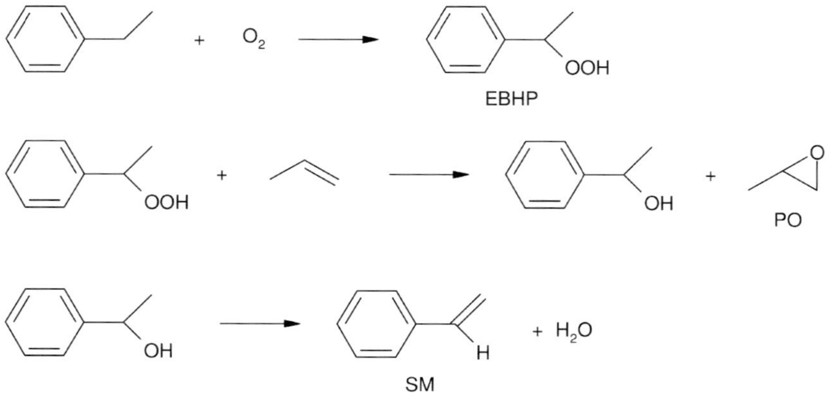

Scheme 6.3 SM technology for PO synthesis.

2.0 tons of $CaCl_2$ are obtained as a coproduct. In addition, there is a large excess of water; therefore, a large volume of wastewater (brine solution) containing the Ca is formed, with about 40 tons being produced per ton of PO. The main by-product is 1,2-dichloropropane (up to 10% yield), which is either sold as a solvent or incinerated together with the other organic chlorinated by-products to produce HCl from the off-gas. The minor by-products are dichloropropanols, which are produced from allyl chloride, itself being produced by a reaction between propene and Cl_2.

The recycling of the solution produced in the dehydrochlorination step to the electrolysis cell of the chloro-alkali facility, and the hydrodechlorination of the chlorinated propane to propane, have been considered as possible options to make the process environmentally more sustainable. However, these processes have not yet been implemented at the industrial level. No new plants using the CHPO technology are expected to be built – because of the large investment required a new plant would not be economically viable. However, existing plants that have been fully depreciated, are technologically up-to-date and are integrated with Cl_2 production are still competitive and will keep operating.

In the hydroperoxidation route, either α-methylbenzyl alcohol or t-butanol are the reaction coproducts; both molecules are dehydrated to yield either a styrene monomer (2.5 ton per ton of PO) or isobutene (2.1 ton per ton of PO), respectively. t-Butanol may also be the direct feed for MTBE production. Balancing the markets for PO and the coproducts has proven difficult with these routes, thus leading to a volatile economic performance over time. Furthermore, the hydroperoxidation routes require relatively large capital investments. Existing hydroperoxidation plants continue to operate and are incrementally improved; however, future investments in these technologies will decline.

6.2
PO-only Routes: Several Approaches for Sustainable Alternatives

Alternative routes that do not produce sizeable quantities of coproducts and that do not use chlorine-based chemistry have already been, or will be, implemented at the commercial level. In April 2003, Sumitomo Chemical commercialized the first "PO-only" plant in Japan, which produces PO by oxidation of propene with cumyl hydroperoxide (the latter being obtained by hydroperoxidation of cumene) without a significant formation of coproducts. Nowadays, the plant located at the Chiba factory, a joint venture between Nihon Oxirane Co and Lyondell, produces around 200 000 tons of PO/year. A second plant was started in May 2009 in Saudi Arabia, a joint project with Saudi Arabian Oil Co.

This process is a variant of the PO/SM process that uses cumene instead of ethylbenzene and recycles the coproduct cumyl alcohol via dehydration to α-methylstyrene and hydrogenation back to cumene (the latter two steps can be combined into a single hydrogenolysis step). Cumene hydroperoxidation technology is well-known for its use in phenol and acetone production.

Much attention has been directed recently towards processing approaches based on the epoxidation of propene using hydrogen peroxide (HP) as the oxidant. In principle, this technology can employ commercial HP, HP produced in an integrated facility or even that generated by the direct combination of hydrogen and oxygen (for an overview of the different technologies for HP production see the review published by Fierro *et al.* [4]). Recently developed technologies include:

- liquid-phase epoxidation with conventionally produced HP (HPPO: hydrogen-peroxide-propene-oxide);
- liquid-phase epoxidation with *in situ* generated HP (*in situ* HPPO).

The estimate for the future PO production market share indicates that by 2012 roughly 18% of all PO production will be based on HPPO technology, corresponding to a total capacity of almost 1.5 million-tons per year, 15% will be based on PO/TBA technology, 29% on PO/SM technology, 34% on CHPO technology and 4% on the Sumitomo cumene/PO process. Current PO production capacities are mainly located in Europe, North America and Asia; there are a few small-capacity plants in the Middle East. However, this region is of interest with regards to the natural gas reserves and the new routes of natural gas transformation, such as the methanol-to-olefins, methanol-to-propene and propane dehydrogenation routes. In fact, nowadays propene is produced mainly in steam-cracking units of naphtha and in FCC plants; 97% of the propene is of oil-based origin, while only 3% is gas-based. In the next few years, this ratio is expected to shift in favor of natural gas.

New approaches under investigation include the direct oxidation of propene with molecular oxygen, eventually in the presence of hydrogen:

- direct oxidation of propene with oxygen (DOPO: direct-oxidation-propene-oxide);
- hydro-oxidation of propene with oxygen and hydrogen (HOPO: hydrogen-oxygen-propene-oxide).

Several companies are working on the direct oxidation of propene; for instance Lyondell is operating a pilot plant in Newtown Square, PA, and intends to commercialize the technology by 2010. Shell Chemical is also working on a direct route to PO production, based on variations of the gold and silver catalysts it uses to make ethene oxide.

Table 6.1 summarizes the newly developing PO production processes.

Evidently, PO synthesis is a compendium of industrial chemistry. In fact, the different approaches and technologies used nowadays for PO synthesis are emblematic not only of the limitations and drawbacks that have burdened the chemical industry, but also of how the discovery of new catalysts and catalytic technologies may lead to the development of more economical and more sustainable processes. Furthermore, it is an example of how the oxidation of an organic substrate can be achieved through quite different approaches, with various oxidants (organic hydroperoxides, HP, molecular oxygen) and catalyst types, either in the liquid or in the gas-phase (Figure 6.2). However, in all the recently developed technologies, the discovery and use of a new heterogeneous catalyst has been the turning point for the successful commercial implementation of the corresponding process.

Table 6.1 Summary of newly developing PO production processes.

Technology	Oxidant	Catalyst: main active component
HPPO and *in situ* HPPO (BASF/Dow/Solvay, Degussa/ Uhde, Lyondell)[a]	H_2O_2 (by the anthraquinone route)	TS-1
HPPO and *in situ* HPPO (Lyondell, BASF, Degussa/Headwaters, Tosoh Co, Hoechst)	H_2O_2 (by direct liquid-phase oxidation of H_2)	Pd/Pt-TS-1
DOPO (Lyondell, Olin, BASF, Dow)	O_2	Ag-CaCO$_3$
HOPO (Dow, Nippon Shokubai, BASF, Dow, Bayer)	H_2O_2 (by gas-phase oxidation of H_2)	Au-Ti silicate

[a]Firstly developed by EniChem.

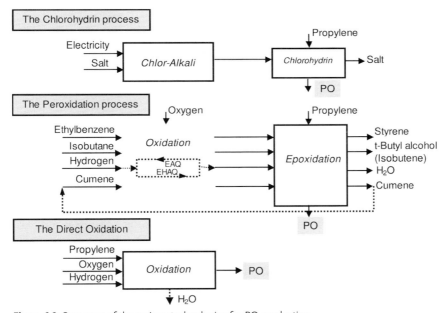

Figure 6.2 Summary of the various technologies for PO production.

6.2.1
The First Industrial PO-Only Synthesis: the Sumitomo Process

The Sumitomo PO-only process consists of the following steps [5] (Scheme 6.4):
1. Oxidation, where cumene is oxidized in air to obtain cumyl hydroperoxide (CHP).
2. Epoxidation, where PO and α,α-dimethylbenzyl alcohol (cumyl alcohol, CA) are obtained from CHP and propene, in the presence of a proprietary heterogeneous epoxidation catalyst.

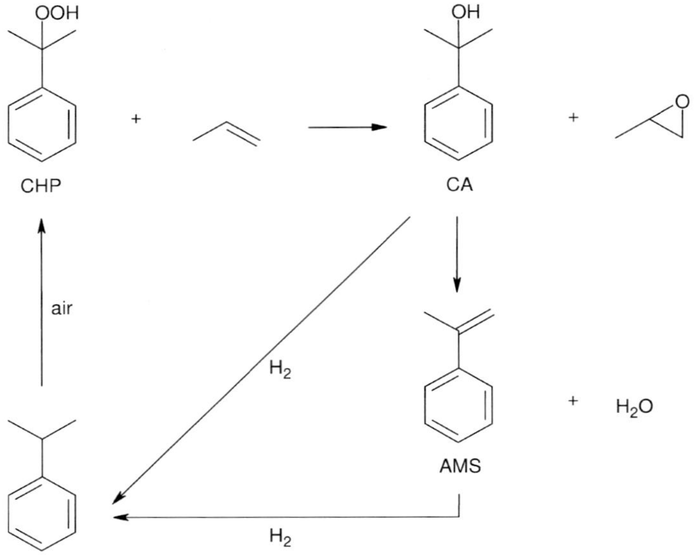

Scheme 6.4 PO-only Sumitomo process.

3. Hydrogenolysis, where the alcohol is first dehydrated to α-methylstyrene (AMS) and then hydrogenated to cumene, in the presence of a hydrogenation catalyst.

Cumene is then purified and recycled for use in the oxidation process.

The salient characteristic of the process is the large cumene inventory, which serves as the oxygen carrier. Owing to the high stability of cumene and its derivatives (CHP and CA), the overall cumene loss is minimal; this has allowed the establishment of a cost-effective cumene cycle for the production of PO. Another key point is the efficient recovery of the reaction heat in the oxidation, epoxidation and hydrogenation processes, minimizing the energy consumed in the separation and purification phases.

The use of CHP for the oxidation of propene has advantages over EBHP (PO/SM technology); as CHP is a stable compound, cumene hydroperoxidation is more selective than ethylbenzene hydroperoxidation. The reaction is an autoxidation process that occurs without a catalyst, and it is at least five-times faster than the oxidation of ethylbenzene; per pass conversion is two-times higher for CHP. The reaction is carried out in an alkaline/aqueous emulsion. Selectivity for CHP with respect to cumene is 95–98% (whereas selectivity for EBHP is 72–77%). The stability of CHP allows the safe handling of concentrations (75–85%) up to five times those of EBHP, the latter being typically handled in concentrations of 10–15%. These factors allow the use of equipment with smaller dimensions, thus meaning lower capital costs and making the recycling of cumene cost-effective. Moreover, the reaction parameters of the process have been optimized, and the role of impurities in the reaction rate and yield have been evaluated. The same operation is not so efficient when performed on ethylbenzene, by styrene hydrogenation. On the other hand, the

PO/SM and PO/TBA processes distribute the capital costs on the two products, while in the CHP-based process the PO alone carries the investment burden, finally resulting in a specific investment (investment/capacity) that may be not much different in the two cases.

In epoxidation, the propene-to-CHP molar ratio is 10:1, the reaction temperature is 60 °C and the pressure is sufficient to maintain propene in the liquid phase. The feed to the epoxidation reactor must contain less than 1% water in order to limit the hydrolysis of PO to glycol. The reaction is catalyzed by a proprietary, silylated, titanium-containing silicon oxide catalyst. The conversion of CHP is greater than 95%. Selectivity for PO based on hydroperoxide is 95%, whereas selectivity based on propene is around 99%. By-products of the reaction are aldehydes, such as acetaldehyde and propionaldehyde, alcohols (methanol and propene glycol), ketones and esters (e.g., acetone and methyl formate). The catalyst fixed-bed is structured into multiple catalyst layers, with heat exchangers in between the layers. This prevents excessive increases in temperature due to the exothermal reaction that would cause both thermal decomposition of the hydroperoxide and consecutive reactions of PO.

The unreacted propene is recovered from the epoxidation reaction solution and recycled, after separation of light hydrocarbons that are potentially present in the fresh propene feed. Distillation is then used to separate the solution into crude PO (light stream) and a CA/cumene solution (heavy stream). To avoid the dehydration of CA at this step, the bottom temperature of the column must not exceed 190 °C. Crude PO is purified by extractive distillation. The heavy stream is fed to the hydrogenolysis reactor; both the cumene/CA solution and hydrogen are fed to a fixed-bed reactor, packed with a spinel-based catalyst, for example, copper-chromite (in some patents, the dehydration- and hydrogenation-steps are separated; in this case, a supported Pd catalyst is preferred). The conversion of alcohol into cumene and water is very high, because alcohol does not easily take part in undesired side-reactions such as dimerization. The effluent stream passes through an oil/water separation and further purification. Before being recycled to the oxidation step, the cumene is purified to eliminate phenols that may inhibit the reaction rate in auto-oxidation, and so are other heavy compounds.

Figure 6.3 shows a simplified potential flow-sheet of the Sumitomo process, as inferred from the vast patent literature dating from 1999 onwards [6]. Here, two different layouts are hypothesized for the section aimed at the purification of PO; in both cases, however, the key separation is an extractive distillation. It is reported that in these columns the highest temperature must not exceed 130 °C, to avoid the formation of heavy compounds.

One milestone in the development of the process was the discovery of a new epoxidation catalyst made of titanium grafted onto silica, in 1998; the catalyst performs outstandingly in the reaction, with a minimal loss of cumene. Ti-silica catalysts have remarkably high epoxidation activity when a high dispersion of highly active tetrahedral titanium in a hydrophobic crystalline silica matrix is achieved [7]. The activity of these Ti-silica catalysts varies greatly according to the peroxides, olefins or solvent systems used.

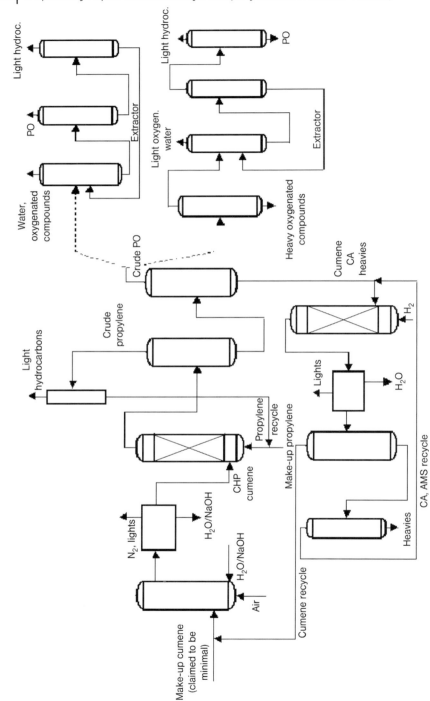

Figure 6.3 Simplified flow sheet of the Sumitomo process for PO synthesis, as inferred from the patent literature [5, 6].

Table 6.2 Comparison of catalytic performance with EBHP versus different heterogeneous Ti-based catalytic systems in propene epoxidation.

Catalyst	Reaction time (h)	EBHP conversion (%)	PO selectivity (%)
Ti-MCM41	0.5	99.2	89.2
Ti-SiO$_2$ (mesop)	1.0	94.9	89.2
TS-1	1.0	32.2	75.6

One peculiarity of the material is its high epoxidation activity with the bulky CHP molecule; as pointed out in a patent [8], the catalyst for the epoxidation of propene with bulky hydroperoxides (either CHP or EBHP) must belong to the class of mesoporous zeotypes, having an average pore size greater than 10 Å, and a specific pore volume greater than $0.2\,cm^3\,g^{-1}$.

As an example, Table 6.2 compares the performance of Ti-MCM-41 [9] with a mesoporous silica-grafted Ti and a microporous Ti silicalite (TS-1) for propene epoxidation with EBHP. With TS-1, the pore diameter is 5.5 Å, which imposes restrictions and limits the size of the molecules to be oxidized; the diffusion of the oxidant is inhibited, and hence the conversion attained is less than with the mesoporous materials. The difference between the mesoporous catalyst and TS-1 is even greater when the epoxidation of propene is carried out with CHP. The affinity for propene is increased by conferring hydrophobic properties to the surface, by means of silylation.

6.2.2
HPPO Processes: HP Generation by Redox Cycles on Organic O Carriers

The use of hydrogen peroxide (HP) as the oxidant in epoxidations is desirable due to its mild reaction conditions and the coproduction of water alone:

$$\text{propene} + H_2O_2 \longrightarrow \text{PO} + H_2O$$

The first attempts to use HP for the epoxidation of propene were reported by Ugine-Kuhlmann, who described the use of HP under anhydrous conditions in an organic medium; the water introduced with the oxidant and formed in the reactor was continuously distilled off, to prevent the deactivation of the metal catalyst. In other processes, developed by Bayer and Degussa, HP was an indirect oxidant, used for the synthesis of peracetic acid and perpropionic acids, the true epoxidizing agents. However, none of these processes went beyond the development stage; in that period, in fact, the hydroperoxidation routes (PO/SM and PO/TBA) were eroding the chlorohydrin market. It was the discovery of TS-1 in the SnamProgetti laboratories at the end of the 1970s that created new perspectives for the application of HP, permitting its direct use in water–methanol mixtures without catalyst deactivation [10].

However, the cost and hazards in shipping standard grades of HP can prove an obstacle for its use in an economically viable epoxidation process. Moreover, the manufacture of HP has other drawbacks, such as the complexity of the process and the proprietary technology. In principle, the problems of cost and logistics can be mitigated by the use of *in situ* produced HP at relatively low concentrations, for instance by means of the direct reaction between hydrogen and oxygen. This option could mean that HP producers will no longer be needed, as only the availability of hydrogen at the industrial site would be required. In this case, the epoxidation process is coupled with the HP-producing process without intermediate concentration or purification. However, this approach presents some drawbacks, for example, the low selectivity to HP based on hydrogen.

An alternative approach is a large co-located HP plant. Integration saves the costs for transport, logistics and buffer tanks at both the HP and the PO plants; this is expected to lead to reduced investments costs and thus to a better, more economic process. The largest plant using this new process with a capacity of 300 000 tons yr^{-1} PO was built in Antwerp, Belgium, by BASF/Dow, and has been in operation since November 2008. A megasized 230 000 tons yr^{-1} plant, three times larger than the largest ever built, for the synthesis of HP using Solvay's high-load anthraquinone technology has been built close to the BASF/Dow HPPO plant. Therefore, integration of the two processes allows a significant reduction of transportation costs, and in general offers opportunities for the reduction of HP production costs; in fact, the cost of HP synthesized in a dedicated plant for the HPPO process turns out to be lower than the HP market price.

Joint ventures and collaborations for the implementation of newly developed HPPO processes have been established between BASF, Dow and Solvay, and between Degussa-Evonik, Headwaters and Krupp-Uhde [11].

6.2.2.1 EniChem Approach: TS-1 Allows the Integration of HP and PO Synthesis

Since the early 1980s, EniChem has been a pioneer in the development of the process, holding a portfolio of patents [10c,d]. The integration of HP synthesis by means of alkylanthraquinone/alkylanthrahydroquinone (RAQ/RAHQ) cycle technology, with PO production, by means of propene epoxidation with HP, is possible because of the peculiar properties of the TS-1 catalyst (Scheme 6.5). TS-1 can selectively epoxidize propene using diluted HP [12]. A water–methanol mixture is the solvent for the epoxidation; the alcohol is necessary to obtain a sufficient reaction rate. Therefore, a cost-saving feature in this process is the fact that the crude HP produced can be used directly in the epoxidation of propene. Moreover, integration of the two processes is also allowed by the easily accomplished separation of propene and PO from the water–methanol mixture. Methanol, after separation and purification, can be recycled to the epoxidation step.

Compared to existing PO process technologies, HPPO is claimed to offer unique benefits from economic and environmental points of view. New PO plants built using HPPO technology are claimed to be more economical because they require up to 25% less capital. The simplicity of the process derives from the fact that there is no need for

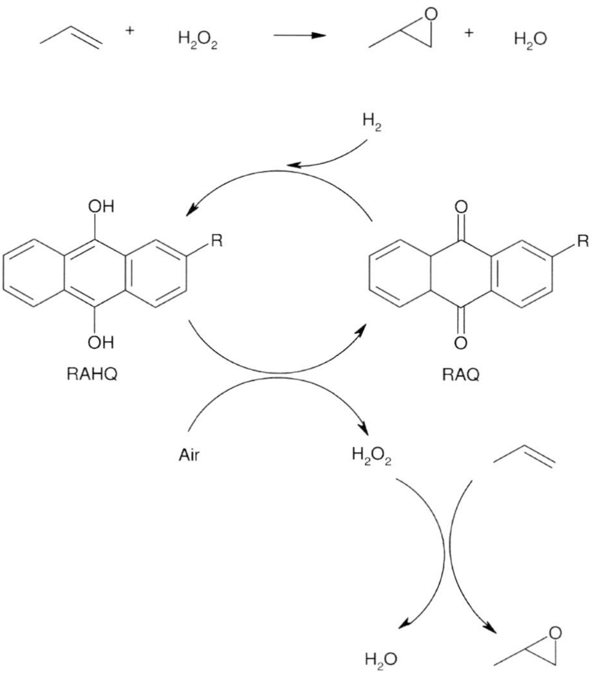

Scheme 6.5 RAQ/RAHQ process for the generation of HP, and integration with propene epoxidation.

additional infrastructure or markets for coproducts, and also because a simpler raw material integration is required. The environmental benefits derive from the reduction of wastewater by up to 70–80% and of energy usage by 35%.

EniChem first developed the concept of process integration for PO production [7f]. The main advantages of the process, compared to the conventional industrial technologies, were reported to be:

1. The high selectivity for PO and the generation of limited amounts of by-products, which can be easily disposed of.

2. The independence from an external supply of HP, since hydrogen and oxygen can be produced at most chemical sites by existing technologies, helping to improve the flexibility of the process in relation to the market, and avoiding the copro-duction of a second chemical.

The reaction is carried out at 50–60 °C, with methanol as solvent (it dissolves both HP and the alkene), and a pressure of 10–15 atm. The reaction is kinetically more favorable in the most polar solvents; water might be even better than methanol: however, in this case, the TS-1 rapidly decays. EniChem chose a mixture of water and methanol as the solvent for the development of the process, as did other companies later on. It is now accepted that the solvent has a role in the partition of the reagents between the external medium and the catalyst pores; however, other factors may overlap with the substrate partition effect. In fact, several lines of evidence exist for the adsorption of protic molecules (i.e., methanol) on Ti sites [13]. The solvation of the olefin by less polar solvents decreases its concentration in the proximity of active sites; the concept of zeolites as solid solvents, introduced by Derouane, is useful in this regard [14].

A PO selectivity on a propene basis as high as 98%, and a yield on a HP basis of around 90%, have been reported (conversion close to 100%, while that of the alkene was lower than 50%) [10e]. The main by-products were obtained due to consecutive reactions involving the epoxide, such as ring opening by methanol (the solvent) to yield the glycol monomethyl ether, or by water to yield the glycol, and glycol ketal. The addition of ppm basic compounds is important for obtaining an almost quantitative selectivity.

The rate of catalyst deactivation is a function of the TS-1 crystal size [12a, 13a]; with larger crystals, the slow diffusion of the epoxide solvolysis products (especially with more hindered products) makes the blockage of pores more likely. In propene epoxidation, polyethers are mainly responsible for this phenomenon, when the catalyst is used in consecutive reaction cycles. The activity of the catalyst can be restored by washing it with a solvent or by calcining it at temperatures higher than 500 °C.

Two different process configurations were proposed by EniChem [15]. The first configuration involves the *in situ* generation of HP, through the classical RAQ/RHAQ route, in the same medium where the epoxidation reaction occurs (*in situ* HPPO). Scheme 6.6 shows the reactions involved.

The first reaction, hydrogenation of the alkylanthraquinone, is catalyzed by Pd. The second, the epoxidation of propene by the HP generated by air oxidation of the RAHQ, is catalyzed by TS-1. This is possible because TS-1 activity is not affected by the polynuclear compounds forming the redox couple, since they do not enter the zeolite cages due to steric hindrance (the average diameter of the channel system of TS-1 and TS-2, with MFI and MEL type structures, respectively, is 0.55 nm; the cross

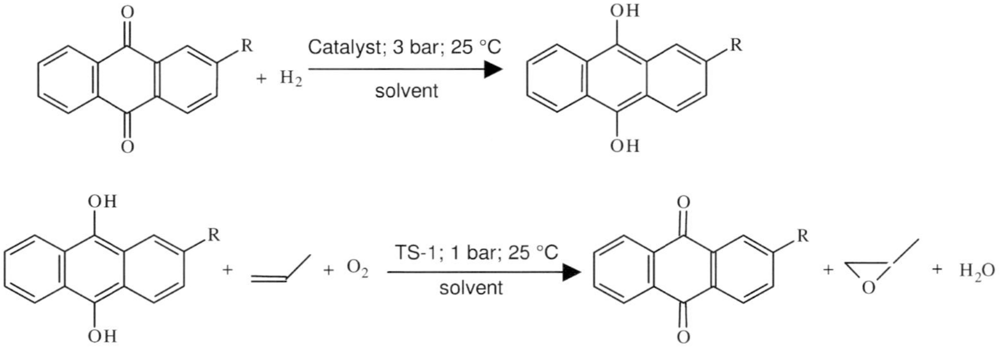

Scheme 6.6 Reactions involved in the EniChem process for propene epoxidation with *in situ* generation of HP.

section of alkylated anthraquinone is greater than 0.6 nm). Therefore, access of these bulky molecules to the Ti active sites is hindered, and potential oxidative degradation or interference with the catalytic processes is prevented.

A mixture of solvents is used: 2-methylnaphthalene (22 vol.%) to dissolve the alkylanthraquinone, a polar compound, preferably methylisobutylcarbynol (68 vol.%) to dissolve the alkylanthrahydroquinone, and methanol (10 vol.%). Methanol is also a co-catalyst, since the rate of reaction is much accelerated in the presence of this solvent. The best yield to PO, based on starting ethylanthraquinone, was 78%, at 30 °C, with 3 atm propene, 2 atm air and 0.31 wt% TS-1 as the catalyst, in a 1.5 h reaction time. For this to happen, autoxidation and epoxidation must occur at the same temperature, that is to say, a moderate temperature, to prevent the degradation of anthraquinone. The disadvantage of the process is that the optimal conditions for the generation of HP are not the same as for the epoxidation reaction.

The second process configuration is also based on process integration, but the RAQ/RAHQ and the epoxidation reactions occur in different reactors. The solvent is a methanol–water mixture, which extracts HP from the organic solvent of the redox process, and which is also the solvent for the solution fed to the epoxidation reaction. Once again, the advantage of this process over a conventional one, in which HP is produced separately, is the elimination of the expensive HP purification and concentration steps. Organic impurities that may have been extracted along with HP are precluded from diffusing inside TS-1 by their relatively large molecular size. The quantity of methanol passing in the working solution is small enough not to interfere with the cycle of reactions in the RAQ process. A yield to PO of 71.3% with respect to charged anthraquinone has been reported [15c]. The RAQ/RAHQ cycle for the production of HP is performed in a mixture of a hydrophobic (e.g., xylene) and a hydrophilic (water + methanol) phase.

The oxidation and reduction steps in the RAQ/RAHQ cycle are performed in two separate reactors. A bubble column is applied for the oxidation of the RAHQ, during which HP is produced. For the Pd-catalyzed hydrogenation of the quinones, a slurry, fixed-bed or monolith reactor can be used. After the reactor and L/L settler, a diluted HP-containing water–methanol stream is finally obtained. After the epoxidation step, crude PO is separated and the water–methanol mixture is returned to the HP synthesis process, thus realizing an efficient process integration.

Figure 6.4 shows a simplified flow-sheet of the two process configurations proposed by EniChem.

6.2.2.2 From the Dream Reaction to the Real Process: the Implemented HPPO Process

The HPPO process has been implemented recently by BASF and The Dow Chemical Company; the first plant has been in operation since November 2008 in Antwerp, Belgium. Dow acquired its PO technology from its 2001 purchase of EniChem's polyurethane business. BASF had been exploring HP-based routes to PO since the mid-1990s; Dow and BASF began collaborating in 2003. The two companies have established a long-term partnership with Solvay SA (supplier of HP to the new facility). The HPPO plant is fed with HP from a new mega-sized plant next to it.

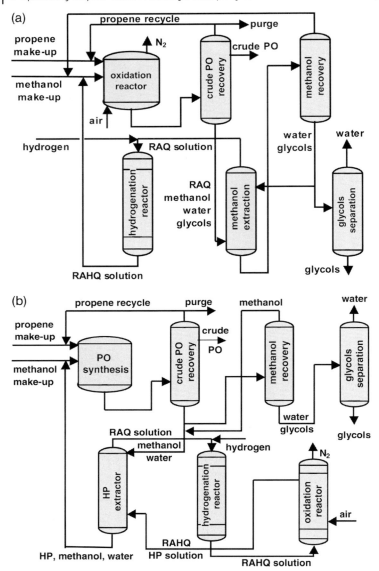

Figure 6.4 Simplified flow-sheet of the integrated *in situ* HPPO (a) and HPPO (b) processes for propene epoxidation with generation of HP, developed by EniChem.

The HP plant, built jointly by Dow, BASF and Solvay, has a capacity of 230 000 tons yr^{-1}, based on the RAQ/RAHQ route for HP production. The 300 000 metric-tons per year HPPO single-train plant, built jointly by BASF and Dow, started up in 2008. In 2007, Dow and Solvay also announced an agreement to create a joint venture for the construction of a HP plant in Map Ta Phut, Thailand. The plant will be the largest

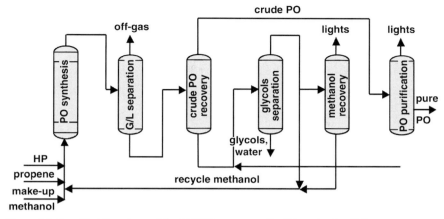

Figure 6.5 Simplified flow-sheet of the BASF/Dow process for PO production [16].

in the world, with a capacity of over 330 000 ton yr^{-1} of HP at 100% concentration. Scheduled to be operational by 2010, it will serve as a raw material source for the manufacture of PO. Dow and BASF are negotiating the construction of a world-scale, 390 000 metric-ton per year PO manufacturing facility in Thailand. Propene will be supplied from the liquids cracker that Dow announced it was building together with The Siam Cement Group in October 2006, which is expected to be fully operational by 2010.

Figure 6.5 shows a simplified flow-sheet of the BASF/Dow PO process [16]. Even though the scheme does not show any propene recycling, it is possible that the process also includes the recycle of the unconverted alkene.

The dilute HP–alcohol solution (HP concentration less than 10%) is introduced in a fixed-bed epoxidation reactor. A Ti silicalite (TS-1) catalyst is used also in this case, to produce PO from propene and HP. The reaction is carried out at 40 °C and 20 atm pressure. Process PO yield is estimated to be around 95 mol.%; by-products are 1,2-propandiol and the ethers formed by the methanolysis of the oxirane ring (1-methoxy-2-propanol and 2-methoxy-1-propanol), which may further react with PO to yield dipropeneglycol monomethylethers. PO may also form propanol hydroperoxides (1-hydroperoxy-2-propanol and 2-hydroperoxy-1-propanol). Other side-reactions such as the decomposition of HP normally occur to a very low extent.

The following achievements are claimed for the newly developed HPPO process in the patents jointly issued by BASF and Dow [17]:

1. An optimal performance in terms of PO yield is obtained when the reaction of PO synthesis is carried out in two steps, with intermediate separation of the PO produced [17b]. The output from the first reactor consists of methanol, water, PO, by-products, unreacted propene and HP. The yield achieved is 85% with a PO selectivity based on HP of 95%. The output is depressurized into a column operating at atmospheric pressure, at a bottom temperature of 69 °C. PO is distilled off together with propene and some methanol, while the bottom product

is fed to a second reactor operating under the same conditions as the first reactor, together with fresh propene. In the second step, the HP conversion achieved is 96%, and the PO selectivity is 96%. The overall HP conversion is 99.4%, and the overall PO selectivity is 95–96%; the PO yield based on HP is 94–95%. When the reaction is carried out in a single reactor, the HP conversion achieved is similar, that is, 98.4%, but the PO selectivity based on HP is 80.3%, with a PO yield of 79%. Overall, the higher propene-to-HP ratio in each reactor leads to a lower propene conversion, but the selectivity for PO with respect to HP is improved.

2. After the epoxidation reactor(s), a first column separates the lights, mainly propene, propane, nitrogen (used as ballast to avoid the formation of flammable mixtures in the epoxidation reactor) and oxygen, the latter deriving from the decomposition of HP. Specifically, the off-gas stream is compressed (16 atm) with cooling (35 °C) and fed to a stripping column under pressure, to obtain an olefin-containing bottom stream (potentially also containing propane) recycled to the epoxidation, and an off-gas stream with a low hydrocarbon content, containing nitrogen, oxygen and further volatile by-products [17c]. The following is an example of the stream outlet composition from the epoxidation reactor with the subsequent separation of the light-boiling compounds [17d]: unreacted propene 0.01 mass%, formaldehyde 0.01%, acetaldehyde 0.03%, PO 9.45%, methanol 71.97%, water 17.54%, glycol ethers 0.43%, propene glycol 0.05%, heavy boilers remainder.

3. PO is separated from methanol by extractive distillation, using water (or in another embodiment, propene glycol) as the extractor. PO is distilled overhead from the extractive column as top stream, while the bottom stream contains methanol and water [17d, f]. The energy integration of the column is one important issue of the patents; the vapors of the top stream are compressed, and the condensation heat is returned to the vaporizer employed in the extractive distillation column.

Degussa-Evonik, currently the world's second largest producer of HP after Solvay (both use the anthraquinone route), with approximately 600 000 metric tons, has investigated both the direct synthesis of HP and PO production with HP. The HPPO process, developed in cooperation with Krupp-Uhde, is available for licensing. To meet the growing demand for PO in the Asian market, the Seoul-based Korean company SKC acquired a license for the Degussa/Uhde process; an HPPO plant with an annual capacity of 100 000 metric tons per year at Ulsan was successfully started up in 2009 [18]. The process uses a titanium silicalite catalyst developed by Degussa, and will be supplied with HP from a conventional anthraquinone plant belonging to Degussa/Headwaters. The joint venture has also acquired a HP plant in Ulsan from the Helsinki-based Kemira, more than doubling its annual capacity from the previous 34 000 metric-tons per year to supply HP to the SKC plant.

In the Degussa/Uhde process [19], the reactor works at 60 °C. The raw PO is purified by extractive distillation using propene glycol that effectively removes impurities such as acetaldehyde. The final purity of PO is 99.97 wt%. The main by-product is propene glycol. The catalyst deactivates due to PO oligomers formation and is regenerated by calcining or treatment with HP solutions. The high selectivity is

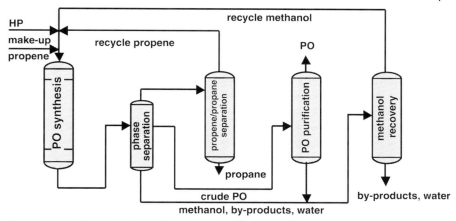

Figure 6.6 Simplified flow-sheet of the Degussa/Uhde process [19].

reached thanks to the moderate reaction temperature; the continuous removal from circulation of substances that can deactivate the catalyst brings about a considerable improvement in the performance and catalyst longevity.

Figure 6.6 shows a simplified flow-sheet of the Degussa/Uhde process [19]. In this process configuration, after the fixed-bed reactor the product mixture, consisting of methanol, water, propene and PO is depressurized; the resulting propene-rich gas phase is compressed, condensed and after separation of propane, it is returned to the reactor. Propane enters the process because chemical-grade propene can be used as the feedstock; to avoid the build-up of the alkane in circulation, which would behave as an inert gas in the reaction, propane is separated from propene. The expanded liquid phase is split into a raw PO stream and a stream consisting of methanol and water. The raw PO is fed to the purification section; purification is performed by extractive distillation using propene glycol that effectively removes impurities such as acetaldehyde. The final purity of PO is 99.97 wt%. The methanol–water mixture from the pre-separation section and the PO purification section is fed to the methanol processing unit, in which methanol is separated and then returned to the reaction section; this separation requires a large amount of energy. The bottom product of this section consists of water and by-products.

The main features of the process, as inferred from the patent literature [20], include:

1. PO synthesis is carried out in a methanol solvent, propene and 40% HP adjusted with ammonia to pH 4.5; pressure is 25 atm [20a]. The feed stream contains 21.5 wt% propene, 57 wt% methanol and 9.4 wt% HP. In some patents, a different feed composition is reported, containing 43 wt% propene, 43 wt% methanol and 8.4 wt% HP (from the feeding of a 60% solution of HP in water, adjusted to pH 4.5 with 1100 ppm ammonia) [20c]. The effect of temperature is shown in Table 6.3, for an upflow feed. By-products are 1-methoxy-2-propanol, 2-methoxy-1-propanol (propene glycol monomethyl ethers) and propene glycol (1,2-propandiol). How-

Table 6.3 Catalytic performance in propene epoxidation with HP in the Degussa/Uhde process [20].

Temperature (°C)	Flow rate (kg h^{-1})	HP conversion (%)	PO selectivity (%)	PO yield based on HP (%)
30	0.35	81	95	77
40	0.55	96	93	89
60	1.8	92	85	78
60	4.1	87	70	61

ever, when the reactor feed is downflow, at 40 °C, $P = 25$ atm and with a flow rate of 0.55 kg h^{-1}, the PO selectivity is 96% and the PO yield with respect to HP is 92%. The maximum recommended temperature in the catalytic bed is 60 °C, whereas the average cooling temperature is 40 °C. The optimum values reported are 97% selectivity for PO and 93% yield, both calculated with respect to HP, with a 96% HP conversion [20d]. The content of alkali metal ions and of bases is another key issue of the process [20e]. The catalyst is regenerated by washing with methanol at a temperature of at least 100 °C [20c].

2. A more detailed configuration of products separation and recovery section can be inferred from patents [20]. The stream exiting the reactor consists of a gas and a liquid phase. The gaseous stream, containing PO, propene, propane, oxygen and the inert gas is first directed to a condensation unit, wherein condensable components like propene, PO, propane and the solvent are partially condensed and recycled to the reactor. In an absorption unit, the gas stream from the phase-separator is subsequently put into contact with the same solvent used in the reaction stage, that is, methanol. The solvent stream loaded with propene, propane and PO is removed from the absorption unit and recycled to the reactor, with the gas stream containing oxygen and the inert gas. The presence of the inert gas prevents the formation of flammable mixtures [20f,h]. The liquid stream from the reactor (Figure 6.7), containing water, the water soluble solvent (methanol), PO, propane and propene, is first directed to a pressure release unit, and then separated into an overhead gas stream containing propene, PO, propane and the organic solvent, whereas the bottom stream contains the remainder of the organic solvent, water and by-products [20i,j]. The methanol–water mixture from the pre-separation section and the PO purification section is fed to the methanol processing unit, in which methanol is separated and then returned to the reaction section; this separation requires a large amount of energy. The bottom product of this section consists of water and by-products.

An alternative scheme for the separation of products is reported in [20j], in which the gas stream leaving the pressure release unit is directly recycled to the reactor. The liquid stream, which contains PO, methanol, water, high-boiling compounds and unconverted HP, as well as some propene and propane, is treated in a pre-evaporation column to obtain an overhead stream and a bottom stream. The latter, containing methanol, water and by-products, is subjected to subsequent purification steps,

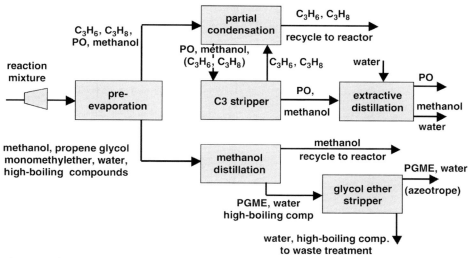

Figure 6.7 Block diagram of the liquid stream separation section of the Degussa/Uhde process for PO production [20].

whereas the former is further treated in a condenser and in a C_3 stripping unit, to finally obtain a hydrocarbon-rich stream that is recycled to the reactor, and a PO-in-methanol–water-rich stream that is subjected to further purification. The following is an example of the composition of the liquid stream exiting the reactor: 20.4 wt% propene, 2 wt% propane, 13.4 wt% PO, 0.1 wt% low-boilers, 47.7 wt% methanol, 15.5 wt% water, 0.5 wt% HP and 1.1 wt% high-boilers.

The bottom product of the pre-evaporation stage (Figure 6.7) can eventually be subjected to hydrogenation in a trickle bed reactor, to purify the solvent recycle stream by eliminating impurities in the form of formaldehyde and acetaldehyde, reducing them to methanol and ethanol, and also to eliminate traces of unconverted HP. Moreover, traces of hydroperoxypropanol and hydroxyacetone are converted into 1,2-propanediol. This allows a considerable decrease in catalyst deactivation in the epoxidation reactor and the improvement of product quality [20k].

6.2.2.3 Other Integrated HPPO Processes

An integrated process has also been reported, but not implemented at a commercial level, by ARCO (now Lyondell) [21]. In this case, the O-donor species is produced by the autoxidation of aryl-substituted secondary alcohols (α-methyl benzyl alcohol) to ketones, and the resulting solution is used to feed the epoxidation reactor. The alcohol is then recovered by hydrogenating the ketone. The main difference with respect to the EniChem process is the higher temperature of the autoxidation process (90–140 instead of 40 °C).

The procedure described involves the initial preparation of the oxidant mixture (containing 5 wt% HP, 1% H_2O, 66% unconverted α-methylbenzyl alcohol, 28% acetophenone, plus minimal amounts of EBHP) by air oxidation of α-methylbenzyl alcohol. The mixture is then fed to an autoclave loaded with the catalyst (TS-1), methanol and propene, and heated at 40 °C (molar ratio olefin/$H_2O_2 = 1.5/1$). A HP conversion of 70% is finally achieved, with 89% selectivity to PO on an HP basis (but conversions of up to 97% have been reported in other examples, with selectivity to PO equal to 84%). After separation of PO, the fraction containing acetophenone is hydrogenated with a Pd/C catalyst at 50 °C to achieve at least 80% conversion of acetophenone to α-methylbenzyl alcohol. It is reported that the integrated process, which uses a crude, unpurified oxidant mixture obtained by air oxidation of the alcohol, provides PO yields equivalent to those obtained using purified HP diluted in a clean alcohol/ketone reaction medium. Thus, the advantage of the integrated process is the reduced costs associated with the purification and concentration of HP. A disadvantage of the process is the deactivation of TS-1 when the process is carried out in a continuous mode, because of the accumulation of oligomeric by-products derived from the secondary alcohol. Figure 6.8 shows the general outline of the process.

In some patents [21b,c], a different integrated process is described, where a secondary-alcohol (isopropanol) is oxidized to an aliphatic ketone with oxygen to generate a stream containing HP. Before being fed to the epoxidation reactor, the ketone is eliminated by separation and hydrogenation; the epoxidation is then carried out, catalyzed by TS-1. The ketone is hydrogenated to the alcohol, and recycled to the autoxidation step. The advantage of separating the ketone is that the accumulation of a ketone peroxide species is prevented, which would otherwise be produced through a reaction between the ketone and HP at epoxidation conditions (conversely, the alcohol is reported not to react with HP). This side-reaction also causes consumption of HP, and complicates the purification and separation steps following epoxidation.

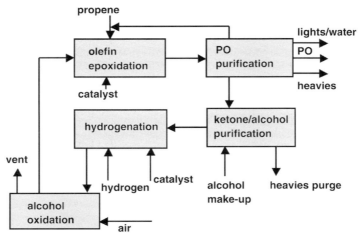

Figure 6.8 Block diagram of the ARCO integrated process for propene epoxidation with *in situ* generation of HP [21].

6.2.3
HPPO and *In Situ* HPPO Processes: HP Generation by Direct Oxidation of H_2 (DSHP)

The price of the HP for PO synthesis is a major factor determining the overall cost-effectiveness of the process. With DHSP, by direct reaction between hydrogen and oxygen, the epoxidation process might be coupled with the HP-producing process without intermediate concentration or purification. On the other hand, the selectivity to HP on a hydrogen basis is still low.

The DSHP is carried out in an alcoholic solvent, and produces dilute solutions of HP. The dilute methanol solution obtained might be suitable for selected organic syntheses. Clearly, the direct synthesis might be aimed at covering a small part of the HP market, also because the major use of HP (over 50% for bleaching paper, wood, and textiles, the remainder being distributed among applications in the fields of detergents, disinfectants and cosmetics) requires aqueous solutions with concentrations of between 30 and 70% [4]. Chemical applications for HP currently constitute less than 10% of all the applications of this product; however, a new, cost-effective HP-inorganic solution could be available for chemical oxidation in industrial applications, for example, in phenol and epichlorhydrin manufacture.

The DSHP, which is catalyzed by C-supported Pd-based systems, may even occur in aqueous solutions, in the presence of strong mineral acids and halide ions, at temperatures in the range 0–25 °C and at 100 atm pressure [22]. DuPont have studied extensively the use of Pt and Pd bimetallic catalysts on silica; the optimal weight ratio of Pt to metal loading on silica was 0.02–0.2; the selectivity to HP obtained was about 70%, with a concentration of HP exceeding 20%. Key promoters were halogens, such as Cl^- or Br^- salts. However, the presence of strong acids may be not compatible with the epoxidation of olefins.

6.2.3.1 Several Technologies for *In Situ* HPPO with TS-1-Supported Pd Catalysts
A process for the epoxidation of propene with *in situ* generation of HP was proposed in the 1990s by the Tosoh Corporation [23]. It was suggested that PO could be made via a direct reaction between hydrogen and oxygen in the presence of propene, using a catalyst made of Pd supported on crystalline titanium silicate, in a flow system. The solvent for the reaction was *t*-butanol, and the reaction temperature was 45 °C. A conversion of 0.8% was reported, with a selectivity for PO of 99%.

ARCO Chem Tech (now Lyondell) has reported catalysts that show enhanced selectivity and yield for PO [24]. Again, the catalyst was made of Pd impregnated over TS-1, but the addition of either NH_4OH, benzothiophene or triphenylphosphine in the reaction medium enabled an increase in the PO yield with respect to propene and H_2. Selectivity was also improved due to the suppression of the propene hydrogenation reaction.

BASF has also claimed the use of metal-modified TS-1 catalysts [25]. Various catalyst compositions have been described, including (i) titanium or vanadium silicates containing rare earth ions [25a] and (ii) titanium or vanadium silicates containing noble metals (Ru, Rh, Pd, Os, Ir, Pt, Re, Au, Ag). In these systems the

Table 6.4 Performance data for BASF catalysts in HPPO with generation of HP by means of H_2 oxidation with O_2 (DSHP) [25][a].

Catalyst	C_3H_6 conversion (%)	Selectivity PO/C_3H_8 (%)	Solvent	Reaction time (h)
0.49% Pd/TS-1	1.8	5.2/94.7	Methanol	17
0.49% Pd/TS-1	1.4	94.0/5.9	Water	3
0.49% Pd/TS-1	1.1	91.1/nr	Water	20

[a]Reaction conditions: 1 g catalyst, 60 mL solvent, gassed with 0.45 $L h^{-1}$ hydrogen for 0.5 h, then feed: 0.17% propene (0.08% when water is the solvent), 18.7% hydrogen, 18.7% oxygen, remainder nitrogen, $T = 45$–$50 °C$, $P = 1$ atm.

metal is (i) highly dispersed, to avoid metal–metal interaction, and (ii) present in at least two different oxidation states, that is, Pd^0, Pd^{+1} or Pd^{+2} [25b]. A procedure for compacting/molding this catalyst has also been reported [25c]. Table 6.4 compiles some relevant results claimed in these patents.

In more recent patents the two steps, that is, HP generation and PO synthesis, are carried out separately, each one with a specific catalyst and optimized reaction conditions [25d]. The gaseous stream exiting the PO synthesis reactor, containing the unconverted propene and the oxygen generated by HP decomposition, can be fed to the HP generation reaction. The reactions of (i) propane dehydrogenation to propene, (ii) HP generation by reaction between hydrogen and oxygen and (iii) PO generation by epoxidation of propene with HP are combined in an integrated process (Figure 6.9) [25e].

In BASF's two-stage integrated approach for PO production, HP is first produced by the direct combination of hydrogen and oxygen in a 6–7 wt% concentration in a

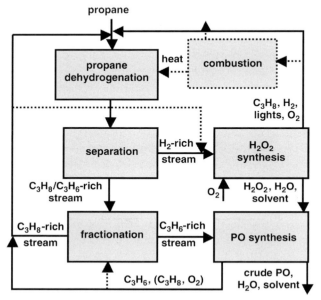

Figure 6.9 Process integration in BASF technology for PO production with DSHP-HPPO.

methanolic or hydro-alcoholic solution. The catalyst system consists of alternately laid corrugated and flat steel nets made from fine wires, and rolled into a cylindrical monolith. This monolith is impregnated with a palladium salt. The reactor vessel is flooded with methanol and hydrogen is bubbled in it as fine droplets from multi-leveled feed points. Enriched oxygen is fed from the bottom of the reactor. The oxygen-to-hydrogen ratio in the reactor is within safe limits. Hydrogen conversion is 76% and selectivity for HP is 82 mol.%. Reaction conditions are 40–50 °C and 50–54 atm; the reactor temperature is maintained by external cooling.

In the second step, a dilute HP–methanol solution is introduced in a fixed-bed epoxidation reactor. Make-up propene, recycled propene and HP from the product purification stage are fed into the reactor. The reaction is catalyzed by titanium silicalite, and takes place at 40–50 °C and 300 psi. HP per-pass conversion is initially 96% but drops down to 63% after 400 hours. PO selectivity is 95 mol.%; propene per-pass conversion is 39.8%. This technology gives capital savings compared to conventional hydroperoxidation technologies; however, it is likely that the operating costs of such a plant are higher than that of the latter.

Hoechst [26] (now Sanofi-Aventis) studied the epoxidation of propene with *in situ* generated HP, using a Ti silicate support with Pt/Pd metals. A very active system was reported, which at 43 °C (methanol and water solvent) converts 25% propene in 2 hours in a batch system, at 60 atm pressure, with a selectivity for PO (with respect to all the organic products formed) of 46%, and therefore a yield (with respect to propene) of 11.7% [26b,c]. The catalyst was made of 1 wt% Pd and 0.02 wt% Pt co-supported on TS-1. In general, the 1% Pd/TS-1 system exhibited a lower PO yield from the reaction between propene and HP than TS-1, due to the ability of the metal to decompose HP. However, it was possible to obtain similar results to those obtained with TS-1 due to a peculiar impregnation and reduction procedure; specifically, the catalyst was prepared by impregnation of TS-1 with Pt and Pd tetramine ligands. The best performance was obtained when the catalyst was reduced simply by treating the sample in a N_2 flow at 150 °C (autoreduction). This was attributed to the fact that this reducing treatment favored the formation of small Pd clusters, and a high fraction of Pd^{2+} species, which seem to play an important role in the reaction. The addition of small amounts of Pt also favors the development of Pd^{2+} species. Performance was greatly improved by the addition of NaBr as a promoter [26a]; with a doped TS-1/Pd/Pt catalyst, a selectivity for PO of 87.3% was achieved at 19.4% propene conversion. Without the NaBr promoter, the selectivity dropped to 34%.

To our knowledge, however, all these technologies have not gone beyond the pilot-plant stage. From an investment point of view, the *in situ* DSHP-HPPO would be the most attractive option, because a single reactor is needed. However, the reaction conditions for the two desired reactions are hardly compatible: the HP synthesis needs low temperature (<20 °C), strong Brønsted acids as co-catalysts and halides as selective catalyst poisons, while the epoxidation needs higher temperatures, usually above 40 °C, strong acids lead to the formation of by-products from PO and bromide can be oxidized to bromine, leading to brominated by-products. Moreover, the precious metal catalyst needed to make HP should not catalyze the hydrogenation of propene to propane. The best catalytic systems found give good selectivities of PO based on propene at moderate to low conversions, but the selectivity of PO based on H_2 is usually very low, $<30\%$. Therefore, to make this approach an improvement in the direction of better sustainability, much higher selectivity should be achieved.

6.2.3.2 DSHP-HPPO Technology Developed by Degussa Evonik/Headwaters

The DSHP has been developed jointly by Degussa-Evonik and Headwaters Nanokinetix, USA [27]. A demonstration unit for the production of several thousand metric tons per year of dilute HP in methanol went on stream in 2006, in Hanau-Wolfgang, Germany; commercial production will begin in 2009. The process operates below the explosion limit, and uses a Pd(Pt) nanocatalyst developed by Headwaters Technology Innovation, denominated NxCat™. It was anticipated that investment costs for the direct synthesis of HP will be approximately one-third to one-half lower than for conventional technologies. The process claims to eliminate all the hazardous reaction conditions and chemicals of the existing process, along with its undesirable by-products, to produce HP more efficiently, cutting both energy usage and costs. As no aqueous HP solution is required for PO production, coupling of direct HP synthesis with the HPPO process offers advantages over conventional PO production [27]. The miniplant set up by Degussa for direct HP synthesis is already coupled with PO production. The two steps are carried out separately, and therefore optimal conditions for the HP and the PO synthesis can be used; HP selectivity based on H_2 is at best 75%.

In 2007, the "Direct Synthesis HP by Selective Nanocatalyst Technology" received the coveted Presidential Green Chemistry Award.

Patents issued by Headwaters Nanokinetix cover the preparation and regeneration of nanosized catalysts for direct HP synthesis [28]. The main aspects reported include:

1. The supported catalyst is made of particles with a top layer of metal atoms with a controlled nearest-neighbor coordination number of 2 [28a,b]. Preferential exposure of specific crystallographic planes of the metal is achieved through the use of templating agents, such as polyacrylic acid. This agent also acts to chemically anchor the reactive catalyst particles to the support. Anchoring reduces the tendency of catalyst particles to become agglomerated, which decreases the available surface area, and also prevents leaching of active atoms into the liquid phase. Very small catalytic particles, less than 5 nm in size, are formed. For example, a supported Pd/Pt catalyst (the support being either C black or TS-1, with metal loading of 0.8 wt%) is prepared by dissolving Pd chloride, sodium polyacrylate or polyacrylic acid and Pt chloride in an acidic aqueous solution. The solution is treated with a H_2 flow, and then used to impregnate the support. Drying and treatment in hydrogen flow at 300 °C completes catalyst preparation. The reaction is carried out in a semi-batch reactor, using a liquid solution consisting of water with 1% H_2SO_4 and 5 ppm NaBr (a reaction promoter); the gas feed consists of 3% H_2 in air. The reactor is maintained at 45 °C and at a pressure of 90–100 atm. After 3 h reaction time, a conversion of 40% of the hydrogen with a selectivity for HP close to the 90% are obtained; the HP concentration is 2.5 wt%. An important role is played by the reaction solvent, because the HP productivity is proportional

to the solubility of hydrogen in the solvent; the latter, however, must also be a solvent for HP. The best results are obtained with methanol, 1% H_2SO_4 and 5 ppm NaBr; after 2 hours reaction, 4.1 wt% concentration of HP is obtained, with a hydrogen conversion of 85.2% and a productivity of approx $880\,g_{HP}\,g_{Pd}^{-1}h^{-1}$, with a catalyst made of C-supported Pd. Results obtained in methanol or in other solvents are better than those obtained with water, even though the addition of methanol or other water-soluble organic additives in a small % improves the HP productivity as compared to water alone [28c].

2. The anchoring of the nanosized active metal onto the support surface is improved when the support is treated so as to attach inorganic acid functional groups to it. This is made by reacting a solid support, for example, silica, with an acid such as p-toluenesulfonic acid in toluene, at the reflux temperature. An organic anchoring agent is then used to anchor metal nanoparticles to the acid-functionalized support, following the procedure described above [28d,e]. The functionalization is optimized by limiting the amount of water present in the support.

3. Staging or sequential feeding of the hydrogen stream into the catalytic fixed-bed reactor, in quantities sufficient to maintain a constant ratio of oxygen-to-hydrogen at the inlet of each of the vessel zones; in such a way, the best selectivity for HP production is obtained and excess oxygen recycling is minimized [28f]. The catalytic fixed bed, made of 0.75% Pd on a carbon support, is subdivided into different sections, and the hydrogen that is fed to the reactor is injected serially in diminishing discrete stages along the length of the reactor, each stage being fed separately before each catalytic section; the liquid and gas streams flow concurrently. The liquid stream consists of methanol, and also contains 1% H_2SO_4 and 5 ppm NaBr. The reactor operates at 45 °C and 27.5 atm. With this method, the best conditions of a local excess of oxygen are obtained (oxygen-to-hydrogen ratio of at least 2 over the entire reactor), and a selectivity for HP of 90%, with a yield of 81% (both calculated with respect to the converted hydrogen) and a HP yield, with respect to the oxygen fed, higher than 50%. This also provides a diminished requirement in terms of the volume of the oxygen recycling stream.

Still several problems remain unsolved to make the DSHP-HPPO process economically viable: (i) safety: the reaction of H_2 with O_2 in the presence of a flammable solvent (methanol) puts high hurdles on safety; (ii) removal of acid and bromide: the Brønsted acid and the bromide needed to produce HP from the elements have to be removed before the generated HP solution can be used for epoxidation; (iii) solvent recycle: after the generated HP solution has been used for epoxidation, the methanol has to be separated and recycled. During work-up some additional by-products are formed: formaldehyde, acetaldehyde, propionaldehyde, methyl formate, dimethoxymethane, 1,1-dimethoxyethane and 1,1-dimethoxypropane. These compounds are difficult to separate (many make azeotropes with methanol), so the recovered methanol will be contaminated. However, even small amounts of aldehydes or formates can poison the Pd or Pd/Au catalyst. Additional equipment needed to solve these problems will increase the investment costs.

6.2.4
An Alternative Approach: Gas-Phase Reaction Between Propene and HP Vapors

A consortium including Degussa, Uhde and academic teams has developed an innovative approach for the gas-phase epoxidation of propene with HP, catalyzed by TS-1; a pilot unit has been designed and engineered at the Degussa site in Hanau-Wolfgang [29a,c]. The BMBF-supported joint project DEMiS (Demonstration project for the Evaluation of Microreaction technology in Industrial Systems) uses micro-structured devices for the gas-phase epoxidation of propene with vapors of HP. The use of the gas phase avoids the need to use a solvent for the reaction, and allows operation at high space–time yields; indeed, the reaction with HP vapors can be carried out under thermal conditions, but the use of catalysts means that milder reaction conditions can be used [29d].

A gas phase process involving HP is extremely challenging, since safe and stable evaporation of HP is required, and precautions have to be taken to minimize the risk of explosions of the gas mixture. These problems were solved using a microstruc-tured reactor. A catalyst activity greater than $80 \, mol_{PO} \, (kg_{cat} \cdot h)^{-1}$ has been reported; activity of this magnitude in a conventional reactor would cause hot spots and increased decomposition of HP.

Microreaction technology allows for better temperature control, and a safer pro-vision and handling of HP both within and outside the range of explosive mixtures with propene and PO. To minimize the decomposition of HP during the evaporation phase, a microstructured falling-film evaporator is used, placed below the mixing device and reaction zone. The reaction is carried out with an excess of propene, at temperatures below 160 °C and a pressure below 1.5 atm. For a propene conversion of 5–20%, the selectivity for PO obtained is around 90%, but the selectivity with respect to HP is 25%. The catalyst is coated onto the reactor module, forming a layer several hundred micrometers thick. One problem encountered was the slow accumulation of deposits on the catalysts, which were derived from the formation of by-products.

This technology is emblematic of the new intensification approach currently used in the chemical industry, which is aimed at developing a safer, more flexible and more economic production of bulk chemicals, with minimal energy expenditure.

6.2.5
An Efficient Alternative Reductant for O$_2$: Methanol

Besides hydrogen, other reductants for O_2 in the liquid-phase epoxidation of propene include carbon monoxide, aldehydes, alcohols and other organic compounds. The reaction proceeds very efficiently with methanol as the reductant, in the presence of Pd and the Ti–Al-MCM-22 catalyst or Pd and peroxo-polyoxometalate catalysts; the latter have been intercalated inside layered double hydroxides to make them heterogeneous [29e,h]. A propene conversion of 47%, with 91.5% selectivity for PO, was obtained at 80 °C [29h].

The high activity of this heterogeneous system stemmed from the highly active Pd particles for effectively generating peroxy intermediates *in situ*, the good epoxidation capability of the peroxo-polyoxometalates in the interlayer region with dilute HP, and the promoting effect of the basic MgAl-type brucite-like layers for olefin epoxidation [29h].

It has been proposed that the methanol molecule reacts with the oxygen molecule rather than with Pd, to form a peroxy intermediate $HOCH_2OOH$, which regenerates the peroxoheteropoly-compound species that is active in propene epoxidation [29i]. As the peroxy intermediate is not stable and eventually decomposes to CO_x and H_2O, a part of the methanol is co-oxidized. Moreover, HP is probably also formed *in situ* in the catalytic system during the reaction and plays an important role in regenerating the active species.

6.2.6
Potential Future Solutions for PO Synthesis: Direct Gas-Phase Oxidation of Propene with Oxygen (DOPO)

In theory, it should be possible to obtain PO through direct oxidation of propene with oxygen, similarly to the industrial production of ethene oxide:

$$\text{propene} + \tfrac{1}{2}\,O_2 \longrightarrow \text{propene oxide}$$

The challenge in the case of propene oxidation is to find a catalyst that gives sufficiently high selectivity for PO by suppressing the competing combustion reaction. The catalysts and reaction conditions employed for the gas-phase epoxidation of ethene with molecular oxygen are not suitable for carrying out epoxidation of higher olefins, such as propene [2a]. In fact, the rate constant for the complete oxidation of PO is negligible, but the rate of the complete oxidation of propene is very high, being almost ten-fold faster than the complete oxidation of ethene. In contrast, the epoxidation reaction of propene is almost ten-times slower than the epoxidation reaction of ethene. All these factors lead to a poor selectivity for PO.

Plant design for the direct oxidation of propene would most likely be based on pure oxygen feed, rather than air, to gain yield advantage and lower capital costs. The minimal purge gas flow in an oxygen-based process makes it economically feasible to use a ballast gas (diluent) other than nitrogen.

The gas-phase epoxidation of propene can be carried out on catalysts substantially similar to those used for ethene epoxidation [30a], when small amounts of either a nitrogen oxide species (NO, NO_2) or a volatile organic chloride (ethyl chloride) are added to the feed stream [30b]. However, the addition of organic chlorides has disadvantages, such as an increase in the cost of raw materials, and the necessity to recover chlorides in the exit stream (these recovery treatments generate corrosive effluents).

This reaction has been studied in depth by ARCO Chem Tech, now Lyondell, which since 1996 has issued many patents on this subject [31]. The catalysts have analogies with those originally reported by Dow, which cited best values of 3.7% propene conversion and 47.2% selectivity for PO at 180 °C (catalyst $Ag/Mg\text{-}SiO_2$) [32a].

ARCO patents describe various approaches for the development of catalysts able to selectively epoxidize olefins other than ethene with oxygen. In all the catalyst formulations claimed, the main active component is supported silver, doped with various components [31a,b]. In the earlier patents, the best results reported were: propene conversion 4.5%, selectivity for PO 59–61%, with a catalyst composition of 54% Ag, 2% K, 0.5% Mo, supported over calcium carbonate. Molybdenum was used to increase the selectivity (but the addition of Mo also caused a decrease in propene conversion).

Several modifications of either the catalyst formulation or the process conditions were then reported, with remarkable improvements in performance being recorded. However, in all cases gas-phase promoters, be they organic halides or NO_x, were necessary to have acceptable conversion and relatively stable performance. Carbon dioxide is also a promoter, since in the presence of CO_2 the selectivity was improved, but the conversion considerably decreased. These features are clearly analogous to those applied in the industrial epoxidation of ethene.

The following improvements to the basic catalyst formulation and to the reaction conditions have been reported:

1. Use of catalyst modifiers, such as gold, rhenium or tungsten [31c,e], promoters for either activity or selectivity, or both. Rhenium is the most effective dopant. However, the system is very sensitive to the concentration of gas-phase promoters, such as NO_x, organic halides and carbon dioxide. In the absence of gas-phase promoters, conversion falls and catalyst deactivation also occurs.

2. Use of a support made of an alkaline earth metal compound [31f]. Tribasic calcium phosphate, calcium carbonate, calcium fluoride, strontium titanate and magnesium aluminate are optimal supports.

3. Addition of inorganic chloride promoters (alkali or alkaline earth chlorides) to the catalyst composition [31g,h]. A catalyst prepared by using potassium chloride gives a much higher activity and selectivity than one prepared with another K^+ salt (e.g., a nitrate). After approximately 20 h on line, this latter catalyst gives a propene conversion of 21%, with selectivity for PO based on propene of 45% (see Table 6.5 for catalyst composition and reaction conditions) in the presence of chloride gas-phase promoters, while in the absence of chloride the selectivity is <4%, with a propene conversion of 4.8%. In contrast, the addition of gas phase promoters does not affect the catalytic performance of the Cl-containing catalyst. Therefore, the promoter effect of halogens is analogous to that obtained by feeding chlorides in the gas phase; in fact, the promotion effect is observed as long as the halogen ions are progressively released from the catalyst (in the form of organic chlorides) into the gas phase. The solid thus acts as a temporary reservoir of halogens. Halogens inhibit the oxidation activity of the catalyst, preventing the overoxidation of propene to carbon oxides, but they also promote homogeneous radical reactions that lead to the selective transformation of propene.

4. The addition of CO_2 in the stream also has an important effect. Further improvement is obtained by treating the catalyst with a CO_2-containing stream after treatment with ethyl chloride. This helps in maintaining the improved perfor-

Table 6.5 Performance of catalysts in the Lyondell DOPO process for propene epoxidation with O_2.

Component, amount (wt%)	C_3 (%)/O_2 (%)/ CO_2 (%)/NO (ppm)/ CHCl (ppm)[a]	T (°C)/ GHSV (h^{-1})	C_3H_6 conversion (%)	PO selectivity (%)
$Ag(43)K(1.7)/CaCO_3$	10/5/0/50/200	250/1200	11	33
$Ag(54)K(2)Mo(0.5)/CaCO_3$	10/5/0/50/200	250/1200	11.2	39.5
$Ag(54)K(2)Mo(0.5)/CaCO_3$	5/5/20/75/200	240/1200	4.5	60
$Ag(52)K(2.1)/CaCO_3$	10/5/10/75/200	245/1200	5.1	51
$Ag(25)Au(4.6)K(1)/CaCO_3$	10/5/10/200/0	250/1200	6	44
$Ag(45)Re(0.37)K(1.3)/CaCO_3$	10/5/0/200/50	250/1200	14	37
$Ag(45)Re(0.37)K(1.3)/CaCO_3$	10/5/11.2/200/50	250/1200	10	51
$Ag(47)K(1.7)/CaCO_3$	10/5/0/200/50	250/1200	11	32
$Ag(47)K(1.7)/CaCO_3$	10/5/11.8/200/50	250/1200	8	42
$Ag(38)K(2)W(5)/CaCO_3$	10/5/25/200/50	260/1200	8.8	53
$Ag(40)K(2)/CaCO_3$	10/5/25/200/50	260/1200	8.7	52
$Ag(38)K(2)W(5)/CaCO_3$	10/5/0/200/50	260/1200	13.3	35
$Ag(40)K(2)/CaCO_3$	10/5/0/200/50	260/1200	12.6	32
$Ag(50)K(1.3)/SrTiO_3$	10/5/0/200/50	250/1200	10	38
$Ag(54)K(1.1)Mo(0.55)$ $Cl(0.5)/CaCO_3$	4/8/0/0/0 —	250/1200 —	21 —	45 —
$Ag(44)K(1.7)Mo(0.44)$ $Cl(0.5)/CaCO_3$	4.2/8.2/0/0/0 —	232/2400 —	9.1 —	51 —

[a]C_3 propene, O_2, CO_2, NO nitrogen oxides, CHCl organic chlorides, PO propene oxide.

mance for longer periods of time [31k]. Moreover, pre-treatment of the catalyst with CO_2 means that CO_2 does not need to be added to the reactor feed stream (but a chlorine source must be fed, anyway), and maintains high selectivity for extended periods of time. The feeding of CO_2 is also fundamental for achieving better selectivity to the epoxide in the case of ethene epoxidation.

5. Alternate feeding of propene and oxygen [31j]. The catalyst is first put in contact with the olefin in the absence of oxygen, and is then rapidly put in contact with oxygen under the reaction conditions; oxygen is pulsed for 20 s. During this time, PO is obtained with great selectivity (>55%); the oxygen is then discontinued for a few minutes, and the oxygen pulse cycle is repeated.

The best selectivity to PO, achieved in the presence of the optimized catalyst formulation and gas-phase compositions, is not greater than 55%. In some cases better selectivity is reported, but for very low conversions of propene. The best yield to PO under steady conditions, under the continuous feeding of the gas-phase promoters, is close to 5–5.5%. However, the productivity reported is still low, which would imply very large reactor volumes to achieve an acceptable epoxide capacity, and a large recycle stream.

A vast scientific literature exists on propene oxidation with O_2 [32]. Catalysts based on supported Ag have been investigated by several groups [32b,d,i]. Moreover, other types of catalytic systems have been studied, for example, Ti/HSZ [32e], Ti/SiO$_2$ [32f], MoO_3/SiO_2 [32g], Ag-MoO_3/ZrO_2 [32h] and Ti-Al-HMS [32i,p]. Table 6.6 compares some of the results reported in the scientific literature.

Table 6.6 Results over various catalytic systems reported in the literature

Catalyst	T (°C)	Propane conversion (%)	Selectivity PO (%)	Reference
Ag/NaCl	390	54.0	26.3	[32i]
Ag/MoO$_3$/NaCl	400	<2.0	53.1	[32k]
Ag/MoO$_3$/NaCl-ZrO$_2$	400	<2.0	66.7	[32q,r]
Ag/Y$_2$O$_3$/K$_2$O-αAl$_2$O$_3$	245	4.0	46.8	[32s]
Ti^{4+}/HSZ	300	78.2	26.1	[32e]
CuO$_x$/K$_2$O-SBA-15	300	~15	~5	[32m]
Cu-SiO$_2$	225	~0.2	~55	[32l]
Ti^{4+}/Al^{3+}-HMS	250	47.8	30.6	[32n,p]
MCM-22	300	90.4	12.5 (55 to oxygenates)	[32o]
MoO$_x$-SiO$_2$	300	17.6	43.6	[32g]
TiO$_2$-SiO$_2$	300	22	21	[32f]
Ag/NaCl-αAl$_2$O$_3$	260	4.4	30.1	[32c]
Ag/NaCl-CaCO$_3$	260	3.6	45	[32c]

In some cases, an important contribution of homogeneous radical reactions to PO formation is postulated [32f]; in practice, an allyl radical is initially formed at the catalyst surface, which then interacts with O$_2$ to form peroxy radicals, leading to the formation of peroxodimers or hydroperoxides. The hydroperoxides epoxidize propylene to PO and the peroxodimers may decompose to form PO.

In the late 1980s, Olin Co patented the use of molten alkali-nitrate salts for the epoxidation of propene with oxygen [33]. The reaction was carried out at 20 atm and 200 °C, and a PO selectivity of 65% was obtained at 15% propene conversion, by flowing a propene–air mixture through the molten salt mixture. The by-products of this reaction were acetaldehyde, acrolein, carbon monoxide and carbon dioxide. Higher selectivity was achieved by recirculation of the aldehyde, or by adding NaOH to the molten salt. Selectivity of 65% at 7% propene conversion has been reported [33f]. Using the same system, propane could also be directly epoxidized to PO, but with a selectivity lower than 15%. Later, it was found that the molten salt was indeed a radical initiator for the homogeneous gas-phase reaction [34]. In fact, relatively good PO yields can be obtained in homogeneous gas-phase epoxidation with oxygen [35]; at temperatures higher than 200 °C and under pressure, radical reactions occur quickly. However, the main problem met in the radical gas-phase reactions is the difficulty in controlling conditions and, hence, the overall performance, in addition to the formation of several oxygenated by-products.

6.2.7
Potential Future Solutions for PO Synthesis: Gas-Phase Hydro-oxidation of Propene with Oxygen and Hydrogen (HOPO)

The hydro-oxidation reaction involves one mole of propene, one mole of oxygen and one mole of hydrogen, forming PO and water:

$$\text{(propene)} + H_2 + O_2 \longrightarrow \text{(propene oxide)} + H_2O$$

Competing side reactions include propene hydrogenation to propane, and propene combustion to carbon dioxide and water. Additional hydrogen and oxygen consumption occurs to form water.

The best systems for this reaction are based on Au^0 nanoparticles supported either over TiO_2 (anatase), or over microporous or mesoporous Ti-silicates (TS-1, Ti-MCM41, Ti-β, Ti-SiO_2, Ti-TUD, silylated titanosilicate). The pioneer in this field was Haruta [36]; many studies concerning novel procedures for the preparation of Au-based systems have since been reported in the literature, including various types of supports [37a,l] and the use of other metal active components, for example, Ag, Pd or Pt [26c, 37m,n].

The best results reported are: 0.090–$0.120\,g_{PO}\,g_{cat}^{-1}\,h^{-1}$ productivity, at a propene conversion close to 8%, PO selectivity above 90% and selectivity based on hydrogen over 20%. Table 6.7 shows several results obtained from the literature. A commercially viable process would probably require a propene conversion >10%, PO selectivity >90% based on propene, and >50% based on hydrogen [36v].

Table 6.7 Performance of Au-based catalytic systems in HOPO^a.

Catalyst	T ($^\circ$C)	C_3H_6/H_2 conversion (%)	Selectivity for PO (%)	Reference
0.98% Au/TiO$_2$ (anatase)	50	1.1/3.2	>99	[36d]
0.20% Au/TiO$_2$(3%)-SiO$_2$	120	2.5/2.6	93	[36d]
1.0% Au/Ti-MCM-41	100	3.1/47	92	[36p]
1%CsCl-1%Au/Ti-MCM-41	100	1.7/4.5	97	[36p]
1.2% Au/Ti-MCM41	100	1.8/38	95	[36l]
1.0% Au/TiO$_2$-SiO$_2$	100	1.5/13	94	[36l]
1% Au/TiO$_2$	70	1.3/ng	>99	[37b]
1% Au/TS-1	125	0.8/ng	>99	[37b]
1% Au/TS-1	200	2.0/ng	50	[37b]
1% Au/TiO$_2$-SiO$_2$	100	1.0/ng	>99	[37b]
0.37%Au/TiO$_2$(1%)-SiO$_2$	150	6.8/36.4	91	[36s]
0.49% Au/TiO$_2$	150	2.4/44.1	38.6	[36s]
0.06% Au/0.09% Na/TiO$_2$(1%)-SiO$_2$	180	1.5/5.1	96	[36s]
Au/0.025% Mg/TiO$_2$(1%)-SiO$_2$	180	7.3/16.8	92.6	[36s]
Au/TiO$_2$(1%)-SiO$_2$	180	5.8/15.9	94.4	[36s]

^{a}Reaction conditions: [36d]: 0.5 g of catalyst, W/F 0.9 g s mL^{-1}, feed composition inert/oxygen/hydrogen/propene = 70/10/10/10 (mol.%). With 0.20 wt% Au/TiO$_2$-SiO$_2$ the catalyst feed composition was 45/10/40/5 (mol.%). [36p]: 0.5 g of catalyst, W/F 0.9 g s mL^{-1}, feed composition inert/oxygen/hydrogen/propene = 70/10/10/10 (mol.%). [36l]: 0.5 g of catalyst, W/F 0.9 g s mL^{-1}, feed composition inert/oxygen/hydrogen/propene = 70/10/10/10 (mol.%). [37b]: 0.3 g of catalyst, W/F 0.55 g s mL^{-1}, feed composition inert/oxygen/hydrogen/propene = 70/10/10/10 (mol.%). [36s]: 1 g of catalyst, W/F 0.72 g s mL^{-1}, feed composition inert/oxygen/hydrogen/propene = 70/10/10/10 (molar ratios); for catalyst 0.06% Au/0.09% Na/TiO$_2$(1%)-SiO$_2$, feed composition inert/oxygen/hydrogen/propene = 55/5/20/20 (mol.%), W/F 0.45 g s mL^{-1}.

A particular feature of the Au-based catalyst is the presence of a chemical interaction between titanium and gold, which confers to the system the ability to generate the active species for olefin epoxidation [36]. In fact, the main advantage of these catalysts is that they can epoxidize propene very selectively (selectivities of $95 + \%$) under mild reaction conditions ($50\,^\circ$C and 1 atm). Higher temperatures are needed when Au is deposited over Ti-silicates, but the latter systems are more stable than those made of Au supported over titanium dioxide, and give a higher PO yield. Modifying the surface of a Ti-Si support by silylation or fluoridation [38] can improve the performance of the catalyst even further.

However, when Au is deposited over other types of supports (silica or alumina), no epoxidation activity is observed [37b]. In addition, neither gold sponge nor titanium dioxide alone are active in the epoxidation of propene; thus, close cooperation between gold and titanium sites exist. For this reason, one major parameter affecting activity is gold dispersion, and hence the contact area between the two components. In fact, the chemistry of preparation of the catalyst, which affects Au dispersion and crystal size, is very important in determining the final catalytic behavior [36q]. Therefore, the use of a support with a surface area optimally higher than $100\,\mathrm{m^2\,g^{-1}}$ and a low titanium content favor the dispersion of both the Au and Ti^{4+} active components, allowing optimal cooperation between the two species [36s]. With Au particles between 2 and 5 nm dispersed over titanium dioxide, no dissociation of H_2 occurs, and the development of the active species is possible, with good selectivity for PO. With smaller Au particles, hydrogenation of propene to propane becomes the predominant reaction [36r]. The hydrogenation, however, is observed only in the presence of oxygen. This is explained by hypothesizing that oxygen can make the small Au particles electron-deficient, such that Au operates more like a hydrogenating metal (i.e., like Pt or Pd). In contrast, with particles bigger than 10 nm, the combustion reaction is favored, because of the reduced Au–Ti interface.

The mechanism proposed involves adsorption of propene onto gold, and the reaction of the adsorbed species with oxygen species (hydroperoxo and peroxo species) formed at the interface between the gold particles and the titanium support, through the reductive activation of oxygen with hydrogen [36j]. Scheme 6.7 shows the reaction mechanism proposed in the literature [37b,g,h].

The rate-determining step is the formation of a peroxide species on Au, whereas the reactive adsorption of propene onto titanium dioxide (catalyzed by the Au

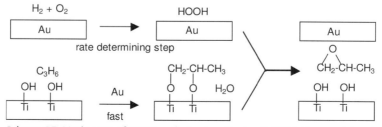

Scheme 6.7 Mechanism of HPPO, with direct synthesis of HOPO. Source: elaborated from [2a].

nanoparticles) to produce a propoxy species on titania is faster. The adsorbed species reacts with the peroxide to generate PO, which desorbs into the gas phase. The O atom left on Ti is consumed by hydrogen dissociatively adsorbed on Au, with the generation of a Ti-OH species and then of water, restoring the Au site. The progressive increase in PO selectivity occurs with a decrease in hydrogen consumption. The fact that gold is relatively unable to dissociatively adsorb molecular oxygen (which would burn hydrogen, and also the hydrocarbons) allows the desired reaction to take place with acceptable selectivity. In general, the hydrogen consumption is high and the selectivity for PO with respect to hydrogen consumed is thus low, at around 10–30%, since most of it is consumed in the production of water. The best hydrogen efficiency reported is no greater than 30%.

Haruta and Oyama proposed a slightly different mechanism [36t, 37s], in which the true intermediate was suggested to be the Ti-hydroperoxo species; the latter is formed by adsorption of O_2 onto Au, with the development of a Au^+-O_2^- species, which then reacts with H_2 to generate HP. HP finally generates the Ti-OOH active species over tetrahedral Ti sites, for the reaction with propene. Scheme 6.8 summarizes the main steps of the mechanism proposed [36t].

Gas-phase promoters, such as trimethylamine, play an important role in the reaction, adsorbing onto Au particles and preventing hydrogen combustion [36v]. In this case, the deactivation was appreciably depressed with a catalyst made of trimethylsilylated $Ba(NO_3)_2$-promoted Au/titanosilicate (Ti/Si 3/100) in the presence of 13–15 ppm trimethylamine as a gaseous promoter. Up to 80% of the high initial activity of the catalyst remained after 5 h of operation (propene conversion roughly 6.5%) with almost constant PO selectivity (about 91%) and H_2 efficiency (about 35%), under the following reaction conditions: 150 °C, atmospheric pressure, feed composition $C_3H_6/O_2/H_2/Ar = 1{:}1{:}1{:}7$, space velocity (SV) $= 4000\,mL\,h^{-1}\,g_{cat}^{-1}$. The PO productivity was approximately $0.08\,g\,g_{cat}^{-1}\,h^{-1}$, that is to say, not very different from the EO space–time yield in industrial ethene epoxidation.

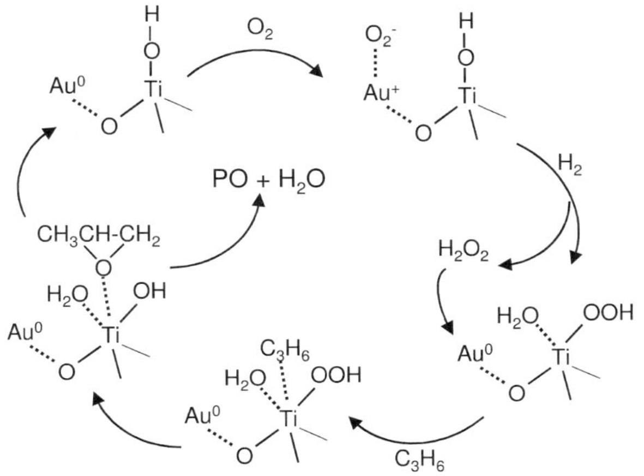

Scheme 6.8 Mechanism of HOPO, with direct synthesis of HP. Source: elaborated from [36t].

Moisture also enhances the catalytic activity by no less than two orders of magnitude [36y]. Water is proposed to play two roles in the reaction, namely in the activation of oxygen and in the decomposition of carbonate. The presence of reducing agents in the reaction environment may also considerably enhance the catalytic performance with regard to both conversion and selectivity. In particular, the preferred gas-phase reductants are CO, NO, 2-propanol and methanol. Improved hydrogen efficiencies can be obtained using CsCl as a promoter [36p], or other alkali or alkaline earth metal ions [39a]. These promoters block acidic sites on the catalyst, which could decompose or oligomerize PO; in general, desorption of PO is favored by an increase in the surface basicity of the catalyst [37i]. Also, alloying the gold with small quantities of platinum greatly improves the hydrogen efficiency [40].

Recently, Bayer reported a promising PO space–time yield of more than $0.2\,g_{PO}$ $g_{cat}^{-1}\,h^{-1}$, with selectivity for PO of 95%, obtained with a propene/oxygen/hydrogen feed of molar composition 30:10:60, at 180 °C and 2 atm pressure [41a]. However, as with the other catalysts, the main problem was a rapid catalyst deactivation. Key issues of Bayer's process include:

1. A catalyst based on Au/(Ag) particles deposited over Ti-containing organic/ inorganic hybrid silicon oxide material, prepared by a sol–gel method. The surface of the catalyst is modified with Si alkyl/alkoxy compounds [41a,c,f,i,k]; therefore, one important feature of the catalyst is its surface hydrophobicity. The material had a DRIFT spectrum characterized by the presence of bands assigned to the hydrocarbon coating (methyl groups), and to Si-H groups. A maximum yield of 10.4% (selectivity 95%) was reported, obtained at 140 °C and 4 atm [41j]. By-products were propionaldehyde (selectivity 2%), acetaldehyde (1%), acetone (<1%), traces of acetic acid, propene glycol and butanedione [41l]. A non-hybrid catalyst, based on nanosized Au particles supported over Ti-containing support, gave a PO yield of 5.8% at 46 °C and feed composition of propene/oxygen/ hydrogen 0.4:0.1:1.3 (molar ratios) [41e].

2. A method for the regeneration of spent catalysts, that includes placing them in contact with water and a diluted hydroperoxide solution [41b].

3. The use of CO as a gas-phase promoter, to limit catalyst deactivation; for example, with a catalyst made of Au with a particle size less than 4 nm, deposited over titanium dioxide, a yield for PO of 1.4% at 10 °C (selectivity >99%) was maintained over 6 h reaction time, whereas in the absence of CO the yield declined over the same period of time [41d].

4. The use of additional catalyst components, for example, Mo^{6+}, as promoters for Au-based catalysts [41g].

The use of a hybrid TS-1, that is to say, a TS-1 in which non-hydrolysable organic ligands have been incorporated while the crystalline structure of TS-1 is largely retained, has also been reported for the epoxidation of propene with aqueous HP [41h]. This structure is reported to possess enhanced hydrophobic properties, despite the relatively large amount of Ti incorporated; this property enhances the adsorption of propene and the desorption of PO, thus limiting the consecutive reaction on PO, and limiting the decomposition of HP.

Both Pd- and Pt-based systems are very active in propene and hydrogen conversion, but both catalyze the hydrogenation of propene. Nevertheless, an outstanding PO yield of 11.7% (selectivity 46%) was obtained on a TS-1 catalyst loaded with 1 wt% Pd and 0.02 wt% Pt, autoreduced under N_2 flow at 150 °C. The increase in PO yield by the addition of minor amounts of Pt to a Pd/TS-1 catalyst was correlated with an increase in the fraction of Pd^{2+} species, which play an important role in the reaction mechanism [26c].

This reaction has also been thoroughly investigated by Dow Chem Co [39]. For example, the presence of Na/Mg in Au/TiO_2 systems allows the catalyst surface to be kept cleaner, thus increasing the turnover of catalytic sites and also decreasing the rate of deactivation [39b]. The selectivity with respect to the hydrogen consumed is low, as indicated by the high H_2O/PO ratio. In most of the Dow patents, propene conversions lower than 1% are reported, selectivity for PO is well above 90%, the H_2O/PO ratio is between 3 and 10, with catalysts based on Au supported over various Ti-containing supports (TS-1, TS-2, Ti-MCM, Ti-β, Ti grafted over silica, etc.), and doped with alkali metal ions, alkaline earth and eventually lanthanides. However, in one patent a propene conversion of 3.4% was attained, with selectivity for PO of 97% [39j], at 140 °C, atmospheric pressure, with a feed composed of 20% propene, 10% oxygen, 10% hydrogen, and overall GHSV $960 \, h^{-1}$. The catalyst was made of Ba/Na-doped Au (average particle size of 5 nm) supported over a Ti-silicate, with 1.5% highly dispersed Ti.

Besides supported Au [39f,i,l,o], other catalysts claimed in Dow patents include Ag/alkali (alkaline earths, lanthanides) over Ti-containing supports. These systems give a propene conversion lower than 1%, high PO selectivity (but lower than that achieved with Au-based systems) and a H_2O/PO ratio much higher than that obtained with Au-based systems [39c,d,h,m,n]. The co-presence of Au, however, remarkably reduces the H_2O/PO ratio, thus increasing hydrogen efficiency [39g,k].

UOP has investigated the synthesis of HP by a reaction between hydrogen and oxygen with a favorable molar ratio close to 1:1, at 30 atm, without inert gases, at high space–time yield and in the presence of a heterogeneous catalyst, inside a microstructured apparatus, which guarantees safe plant operation within the explosive regime. UOP's interest is to have this direct HP synthesis route in the framework of PO manufacture [42]. Microchannel reactors are intrinsically safe, because thermal runaway and uncontrolled self-accelerating radical-chain propagation do not occur therein. In fact, the flame arrestor effect that quenches chain-growth by chain termination at the channel walls is favored because of the large interfaces and short diffusion distances. A space–time yield of $2 \, g_{HP} \, g_{cat}^{-1} \, h^{-1}$ is claimed, which is considerably higher than that achieved with conventional reactors.

Integration of the processing of propene epoxidation with that of propane dehydrogenation should in principle lead to economic advantages [36s]. In the integrated process, oxygen is added at the outlet from the dehydrogenation reactor, in the presence of an Au-based catalyst, to produce a gas that contains PO, unreacted propane, propene, hydrogen and oxygen. The epoxide is separated from the other gases. Before recycling the remaining gas to the dehydrogenation reactor, oxygen is eliminated from the gas mixture by letting it react with part of the hydrogen. In this way, hydrogen does not accumulate in the recycling loop. The advantage of this process configuration is that there is no need to separate hydrogen from the other gases.

Catalyst deactivation is one problem in the systems described in the literature. Deactivation is caused by the consecutive oxidation of the propoxy species to carboxylates, and to oligomerization of PO with accumulation of oligomerized and oxidized PO by-products around the gold nanoparticles. Other major hurdles for the industrial application of this process are the low propene conversion, the lower activity of the successively regenerated catalyst, and, especially, the low H_2 efficiency. A higher hydrogen selectivity will be needed for better process economics to avoid excessive costs for hydrogen feed.

6.2.8
Alternatives for Gas-Phase PO Synthesis

6.2.8.1 Gas-Phase Oxidation with N_2O
Although N_2O is still an expensive oxidant, the chemistry concerning its application in selective oxidation catalysis has attracted much attention. Actually, many studies have shown that N_2O can be used as an efficient oxidant for the selective oxidation of hydrocarbons such as benzene to phenol, propane to propene, and methane to methanol.

The direct oxidation of propene with N_2O can be carried out with Fe_2O_3 dispersed over SiO_2 (with an Fe content <0.1 wt%), or Fe supported over mesoporous silica (SBA15, MCM-41), with an Fe content of about 1 wt% [43]. The mesoporous silica may allow high and homogeneous dispersion of FeO_x species because of its large surface area and ordered mesoporous channel. The reaction occurs at temperatures higher than 350 °C and with a large excess of N_2O. The dispersed iron species residing within the pore structure are responsible for generating the electrophilic oxygen species upon nitrous oxide decomposition.

Duma and Hönicke were the first to report the successful use of N_2O in propene epoxidation. A PO yield of 5% was obtained over silica-supported iron oxide catalysts promoted with Na ions [43b]. The pore shape and diameter of the support as well as iron oxide dispersion are crucial parameters in the reaction [43b,c]. Doping with alkali metal may also considerably affect the Fe dispersion, and favor epoxidation over allylic oxidation [43f]. Further modification by boron can also significantly enhance the catalytic performance of the K-doped FeO_x/SBA-15 catalyst [43g].

The reaction pattern includes the formation of PO, its consecutive isomerization to propanal, acetone and allyl alcohol on acidic sites and combustion [43a]. Propanal and acrolein are also primary products. The formation of lower alkanes, alkenes, acetaldehyde and methanol results from cracking and oxidative C—C bond cleavage of propene and products. Additional side-reactions may occur in the gas phase, including radical-type oxidation of propene to acrolein, hexadiene and other by-products. Alkyl dioxanes and alkyl dioxolanes may form via dimerization reactions of PO on acidic catalysts. Indeed, major by-products are heavy compounds that accumulate on the catalyst and finally lead to coke; when taking into account these compounds, the true selectivity for PO may be well below 50% [43e].

The best yield reported in the literature is 13.3%, with selectivity of about 60% obtained with silica-supported K-promoted iron oxide catalysts modified by amines [43c]. The same catalyst is inactive in propene oxidation with air. However, the use of ammonia/air mixtures leads to a considerably enhanced conversion with respect to air only, with 60% selectivity for the epoxide. This observation suggests a mechanism whereby ammonia is first oxidized to nitrous oxide, which subsequently produces the active oxygen species for epoxidation.

6.2.8.2 Gas-Phase Oxidation with O_3

A novel, non-catalytic homogeneous route has been proposed, in which O_3 is co-fed together with NO_2, to generate *in situ* the true oxidant for propene, a N^V-oxide species (e.g., NO_3, N_2O_6 or N_2O_5, the latter being produced through a reaction between NO_2 and NO_3) [44].

Results comparable to those achieved for ethene epoxidation were obtained; for instance, at 10 mbar total pressure, at 180 °C and residence time of about 8 ms, with a reactor inlet feed containing 10 mol.% propene, 55% O_2, 3% O_3 and 7% NO_2, propene conversion was 30%, with 95–98% selectivity for PO [44a]. The major by-product was acetaldehyde, and there was minimal formation of CO_2. Optimization of the process parameters led to a further enhancement in performance; in the absence of a dilution gas, for a propene conversion of 91.8%, a selectivity for PO of 93.3% was recorded [44b]. An optimal O_3 utilization of 95%, with PO selectivity of 80.5%, was achieved by limiting the propene conversion to 15%, at 300 °C, 50 mbar total pressure and 1.6 ms residence time. Under the latter conditions, the productivity of PO was 3640 g $L_{reactor}^{-1}$ h^{-1}, which is higher than that achieved for industrial EO production.

In the reaction, NO_2 is not consumed, and it thus acts as a regenerable source for the O-transfer species; in the absence of NO_2, simple ozonolysis takes place and there is no formation of PO:

$$NO_2 + O_3 \rightarrow N^V\text{-oxide} + O_2$$

$$N^V\text{-oxide} + C_3H_6 \rightarrow PO + xNO_2$$

The reaction conditions and the expensive oxidant used are obvious drawbacks of this approach. However, the excellent PO productivity, the good yield achieved, the absence of catalysis costs and the associated problems related to catalyst deactivation and catalytic performance stability are all attractive features of this alternative approach for PO production.

6.2.9
The Ultimate Challenge: Direct Oxidation of Propane to PO

A one-stage process for the manufacture of PO from propane instead of propene would have substantial economical advantages. In one patent, a catalyst composed of $Ag/Cl/NaNO_3/La/Cr/BaCO_3$ is claimed that gives 10% propane conversion with a PO selectivity of 8% at 480 °C, resulting in a PO space–time-yield of 0.002 $g_{PO} g_{cat}^{-1} h^{-1}$; however, this catalyst deactivates very rapidly [45a].

An alternative approach is to combine the two steps, that is, the oxidative dehydrogenation of propane to propene and the epoxidation of propene to PO, in a single reactor, with two sequential catalytic beds, but with similar reaction conditions for the two steps [45b]. The problem is that most of the active and selective catalysts in propane ODH operate at temperatures that are too high for propene epoxidation.

In a recent paper, Oyama *et al.* [45b] documented an Au-TiO_2 catalyst that was able dehydrogenate propane to propene in the presence of H_2 and O_2, with good selectivity at 170 °C. A dual-bed catalyst, made of Au-TiO_2 and Au-TS-1, gave 2% propane conversion with propene selectivity of 57% and PO selectivity of 8%, when a propane/oxygen/hydrogen mixture was fed at 170 °C. The catalysts showed minimal deactivation and maintained their conversion and selectivity levels for 12 h. With a catalyst made of Au on TS-1, on the other hand, propane was transformed with great selectivity into acetone and 2-propanol [45c,d].

6.3
Conclusions

In PO synthesis, some options investigated, besides being more favorable from the economic standpoint, are also more sustainable in the sense that they adopt concepts like process integration or avoid the synthesis of coproducts. The successful industrial implementation of HPPO technology is emblematic of the steps that the chemical industry has been taking in the last ten years towards a more sustainable chemistry. Other options, finally, will hardly be more sustainable, due to the worse selectivity and the worse raw material usage, although being perhaps economically convenient because, for example, of the lower investment costs. Others still need considerable improvements to become economically viable, like the DOPO. The latter, however, if the target yield and space–time-yield were achieved, would be an outstanding example of better sustainability.

Many problems remain to be solved for these fascinating processes that directly oxidize propene with green oxidants in the gas phase: (i) the rapid deactivation of the catalyst, due to the accumulation of heavy compounds, precursors for coke formation; (ii) the need for gas-phase promoters, for example, NO, or chlorocarbons, which act to moderate activity and enhance PO selectivity; even in the very first patents issued in this field, this was claimed to be a key feature for optimal performance [46]; and (iii) the low space–time-yield achieved, due to the low conversion of propene and/or the low residence time.

For conversions lower than 5%, very high selectivity for PO based on propene can be obtained (e.g., higher than 90%) with the O_2/H_2 mixture (HOPO), whereas in the presence of O_2 alone the selectivity is not higher than 50–60% even at very low propene conversion. In general, yields for the direct oxidation of propene are lower than 5%. As shown clearly in [43a], if all the results achieved in the gas-phase epoxidation of propene with various oxidants, that is, O_2, $O_2 + H_2$, HP vapors or N_2O, are compiled in a cumulative plot of PO selectivity versus propene conversion, a limit curve can easily be drawn up, which seems to indicate that the conditions needed to increase propene conversion are not compatible with good PO selectivity. Moreover, selectivity to PO with respect to hydrogen is still too low.

Presently, the scientific community is making every effort to study innovative catalytic materials that combine morphological features suited to a fast counter-diffusion of PO, surface properties designed to favor the desorption of PO and limited side-reactions deriving from acid or base-catalyzed reactions – especially those leading to heavy product accumulation – as well as to stabilizing metal nanoparticles. The ultimate aim is to favor those synergetic phenomena deriving from the interaction between different catalyst components, multifunctionality being one of the essential properties for catalytic materials in gas-phase propene epoxidation with a feed containing oxygen and a reducing component.

Acknowledgement

Dr. Henrique Teles, BASF, is gratefully acknowledged for helpful suggestions and inspiring discussion.

References

1 (a) Tullo, A.H. and Short, P.L.(2006) *Chem. & Eng. News*, **84** (41), 22; (b) http://www.researchandmarkets.com/reports/29824; (c) http://www.icis.com/v2/chemicals/9076450/propylene-oxide/uses.html; (d) http://www.prnewswire.co.uk/cgi/news/release?id=60915. All web sites were accessed on May 2009.

2 (a) Nijhuis, T.A., Makkee, M., Moulijn, J.A. and Weckhuysen, B.M. (2006) *Ind. Eng. Chem. Res.*, **45**, 3447; (b) Hoelderich, W. (2000) *Appl. Catal. A*, **194–195**, 487; (c) Monnier, J.R. (2001) *Appl. Catal. A*, **221**, 73.

3 (a) Kollar, J. (1966) US Patent 3,350,422 (assigned to Halcon International); (b) Joustra, A.H., De Bruijn, W., Drent, E.

and Reman, W.G. (1989) Eur Patent 345,856 (assigned to Shell); (c) Buijink, J.K.F., van Vlaanderen, J.J.M., Crocker, M. and Niele, F.G.M. (2004) *Catal. Today*, **93**, 199; (d) Clerici, M.G., Ricci, M. and Strukul, G. (2007) *Metal Catalysis in Industrial Organic Processes* (eds G.P. Chiusoli and P.M. Maitlis), Royal Society Chemistry Publishing, Oxford, p. 23.

4 Campos-Martin, J.M., Blanco-Brieva, G. and Fierro, J.L.G. (2006) *Angew. Chem. Int. Ed.*, **45**, 6962.

5 (a) Tsuji, J., Yamamoto, J., Ishino, M. and Oku, N. (2006) *Sumitomo Kagaku*, **1**, 4; (b) Tsuji, J., Ishino, M. and Uchida, K. (2000) US Patent 6,160,137 (assigned to Sumitomo Chem); (c) Oku, N. and Seo, T.

(2003) US Patent 6,646,138 (assigned to Sumitomo Chem); (d) Tsuji, J. and Ishino, M.US Patent Appl. 2006/293531; (e) Tsuji, J.US Patent Appl. 2005/085647; (f) Tsuji, J. and Ishino, M.US Patent Appl. 2004/254386.

6 (a) Goto, S., Shinohara, K. and Katao, M. (2006) US Patent 7,030,254 (assigned to Sumitomo Chem); (b) Ito, K. and Katao, M. Jap Patent 2003/261551 (assigned to Sumitomo Chem); (c) Ito, K. and Katao, M. Jap Patent 2003/160519 (assigned to Sumitomo Chem); (d) Oku, N., Nakayama, T. and Shinohara, K. (2005) Eur Patent 1,498,414 (assigned to Sumitomo Chem); (e) Goto, S., Shinohara, K. and Katao, M. (2004) Eur Patent 1,420,014 (assigned to Sumitomo Chem); (f) Tsuji, J. and Yamamoto, Y. (2003) US Patent 6,639,085 (assigned to Sumitomo Chem).

7 (a) Tatsumi, T. (1995) *Catalysts Catal.*, **37**, 598; (b) Vayssilov, G.N. (1997) *Catal. Rev.-Sci. Eng.*, **39**, 209; (c) Chen, L.Y., Chuah, G.K. and Jaenicke, S. (1998) *J. Mol. Catal. A*, **132**, 281; (d) Corma, A., Fomes, V., Pergher, S.B., Maesen, Th.L.M. and Buglass, G. (1998) *Nature*, **396**, 353; (e) Tatsumi, T. (2005) *Catalysts Catal.*, **47**, 219; (f) Clerici, M.G., Bellussi, G. and Romano, U. (1991) *J. Catal.*, **129**, 159; (g) Thiele, G.F. and Roland, E. (1997) *J. Mol. Catal. A*, **117**, 351.

8 Tsuji, J., Yamamoto, J., Corma, A. and Rey Garcia, F. (2001) US Patent 6,211,388

9 Corma, A., Perez Pariente, J. and Villalba, N. (1998) US Patent 5,783,167

10 (a) Taramasso, M., Perego, G. and Notari, B. (1983) US Patent 4,410,501 (assigned to SnamProgetti); (b) Neri, C., Anfossi, B., Esposito, A. and Buonomo, F. (1986) Eur Patent 100,119 (assigned to EniChem); (c) Paparatto, G., Forlin, A. and Tegon, P. (2000) Eur Patent 1,072,599 (assigned to EniChem); (d) Clerici, M.G. and Ingallina, P. (1993) US Patent 5,221,795 (assigned to Enichem); (e) Buonomo, F., Neri, C., Anfossi, B. and Esposito, A. (1989) US Patent 4,833,260 (assigned to ANIC SpA).

11 (a) Tullo, A.H. (2004) *Chem. Eng. News*, **82** (36), 15; (b) http://www.basf.com/group/

pressrelease/P-07-351; (c) Tullo, A.H. (2006) *Chem. Eng. News*, **84** (41), 22–23; (d) http://www.basf.com/group/pressrelease/P-09-154; (e) http://www.dow.com/propylenoxide/news/20060927a.htm; If (f) Tullo, A. (2005) *Chem. Eng. News*, **83**, 7. All web sites were accessed in May 2009.

12 (a) Clerici, M.G. and Ingallina, P. (1993) *J. Catal.*, **140**, 71; (b) Lane, B.S. and Burgess, K. (2003) *Chem. Rev.*, **103**, 2457.

13 (a) Clerici, M.G. (2005) Oxidation and functionalization: classical and alternative routes and sources. Proceedings of the DGMK/SCI Conference, DGMK Tagungsbericht 2005-2, p. 165; (b) Clerici, M.G. (2001) *Topics Catal.*, **15**, 257.

14 (a) Derouane, E.G. (1998) *J. Mol. Catal.*, **134**, 29; (b) Langhendries, G., De Vos, D.E., Baron, G.V. and Jacobs, P.A. (1999) *J. Catal.*, **187**, 453.

15 (a) Clerici, M.G. and Ingallina, P. (1998) *Catal. Today*, **41**, 351; (b) Clerici, M.G. and Ingallina, P. (1993) US Patent 5,252,758 (assigned to Enichem); (c) Clerici, M.G., De Angelis, A. and Ingallina, P. (1998) US Patent 5,817,842 (assigned to Enichem).

16 http://www.corporate.basf.com/basfcorp/img/presse/konferenzen/antwerpen/BASF_HPPO/. Accessed in January 2008. Technical_Backgrounder_e.doc.

17 (a) Gobbel, H.G., Schultz, H., Schultz, P., Patrascu, R., Schultz, M. and Weidenbach, M.US Patent Appl 2006/0161010; (b) Bassler, P., Harder, W., Resch, P., Rieber, N., Ruppel, W., Teles, J.H., Walch, A., Wenzel, A. and Zehner, P. (2002) US Patent 6,479,680; (c) Babler, P., Gobbel, H.G., Teles, J.H. and Rudolf, P. US Patent Appl 2006/058539; (d) Schultz, H., Schultz, P., Patrascu, R., Schultz, M. and Weidenbach, M.Eur Patent 1,778,659 (assigned to BASF and The Dow Chem Co); (e) Goebbel, H.G., Schultz, H., Schultz, P., Patrascu, R., Schultz, M. and Weidenbach, M.US Patent Appl 2007/0238888; (f) Patrascu, R., Astori, S. and Weidenbach, M.WO Patent Appl 2004/083196 (assigned to Dow Global).

18 Ullrich, N., Kolbe, B. and Bredemeyer, N. (2007) *ThyssenKrupp Techforum*, **1**, 39.

19 Bredemeyer, N., Langanke, B., Ullrich, N., Haas, Th., Hofen, W. and Jaeger, B.(October 2005) Proceedings of the DGMK/SCI on "Oxidation and Functionalization: Classical and alternative routes and sources", Milan, DGMK Tagungsbericht 2005-2, p. 179.

20 (a) Haas, T., Hofen, W., Sauer, J. and Thiele, G. (2002) Eur Patent 1,373,235 and 1,373,236 (assigned to Degussa AG and Uhde); (b) Haas, T., Hofen, W., Sauer, J. and Thiele, G. (2003) US Patent 6,600,055 (assigned to Degussa AG and Uhde); (c) Haas, T., Brasse, C., Woll, W., Hofen, W., Jaeger, B., Stochniol, G. and Ullrich, N. (2005) US Patent 6,878,836 (assigned to Degussa AG and Uhde); (d) Haas, T., Hofen, W., Thiele, G., Pilz, S. and Woell, W. (2003) US Patent 6,608,219 (assigned to Degussa AG and Uhde); (e) Brasse, C., Haas, T., Hofen, W., Stochniol, G., Thiele, G. and Woell, W. (2003) Eur Patent 1,556,366 (assigned to Degussa and Uhde); (f) Hofen, W., Thiele, G., Haas, T., Woell, W., Kampeis, P. and Kolbe, B. (2003) Eur Patent 1,485,366 (assigned to Degussa AG and Uhde); (g) Hofen, W., Thiele, G., Haas, T., Woell, W., Kampeis, P. and Kolbe, B. (2004) US Patent 6,720,436 (assigned to Degussa AG and Uhde); (h) Hofen, W. and Thiele, G. (2003) US Patent 6,624,319 (assigned to Degussa AG and Uhde); (i) Hofen, W., Thiele, G. and Moller, A. (2003) US Patent 6,646,141 (assigned to Degussa AG and Uhde); (j) Haas, T., Hofen, W., Thiele, G. and Woell, W. (2003) WO Patent Appl 03/018567 (assigned to Degussa AG and Uhde); (k) Berges, J., Brasse, C., Eickhoff, H., Haas, T., Hofen, W., Kampeis, P., Moroff, G., Pohl, W., Stochniol, G., Thiele, G., Ullrich, N. and Woell, W. (2003) Eur Patent 1,499,602 (assigned to Degussa and Uhde).

21 (a) Zajacek, J.G. and Crocco, G.L. (1993) US Patent 5,214,168 (assigned to ARCO Chem Tech); (b) Zajacek, J.G., Jubin, J.C. and Crocco, G.L. (1995) US Patent 5,384,418 (assigned to ARCO Chem Tech); (c) Zajacek, J.G., Jubin, J.C. and Crocco, G.L. (1997) US Patent 5,693,834 (assigned to ARCO Chem Tech).

22 (a) Gosser, L.W. (1987) US Patent 4,681,751 (assigned to DuPont); (b) Gosser, L.W. (1989) US Patent 4,832,938 (assigned to DuPont)

23 (a) Sato, A. and Miyake, T. (1992) JP Kokai 4-352771 (assigned to Tosoh); (b) Sato, A., Miyake, T. and Saito, T. (1992) *Shokubai (Catalyst)*, **34** (2), 132; (c) Sato, A., Oguri, M., Tokumaru, M. and Miyake, T. (1996) JP Patent H8 269029 (assigned to Tosoh); (d) Sato, A., Oguri, M., Tokumaru, M. and Miyake, T. (1996) JP Patent H8 269030 (assigned to Tosoh); (e) Tatsumi, T., Yuasa, K. and Tominaga, H. (1992) *J. Chem. Soc., Chem. Commun.*, 1446.

24 (a) Dessau, R.M., Kahn, A.P. and Grey, R.A. (1999) US Patent 6,005,123 (assigned to ARCO Chem. Tech); (b) Dessau, R.M., Kahn, A.P., Grey, R.A., Jones, A. and Jewson, J.D. (1999) US Patent 6,008,388 (assigned to ARCO Chem Tech).

25 (a) Müller, U., Schulz, M., Gehrer, E., Grosch, G.H., Harder, W. and Dembowski, J. (1997) WO Patent Appl 97/25143 (assigned to BASF); (b) Dembowski, J., Bassler, P., Ellen, K., Kohl, V., Fischer, M., Harder, W., Mueller, U., Rieber, N. and Lingelbach, P. (1999) US Patent 5,859,265 (assigned to BASF); (c) Grosch, G.H., Mueller, U., Rieber, N., Schulz, M. and Wuerz, H. (1999) US Patent 6,008,389 (assigned to BASF); (d) Schindler, G.P., Walsdorff, C., Korner, R. and Gobbel, H.G.US Patent Appl. 2007/0004926 (assigned to BASF and The Dow Chem Co); (e) Bendewr, M., Zehner, P., Machhammer, O., Mueller, U., Harth, K., Schindler, G.P. and Junicke, H. (2007) US Patent 7,173,143 (assigned to BASF).

26 (a) Hoelderich, W. (1998) German Patent DE 98-19845975; (b) Laufer, W., Meiers, R. and Hölderich, W. (1999) *J. Mol. Catal. A*, **141**, 215; (c) Meiers, R., Dingerdissen, U. and Hölderich, W. (1998) *J. Catal.*, **176**, 376.

27 (a) http://www.degussa-award.com/
degussa/en/press/news/details?
News ID=1475 (accessed in May 2009);
(b) Brasse, C. and Jaeger, B. (2006) *Degussa
Sci. Newsetter*, **17**, 4.

28 (a) Zhou, B., Rueter, M. and Parasher, S.
(2006) US Patent 7,011,807 (assigned to
Headwaters Nanokinetix); (b) Zhou, B. and
Rueter, M. (2006) US Patent 7,045,479
(assigned to Headwaters Nanokinetix);
(c) Rueter, M., Zhou, B. and Parasher, S.
(2006) US Patent 7,144,565 (assigned to
Headwaters Nanokinetix); (d) Parasher, S.,
Rueter, M. and Zhou, B. (2006) US Patent
7,045,481 (assigned to Headwaters
Nanokinetix); (e) Rueter, M., Parasher, S.,
Zhang, C. and Zhou, B.US Patent Appl.
2007/0231248; (f) Rueter, M.US Patent
Appl 2006/0002847.

29 (a) Plettig, M., Döring, H., Dietzsch, E.,
Schwartz, T., Klemm, E., Markowz, G.,
Becker, F., Albrecht, J., Schütte, R.,
Schirrmeister, St. and Caspary, K.J. (2005)
in Proceed. DGMK/SCI Conference on
"Oxidation and Functionalization:
Classical and Alternative Routes and
Sources", Milan, p. 177; (b) Markowz, G.,
Schirrmeister, St., Albrecht, J., Becker, F.,
Schütte, R., Caspary, K.J. and Klemm, E.
(2005) *Chem. Eng. Technol.*, **28**, 459; (c)
Schütte, R., Markowz, G., Esser, P., Balduf,
T., Thiele, G. and Hasenzahl, S. (2001) Eur
Patent 1,118,613 (assigned to Degussa
AG); (d) Nagiyev, T.M., Nagiyeva, Z.M. and
Mustafayeva, Ch.A. (1991) *Petrol. Chem.*,
31 (5), 670; (e) Murata, K., Liu, Y., Mimura,
N. and Inaba, M. (2003) *J. Catal.*, **220**, 513;
(f) Liu, Y., Murata, K. and Inaba, N. (2004)
Chem. Commun., 582; (g) Liu, Y., Murata,
K. and Inaba, M. (2004) *Green Chem.*, **6**,
510; (h) Liu, Y., Murata, K., Hanaoka, T.,
Inaba, M. and Sakanishi, K. (2007)
J. Catal., **248**, 277; (i) Liu, Y., Murata, K.,
Inaba, M. and Mimura, N. (2005) *Appl.
Catal. B*, **58**, 51; (j) Liu, Y., Murata, K. and
Inaba, M. (2004) *Catal. Lett.*, **93**, 109.

30 (a) Hayden, P. and Sampson, R.J. (1977)
US Patent 4,007,135 (assigned to ICI LTD);
(b) Hayden, P., Clayton, R., Ramforth, J.R.

and Cope, A.F. (1995) US Patent 5,387,751
(assigned to ICI PLC).

31 (a) Rangasamy, P., Kahn, A.P. and Gaffney,
A.M. (1997) US Patent 5,625,084 (assigned
to ARCO Chem Tech); (b) Rangasamy, P.,
Kahn, A.P. and Gaffney, A.M. (1997) US
Patent 5,686,380 (assigned to ARCO Chem
Tech); (c) Pitchai, R., Kahn, A. and Gaffney,
A.M. (1997) US Patent 5,703,254 (assigned
to ARCO Chem Tech); (d) Gaffney, A.M.
(1999) US Patent 5,864,047 (assigned to
ARCO Chem Tech); (e) Kahn, A.P. and
Gaffney, A.M. (1999) US Patent 5,861,519
(assigned to ARCO Chem. Tech);
(f) Rangasamy, P., Kahn, A. and Gaffney,
A.M. (1998) US Patent 5,763,630 (assigned
to ARCO Chem Tech); (g) Cooker, B.,
Gaffney, A.M., Jewson, J.D. and Onimus,
W.H. (1998) US Patent 5,780,657 (assigned
to ARCO Chem Tech); (h) Jewson, J.,
Cooker, B., Gaffney, A.M. and Onimus,
W.H. (1999) US Patent 5,965,480 (assigned
to ARCO Chem Tech); (i) Cooker, B.,
Gaffney, A.M., Jewson, J.D., Kahn, A.P.
and Pitchai, R. (1998) US Patent 5,770,746
(assigned to ARCO Chem Tech);
(j) Gaffney, A.M., Jones, C.A., Pitchai, R.
and Kahn, A.P. (1997) US Patent 5,698,719
(assigned to ARCO Chem Tech);
(k) Jewson, J.D., Kahn, A.P., Cooker, B. and
Gaffney, A. (1999) US Patent 5,856,534
(assigned to ARCO Chem Tech).

32 (a) Bowman, R.G. (1989) Eur Patent
318,815 (assigned to Dow Chem Co);
(b) Oyama, S.T., Murata, K. and Haruta, M.
(2004) *Shokubai*, **46**, 3; (c) Lu, J., Bravo-
Suarez, J.J., Haruta, M. and Oyama, S.T.
(2006) *Appl. Catal. A*, **302**, 283; (d) Lu, J.,
Bravo-Suarez, J.J., Takahashi, A., Haruta,
M. and Oyama, S.T. (2005) *J. Catal.*, **232**,
85; (e) Murata, K. and Kiyozumi, Y. (2001)
Chem. Commun., 1356; (f) Mimura, N.,
Tsubota, S., Murata, K., Bando, K.K.,
Bravo-Suarez, J.J., Haruta, M. and Oyama,
S.T. (2006) *Catal. Lett.*, **110**, 47; (g) Song,
Z.X., Mimura, N., Bravo-Suarez, J.J., Akita,
T., Tsubota, S. and Oyama, S.T. (2007) *Appl.
Catal. A*, **316**, 142; (h) Jin, G., Lu, G., Guo,
Y., Guo, Y., Wang, J. and Liu, X. (2004)

Catal. Today, **93–95**, 173; (i) Lu, G. and Zuo, X. (1999) *Catal. Lett.*, **58**, 67; (j) Lu, J., Luo, M., Lei, H. and Li, C. (2002) *J. Catal.*, **211**, 552; (k) Jin, G.J., Lu, G., Guo, Y., Guo, Y., Wang, J. and Liu, X. (2003) *Catal. Lett.*, **87**, 249; (l) Vaughan, O.P.H., Kyriakou, G., Macleod, N., Tikhov, M. and Lambert, R.M. (2005) *J. Catal.*, **236**, 401; (m) Chu, H., Yang, L., Zhang, Q. and Wang, Y. (2006) *J. Catal.*, **241**, 225; (n) Liu, Y., Murata, K., Inaba, M. and Mimura, N. (2006) *Appl. Catal. A*, **309**, 91; (o) Murata, K., Liu, Y., Mimura, N. and Inaba, M. (2003) *Catal. Comm.*, **4**, 385; (p) Liu, Y., Murata, K., Inaba, M. and Mimura, N. (2003) *Catal. Lett.*, **89**, 49; (q) Jin, G., Lu, G., Guo, Y., Guo, Y., Wang, J., Liu, X., Kong, W. and Liu, X. (2004) *Catal. Lett.*, **97**, 191; (r) Jin, G., Lu, G., Guo, Y., Guo, Y., Wang, J., Kong, W. and Liu, X. (2005) *J. Mol. Catal A*, **232**, 165; (s) Yao, W., Guo, Y., Liu, X., Guo, Y., Wang, Y., Wang, Y., Zhang, Z. and Lu, G. (2007) *Catal. Lett.*, **119**, 185.

33 (a) Pennington, B.T. and Fullington, M.C. (1990) US Patent 4,943,643 (assigned to Olin Co); (b) Pennington, B.T. (1989) US Patent 4,883,889 (assigned to Olin Co); (c) Pennington, B.T. (1988) Eur Patent 268,870 (assigned to Olin Co); (d) Pennington, B.T. (1988) US Patent 4,885,374 (assigned to Olin Co); (e) Meyer, J.L. and Pennington, B.T. (1991) US Patent 4,992,567 (assigned to Olin Co); (f) Fullington, M.C. and Pennington, B.T. (1992) WO Patent Appl 92/09588.

34 Nijhuis, T.A., Musch, S., Makkee, M. and Moulijn, J.A. (2000) *Appl. Catal. A*, **196** (2), 217.

35 (a) Stark, M.S. and Waddington, D.J. (1995) *Int. J. Chem. Kinet.*, **27** (2), 123; (b) Dagaut, P., Cathonnet, M. and Boettner, J.C. (1992) *Combust. Sci. Technol.*, **83**, (4–6), 167.

36 (a) Kobayashi, T. and Haruta, M. (1989) US Patent 4,839,327 (assigned to Ministry of International Trade & Industry, and to Agency of Industrial Science & Technology); (b) Haruta, M., Tsubota, S., Kobayashi, T. and Nakahara, Y. (1990)

US Patent 4,937,219 (assigned to Ministry of International Trade & Industry, and to Agency of Industrial Science & Technology); (c) Qi, C., Okumura, M., Akita, T. and Haruta, M. (2004) *Appl. Catal. A*, **263** (1), 19; (d) Hayashi, T., Tanaka, K. and Haruta, M. (1998) *J. Catal.*, **178**, 566; (e) Haruta, M. (1997) *Catal. Today*, **36**, 153; Haruta, M. (1997) *Catal. Surv. Jpn.*, **1**, 61; (f) Haruta, M., Uphade, B.S., Tsubota, S. and Miyamoto, A. (1998) *Res. Chem. Intermed.*, **24**, 329; (g) Haruta, M., Kalvachev, Y.A., Tsubota, S., Hayashi, T. and Wada, M. (1999) Eur Patent 916,403 (assigned to Director-General of the Agency of Industrial Science and Technology, and to Nippon Shokubai Co); (h) Haruta, M. (1999) US Patent 5,929,258 (assigned to Agency of Industrial Science and Technology, and Nippon Shokubai Co); (i) Hayashi, T., Wada, M., Haruta, M. and Tsubota, S. (1996) Eur Patent 709,360 (assigned to Agency of Industrial Science & Technology, and Ministry of International Trade & Industry); (j) Hayashi, T., Tanaka, K. and Haruta, M. (1998) *J. Catal.*, **178**, 566; (k) Kalvachev, Y.A., Hayashi, T., Tsubota, S. and Haruta, M. (1997) *Stud. Surf. Sci. Catal.*, **110**, 965; (l) Kalvachev, A.Y., Hayashi, T., Tsubota, S. and Haruta, M. (1999) *J. Catal.*, **186**, 228; (m) Uphade, B.S., Tsubota, S., Hayashi, T. and Haruta, M. (1998) *Chem. Lett.*, 1277; (n) Sinha, A.K., Seelan, S., Akita, T., Tsubota, S. and Haruta, M. (2003) *Appl. Catal. A*, **240**, 243; (o) Sinha, A.K., Seelan, S., Tsubota, S. and Haruta, M. (2004) *Angew. Chem. Int. Ed.*, **43**, 1546; (p) Uphade, B.S., Okumura, M., Tsubota, S. and Haruta, M. (2000) *Appl. Catal. A*, **190**, 43; (q) Haruta, M. and Daté, M. (2001) *Appl. Catal. A*, **222**, 427; (r) Haruta, M. (1997) *Stud. Surf. Sci. Catal.*, **110**, 123; (s) Hayashi, T., Wada, M., Haruta, M. and Tsubota, S. (2000) US Patent 6,034,028 (assigned to Agency of Industrial Science and Technology and to Nippon Shokubai Co); (t) Bravo-Suarez, J.J., Bando, K.K., Lu, J., Haruta, M., Fujitani, T. and Oyama, S.T. (2008) *J. Phys.*

Chem. C, **112**, 1115; (u) Sinha, A.K., Seelan, S., Okumura, M., Akita, T., Tsubota, S. and Haruta, M. (2005) *J. Phys. Chem. B*, **109**, 3956; (v) Chowdhury, B., Bravo-Suarez, J.J., Date, M., Tsubota, S. and Haruta, M. (2006) *Angew. Chem. Int. Ed.*, **45**, 412; (w) Haruta, M. (2003) *Chem. Rec.*, **3**, 75; (x) Sinha, A.K., Seelan, S., Tsubota, S. and Haruta, M. (2004) *Top. Catal.*, **29**, 95; (y) Daté, M., Okumura, M., Tsubota, S. and Haruta, M. (2004) *Angew. Chem. Int. Ed.*, **43**, 2129.

37 (a) Stangland, E.E., Stavens, K.B., Andres, R.P. and Delgass, W.N. (2000) *J. Catal.*, **191**, 332; (b) Nijhuis, T.A., Huizinga, B.J., Makkee, M. and Moulijn, J.A. (1999) *Ind. Eng. Chem. Res.*, **38**, 884; (c) Sivadinarayana, C., Choudhary, T.V., Daemen, L.L., Eckert, J. and Goodman, D.W. (2004) *J. Am. Chem. Soc.*, **126** (1), 38; (d) Nijhuis, T.A., Visser, T. and Weckhuysen, B.M. (2005) *Angew. Chem. Int. Ed.*, **44**, (7), 1115; (e) Chretien, S., Gordon, M.S. and Metiu, H. (2004) *J. Chem. Phys.*, **121**, (8), 3756; (f) Yap, N., Andres, R.P. and Delgass, W.N. (2004) *J. Catal.*, **226**, 156; (g) Nijhuis, T.A., Visser, T. and Weckhuysen, B.M. (2005) *J. Phys.Chem. B*, **109**, 19309; (h) Nijhuis, T.A., Gardner, T.Q. and Weckhuysen, B. (2005) *J. Catal.*, **236**, 153; (i) Zwijnenburg, A., Makkee, M. and Moulijn, J.A. (2004) *Appl. Catal. A*, **270**, 49; (j) Nijhuis, T.A. and Weckhuysen, B.M. (2006) *Catal. Today*, **117**, 84; (k) Taylor, B., Lauterbach, J. and Delgass, W.N. (2005) *Appl. Catal. A*, **291**, 188; (l) Stangland, E.E., Taylor, B., Andres, R.P. and Delgass, W.N. (2005) *J. Phys.Chem. B*, **109**, 2321; (m) Wang, R., Guo, X., Wang, X., Hao, J., Li, G. and Xiu, J. (2004) *Appl. Catal. A*, **261**, 7; (n) Meiers, R. and Hoelderich, W.F. (1999) *Catal. Lett.*, **59**, 161; (o) Wells, D.H. JJr, Delgass, W.N. and Thomson, K.T. (2004) *J. Am. Chem. Soc.*, **126**, 2956; (p) Wells, D.H. JJr, Delgass, W.N. and Thomson, K.T. (2004) *J. Catal.*, **225**, 69; (q) Wells, D.H. JJr, Joshi, A.M., Delgass, W.N. and Thomson, K.T. (2006) *J. Phys. Chem. B*, **110**, 14627; (r) Taylor, B.,

Lauterbach, J., Blau, G.E. and Delgass, W.N. (2006) *J. Catal.*, **242**, 142; (s) Lu, J.Q., Zhang, X., Bravo-Suarez, J.J., Tsubota, S., Gaudet, J. and Oyama, S.T. (2007) *Catal. Today*, **123**, 189; (t) Sacaliu, E., Beale, A.M., Weckhuysen, B.M. and Nijhuis, T.A. (2007) *J. Catal.*, **248**, 235.

38 Weisbeck, M., Schild, C., Wegener, G. and Wiessmeier, G. (2004) US Patent 6,734,133 (assigned to BASF).

39 (a) Kuperman, A., Bowman, R.G., Hartwell, G.E., Schoeman, B.J., Tuinstra, H.E. and Meima, G.R. (2001) US Patent 6,255,499 (assigned to Dow Chemical Co); (b) Bowman, R.G., Womack, J.L., Clark, H.W., Maj, J.J. and Hartwell, G.E. (1998) WO Patent Appl 98/00413 (assigned to The Dow Chem Co); (c) Bowman, R.G., Clark, H.W., Kuperman, A., Meima, G.R. and Hartwell, G.E. (1997) Eur Patent 1,314,473 (assigned to The Dow Chem Co); (d) Bowman, R.G., Clark, H.W., Kuperman, A., Melma, G.R. and Hartwell, G.E. (1999) WO Patent Appl 99/00188 (assigned to The Dow Chem Co); (e) Clark, H.W., Bowman, R.G., Maj, J.J., Bare, S.R. and Hartwell, G.E. (1999) US Patent 5,965,754 (assigned to The Dow Chem Co); (f) Bowman, R.G., Womack, J.L., Clark, H.W., Maj, J.J. and Hartwell, G.E. (2000) US Patent 6,031,116 (assigned to The Dow Chem Co); (g) Bowman, R.G., Clark, H.W., Kuperman, A., Hartwell, G.E. and Meima, G.R. (1999) Eur Patent 1,140,883 (assigned to The Dow Chem Co); (h) Bowman, R.G., Kuperman, A., Clark, H.W., Hartwell, G.E. and Meima, G.R. (2001) US Patent 6,323,351 (assigned to The Dow Chem Co); (i) Bowman, R.G., Womack, J.L., Clark, H.W., Maj, J.J. and Hartwell, G.E. (2001) US Patent 6,309,998 (assigned to The Dow Chem Co); (j) Kuperman, A., Bowman, R.G., Clark, H.W., Hartwell, G.E., Schoeman, B.J., Tuinstra, H.E. and Meima, G.R. (2001) US Patent 6,255,499 (assigned to The Dow Chem Co); (k) Meima, G.R., Clark, H.W., Bowman, R.G., Kuperman, A. and Hartwell, G.E. (2003) US Patent 6,646,142 (assigned to Dow Global Technologies); (l) Bowman, R.G., Womack, J.L., Clark, H.W., Maj, J.J. and Hartwell, G.E.US Patent Appl 2002/0052290; (m) Bowman, R.G., Kuperman, A., Clark, H.W., Hartwell, G.E. and Meima, G.R.US Patent Appl 2002/161250; (n) Bowman, R.G., Kuperman, A., Clark, H.W., Hartwell, G.E. and Meima, G.R.US Patent Appl 2003/144141; (o) Bowman, R.G., Kuperman, A., Clark, H.W., Hartwell, G.E. and Meima, G.R.US Patent Appl 2004/176620; (p) Siler, S.J. and Henry, J.D.WO Patent Appl 2005/102,525 (assigned to Dow Global Technologies).

40 Makkee, M., Zwijnenburg, A. and Moulijn, J.A. (2003) WO Patent Appl 03/062196 (assigned to Huntsman).

41 (a) Weisbeck, M., Heinen, M.Th., Schmitt, J., Wegener, G. and Dugal, M. (2003) US Patent 0,134,741; (b) Weisbeck, M., Dorf, E.U., Schild, C., Lücke, B., Dilke, H. and Schülke, U. (1999) Eur Patent 1,051,256 (assigned to Bayer Aktiengesellschaft); (c) Weisbeck, M., Kraus, H. and Wegener, G. (2000) Eur Patent 1,196,242 (assigned to Bayer Aktiengesellschaft); (d) Weisbeck, M., Wegener, G., Dilke, H., Schülke, U. and Lücke, B.US Patent Appl 2002/0115894. (e) Weisbeck, M., Dorf, E.U., Wegener, G., Schild, C., Lücke, B., Dilke, H. and Schülke, U. (2003) US Patent 6,548,682 (assigned to Bayer Aktiengesellschaft); (f) Weisbeck, M., Wegener, G., Arlt, W. and Puppe, L.US Patent Appl 2003/0148885; (g) Dugal, M., Weisbeck, M. and Wegener, G.US Patent Appl 2004/0133019; (h) Weisbeck, M., Heinen, M.Th., Schmitt, J., Wegener, G. and Dugal, M.US Patent Appl 2004/0229747; (i) Weisbeck, M., Schild, C., Wegener, G. and Wiessmeier, G. (2004) US Patent 6,753,287 (assigned to Bayer Aktiengesellschaft); (j) Weisbeck, M., Wegener, G., Wiessmeier, G. and Vogtel, P. (2006) US Patent 6,995,113 (assigned to Bayer Aktiengesellschaft); (k) Weisbeck, M., Schild, C., Wegener, G. and Wiessmeier, G. (2004) US Patent 6,734,133 (assigned to Bayer Aktiengesellschaft); (l) Schummer, G., Zurlo, C., Woynar, H., Weisbeck, M., Wegener, G. and Hallenberger, K.US Patent Appl 2003/0031624.

42 Pennemann, H., Hessel, V. and Löwe, H. (2004) *Chem. Eng. Sci.*, **59**, 4789.

43 (a) Ananieva, E. and Reitzmann, A. (2004) *Chem. Eng. Sci.*, **59**, 5509; (b) Duma, V. and Hönicke, D. (2000) *J. Catal.*, **191**, 93; (c) Horvath, B., Hronec, M. and Glaum, R. (2007) *Top. Catal.*, **46**, 129; (d) Costine, A., O'Sullivan, T. and Hodnett, B.K. (2006) *Catal. Today*, **112**, 103; (e) Thömmes, T., Zürcher, S., Wix, A., Reitzmann, A. and Kraushaar-Czarnetzki, B. (2007) *Appl. Catal. A*, **318**, 160; (f) Zhang, Q., Guo, Q., Wang, X., Shishido, T. and Wang, Y. (2006) *J. Catal.*, **239**, 105; (g) Yang, S., Zhu, W., Zhang, Q. and Wang, Y. (2008) *J. Catal.*, **254**, 251.

44 (a) Berndt, T., Böge, O. and Heintzenberg, J. (2003) *Ind. Eng. Chem. Res.*, **42**, 2870; (b) Berndt, T. and Böge, O. (2005) *Ind. Eng. Chem. Res.*, **44**, 645.

45 (a) Mul, G., Asaro, M.F., Hirschon, A.S. and Wilson, R.B. JJr, (2003) US Patent 6,509,485 (assigned to SRI International); (b) Bravo-Suárez, J.J., Bando, K.K., Lu, J., Fujitani, T. and Oyama, S.T. (2008) *J. Catal.*, **255**, 114; (c) Bravo-Suarez, J.J., Bando, K.K., Akita, T., Fujitani, T., Fuhrer, T.J. and Oyama, S.T. (2008) *Chem. Commun.*, 3272; (d) Bravo-Suarez, J.J., Bando, K.K., Fujitani, T. and Oyama, S.T. (2008) *J. Catal.*, **257**, 32.

46 Thorsteinson, E.M. (1991) Canadian Patent 1,282,772.

7
Synthesis of Adipic Acid:
On the Way to More Sustainable Production

Fabrizio Cavani and Stefano Alini

7.1
Introduction: The Adipic Acid Market

Adipic acid (AA) is an important chemical intermediate [1]. The production of Nylon-6,6 (for fibers and resins) accounts for 63% of AA consumption; the fibers are used in applications such as fishing lines, tires, carpets, home furnishings and in tough fabrics for parachutes, backpacks, luggage and business cases. Nylon engineering resins are chiefly used in electrical connectors, auto parts and items such as self-lubricating bearings, gears, and cams.

The major non-nylon uses of AA are in polyester polyols (for polyurethane resins, 25% of AA production), in plasticizers (7%: dioctyl adipate, diisodecyl adipate, etc. for vinyl chloride, nitrocellulose and cellulose acetate polymers), resins (2%: unsaturated polyesters) and 3% for miscellaneous applications, such as a food ingredient in gelatins, and as a component in cosmetics, pharmaceuticals, fertilizers, paper, cements, waxes, and so on.

Currently, the overall growth for AA is close to 3% per year, while demand is growing faster and is forecast to be around 5–6% per year by 2010. The expected growth is 12% in Asia-Pacific, around 8–9% in Asia/Middle East, while remaining stable in Western Europe and in the USA. The most rapidly growing sector is that of nylon resins, which has grown by 8–10% per year during the past decade. However, the market anticipates that the substitution of nylon engineering plastics into automobile parts is becoming saturated. Moreover, with 62% of the AA take, a slowdown in nylon fiber production is retarding the AA market. Nylon fibers are highly sensitive to the performance of the housing sector, which has been growing slowly for some years now, by about 1.7% on an annual basis. Nevertheless, despite the new plants being installed in China, capacity will fall short by 2010, because of the poor profitability that has hindered investments.

In 2006, the global AA capacity was around 2.8 million metric tons per year; nearly 1 M-ton yr^{-1} was produced in Western Europe, 1 M-ton yr^{-1} in the USA (where 62% of the total global production is consumed), 0.5 M-ton yr^{-1} in Asia (excluding Japan), with the remainder being produced in Japan, Canada and Latin

Sustainable Industrial Processes. Edited by F. Cavani, G. Centi, S. Perathoner, and F. Trifiró
Copyright © 2009 WILEY-VCH Verlag GmbH & Co. KGaA, Weinheim
ISBN: 978-3-527-31552-9

Table 7.1 Main producers of AA [1a].

Producer	Capacity (million tons per year)
Inolex Chem (USA)	0.02
Solutia (USA)	0.40
BASF (Germany)	0.26
Asahi Kasei Corporation (Japan)	0.17
Azot Severodonetsk (Ukraine)	0.03
Lanxess AG (Germany)	0.07
Invista (Koch Ind) (USA, Canada, Singapore, UK)	1.09
(in 2004 Koch acquired the DuPont textile business)	
Rhodia (France, Brazil, Korea)	0.54
Radici Chimica (Italy, Germany)	0.15
Rivneazot	0.03
China Shenma Group (China)	0.11 (planned 0.15)
PetroChina Liaoyang Petrochemical (China)	0.14 (planned 0.30 by 2009)
Xinjiang Dushanzi Tianli (China)	0.08
Taiyuan Chemical Industry (China)	0.05 (planned)
Shandong Bohui Chemical Industry Group (China)	0.08
Shandong Hongye Chemical Industry Group (China)	0.10

America. Notably, in 2002 the share was quite different: 42% of the global production was in the USA, 40% in Western Europe and 13% in Asia-Pacific, while the other regions accounted for the remaining 5%. The share is foreseen to shift again, in favor of China; in fact, at the end of 2007, a new production unit was started up in China, and two more units were installed in the same country in 2008.

Between 2002 and 2005, the AA market suffered from overcapacity. In 2006, planned and unplanned shutdowns in Europe and Asia kept supply tight, while demand for nylon and non-nylon sectors was strong. In addition, geographic demand has been changing rapidly in recent years; while a few years ago more than 60% of the global production was consumed in the USA, nowadays demand is split, with approximately 33% of the demand coming from Western Europe, 32% from the USA, 20% from Asia-Pacific, and 6% from Japan. In 2006, the USA was the largest net exporter of AA, whereas Asia was the largest net importer.

Table 7.1 summarizes the main producers of AA, and the installed capacity.

7.2
Current Technologies for AA Production

The industrial process accounting for the total AA production is the oxidation of either a mixture of cyclohexanol and cyclohexanone, the so-called KA Oil (ketone/alcohol oil or Ol/One), or of cyclohexanol [2]. These compounds can be produced by different methods, the principal one being the oxidation of cyclohexane with air.

7.2.1
Two-Step Transformation of Cyclohexane to AA: Oxidation of Cyclohexane to Ol/One with Air

Cyclohexane is obtained either by the hydrogenation of benzene, or from the naphtha fraction in small amounts. Its oxidation to the KA Oil dates back to 1893 and was first industrialized by DuPont in the early 1940s. Oxidation is catalyzed by Co or Mn organic salts (e.g., naphthenate), at between 150 and 180 °C and 10–20 atm. Indeed, this reaction is a two-step process (an oxidation and a deperoxidation step), and two variants are currently in use [2, 3]. The oxidation step can be performed with or without a catalyst. The deperoxidation step always uses a catalyst (Co(II) or NaOH). The overall performance of both variants is almost identical, although the selectivity in the individual steps may be different. For example, in a first reactor, cyclohexane is oxidized to cyclohexylhydroperoxide; the concentration of the latter is optimised by carrying out the oxidation in passivated reactors and in the absence of transition metal complexes, in order to avoid the decomposition of the hydroperoxide. In fact, the synthesis of the hydroperoxide is the rate-limiting step of the process, and, on the other hand, alcohol and ketone are more reactive than cyclohexane. The decomposition of the hydroperoxide is then carried out in a second reactor, in which the catalyst amount and reaction conditions are optimised, thus allowing the Ol/One ratio to be controlled. The per-pass conversion is 5–7%, to limit the consecutive reactions; in fact, alcohol and ketone are more reactive towards oxygen than cyclohexane. Selectivity to the KA Oil is around 75–80% (product weight ratio Ol/One 60/40); by-products are carboxylic acids (6-hydroxyhexanoic, *n*-butyric, *n*-valeric, succinic, glutaric and adipic among others), and cyclohexyl hydroperoxide. Unconverted cyclohexane is recycled [2a, c]:

$$2 \bigcirc + 3/2\ O_2 \longrightarrow \bigcirc\!\!\!-OH + \bigcirc\!\!\!=O + H_2O$$

$$\text{cyclohexane} + \frac{1}{2}O_2 \rightarrow \text{cyclohexanol} \quad \Delta H_R -47.4\ \text{k cal mol}^{-1}$$

$$\text{cyclohexane} + O_2 \rightarrow \text{cyclohexanone} \quad \Delta H_R -98\ \text{k cal mol}^{-1}$$

The chemistry of the reaction is an homolytic autoxidation. Cyclohexanol is formed from the RO$^{\bullet}$ radical (R = C_6H_{11}). At high concentrations (>3 mmol dm^{-3}), Co may act as an initiator [2g], with a direct attack on cyclohexane (this reaction does not occur with Mn, in agreement with the lower electrochemical potential of the Mn(II)/Mn(III) redox couple) [3b]:

$$\text{RH(cyclohexane)} + \text{Co(III) (or In}^{\bullet}) \rightarrow \text{R}^{\bullet}\text{(cyclohexyl radical)} + \text{Co(II)}$$
$$+ H^+ \text{ (or InH)}$$

$$R^{\bullet} + O_2 \rightarrow ROO^{\bullet}$$

$$ROO^{\bullet} + RH \rightarrow ROOH \text{ (cyclohexylhydroperoxide)} + R^{\bullet}$$

The main role of Co is to accelerate the autoxidation reaction by decomposition of the intermediate hydroperoxide, ROOH, thus leading to the formation of alkoxyl or peroxyl radicals (Haber–Weiss mechanism):

$$ROOH + Co(II) \rightarrow RO^\bullet + Co(III) + OH^- \text{ (dominating in non-polar solvents)}$$

$$ROOH + Co(III) \rightarrow ROO^\bullet + Co(II) + H^+ \text{ (present in polar solvents)}$$

which corresponds to:

$$2ROOH \rightarrow RO^\bullet + ROO^\bullet + H_2O$$

These reactions increase the rate of free radical formation in the chain reaction, explaining the global catalysis of autoxidation.

At high concentrations of Co(II) it competes with cyclohexane for the alkylperoxy radical:

$$ROO^\bullet + Co(II) \rightarrow ROOCo(III)$$

and therefore a termination step would occur instead of a propagation step [3a]. Consequently, under these conditions the catalyst may become an inhibitor.

The predominant reaction for the formation of cyclohexanol and cyclohexanone is the Russell mechanism of decomposition of secondary cyclohexylperoxy radicals, which first yields the product of coupling and then reacts by a non-radical, six-center 1,5-H-atom shift (termination of the radical-chain sequence):

$$2ROO^\bullet \rightleftarrows ROOOOR$$

$$ROOOOR \rightarrow C_6H_{11}OH + C_6H_{10}O + O_2$$

Another possibility for the formation of cyclohexanol (ROH) and radical propagation is:

$$RO^\bullet + RH \rightarrow ROH + R^\bullet$$

or by homolytic scission of the hydroperoxide:

$$ROOH \rightarrow RO^\bullet + {}^\bullet OH$$

$$RO^\bullet + ROOH \rightarrow ROH + ROO^\bullet$$

Moreover, cleavage of an α C–H bond can yield cyclohexanone directly:

$$ROOH \rightarrow C_6H_{10}O + H_2O$$

Cyclohexanol may also undergo consecutive oxidation to cyclohexanone.

Recently, however, Hermans et al. [2h–m] have revisited the mechanism of cyclohexane autoxidation, and found that indeed the most efficient mechanism for chain propagation is not:

$$ROO^\bullet + RH \rightarrow ROOH + R^\bullet$$

but instead:

$$ROO^\bullet + ROOH \rightarrow R_{-\alpha H} \cdot OOH + ROOH \ (R_{-\alpha H} = Q = C_6H_{10})$$

and that the quicker mechanism for the formation of cyclohexanone and cyclo-hexanol is not:

$$2ROO^\bullet \rightarrow ROH + Q = O + O_2$$

but:

$$R_{-\alpha H}{}^\bullet OOH \rightarrow Q = O + {}^\bullet OH$$

$$RH + {}^\bullet OH \rightarrow R^\bullet + H_2O$$

$$ROOH + R^\bullet \rightarrow ROH + RO^\bullet$$
$$(RO^\bullet = \text{the resonance-stabilized ketonyl radical})$$

and also:

$$RO^\bullet + RH \rightarrow ROH + R^\bullet$$

This propagation of the hydroperoxide is of great importance because it produces both the ketone and alcohol from the start of the reaction. As soon as Q=O (cyclohexanone) is formed, an autoacceleration effect is observed, due to the reaction:

$$ROOH + Q = O \rightarrow RO^\bullet + Q_{-\alpha H^\bullet} = O + H_2O$$

in which an $^\bullet OH$ radical abstracts a weakly bonded αH atom from the ketone, which also is an efficient bimolecular initiation reaction.

It was also proposed that starting from $RO^\bullet$, ring opening may proceed efficiently.

$$RO^\bullet \rightarrow {}^\bullet CH_2 - (CH_2)_4 - CHO$$

Figure 7.1 shows a simplified flow sheet of the process of cyclohexane oxidation with air. The reaction is carried out in three in-series reactors, with cyclohexane being fed in the first reactor and air being distributed to the three reactors. This operation allows a better control of the reaction and improves safety.

A variation of this step currently adopted in several plants, is characterized by the addition of substantial quantities of anhydrous meta-boric acid as a slurry in

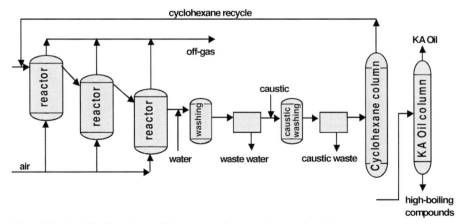

Figure 7.1 Simplified flow-sheet of the process for the oxidation of cyclohexane with air.

cyclohexane to the first of a staged series of oxidation reactors (Bashkirov Oxidation). No other catalyst is necessary. Boric acid reacts with cyclohexanol to give a borate ester that stabilizes the product and reduces its tendency to be oxidized further to form either cyclohexanone or degradation products. This results in a higher once-through conversion (up to 10 or even 15%), and selectivity (90%, with a molar ratio Ol/One of around 10). The borate ester formed is easily hydrolyzed by hot water to boric acid and cyclohexanol. After distillation, a 99.5% Ol/One mixture is obtained.

The cyclohexane volatilizes immediately when depressurized from reaction conditions (e.g., 155 °C, 8 atm) to atmospheric conditions. This leads to potentially dangerous operations; for instance, in June 1974 an accident occurred at Flixborough, England, with several casualties and injuries. Because of the inadequate support of a flexible bellow-type piping between two reactors, installed to replace a reactor that had been removed for maintenance, a rupture occurred, and an estimated 30 tons of cyclohexane volatilized and formed a large vapor cloud. The cloud was ignited by an unknown source approximately 45 seconds after release.

The use of oxygen instead of air might lead to improvements, such as a higher rate of oxidation and an increased productivity, the use of milder reaction conditions, a lower capital and operating energy costs, lower purge streams and a reduced cost for the treatment of vent gas. In fact, several companies already use O_2 instead of air. Safety concerns are greater than with air operation, but the safety issues have been solved.

A special reactor design, called the liquid oxidation reactor (LOR) allows the use of oxygen or oxygen-enriched air [3e–g]. The LOR has a special enclosure around the stirrer where most of the oxygen is fed and consumed. The remaining oxygen passes into the vapor space and is diluted with nitrogen before being vented. Although the formation of flammable compositions in the overhead space is unlikely, because of the high oxygen conversion, there are still concerns regarding the potential explosions that may occur inside the vapor bubbles in the reactor.

A paper by the China Petrochemical Dev Co [3h] reported that the use of pure oxygen for cyclohexane oxidation leads to an increased yield and selectivity to Ol/One with respect to the traditional air-based technology, under inherently safe conditions. The latter are achieved by the addition of water, which avoids the formation of flammable mixtures in the overhead vapor space and in the vapor bubbles. In fact, cyclohexane and water form a minimum-boiling azeotrope, the vapor pressure of which is higher than that of cyclohexane. The increased vapor pressure acts as an inert component.

7.2.2
Alternatives for the Synthesis of Ol/One

Two variants are currently employed for the synthesis of Ol/One: (i) the hydrogenation of phenol to cyclohexanol and cyclohexanone and (ii) the hydration of cyclohexene to cyclohexanol; cyclohexene is synthesized by the selective hydrogenation of benzene. Cyclohexanol is then oxidized to AA using the same process as that employed for the nitric acid oxidation of the KA Oil.

The hydrogenation of phenol has been adopted by Solutia and Radici. This process has some advantages, particularly for smaller scale manufacturers and for companies that are large-scale manufacturers of phenol. The equipment needed for KA Oil manufacture from phenol is less complex and the process is safer than that based on cyclohexane oxidation, resulting in reduced investment costs. Moreover, current hydrogenation technology allows one to directly obtain a mixture of cyclohexanol and cyclohexanone with the desired ratio. By increasing the percentage of ketone it is possible to save hydrogen in this step and nitric acid during oxidation. The hydrogenation process is very selective and the final product is extremely pure, if compared with KA Oil stemming from cyclohexane oxidation. This could render AA purification simpler.

The hydrogenation of benzene to cyclohexene, followed by the hydration of cycloolefin, was developed by Asahi, and is currently employed by this company and some Chinese producers as the first step in the manufacture of AA; in 1990 Asahi built a plant with a capacity of 60 000 tons yr^{-1}. The partial hydrogenation reaction product is a mixture of unreacted benzene, cyclohexene and by-product cyclohexane. Figure 7.2 shows a simplified flow sheet of the Asahi process.

Original work done at the University of Delft [4a, b] disclosed the possibility of performing the selective hydrogenation of benzene to cyclohexene. Thermodynamically, total hydrogenation to cyclohexane is much more favorable than partial hydrogenation; therefore, under normal conditions, it is very difficult to stop the reaction at the monoolefin. This method is based on a catalyst consisting of Pt or Ru powder, coated with a layer of an aqueous solution of zinc sulfate. The reaction is performed in bulk benzene; since the catalyst is surrounded by the aqueous phase, those organic molecules that are better soluble in the aqueous phase are preferentially hydrogenated. Cyclohexene is less soluble in the aqueous phase than benzene, and hence as it forms it migrates preferentially to the organic phase, preventing further hydrogenation [4c]. Cyclohexene is obtained with 80% selectivity (the remainder being cyclohexane), at 70–75% benzene conversion.

Asahi has been investigating solvent mixtures suitable for extractive distillation to make the separation of cyclohexene practical [5]. Cyclohexene is hydrated to cyclohexanol on a ZSM-5 based catalyst. With respect to the conventional process from

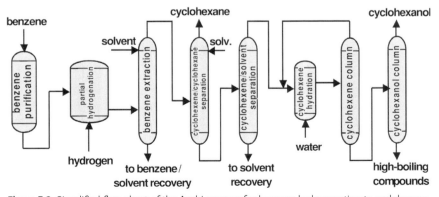

Figure 7.2 Simplified flow sheet of the Asahi process for benzene hydrogenation to cyclohexene.

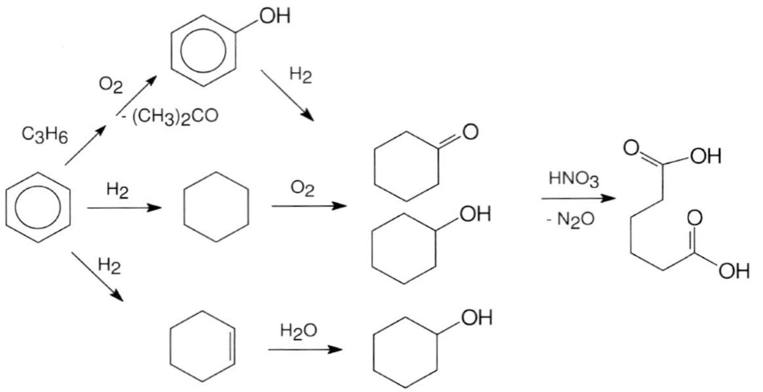

Scheme 7.1 Summary of the current processes for AA production, starting from benzene.

cyclohexane, the theoretical consumption of hydrogen is reduced by one-third, and fewer by-products are formed.

Scheme 7.1 summarizes the various industrial processes currently employed for AA production from benzene.

7.2.3
Alternative Homogeneous Catalysts for Cyclohexane Oxidation to Ol/One

In the literature, homogeneous catalysts other than Co complexes have been reported for the oxidation of cyclohexane to Ol/One with either oxygen (air) or hydroperoxides, sometimes offering selectivity comparable to that achieved with the Co-based catalyst. The aim is to find catalysts that permit high cyclohexane conversion while maintaining high selectivity to cyclohexanol and cyclohexanone. Indeed, with oxygen, a coreductant (typically an aldehyde) that is more readily oxidizable by oxygen into peroxide than the substrate is often added to the reaction medium. In this way, the substrate is not oxidized by oxygen, but will react with the resulting peroxide through a homolytic or a heterolytic mechanism. Often, the reactivity is greatly affected by the presence of additives [6a].

Examples of alternative systems investigated in the literature include:

1. Keggin-type polyoxometalates, POMs, can act either as redox-type oxidants towards the substrate, furnishing nucleophilic O^{2-} species that are then reversibly reoxidized by molecular oxygen under mild conditions (P/Mo/V POMs), or may catalyze radical autoxidation, or may even activate hydroperoxides, depending on their composition. For instance, Fe_2Ni-substituted P/W POMs facilitate the oxidation of cyclohexane to Ol/One with air at atmospheric pressure [6b–d], whereas Co(Fe)-substituted P/W POMs catalyze the oxidation with t-BuOOH [6e]. Fe-substituted polyoxotungstates are active catalysts for the oxidation of cyclohexane with oxygen at atmospheric pressure and 120 °C with high selectivity to Ol/One, although with low conversion [6f]. A Pt/C-P/W polyoxometalate mixed heterogeneous–homogeneous system catalyzes the oxidation in the presence of a H_2–O_2 mixture, at 35 °C, in acetonitrile solvent, via intermediate formation of hydrogen peroxide (HP). In the

absence of the polyoxometalate, the Pt/C catalyst rapidly converts the H_2–O_2 mixture into water, without yielding organic products [6g].

2. Cu-based systems have been widely investigated as catalysts for the oxidation of cyclohexane under relatively mild conditions, using various oxidants like HP, t-BuOOH, peracetic acid and oxygen. Catalysts are based on Cu(I) and Cu(II) complexes, with various types of ligands [6h–r]. However, most of them still provide very low yields and low selectivity, require rather expensive and environmentally unfriendly components, are active only in the presence of various additives or coreductants or involve complicated syntheses, thus being inaccessible on a large scale. Nevertheless, a remarkable 39% yield (overall to Ol/One) was obtained with polynuclear Cu triethanolamine complexes [6p–q] in the oxidation of cyclohexane with aqueous HP in acidic medium (liquid biphasic catalysis) at room temperature in acetonitrile at atmospheric pressure. The highest yield to Ol/One reported (69%), however, was obtained with a bis-(2-pyridylmethyl)amine Cu(II) complex and with HP as the oxidant [6t], in acetonitrile solvent under mild conditions. Yields achieved with O_2, on the other hand, are usually lower than 5% [6l, s].

3. Various iron salts and mononuclear Fe or binuclear Fe complexes with a N,O environment, biomimetic to methane monooxygenase complexes, have been applied to the oxidation of cyclohexane with various oxidants [6u,v,7a–g], but their catalytic activity is usually modest, with the exception of a hexanuclear Fe(III) compound derived from p-nitrobenzoic acid, which gives the highest total yield to Ol/One of about 30% [7a]. Moreover, most of these complexes are often unstable and very expensive. A hexanuclear heterotrimetallic Fe/Cu/Co complex bearing two $Cu(\mu\text{-}O)_2Co(\mu\text{-}O)_2Fe$ cores, prepared by self-assembly, oxidizes cyclohexane with aqueous HP, with a maximum yield to Ol/One of 45%, virtually total selectivity to the two compounds, and preferred formation of cyclohexanol [7h]. The remarkable activity of the Fe/Cu/Co cluster was associated with the synergic effect of the three metals.

4. Vanadium-based complexes and salts have been used for oxidation either with oxygen in the presence of coreductants [7i], with peroxyacetic acid in acetonitrile [7j], or with HP/O_2 combined oxidants [7k–m].

5. The addition of an alkyl nitrite (e.g., isoamyl nitrite) as a promoter in the aerial oxidation of cyclohexane in the presence of a Co/Mn catalyst led to an improvement in the conversion and in the selectivity to Ol/One and AA, compared to the unpromoted reaction [7n]. For instance, at 120 °C and 9 atm oxygen pressure, 9.7% cyclohexane conversion was obtained in 3 h reaction time, with selectivity to Ol/One of 65% and to AA of 35%. In the absence of the nitrite, the conversion was less than 4%.

7.2.4
Two-Step Transformation of Cyclohexane to AA: Oxidation of Ol/One to AA with Nitric Acid

The second step of the process from cyclohexane is the oxidation of the KA Oil with a large excess of 50–65% HNO_3 (molar ratio HNO_3/Ka Oil at least 7/1), and with a

Cu(II) and ammonium metavanadate catalyst [2a, b]. The formal stoichiometries are as follows (indeed, the reaction leads not only to N_2O, but also to the formation of NO and NO_2):

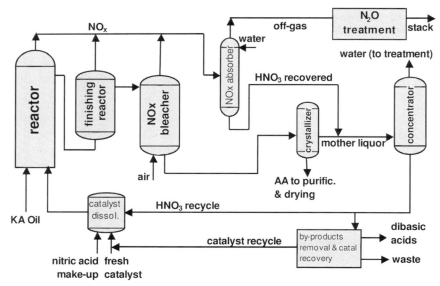

cyclohexanone to AA, $\Delta H_R = -172\ \text{kcal mol}^{-1}$
cyclohexanol to AA, $\Delta H_R = -215\ \text{kcal mol}^{-1}$

The reaction is carried out in two (or more) in-series reactors, the first one operating at 60–80 °C, the second one being maintained at 90–100 °C and a pressure of 1–4 atm. The molar yield obtained for total KA Oil conversion is as high as 95%; the by-products are glutaric (selectivity 3%) and succinic (selectivity 2%) acids.

Figure 7.3 shows a simplified flow sheet of the process of KA Oil oxidation, with the main process units.

The reaction mechanism was discussed in detail by van Asselt and van Krevelen in 1963 [2e, f]; a complete analysis of the mechanism has also been reported ([2b] and

Figure 7.3 Simplified flow sheet of the KA Oil oxidation process.

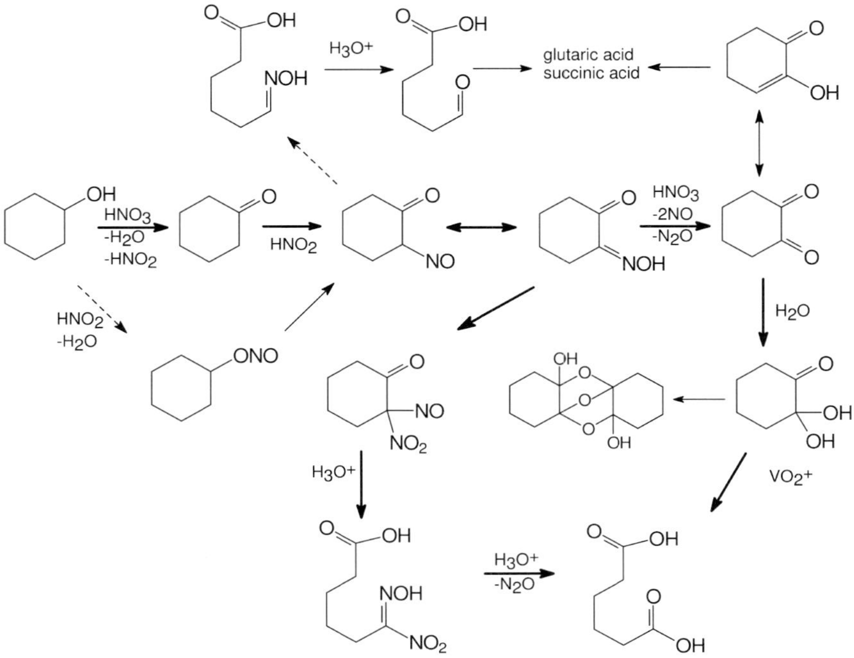

Scheme 7.2 Main reactions involved in the mechanism of KA Oil oxidation to AA with nitric acid. Source: elaborated from [2b, e, f].

references therein) (Scheme 7.2). Cyclohexanol is first oxidized to cyclohexanone, followed by nitrosation of cyclohexanone by nitrous acid (HNO_2) to produce 2-nitrosocyclohexanone; the latter may undergo various transformations.

In the presence of HNO_2, the nitrosoketone may be hydrolyzed to the α-diketone and hydroxylamine (Claisen–Manasse reaction), via intermediate formation of the ketoxime tautomer. Alternatively, the oxidation of the oxime by the stronger oxidant nitric acid also yields the α-diketone, with co-formation of NO_x (the reaction shown in Scheme 7.2). Finally, the diketone is oxidized to AA; oxidation may occur directly either by the nitric acid (with the possible side co-formation of succinic acid and oxalic acid) or, with a suitable vanadium concentration, by VO_2^+; in the latter case, the reaction is very selective. HNO_3 rapidly reoxidizes the resulting reduced V species, VO^{2+} or VO^+, back to VO_2^+. In this step, nitric acid is reduced to the regenerable oxides NO and NO_2. Cu(II) helps to limit multiple nitrosation of cyclohexanone and the formation of glutaric acid. Only one mole of nitric acid is consumed per mole of cyclohexanone via this route. 2-Nitrosocyclohexanone may eventually yield 2-nitro-cyclohexanone in strongly oxidizing solutions.

The main pathway, however, is the reaction of 2-nitrosocyclohexanone to yield 2-nitro-2-nitrosocyclohexanone; the latter is hydrolyzed to 6-nitro-6-hydroximinohex-anoic acid (adipomononitrolic acid) that undergoes oxidative hydrolysis to AA via intermediate adipomonohydroxamic acid; in this case, HNO_3 is completely reduced to N_2O. Therefore, this pathway consumes two moles of nitric acid per mole of cyclohexanone.

Nitrosation may potentially also occur on cyclohexanol; in fact, cyclohexanol can be oxidized at much lower temperatures than cyclohexanone. The active reactant is HNO_2; therefore, in this case, the first product of cyclohexanol oxidation is cyclohexyl nitrite. The latter is then rearranged into 2-nitrosocyclohexanone, which is also the key intermediate in the main reaction pathway involving cyclohexanone.

The nitric acid oxidation of cyclohexanol at low temperatures, for example, 10–15 °C, leads to the formation of both adipomononitrolic acid and the hemihydrate of cyclohexadione (predione) rather than the α-diketone. The predione is oxidized to AA in the presence of metavanadate.

For many years industries have shown an interest in the development of a new process that does not produce nitrogen oxides; this would not only be more environmentally sustainable than the current process but it would also not require denitration equipment.

7.2.5
Environmental Issues in AA Production

The major environmental concern for nitric acid oxidation of either cyclohexanol or the KA Oil is the emission of the greenhouse gas nitrous oxide, N_2O. The emission factor is estimated to be around 300 kg of N_2O per ton of AA; indeed, the real value is between 260 and 330 (around 0.75–1.2 mol of N_2O per mole of AA), depending on the amount of catalyst used and the KA Oil composition. For a global AA production of 3.0 M-ton yr^{-1}, the total amount of N_2O produced is estimated to be about 0.9 M-ton yr^{-1}. However, nowadays less than 5% of total emission of N_2O into the atmosphere is from anthropogenic origin, and of this 5% only a small fraction is due to AA production, because almost all the AA producers have installed N_2O decomposition units.

HNO_2 is an intermediate of the reduction of nitric acid; the surplus HNO_2 produced by various reactions eventually decomposes into NO and NO_2, and leaves the reactors in the gas stream. However, NO and NO_2 (about 1 mol NO_x per mole of AA) can be almost completely recovered, yielding a nitric acid solution that can be reused in the oxidation process. Indeed, some companies simply co-feed the off-gas of the AA plant to a large existing nitric acid plant.

The most efficient method for NO_x recovery is the water adsorption in a multistage column. The gaseous stream containing nitrogen oxides is compressed in the presence of excess oxygen and sent to an adsorption column. Here, nitrogen oxides are recovered as nitric acid according to the following reactions:

$$2NO + O_2 \rightarrow 2NO_2$$

$$3NO_2 + H_2O \rightarrow 2HNO_3 + NO$$

Low temperature and pressure are the key factors in the adsorption process (owing to the poor solubility of nitrogen oxides, NO in particular), since they contribute considerably to operative and investment costs. Nowadays it is possible, with a proper design, to achieve almost complete recovery of NO_x in compliance with the current limits of the law (500 mg Nm^{-3} of NO_x expressed as NO_2), without any post-treatment. N_2O, however, cannot be reconverted into nitric acid and so it is subjected to downstream treatment.

N_2O contributes to the greenhouse effect and global warming, due to its strong IR absorption. The estimated atmospheric lifetime of N_2O is about 150 years, and the estimated impact of N_2O is a 6% increase in the ozone depletion rate. The global generation of N_2O, ascribed to both anthropogenic and natural sources, is estimated to be around 26 M-ton yr^{-1} [8a]. The concentration of N_2O in the air is approximately 300 ppb at present; however, the rate of N_2O generation now exceeds the global rate of N_2O destruction, such that the concentration of N_2O is growing at the rate of 0.2% per year [8b].

To evaluate the impact of different greenhouse gases, the IPCC introduced the global warming potential [8c] (GWP), which is a quantified measure of the impact of each gas compared to carbon dioxide, used as a reference, in a defined time and per unit of mass. The relationship between gigagrams of gas and teragrams of equivalent CO_2 is the following:

$$Tg\ CO_2\ eq. = (Gg\ gas) \times GWP \times (Tg/1000\ Gg)$$

According to this criterion, considering the lifetime of N_2O in the atmosphere and a period of 100 years, it has a GWP 310-times higher than that of CO_2.

In 1990, AA production was the largest source of industrial N_2O emissions. As of 1999, all major AA producers had implemented N_2O abatement technologies and, therefore, this source has decreased substantially. N_2O abatement is estimated to have improved from approximately 32% in 1990 to approximately 90% in 2000. As a result, more than 90% of the N_2O generated at AA plants is now decomposed before being released into the environment.

7.2.6
Technologies for N_2O Abatement

Different options are available for the abatement of nitrous oxide: (i) N_2O decomposition in boilers (thermal destruction, efficiency higher than 98%), (ii) conversion of N_2O into recoverable NO and (iii) catalytic dissociation of N_2O to N_2 and O_2 (efficiency higher than 90–95%) [9].

The N_2O content in gaseous emissions from AA plants can be as high as 65%, but more frequently it may reach 40% of the volume. In general, these gases also contain water, oxygen and carbon dioxide. The water content is usually at the saturation level for the gas in question – the exact amount being a function of system temperature and working pressure – but the maximum content is typically 1–2% of the volume. Oxygen is always present, generally ranging from 5 to 15% of the volume. Traces of volatile organic compounds and of NO_x (up to 2000 ppm) may also be present.

Nowadays, Asahi, BASF, Bayer, Invista, Rhodia, Radici and Solutia use catalytic or thermal processes to destroy N_2O. Recovery of waste heat from the exothermic abatement reactions is more effective with thermal systems due to their higher operating temperatures, but 60% of the operating costs may be recovered through steam generation.

7.2.6.1 Catalytic Abatement

Catalytic decomposition can be considered the simplest removal method for N_2O because it does not need any additional chemical compounds. The catalytic decomposition of nitrous oxide is highly exothermic ($-19.6\,\text{kcal mol}^{-1}$):

$$N_2O \rightarrow N_2 + \tfrac{1}{2}O_2$$

It is a redox mechanism [9l, m], and includes the following steps:

1. $N_2O + * \rightleftarrows N_2O^*$
2. $N_2O^* \rightarrow N_2 + O^*$
3. $2O^* \rightleftarrows O_2 + 2^*$
4. $N_2O + O^* \rightarrow N_2 + O_2 + *$

where $*$ is the active site of the catalyst.

The reaction of N_2O with the active site of the catalyst is envisaged as a donation of electronic charge from the catalyst to the antibonding molecular orbital of N_2O, thus destabilizing the N—O bond and leading to scission.

The various processes for the catalytic reaction are similar. The factor that makes the difference is the choice of catalyst, which in turn affects the temperature regime needed to trigger the decomposition of nitrous oxide. In the literature, numerous works illustrate the several classes of catalysts appropriate for this reaction [9a, k]: noble metals (Pt, Au), pure or mixed metal oxides (spinels, perovskite-types, oxides from hydrotalcites), supported systems (metal or metal oxides on alumina, silica, zirconia) and zeolites.

Zeolitic systems are very active at low temperatures but they also have disadvantages related to their hydrothermal stability and the possibility of inhibition or poisoning by different compounds. These drawbacks drastically limit the industrial applications of these catalysts. Rhodium-supported systems are also active at low temperatures and low N_2O concentration, but at high temperatures and in the presence of O_2 the noble metal is oxidized. Furthermore, the high cost of Rh may prove to be a limit for industrial applications.

Owing to their crystal structure, which can contain various metal ions and can stabilize unusual and mixed valence states of active ions, mixed oxide catalysts have been thoroughly investigated, and have been found to be better catalytic materials than other systems.

Some examples of catalysts belonging to the above-mentioned categories are described in several patents [8]. Unfortunately, most systems are unable to trigger the reaction below 300 °C. Since the reaction is carried out in fixed-bed reactors on N_2O-laden industrial emissions, in the absence of an adequate control system, marked increases in temperature (up to and beyond 1000 °C) can take place on the catalytic bed. The exothermic nature of the reaction, therefore, gives rise to a series of problems, for example: (i) the sintering of catalysts or their supports, reducing catalytic activity and shortening lifetime; (ii) high investment costs for the careful

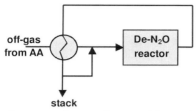

Figure 7.4 Schematic diagram of the catalytic abatement of nitrous oxide.

selection of special heat-resistant materials used in the construction of reactor bodies; and (iii) difficulties in complying with environmental regulations: the higher the temperature on the catalytic bed, the greater the risk of forming higher nitrogen oxides (NO_x), which are subject to strict environmental laws.

During the process, part of the treated and cooled gas is recycled back to dilute the N_2O stream stemming from AA production, thus avoiding high local temperature rises and leveling out the heat production over the entire bed length (Figure 7.4). Typical inlet reactor temperatures are about 450–500 °C, and 700–800 °C at the outlet, considering an inlet concentration of N_2O of around 12 vol.%.

The process developed by Radici includes a multi-bed reactor (Figure 7.5) [10]. The gaseous flow containing N_2O is subdivided into three flows that are fed separately to the three catalytic layers. The flow from the final catalytic bed exhibits a residual nitrous oxide content below 500 ppm and is subdivided into two flows, one being vented to the atmosphere and one being mixed with the feed to the first catalytic bed and re-circulated into the nitrous oxide decomposition process.

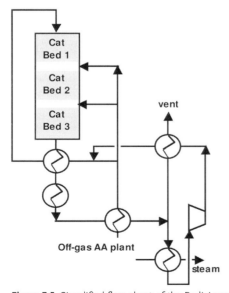

Figure 7.5 Simplified flow sheet of the Radici process for N_2O catalytic decomposition.

The technology developed by Radici has the following advantages: (i) it allows for a better control of the temperature of catalytic layers that operate at the maximum desired temperature; (ii) it provides for maximum heat recovery from the decomposition reaction, thus rendering an environmental protection process that is economically advantageous; (iii) it minimizes the quantity of gaseous diluting agent required, thus reducing costs arising from devices (installation, etc.) for injecting the diluting flow into the system; (iv) it allows for an operating temperature range that minimizes the formation of NO_x, thus making the installation of a downstream NO_x abatement system unnecessary; (v) it minimizes the amount of energy required for triggering and maintaining the reaction, resulting in reduced fuel consumption in the start-up phase and greater heat recovery efficiency during normal operation; and (vi) it allows the start-up of the reaction to take place at the lowest temperature permitted, reducing the need for highly active, but expensive, catalysts.

Apart from the environmental fallout (elimination of emissions of greenhouse gases) this process allows one to obtain industrially useful energy in the form of medium-pressure steam (0.16 ton$_{steam}$ per ton$_{AA}$). All this occurs in the absence of fuel consumption (e.g., methane) and with a negligible use of power for the movement of process fluids.

7.2.6.2 Thermal Abatement

Thermal destruction can be carried out using two different approaches that differ substantially, depending on whether the decomposition reaction takes place at oxidizing conditions or as a result of burning in a reducing flame. In the first case, the process is designed to maximize the conversion of nitrous oxide into higher oxides (NO and NO_2). The reaction is thermodynamically unfavored at low temperatures. Once nitrogen oxides have been adsorbed in water, a modest recovery of nitric acid can be achieved, which can be recycled in the process upstream.

The second approach for thermal destruction is realized at sub-stoichiometric combustion conditions, so as to minimize the formation of NO_x, limiting the latter to values compatible with the environmental regulations. In this case, an efficient recovery of the heat with generation of steam is possible. Thermal abatement is carried out using heat burners. The reducing atmosphere is generated by adding an excess of methane to the gas mixture fed to the burner (containing O_2 and N_2O). In addition to the normal products of combustion (carbon dioxide and water vapor –

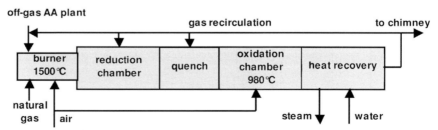

Figure 7.6 Simplified flow sheet of the N_2O thermal decomposition process.

together with nitrogen), the excess methane produces an unburnt share (carbon monoxide and hydrogen) whose concentration in the gas flow is the driving force for the reduction of nitrogen oxides to elemental nitrogen.

The following reactions take place in the reducing section:

$$4N_2O + CH_4 \rightarrow CO_2 + 2H_2O + 4N_2$$

$$4NO + CH_4 \rightarrow CO_2 + 2H_2O + 2N_2$$

$$2O_2 + CH_4 \rightarrow CO_2 + H_2O$$

$$H_2O + CO_2 + 2CH_4 \rightarrow 3CO + 5H_2$$

$$N_2O \rightarrow \tfrac{1}{2}O_2 + N_2$$

Figure 7.6 shows a simplified outline of the reductive thermal abatement process. The off-gas derived from AA manufacturing enters the burner and is mixed with methane and air for combustion, and, as a result of the reactions reported above, the temperature reaches 1500 °C. The gases exit from the burner and pass through the reduction chamber in which the temperature is lowered to about 1200 °C by recycling a portion of the cold tail gas (200 °C approx.), maintaining the gas at high temperatures for a short time, that is, for a few seconds. Downstream of this reduction chamber, gases are cooled with gas recirculation to lower the temperature and prevent the next stage of the oxidation of carbon monoxide and hydrogen (oxidation chamber) from exceeding 980 °C, to keep the concentration of nitrogen oxides at low level upon leaving the plant (thermal NO_x). Once combustion is complete, fumes enter the heat recovery section that consists primarily of a boiler to produce steam at medium pressure.

The thermal nitrous oxide decomposition can reach abatement conversions of approximately 95% on an ongoing basis, giving a final concentration of less than 500 mg Nm^{-3} in the flue gas. The thermal decomposition consumes about 95 Nm^3 of methane per ton_{AA}, with a recovery of approx 1.5 ton_{steam} per ton_{AA}: this corresponds to a yield in the production of steam that is greater than 100%, after having taken into account the contribution of thermal nitrous oxide. The economic limits of this technology are determined by both the cost of natural gas and the value of the steam generated.

7.2.7
N_2O: From a Waste Compound to a Reactant for Downstream Applications

As an alternative, N_2O can be recovered from the off-gas in pure form, either for selling or for use as an oxidant in some downstream processes. Within this context, an innovative solution has been developed by Solutia, together with the Boreskov Institute of Catalysis, in which N_2O is the oxidant used for the hydroxylation of benzene to phenol in the presence of a ZSM-5 catalyst exchanged with Fe(III) [11]. Phenol can then be hydrogenated to yield cyclohexanol, hence completing the N_2O cycle (Scheme 7.3). A growing number of gas-phase applications are being investigated in which N_2O is the co-reactant, for example, the oxidation of methane to methanol, the epoxidation

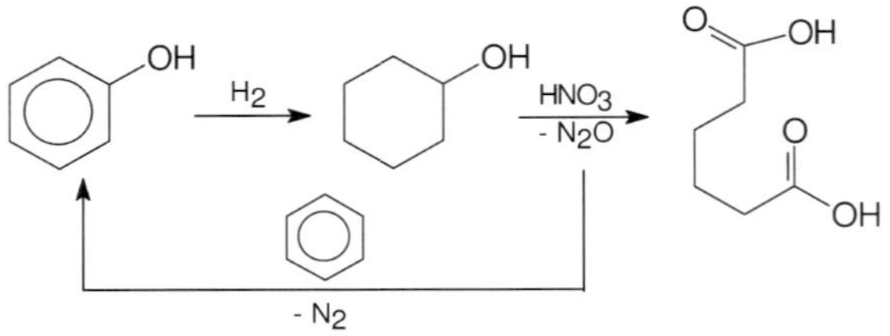

Scheme 7.3 Integration in the BIC/Solutia process.

of olefins, the oxidative dehydrogenation of alkanes to alkenes, amongst others [11h]. Within this context, it is useful to remember that the cost of N_2O is more than ten times that of oxygen; clearly, from a cost standpoint, it is difficult to justify new commodity chemical processes using purposely produced N_2O as the oxygen source. However, for fine chemical applications such as pharmaceuticals and agrochemicals the cost may well be justified. In addition, any downstream process that may use the N_2O-containing stream – possibly without any further treatment, and thus realizing efficient process integration – may constitute a valid alternative to the disposal of that stream [11h].

In the BIC/Solutia process, using a zeolite that contains only small amounts of iron, benzene can be oxidized to phenol with a selectivity of over 95% at around 300 °C, but N_2O selectivity is lower than 95% [11e].

Despite the remarkable performance of this innovative process, which was scaled up to the pilot unit, the BIC/Solutia process has not yet been put into commercial operation. This may be due to rapid catalyst deactivation caused by tar deposition, to low efficiency with respect to N_2O and to the poor economics of such small-sized phenol plants, suited to balance the AA process unit, as compared to bigger, traditional plants for phenol production. An alternative option would be to build large phenol plants and produce N_2O separately. For instance, a patent belonging to Solutia [11g] discusses the use of a Bi/Mn/Al/O catalyst for the oxidation of ammonia to nitrous oxide. N_2O selectivity is reported to be about 92% at 99.2% conversion. The cost of N_2O is projected to be about 25% that of H_2O_2 [11h].

Concerning other reactions that make use of nitrous oxide as the oxidant, a process developed by BASF is worth mentioning, which involves the synthesis of cyclododecanone, a raw material for Nylon-12 and Nylon 6–12 [12]. Cyclodode-canone is oxidized to 1,12-dodecanedioic acid by oxidative cleavage with nitric acid. It is currently made by a five-step sequence from cyclododecatriene. The first step of the new process is the synthesis of cyclododecatriene from butadiene, which is then reacted with nitrous oxide, to produce cyclododecadienone, which is finally hydrogenated to cyclododecanone. The yield to cyclododecanone is significantly higher than with the conventional process, waste is reduced and investment costs are lower. BASF already has a contracted customer and is building a plant due to come on stream in 2009.

7.3
Alternatives for AA Production

7.3.1
Oxidation of KA Oil with Air

The second step of the KA Oil process can be carried out with oxygen as the oxidant, in place of nitric acid, and catalytic amounts of Co and Mn acetate, at 70–80 °C, in acetic acid solvent 13. The fact that N_2O is not generated and that the dedicated HNO_3 plant is not necessary (when a large-scale HNO_3 plant is not already available) may make the air-based process an alternative to the process currently employed. However, the potential use in industrial applications is still unclear, due to the lower yield achieved than that with nitric acid, and also due to the lower quality of the AA obtained. Moreover, using acetic acid as the solvent leads to severe corrosion problems, particularly in combination with the Mn and Co salts.

In this reaction it is important to achieve 100% conversion of the reactant, because recycling of unconverted KA Oil complicates the process. Yields to AA reported in most of the patent literature are not higher than 70% at high Ol/One conversion, with an overall yield to diacids close to 80%.

AA is probably generated via the intermediate formation of 2-hydroxycyclohexanone and 6-oxohexanoic acid. The mechanism may potentially include the formation of the enol tautomer of cyclohexanone, favored by the presence of an acid (Scheme 7.4; this may also be an alternative mechanism for cyclohexanone activation). The cyclohexen-1-ol formed is then oxidized to 2-hydroperoxycyclohexanone. The hydroperoxide generates 2-hydroxycyclohexanone, which is then cleaved to 6-oxohexanoic acid. The latter is then converted into AA via monoperoxyadipic acid; this step is eventually catalyzed by cobalt. Scheme 7.5 shows the steps in the mechanism of reaction.

Side-reactions lead to the formation of lighter acids. For instance, monoperoxyadipic acid can decarboxylate to yield the pentanoic acid radical, precursor of the by-product valeric acid. The same C_5 radical may react with O_2 to yield 5-oxopentanoic acid, which is then oxidized to monoperoxyglutaric acid, a precursor of glutaric acid. An analogous mechanism starting from the butanoic acid radical may yield the by-product succinic acid. Azelaic acid may form by the coupling of radical species (e.g., between the butanoic acid radical and the pentanoic acid radical), whereas the dimerization of the pentanoic acid radical may yield the by-product sebacic acid.

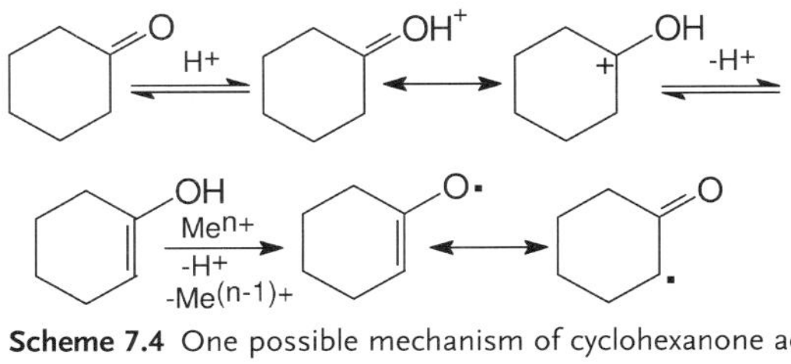

Scheme 7.4 One possible mechanism of cyclohexanone activation via enol tautomer.

$$Co3+ + HBr \ (init) \longrightarrow Co2+ + Br^* + H+$$
$$HBr + O2 \longrightarrow Br^* + HO2^*$$

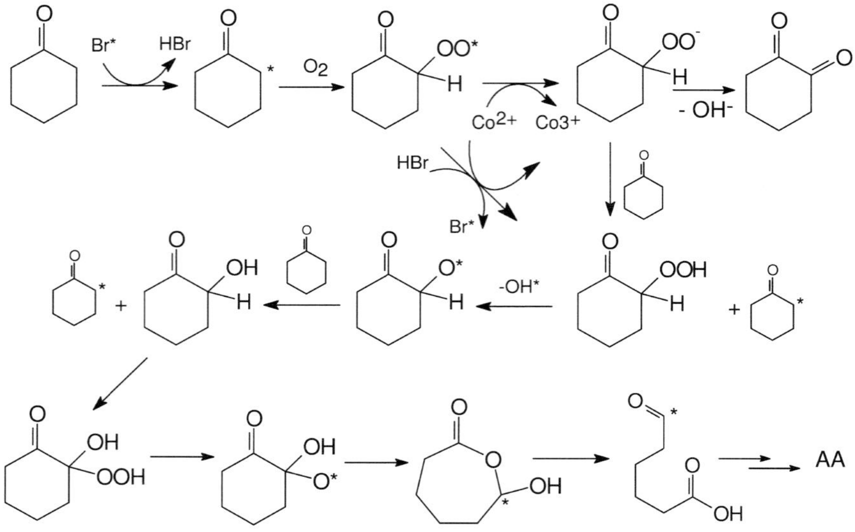

Scheme 7.5 Mechanism for the oxidation of cyclohexanone with O_2.
Source: elaborated from [130].

Researchers from Asahi have ascertained the industrial feasibility of a process based on a Co/Mn catalyst and either pure oxygen at atmospheric pressure or nitrogen-diluted air at 12 atm (to avoid explosion hazards), with water and acetic acid as the solvent [130]. The authors describe in detail the main features of the process and report a detailed study of the reaction mechanism.

Specifically:

1. The combination of a $Mn(OAc)_2$ catalyst and a $Co(OAc)_2$ catalyst (optimum Mn/Co ratio, 1:1) is effective for improving AA selectivity in the liquid-phase oxidation of cyclohexanone with oxygen. With pure oxygen, at atmospheric pressure, a selectivity to AA as high as 77% at total conversion of cyclohexanone is obtained, which is higher than that previously reported in the patent literature. The selectivity to glutaric acid is 12% and that to succinic acid is 2%, to oxoacids 2%, for an overall selectivity to acids of 93%. Selectivity to AA shows a maximum at 70 °C.
2. The water concentration is less than 20%, and does not affect the reaction. The use of an acid, such as *p*-toluenesulfonic acid, accelerates the oxidation reaction (Scheme 7.4).

The process scheme proposed by Asahi includes three in-series high-pressure, continuous-stirred, tank reactors (Figure 7.7).

To avoid explosion hazards, air containing less than 10% oxygen (at best, around 5%) was used. Under these conditions, the selectivity to AA was 73%, with a

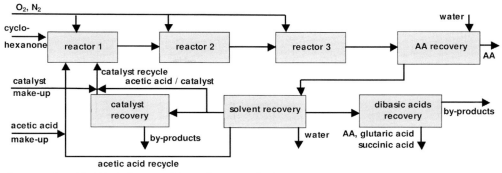

Figure 7.7 Simplified block-diagram of the Asahi process for the oxidation of KA Oil with nitrogen-diluted air. Source: elaborated from [130].

cyclohexanone conversion of 99.3%. The first two reactors operate at 70 °C, the third at over 70 °C. More than 90% of the cyclohexanone is consumed in the first two reactors; the remaining cyclohexanone and the intermediates are completely consumed in the third reactor at elevated temperatures. The reaction solution is fed into the AA recovery section, where AA is purified by recrystallization. Three-stage recrystallization is necessary, because the selectivity of AA is lower than that achieved with nitric acid oxidation, which uses a two-stage recrystallization. Water (from the solvent recovery system) is added in the AA recovery section to wash the AA crystals and to dissolve them for recrystallization. After removing the AA, the solution is moved to the solvent recovery section, where acetic acid is recovered by distillation. Part of the concentrated solution in the solvent recovery section is moved to the DBA (dibasic acid) recovery, aimed at removing the by-products and the remaining AA. The other part of the concentrated solution is moved to the catalyst recovery section, in which part of the by-products is also removed. Acetic acid and the catalysts are recycled to the first reactor.

The technology described by Asahi is very similar to that used for the synthesis of terephthalic acid by *p*-xylene oxidation. The selectivity of AA and the total selectivity of dicarboxylic acids are still lower than those in the nitric acid process [130].

Comprehensive reviews on the catalytic homolytic oxidative C–C bond cleavage of ketones with either oxygen or air have been published by Brégeault [13l, m]. Several homogeneous systems are active and selective in this reaction; vanadium-containing heteropolyacids with the Keggin structure are amongst the most efficient for the aerial oxidation of cyclohexanone [14]. These compounds are soluble either in aqueous or in organic mediums, depending on their composition. For example, at 70 °C, with 1 atm O_2, an acetic acid–water solvent mixture (volume ratio 9:1) and a Keggin-type $H_7(PMo_8V_4O_{40})12H_2O$ catalyst, after a 7-h reaction time 99% conversion of cyclohexanone was achieved, with 50% yield to AA, 19% to glutaric acid and 3% to succinic acid [14a]. A slightly higher yield to AA (54%) was obtained by using a V_2-heteropoly-compound, in acetonitrile–methanol solvent, in 24-h reaction time, at 60 °C, with 98% cyclohexanone conversion. The reactivity of the P/Mo/V polyoxometalates was greatly affected by the composition, for example, the number of V ions incorporated in the Keggin unit and the cation type. At below 90 °C, the mechanism included O^{2-} transfer from the compound to the organic substrate [14l]. Owing to their high reactivity, their low-temperature redox reversibility and structural flexibility, catalysts based on heteropoly-compounds are potentially promising for several oxidation reactions with O_2 under mild reaction conditions [14i].

Fe(III), Ce(IV), Ru(II) and monomeric V species (e.g., [VO{O-i-Pr}$_3$]) also lead to cyclohexanone conversions that are as good as the heteropoly-compounds in this class of reactions, but with lower selectivity to the diacids. However, better performance is obtained when the reaction is catalyzed by Cu(NO$_3$)$_2$ [13l]. At 110 °C and 8 h reaction time, 95% cyclohexanone conversion with 72% yield to AA, 8% to glutaric and 10% to succinic acid were obtained in an acetic acid–water solvent. A similar performance was reported with Mn(OAc)$_2$ in acetic acid–CF$_3$COOH solvent, at 65 °C after a 3-h reaction time: 99.8% conversion, 75% yield of AA, 9% to glutaric acid and 1% to succinic acid [14j].

High yields to AA were obtained when a Co/Mn cluster complex was used, which was superior to the individual Co and Mn acetates [14k]; at 90 °C, and 37 atm pressure, in acetic acid and water solvents, the oxidation of cyclohexanone with air gave complete conversion and 76.3% yield to AA. The authors suggested that 1-hydroxo-cyclohexen-2-one, the tautomeric form of 1,2-cyclohexandione, is the precursor for the formation of the glutaric and succinic acid by-products. Excellent yields were also reported [14m] for the oxidation of cyclohexanone using Mn(NO$_3$)$_2$ and Co(NO$_3$)$_2$ (molar ratio 1:1) in the presence of oxygen and catalytic quantities of nitric acid at atmospheric pressure. The conversion was 97.5% and the selectivity to AA was 93.4%.

Table 7.2 summarizes some results obtained with various catalyst types for the KA Oil oxidation with oxygen [13–15].

Table 7.2 Summary of catalysts used for the aerial oxidation of cyclohexanol (Ol) and cyclohexanone (One).

Catalyst	Reactant	Solvent, T (°C)	Conversion (%)	Selectivity AA (%)	Reference
Pt/charcoal	Ol	Water, 150 °C	100	50	[15b]
Pt/carbon/monolith	One	Water, 140 °C	100	21	[15d]
Mn(OAc)$_2$/Co(OAc)$_2$	Ol/One	Acetic acid, 70 °C	100	77	[13o]
Mn(OAc)$_2$	One	Acetic acid/ CF$_3$COOH, 65 °C	99.8	75	[13j]
Mn(NO$_3$)$_2$/Co(NO$_3$)$_2$	One	Acetic acid + catalytic HNO$_3$, 40 °C	97.5	93.4	[14m]
H$_7$PMo$_8$V$_4$O$_{40}$	One	Acetic acid/water, 70 °C	99	51	[14a]
H$_5$PMo$_{10}$V$_2$O$_{40}$	One	Acetonitrile/methanol, 60 °C	98	55	[14h]
Co/Mn cluster	One	Acetic acid + water (MEK), 100 °C	97.6	86.6 (wt)	[14k]

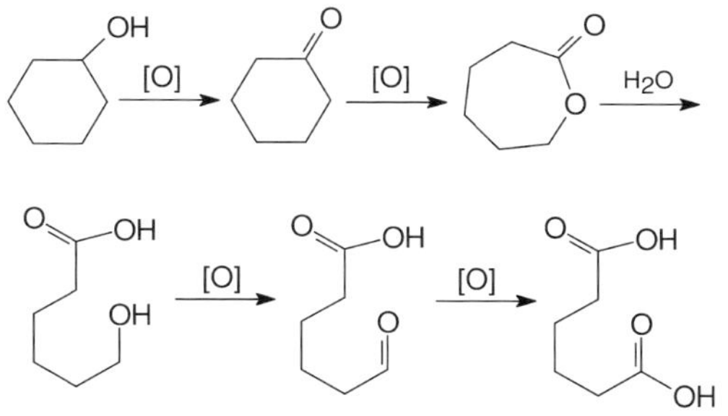

Scheme 7.6 Mechanism of cyclohexanol oxidation to AA with HP.
Source: elaborated from [15e].

The oxidation of Ol/One has also been investigated using aqueous HP as the oxidant, with either homogeneous or heterogeneous catalysts. For instance, the oxidation of the ketone with HP under homogeneous conditions using acetic acid or *t*-butanol as the solvent gave about a 50% yield of AA [15f]. H_2WO_4 afforded the transformation of the ketone with a 91% isolated yield of AA, and that of cyclohexanol with an 87% AA yield, by using a 3.3-molar 30% HP at 90 °C for 20 h reaction time, without the use of solvents [15e]. Remarkably, there was no unproductive decomposition of HP. The mechanism proposed involves the Baeyer–Villiger type oxidation to ε-caprolactone, followed by ring opening to yield 6-hydroxyhexanoic acid (Scheme 7.6).

This ketone conversion into dicarboxylic acid by HP is applicable to five- to eight-membered cyclic ketones. The acidic nature of the catalyst is essential, since Na_2WO_4 did not catalyze the reaction. H_2WO_4 was the precursor of the true active species, $H_2[WO(O_2)_2(OH)_2]$, which is soluble in water. The crystalline product was recovered by filtration (AA precipitated during cooling) followed by drying in air. The aqueous phase of the reaction mixture could be reused with 60% HP to give AA in 71% yield.

7.3.2
Direct Oxidation of Cyclohexane with Air

Compared to the traditional technology, the direct oxidation of cyclohexane with air or oxygen, also called the One Step AA process, should lower the total investment cost. This is due to the following differences:

- elimination of one oxidation step;
- elimination of nitric acid production, handling, recovery, purification and recycle units;
- simplification of the air abatement system by eliminating N_2O and NO_x emissions;
- simplification of the wastewater treatment by elimination of nitrates.

7.3.2.1 Homogeneous Autoxidation of Cyclohexane Catalyzed by Co, Mn or Cu
Many patents and papers [16–22] describe catalysts and process configurations for the single-step aerobial oxidation of cyclohexane to AA, catalyzed by homogeneous

Co-, Cu-, Mn- or Fe-based complexes, including biomimetic systems. Therefore, the same catalyst type that is used for the oxidation of cyclohexane to KA-Oil also oxidizes this reactant to AA, depending on reaction conditions and the cyclohexane conversion achieved. Indeed, the Asahi Chem Co had already developed the process in the 1940s, using Co acetate catalyst and acetic acid as the solvent, under 30 atm of O_2 at 90–100 °C [16]. The best selectivity to AA was 75%, the main by-product being glutaric acid, with a cyclohexane conversion of 50–75%. Asahi succeeded in achieving high conversion of cyclohexane through the use of a relatively high concentration of Co(III) acetate combined with acetaldehyde or cyclohexanone which served as a promoter [16].

A review by Schuchardt *et al.* thoroughly analyses the various catalytic systems reported in the literature up to 2000, both homogeneous and heterogeneous ones, and those that use oxidants other than oxygen [e.g., HP or *t*-butyl hydroperoxide (*t*-BuOOH)] [2c]. The mechanism involves the formation of cyclohexanol via the cyclohexyl radical and cyclohexyl hydroperoxide. According to the Haber–Weiss mechanism, cyclohexyl hydroperoxide decomposes into alkoxy and alkyloxy radicals (Section 7.2.1). Cyclohexanol is finally oxidized to cyclohexanone. A similar mechanism may occur at the α-C, affording 1,2-cyclohexanedione, which is finally cleaved to AA. Oxidation of the intermediately formed cyclohexanone to AA then occurs through a mechanism similar to that illustrated in Scheme 7.5.

Many companies have studied the optimization of catalyst composition and process conditions in order to improve the performance of the reaction and the economics of the process. In the Gulf process, the reaction is carried out at 90–100 °C, with a Co(III) acetate catalyst and acetic acid as the solvent [17]. The molar selectivity is around 70–75%, for a cyclohexane conversion that can be as high as 80–85%. The high concentration of Co(III) acetate used also favors the direct reaction of the cation with cyclohexane, generating the cyclohexyl radical. In fact, in Gulf patents the reaction is reported to occur in a critical amount of Co(III) (25–150 mmoles per mole of cyclohexane). The catalyst is activated during the initial induction period, and water is also added in the initial stage to enhance the selectivity to AA, but the rate of production decreases because the induction period increases.

In Amoco patents [18b], the addition of controlled amounts of water after the initiation of the oxidation reaction is claimed to be a tool to obtain a better yield to AA. The best yield achieved was 88% (based on the identifiable compounds) at 98% cyclohexane conversion, with a Co(II) acetate catalyst, an acetic acid solvent, at 95 °C and 70 atm air pressure. It is reported that water, if present during the induction period, depletes the concentration of free radicals; in the absence of water, the yield was remarkably lower. These results are comparable to those attained by the air/nitric acid two-step oxidation of cyclohexane.

In patents issued by Redox Technologies, oxidation is reported to occur more efficiently if high concentrations of cyclohexane are used (AcOH/cyclohexane weight ratio between 50:50 and 15:85, instead of 80:20), and if the conversion of the reactant is kept below 30%, instead of more than 70% [19]. A low concentration of catalyst is recommended. An 88% AA selectivity at 21% cyclohexane conversion is obtained in an acetic acid solvent with a $Co(OAc)_2$ catalyst (cyclohexane/Co ratio higher than 150), at 105 °C and with 14 atm pressure of O_2-enriched air flow, for 45 min reaction time; acetaldehyde is added as an initiator [19a]. In Reference [19b], the purification stage of the final mixture is disclosed. The treatment consists of cooling the product mixture to bring about the precipitation of the diacid followed by separation by filtrating the diacid from the two liquid phases, a nonpolar one that is recycled and a polar one that is also recycled after an optional hydrolysis and separation of an additional amount of diacid.

Various bicomponent catalytic systems have been suggested for this reaction, including: (i) Co-Fe [18c, d], which affords 80% selectivity to AA by restricting the conversion to below 30%; (ii) Co-Mn [20a]; (iii) Co-Zr [18e, f]; and (iv) Co-Cr [20b].

Rhone Poulenc (now part of Rhodia) [20c] disclosed a process for recycling the catalyst, including treating the reaction mixture by extraction of the by-products glutaric and succinic acid. For example, with a Mn/Cr catalyst, at 105 °C, 100 bar air pressure, in acetic acid solvent, 11.3% cyclohexane conversion is obtained in 170 min reaction time, with 10.2% selectivity to cyclohexanol and 65.5% to AA [20n]. One innovative aspect of recent Rhodia patents [20d–l] is represented by a lipophilic catalyst that can be separated easily from the reaction mixture and recycled together with the unconverted cyclohexane. At the end of the reaction, water is added to the solution to solubilize the AA. The reactant, the catalyst and the intermediate products cyclohexanol and cyclohexanone form an immiscible phase and are recovered and recycled (Figure 7.8). AA is crystallized from the aqueous solution by cooling. Thus, the innovation is the use of a lipophilic acid, for example, 4-*tert*-butylbenzoic acid, in place of acetic acid; the stronger the acid, the faster the oxidation rate. At 130 °C, with an optimal amount of carboxylic acid of 10–12 wt%, 100 wppm of Mn(II), 1 mol% of cyclohexanone (with respect to cyclohexane) as a radical initiator and 100 bar air, the best performance was obtained in terms of AA selectivity (around 33%) and productivity, with a cyclohexane conversion of 10%. The addition of Co(II) as a co-catalyst remarkably improved the oxidation rate, and the selectivity to AA (about 50%). By-products were glutaric acid, succinic acid, cyclohexanone, cyclohexanol, 3-hydroxyadipic acid and formic acid. Under continuous conditions, 10.4% conversion was obtained with 56% selectivity to AA; after recycling of the organic phase, the selectivity increased to 70.6%, also because of the conversion of intermediates into AA. An outstanding value of 95 $g_{AA}\,L^{-1}\,h^{-1}$ was obtained.

During the 1990s, the Twenty First Century Corporation in a joint venture with RPC Inc. carried out an extensive laboratory experimental program aimed at the development of a new oxidation technology. The core of the technology is the design

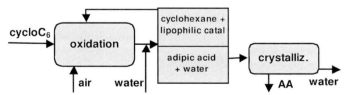

Figure 7.8 General diagram illustrating the Rhodia approach for cyclohexane oxidation. Source: elaborated from [20f].

of a new concept for the oxidation reactor. The unit uses oxygen rather than air (used in conventional units), which significantly reduces the size of the air abatement systems. The solution of cyclohexane and cobalt catalyst dissolved in acetic acid is nebulized through a series of nozzles within the reactor. In the reaction chamber there is a very high liquid–gas interface and condensation can be an excellent factor for controlling the reaction temperature. This is achieved by feeding the solution through the bottom nozzles of the reactor at a temperature lower than the solution fed through the upper line. Conversion is controlled by operating on different variables, such as the concentration of catalyst and reactants, the hold-up time of the liquid feed and the size of droplets. The reactor is equipped with a cooling system for the removal of the heat of reaction. The off-gas containing excess of oxygen, nitrogen and VOCs is cooled, compressed and recirculated. A solvent recovery system removes water, recycles the solvent (acetic acid) and cyclohexane. The reaction also yields over-oxidized product and adipic esters. Esters react with water in the presence of a catalyst, converting them into AA and a small amount of cyclohexanone and cyclohexanol. AA is again separated in recovery crystallization and recycled to an AA recovery section. The combination of the single oxidation and hydrolysis boosts yields to levels comparable with conventional technology.

Many patents have been filed by the Twenty First Century Corporation on reactor technology with reactant atomization [21a, b], AA recovery and solvent separation [21c], catalyst handling and recovery [21d] and the control of temperature and pressure [21e–g]. Figure 7.9 shows a simplified block diagram of the process.

Fluor Daniel [21h, i] have performed a conceptual economic analysis of the Twenty First Century Corporation/RPC Inc technology by comparing a theoretical grass roots facility of comparable capacity to an existing global scale unit. The results show that the new process could reduce the capital cost by 33% and the operating costs by nearly 23% with respect to conventional technologies.

An application combining the direct oxidation of cyclohexane with air and the use of HP as a finishing oxidant, to complete the oxidation of by-products separated during alkaline washing (see Figure 7.1), has been proposed by Sumitomo [21j]. The wastewater containing hydroxycaproic acid is made to react with 30% HP in the presence of sulfuric acid and a tungsten catalyst, with a 95% yield to AA. This technology allows an increase in the yield and a reduction in the cost of waste

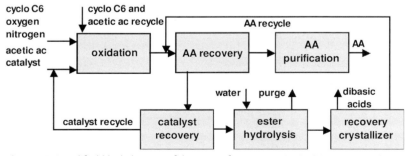

Figure 7.9 Simplified block diagram of the twenty-first century Co/RPC Inc one-step AA process.

treatment. The consumption of HP is limited to the oxidation of a partially oxygenated by-product, and the incidence of its processing costs is reduced.

Despite the moderate yields to AA achieved by the various technologies developed, the One-Step AA process has not yet been implemented at a commercial level, because of the high corrosivity of acetic acid and the high energy demand for solvent recycle and product workup. Moreover, the per-pass conversion and selectivity is generally lower than the two-step process from cyclohexane. Finally, traces of acetic acid are left in the product even after repeated crystallization steps, which lower the quality of the product and limit its use in polyamide synthesis.

7.3.2.2 Heterogeneous Catalysis for Cyclohexane Oxidation to Either Ol/One or AA (Various Oxidants Included)

Several Co-based heterogeneous catalysts have been investigated for the oxidation of cyclohexane [23]. For instance, Co framework-substituted aluminophosphate microporous molecular sieves (CoAPO or CoAlPO) are the heterogeneous analogues of homogeneous Co complexes [23d, f, i]. With CoAPO-15, under oxygen pressure, at 145 °C, keeping the conversion at 3.5% gives the same distribution of products (i.e., selectivity to cyclohexanol and cyclohexanone) as the industrial process [23e]. With a CoAPO-36, Sankar *et al.* [23f] obtained a very high selectivity to Ol/One, at 9.6% cyclohexane conversion ($T = 150$ °C). In this case, the activity proved to be truly heterogeneous. In fact, some authors pointed out that with these catalysts the catalytic activity may indeed be due to the leaching of Co from the structure into the solution, as also occurs with Co-containing dealuminated beta zeolites [23g, h]. Such leaching did not occur either in the case of Co(II)-substituted hexagonal mesoporous aluminophosphate, Co-HMA [23i], which catalyzed the oxidation of cyclohexane to Ol/One with air (MEK as initiator), or HP or *t*-BuOOH.

A Fe framework-substituted AlPO (FeAlPO-31) gave a 65% yield of AA [24a, b]; a high selectivity (81%) at 88% cyclohexane conversion was obtained at 110 °C using acetylperoxyborate (APB), which is a convenient solid source of peracetic acid and HP [24c, d], the by-products being cyclohexanone and lighter dicarboxylic acids. Worse performance was shown by Co(Mn)AlPO-36 [23f] and Fe(Mn)AlPO-5 [24e], with cyclohexanol and cyclohexanone being the principal products in this case. The FeAlPO-31 catalyst, in fact, had the optimal pore diameter (5.4 Å) to enable the more sinuous of the reaction intermediates formed within the pores, such as AA, to diffuse out readily. Larger compounds formed during the reaction are trapped inside on account of their retarded diffusion. Thus, the intermediates cyclohexanol and cyclohexanone, formed from cyclohexyl hydroperoxide, or 2-hydroxycyclohexanone and 1,2-cyclohexanedione, are held near the active site until oxidation proceeds further to yield the more mobile AA. Therefore, the constrained environment favors the formation of the final product.

With MnAPO-5, the activity in the aerobial oxidation of cyclohexane to Ol/One increased linearly with the number of redox-active sites, since the reaction involves cycling between Mn(II) and Mn(III). Therefore, the reaction requires cation sites able to reversibly form charge-balancing cationic species [24f].

Besides MeAPO, several heterogeneous systems have been proposed for the oxidation of cyclohexane with various oxidants [2c], but almost all of them yield Ol/One as the main reaction products, with AA being only a minor product. Heterogeneous catalysts can be either oxides or metal cations and complexes incorporated on inorganic matrixes, such as active carbon, zeolites, aluminophosphates or conventional supports such as alumina and silica. The activity of these systems is greatly affected by the choice of solvent, which determines the polarity of the medium. In addition, the hydrophobicity of the support is important, since a hydrophobic environment rapidly expels the oxidized products from the reaction zone. When oxygen is used as the oxidant, these systems often need small amounts of hydroperoxides as co-catalysts.

Examples of heterogeneous catalysts for the oxidation of cyclohexane to Ol/One or to AA with various oxidants include [25–27]:

1. Nano Au particles supported on mesoporous materials (SBA-15, MCM-41), which catalyze the oxidation of cyclohexane to Ol/One with oxygen, without solvent (conv. 18%, selectivity 93%) [25a]. Similar performance is obtained when Au is dispersed over other supports, for example, silica, alumina or titania [25b, c, d].

2. Copper or iron phthalocyanines encapsulated in X or Y zeolites [25g], which catalyze the oxidation of cyclohexane to Ol/One and to AA with oxygen (in the presence of small amounts of t-BuOOH) at near-ambient conditions. The catalyst remains in the solid phase throughout the reaction, and can be easily filtered off. Moreover, the solvent type affects performance; best selectivity to AA (41%) is achieved with methanol [25g], at 12.7% cyclohexane conversion, with a halogen-substituted phthalocyanine of Fe encapsulated in an X zeolite. Cyclohexanone and cyclohexandione are hypothesized to be the intermediate compounds of the reaction. Incorporation of the zeolite-encapsulated Fe phthalocyanine inside a polymer matrix can serve to enhance catalyst stability and limit leaching phenomena [25h].

3. Ce-exchanged Y zeolite, which catalyzes the oxidation of cyclohexane with t-BuOOH, giving 10% reactant conversion at 90 °C with good selectivity to products of partial oxidation [26a, b]. Alumina-supported Ce(IV) oxide has been used for the aerial oxidation of cyclohexanone and the co-oxidation of cyclohexane and cyclohexanone, in an acetic acid solvent, at 15 atm pressure and 110 °C [26c]. A yield of 66% to AA was obtained, with a cyclohexane conversion of 38% and cyclohexanone conversion of 41%. Interestingly, these authors also reported the formation of ε-caprolactone in 15% yield; the lactone was not an intermediate in AA formation. The authors hypothesized that the lactone forms by rearrangement of the intermediate 1,1-dihydroxycyclohexane, the latter being formed by support-mediated hydroxylation of protonated cyclohexanone, and that the mechanism was not a free-radical type. ε-Caprolactone also formed in the aerial oxidation of cyclohexanone catalyzed by the Ce-exchanged X zeolite [26d].

4. Titanium substituted hexagonal mesoporous aluminophosphate (Ti-HMA) molecular sieves, which exhibit excellent performance in the oxidation of cyclohexanone to Ol/One with HP (yielding mainly cyclohexanol), or t-BuOOH (mainly cyclohexanone) or oxygen (mainly cyclohexanol). Similar performance is obtained

with Ti-MCM-41 [27a]. The use of an acetic acid solvent, however, may favor the leaching of Ti from the molecular sieve [27b]; acetic acid forms peracetic acid, which acts as the true oxidant.

5. Other catalysts investigated include Cu, Cr and Ce silicates, with t-BuOOH as the oxidant. The conversion achieved, however, is generally lower than 5% under the conditions investigated [27c–f]. A high activity was instead obtained with Cr-MCM-41 [27g], with HP in an acetic acid solvent; in this case, peroxy acetic acid also had a role in reaction acceleration.

Table 7.3 compares the results of the several heterogeneous catalysts reported in the literature for oxidation with HP, or t-BuOOH or oxygen.

The use of supercritical (SC) CO_2 or almost-SC (two-phase regime) conditions and heterogeneous catalysts should in principle positively affect the selectivity of partial oxidation reactions, owing to enhancement in mass transfer rates. Various authors have investigated the effect of adding CO_2 to the liquid phase, with respect to the catalytic performance in cyclohexane oxidation with oxygen [29]. In some cases, the passage from a single liquid phase to a two-phase fluid system, because of CO_2 addition, caused an increase in the rate, because the concentration of cyclohexane was higher near the catalyst when the system was split into liquid and vapor phases, while the transport of oxygen between the phases remained high [29b, c]. Also, in some cases, an increase of selectivity to Ol/One was observed with CO_2 addition [29c] – attributed to the increased ability of the products to rapidly diffuse out of the solid catalyst, owing to carbon dioxide's low viscosity, thus preventing by-product formation. However, in general, the yields obtained were not higher than those obtained under conventional liquid-phase conditions.

7.3.2.3 *N*-Hydroxyphthalimide as the Catalyst for the Oxidation of Cyclohexane to AA with Oxygen

One notable achievement is the process that makes use of N-hydroxyphthalimide (NHPI) as the catalyst for the aerial oxidation of cyclohexane [30]. NHPI is a cheap, nontoxic catalyst easily prepared by the reaction of phthalic anhydride and hydroxylamine. It acts as a precursor of the phthalimido-N-oxyl (PINO) radical, which is the effective abstracting species in the free radical process (Scheme 7.7).

A $30 \, \text{ton} \, \text{yr}^{-1}$ pilot plant has been constructed by Daicel Chemical Industry to produce AA with this latter technology; the yield of AA is reported to be greater than 80% [30r]. Typically, a 90% conversion of cyclohexane is obtained after 6 h reaction at 100 °C and 10 atm, with 76% selectivity to AA, and with oxygen as the oxidant. The reaction needs a cocatalyst, for example, $Mn(acac)_2$ and $Co(acac)_2$ (the two salts give a synergic effect), and is carried out in an acetic acid solvent (or acetonitrile), because of the limited solubility of NHPI in nonpolar solvents such as hydrocarbons. Notably, the reaction may proceed even in the absence of the solvent by the use of a lipophilic NHPI derivative. For example, 4-laulyloxycarbonyl-N-hydroxyphthalimide is an efficient catalyst for the oxidation of cyclohexane under solvent-free conditions [31f]. The by-product formed in the largest quantities is cyclohexanone, the intermediate formed in autoxidation.

Table 7.3 Summary of literature reporting on the heterogeneous oxidation of cyclohexane with various oxidants.

Catalyst	Oxidant, solvent	T (°C), P (atm), time (h)	CycloC$_6$ conversion (%)	Main product selectivity (%)[a]	Reference
Co-AlPO-5	Air, no solvent	130, 15, 16	11	Ol/One 80	[23f]
Co-AlPO-36	Air, no solvent	130, 15, 16	57	Ol/One 83	[23f]
Co-HMA	HP (+ MEK initiator), acetic acid	100, 1, 12	99	Ol/One 98	[23i,k]
Co-MCM-41	HP (+ MEK initiator), acetic acid	100, 1, 12	69	Ol/One 96	[23j]
Co-APO-5	HP (+ MEK initiator), acetic acid	100, 1, 12	79	Ol/One 94	[23j]
Co$_3$O$_4$	O$_2$ (+ t-BuOOH)	120, 10, 6	8	Ol/One 89	[23l]
K$_7$HCoW$_{11}$Co(H$_2$O)O$_{39}$ -hydrotalcite	O$_2$, no solvent	130, 5, 24	6	Ol/One 89	[23m]
Co (metal)	O$_2$, no solvent	25, 40, 15	41	Ol/One 80	[23n]
Co-kenyaite (layer silicate)	O$_2$, no solvent	130, 5, 24	10	Ol/One 85	[23o]
Co-AlPO-5	Air (+ t-BuOOH), no solvent	130, 15, 24	12	Ol/One 88, AA 8	[24e]
Co-APO-5	O$_2$, acetone	125, 10, 2.5	15	Ol/One 75	[23d]
Fe-AlPO-31	Air, no solvent	100, 15, 24	7	AA 65	[24a,b]
Fe-AlPO-5	Air, no solvent	100, 15, 24	5	Ol/One 77	[24a,b]
Fe-AlPO-31	Acetylperoxyborate, water	110, 1, 16	89	AA 81	[24c,d]
Fe-AlPO-5	Air, no solvent	130, 15, 24	7	Ol/One 51, AA 31	[24e]
Fe-AlPO-5	Air (+ t-BuOOH), no solvent	130, 15, 24	20	Ol/One 54, AA 33	[24e]
Fe/phthalocyanine – Y zeolite	t-BuOOH, acetone	25, 1, 150	24	Ol/One 95 (mainly One)	[24h]
Fe-HMA	HP (+ MEK), acetic acid	100, 1, 12	89	Ol/One 90	[24i,j]
Fe-HMA	O$_2$ (+ MEK), acetic acid	100, 1, 12	63	Ol/One 96	[24i]
Fe$_2$O$_3$-TiO$_2$	O$_2$ (+ isobutyraldehyde, acetic acid), no solvent	70, 1, 16	26	Ol/One 90	[24g]
Mn-AlPO-5	Air (+ t-BuOOH), no solvent	130, 15, 24	17	Ol/One 63, AA 29	[24e]
Au-SBA-15	O$_2$, no solvent	150, 10, 6	18	Ol/One 93	[25a]
Au-MCM-41	O$_2$, no solvent	150, 10, 6	16	Ol/One 76	[25b]

Au-Al$_2$O$_3$	O$_2$, no solvent	150, 15, 3	13	Ol/One 85	[25c]
Au-SiO$_2$	O$_2$, no solvent	150, 15, 3	9	Ol/One 83	[25d]
Au-graphite	O$_2$ (+t-BuOOH), no solvent	70, 3, 17	6	Ol/One 17	[25e]
Au-CeO$_2$	O$_2$ [+ 2,2'-azobis(isobutyronitrile)], no solvent	120, 15, 24	21	Ol/One 90	[25f]
Cu(Fe)//phthalocyanine – X(Y) zeolite	t-BuOOH, acetonitrile	70, 1, 8	22	Ol/One 64	[25g]
Cu(Fe)//phthalocyanine – X(Y) zeolite	Air (+ t-BuOOH), methanol	70, 53, 8	13	Ol/One 50, AA 41	[25g]
Ce-Y	t-BuOOH, no solvent	70, 1, 24	8	Ol/One 49	[26a,b]
CeO$_2$-Al$_2$O$_3$	Air, acetic acid	110, 15	38	AA 84	[26c]
Ti-HMA	HP (+ MEK), acetic acid	100, 1, 12	90	Ol/One 95 (mainly Ol)	[27a]
Ti-MCM-41	HP (+ MEK), acetic acid	100, 1, 12	88	Ol/One 99 (mainly Ol)	[27a]
Ti-HMA	O$_2$ (+ MEK), acetic acid	100, 1, 12	47	Ol/One 98	[27a]
TS-1	HP, acetic acid	60, 1, 4	16	Ol/One 79	[27b]
Ti-MMM-1	HP, acetone	95, 1, 8	9	Ol/One 90	[27h]
TAPO-5	HP, acetone	80, 1,5	4	Ol/One 100 (mainly Ol)	[27i]
Ti-β	HP, 2-butanone	96, 1, 4	47	AA 54	[27j]
VO$_x$-SiO$_2$	O$_2$, no solvent	175, 10, 20	16	Ol/One 71	[28a]
(VO)$_2$P$_2$O$_7$	HP, CH$_3$CN	65, 1, 20	91	Ol/One 100	[28b]
V-HMA	HP (+ MEK), acetic acid	100, 1, 12	95	Ol/One 95 (mainly Ol)	[28c]
V-MCM-41	HP (+ MEK), acetic acid	100, 1, 12	94	Ol/One 92	[28c]
V-APO-5	HP (+ MEK), acetic acid	100, 1, 12	22	Ol/One 86	[28c]
Cr-MCM-41	HP (+ MEK), acetic acid	100, 1, 12	99	Ol/One 93 (mainly Ol)	[27g]
Cr-MCM-41	O$_2$ (+ MEK) acetic acid	100, 1, 12	86	Ol/One 97 (mainly Ol)	[27g]
Cr-MCM-41	t-BuOOH, acetonitrile	80, 1, 20	92	Ol/One 95 (mainly One)	[28d]
(Cr,V)-APO-5	HP, acetone	60, 1, 40	10	Ol/One 90	[28e]
Cr-HMA	HP (+ 2-butanone), acetic acid	100, 1, 12	94	Ol/One 100	[28f]
Cr-MCM-41	HP (+ 2-butanone), acetic acid	100, 1, 12	96	Ol/One 98	[28f]
Al-todokorite	t-BuOOH, t-butanol	60, 1, 40	18	Ol/One 76	[28g]

a In many cases, small amounts of AA are formed (selectivity less than 10%) together with Ol/One, especially at high cyclohexane conversion.

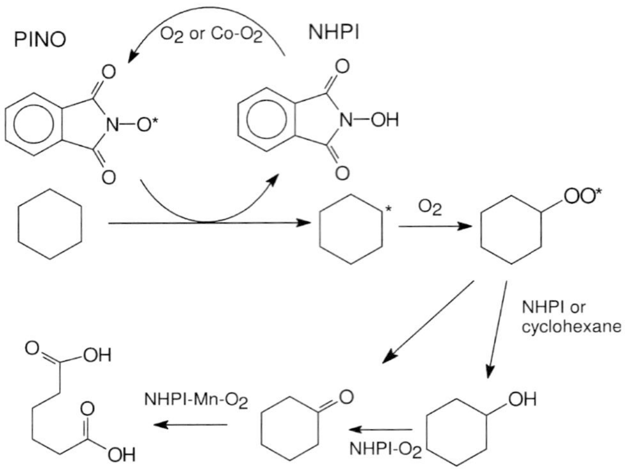

Scheme 7.7 Catalytic oxidation of cyclohexane to AA with the NHPI catalyst. Source: elaborated from [30b,s].

Table 7.4 compiles some of the results reported by Ishii *et al.* [30b]. Professor Ishii and Daicel Chem were awarded the 2003 Green and Sustainable Chemistry Prize for their efforts in the development of NHPI-based industrial processes with low environmental impact.

The best result reported in the open literature is of 73% conversion with 73% selectivity to AA, obtained at normal O_2 pressure, in acetic acid with 1 mol% Mn (acac)$_2$, the by-products being glutaric acid (9%), succinic acid (6%), cyclohexyl acetate (2%) and cyclohexanol (1%) [30b]. The generation of the PINO from NHPI (Scheme 7.7) with oxygen is assisted by the Co(II) species; therefore, the addition of a small amount of Co(OAc)$_2$ enhances the oxidation process. In contrast, if the reaction is performed in an acetonitrile solvent, with a Co(OAc)$_2$ catalyst at 75 °C, the main product is cyclohexanone (78% selectivity at 13% cyclohexane conversion).

The use of NHPI allows an increase in reaction rate but has no effect on the selectivity [30p]. Drawbacks of the reaction are the high amount of NHPI required, and its decay to phthalimide, which cannot be easily recycled to NHPI. The process

Table 7.4 Results for the oxidation of cyclohexane with O_2 and NHPI [30b][a].

Co-catalyst	T (°C)	Time (h)	Cyclohexane conversion (%)	AA selectivity (%)	Cyclohexanone selectivity (%)
Mn(acac)$_2$	100	6	36	46	20
Mn(acac)$_2$	100	20	73	73	Trace
Co(OAc)$_2$	100	6	40	34	36
Co(acac)$_2$	75	6	13	13	78
Cu(acac)$_2$	100	6	14	12	71

[a]Reaction conditions: 3 mmol cyclohexane, 1 or 10 mol% NHPI, 7 mL acetic acid, $PO_2 = 1$ atm.

also looks very complex, in view of the reaction mixture used and the workup needed. Attempts to make NHPI heterogeneous have been reported recently [30u].

NHPI has been introduced as an effective system for C−H activation by hydrogen abstraction on several different substrates [30i,j,s,31a–e]. In 2001, Daicel commercialized the process used to synthesize dihydroxyadamantane, and has carried out pilot trials not only for the oxidation of cyclohexane but also for the oxidation of p-xylene to terephthalic acid. In the latter case, the advantage lies in being able to avoid using special anticorrosive metals currently required in the production of terephthalic acid because of the use of bromine. NHPI can also be used as a catalyst for the in situ production of hydroperoxides, reactants for epoxidation and for the oxidation and ammoxidation of cyclohexanone to caprolactone and caprolactam, respectively.

An interesting example of the application of NHPI is the synthesis of lactones with the in situ generation of HP. Ishii et al. [31g] have reported that in the oxidation of the KA Oil mixture with air, in the presence of both NHPI and small amounts of initiator AIBN, the alcohol was converted into a mixture of cyclohexanone and HP via the intermediate formation of 1-hydroxy-1-hydroperoxycyclohexane; the reaction was performed in an acetonitrile solvent at 70 °C. The addition of catalytic amounts of InCl₃ in the second step led to the reaction between cyclohexanone and HP, with the formation of ε-caprolactone, at 25 °C. The conversion was 23%, and the selectivity to the lactone was 57% (Scheme 7.8).

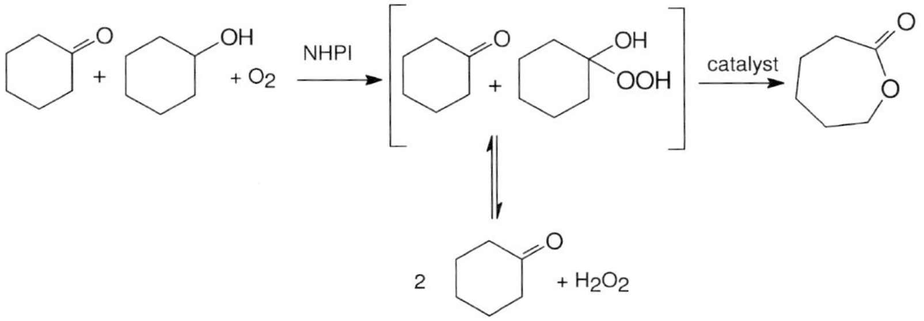

Scheme 7.8 A new strategy for the Baeyer–Villiger oxidation of KA Oil.
Source: elaborated from [31g].

Other examples of nitroxyl radicals such as TEMPO [31d, j] have been used successfully in several examples of environmentally friendly liquid-phase oxidations with oxygen. Sheldon et al. have reported on the use of N-hydroxysaccharin as an alternative to NHPI in the oxidation of cycloalkanes to dicarboxylic acids [31h]. Other examples include the aerobial oxidation in the presence of NHPI, o-phenanthroline and bromine, in an acetonitrile/CCl₄ solvent and in the absence of metals, at 100 °C. The selectivity was 75% to AA and 22% to cyclohexanone, at 48% cyclohexane conversion [31i].

7.3.3
Butadiene as the Starting Reagent

BASF has developed the synthesis of AA by the methoxycarbonylation of butadiene [32a]; the catalyst is based on Co. The process takes place in three steps (Scheme 7.9).

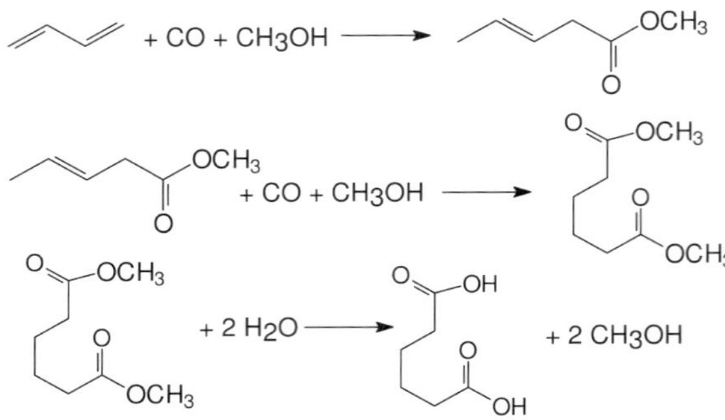

Scheme 7.9 Methoxycarbonylation of butadiene to AA.

The first step is the reaction between butadiene, methanol and carbon monoxide to generate methyl-3-pentenoate with 98% molar selectivity; it occurs at 120–140 °C and high pressure (>300 atm). A fairly high concentration of $HCo(CO)_4$ and pyridine is used to ensure rapid carbonylation of butadiene, thus preventing typical side reactions like dimerization and oligomerization. In the second step, the intermediate is made to react again with methanol and CO to produce dimethyl adipate; the concentration of the pyridine ligand must be low because it has an inhibitory effect on the hydroalkoxycarbonylation. The temperature is 150–170 °C and the pressure is 150–200 atm. The dimethyl adipate is finally hydrolyzed to AA, with an overall yield based on butadiene of 72%, and an AA purity of 99.9%. The major by-products of the process are methyl valerate and the dimethyl esters of C_4 dicarboxylic acids, which could eventually find markets in various applications. Several BASF patents, issued after 1977, describe the technology; BASF, however, announced that it would not proceed with the implementation of this C_4-based technology in the foreseeable future [32b].

A variant of this process, studied by DuPont and DSM [32c], includes the hydrocarboxylation (hydroxycarbonylation) of butadiene with carbon monoxide and water; this technology offers potential savings in raw material costs. The reaction primarily yields 3-pentenoic acid using a palladium/crotyl chloride catalyst system, with a selectivity of 92%. Further conversion of pentenoic acids by reaction with carbon monoxide and methanol and a palladium/ferrocene/phosphorous ligand catalyst has demonstrated a selectivity to dimethyl adipate of 85%; the latter is finally hydrolyzed to AA. The main problem in this reaction is the propensity of pentenoic acid to undergo acid-catalyzed cyclization to γ-valerolactone; one way to circumvent the problem is to carry out the hydrocarboxylation of pentenoic acid using the γ-valerolactone as the solvent.

One further variant of the process has been studied by Rhone Poulenc [32d] and by Shell Chemicals [32f–h] with the development of a new catalyst capable of converting butadiene, carbon monoxide and water into AA (Scheme 7.10). First, butadiene is carboxylated to a mixture of isomeric pentenoic acids, at 105 °C and 80 bar CO. In a second step, the same catalyst catalyzes the carboxylation of the pentenoic acids to AA. Shell patents claim that mono-olefins are not carboxylated and thus allow the use

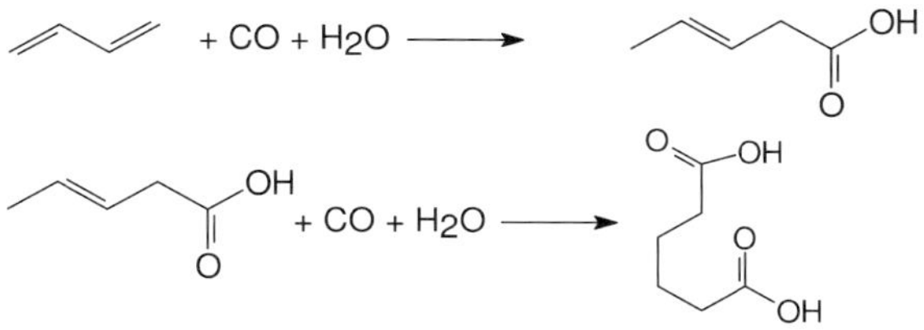

Scheme 7.10 Hydroxycarbonylation of butadiene to AA.

of a crude C$_4$ fraction as the starting material instead of pure butadiene. The catalyst system used consists of Pd(OAc)$_2$, a diphosphine ligand with an *ortho*-xylyl bridge and two P(*t*-Bu)$_2$ groups, and a carboxylate anion. Pentenoic acid acts as the solvent in both steps of the reaction. The reaction rate, selectivity and the stability of the catalytic system are influenced by several factors such as CO pressure, temperature, water content, ligand/Pd ratio and butadiene concentration.

Currently, the new Shell process is at the laboratory scale. The main limit for the industrialization is the use of a complex homogeneous catalyst, composed of precious metals and a phosphine ligand. Phosphine ligands are expensive chemicals and the recovery system must avoid catalyst losses during the AA purification and from the heavy ends purge.

Direct hydroxycarbonylation can be advantageous with respect to the more traditional approach of butadiene hydroformylation to yield 1,6-hexanedial (adipaldehyde), via intermediate pentenal, catalyzed by a metal organophosphorus ligand complex catalyst [32i]. In this latter process, the dialdehyde can either be transformed into ε-caprolactone [32j] or oxidized to AA.

In an alternative synthetic pathway, methyl 5-formylvalerate can be prepared from butadiene via sequential methoxycarbonylation to methyl 3-pentenoate, isomerization to methyl 4-pentenoate, and hydroformylation to methyl 5-formylvalerate. The valerate can then be oxidized to monomethyladipate, which in turn is hydrolyzed to AA (Scheme 7.11).

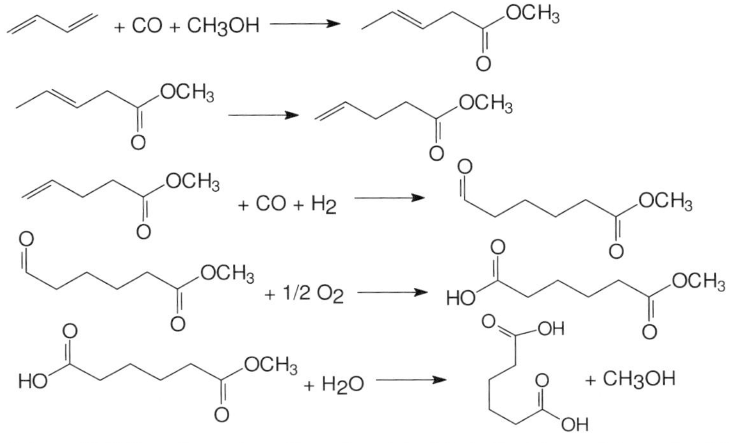

Scheme 7.11 Transformation of butadiene into AA via the intermediate methyl 5-formylvalerate.

The methyl formylvalerate can also be an intermediate for the synthesis of caprolactam [32k]. For instance, in Reference [32l], butadiene is carbonylated in the liquid phase in the presence of an alkanol, CO, palladium and a bidentate organic phosphorus, antimony or arsenic ligand. The bridging group of the ligand, which connects the two phosphorus atoms, consists of a bis(η-cyclopentadienyl) coordination group of Fe. By performing the air oxidation of methyl 5-formylvalerate at high pressure (e.g., 50 atm) and at a relatively moderate temperature (50 °C) in the absence of a catalyst, commercial rates of conversion (>98%) can be obtained with optimum selectivity (>98%) to the desired product [32m].

As proposed in [33a–h], the metal-catalyzed carboxylation of alkanes by CO is an alternative approach, using catalysts such as $Pd(OAc)_2$, $Cu(OAc)_2$, Mg, $CaCl_2$ and Co $(OAc)_2$ in trifluoroacetic acid (TFA), with a peroxodisulfate salt, for example, $K_2S_2O_8$, as the oxidant, at 80 °C. Amongst the most active catalysts for this reaction are V-complexes, which have been applied with various N,O-polydentate ligands of (hydroxyimino)dicarboxylate and aminoalcohol types, in particular synthetic amavadine and its models [33i]. In recent papers, the single-pot catalytic carboxylation of light alkanes by CO has been carried out with group 5–7 metal oxides, in a trifluoroacetic acid solvent, at 80 °C and with $K_2S_2O_8$. For instance, a 16% yield to hexanoic acid from n-pentane was obtained using V_2O_4 as the catalyst [33g, j].

7.3.4
Dimerization of Methyl Acrylate

Methyl acrylate can be dimerized to give a molecule that can be hydrogenated to dimethyl adipate; the latter, in turn, can be hydrolyzed to yield AA. Methyl acrylate is synthesized by esterification of acrylic acid, which is obtained by the two-step oxidation of propylene. However, the overall scheme requires several reaction steps, and investment requirements may be large.

7.4
Emerging and Developing Technologies for AA Production

7.4.1
An Alternative Raw Material for AA Synthesis: Cyclohexene

Cyclohexene can be synthesized by partial hydrogenation of benzene, by partial dehydrogenation of cyclohexane, or by dehydrohalogenation of cyclohexyl halides. The hydrogenation of benzene is the most viable route; in the Asahi process, cyclohexene is obtained with a 60% yield and 80% selectivity, with the remainder being converted into cyclohexane. Asahi has developed a process for the addition of water to cyclohexene to produce cyclohexanol that can then be oxidized to AA using the conventional nitric acid oxidation.

Scheme 7.12 summarizes the possible synthetic pathways from cyclohexene. Many of them involve at least two steps and are thus unlikely to ever lead to a competitive process.

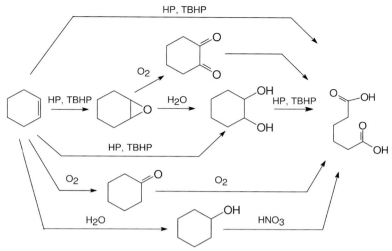

Scheme 7.12 Summary of possible synthetic pathways to AA from cyclohexene.

7.4.1.1 Single-Step Oxidation of Cyclohexene to AA

The catalytic system consisting of tungstate in which the W-peroxide species is the active site for oxidation is very efficient for the oxidative cleavage of alkenes in organic solvents with HP [34a–k]. For instance, Keggin P/W heteropolyanions are precursors of the true catalytic complexes $[PO_4\{W_2O_2(\mu\text{-}O_2)_2(O_2)_2\}_2]^{3-}$ and $[HPO_4\{W_2O_2(\mu\text{-}O_2)_2(O_2)_2\}]^{2-}$. Systems based on either Venturello's or Ishii's conditions, using phase-transfer catalysis in organic solvents, involve similar oxoperoxo anions [34b, e, g].

Noyori has described the efficient oxidation of cyclohexene with aqueous 30% hydrogen peroxide, in the presence of small amounts of Na_2WO_4 and $CH_3(n\text{-}C_8H_{17})_3HSO_4$ as a phase-transfer catalyst (PTC), in the absence of solvents (molar ratio olefin:W:PTC = 100:1:1), at 75–90 °C [34h, i]. Cyclohexene was converted directly into shiny, colorless, analytically pure crystalline AA in almost quantitative yield. This direct conversion used only a 4.4 molar amount of HP per mol of cyclohexene as the oxidant (the theoretical molar amount needed is 4.0). The aqueous phase of the reaction mixture could be reused with a renewed PTC and 30% HP. The reaction did not occur with H_2WO_4 in water, because cyclohexene is not soluble in water [34l] (indeed, in one patent, oxidation is carried out with 60% HP and H_2WO_4 [34m]). In fact the reaction medium is formed by two immiscible liquid phases: (i) an aqueous solution containing the HP and the tungstate and (ii) an organic cyclohexene phase in which the lipophilic trioctylmethylammonium hydrogen sulfate (the phase-transfer catalyst) is dissolved. The tungstate reacts with HP in the aqueous phase, generating the anionic peroxo species that is extracted in the organic phase by the quaternary ammonium cation. The reaction occurs in the organic phase between the peroxo species and cyclohexene, restoring the reduced form of the catalyst, which is again separated between the two phases.

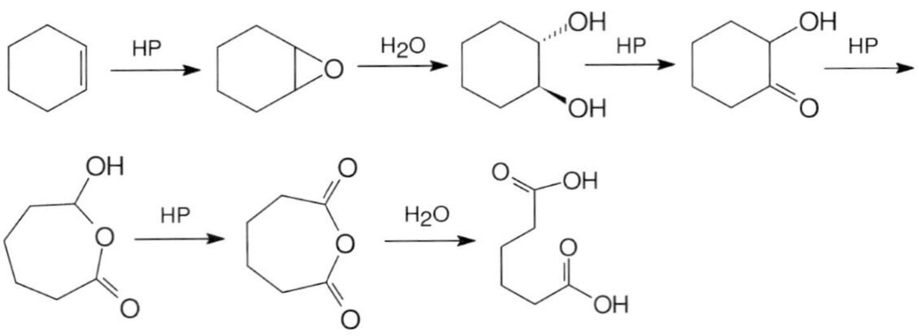

Scheme 7.13 Reaction network in the direct oxidative cleavage of cyclohexene to AA. Source: elaborated from [34h].

The reaction mechanism, illustrated in Scheme 7.13, does not include cyclohexandione as the reaction intermediate; moreover, the Baeyer–Villiger ring enlargement on cyclohexanone was negligible.

In a similar manner, cyclohexanol can be converted into AA with a 87% yield, using a 4.4 molar amount of 30% HP and H_2WO_4 at 90 °C, in 20 h [34n, o].

As pointed out [13l], these systems, although quite efficient, may present problems due both to the oxidative degradation of the organic ligands in ammonium salts and some catalyst instability, with the latter being caused by the fact that the O-transfer from the peroxo active species may lead to its structural collapse and the formation of less active species.

An improvement in reaction conditions was reported whereby a peroxytungstate–organic complex was used as a highly efficient catalyst for the same oxidation with no phase-transfer catalyst [34p]. It was reported that the presence of organic acids as ligands, instead of the quaternary ammonium salt, greatly improved performance. In fact, cyclohexene conversion and the selectivity to AA were 97.3 and 38.9%, respectively, when only peroxytungstate was present in the reaction media. However, the selectivity could be greatly enhanced when both peroxytungstate and organic acids were present in the reaction media [34p, q]. The best catalytic performance was achieved with oxalic acid: complete conversion of cyclohexene, and a 96.6% selectivity to AA in 24 h and a HP/cyclohexene molar ratio of 4.4:1. The main by-products were cyclohexanediol and pentanedioic acid. When the peroxytungstate was complexed with organic acids, the resulting catalyst system not only had the capability of carrying active oxygen species but it also became oleophilic; in fact, since the peroxytungstate was water-soluble but immiscible with cyclohexene it must dissolve into the oil phase before initiating oxidation of the alkene.

Various authors have reported on the use of amphiphilic quaternary ammonium tungstophosphate $Q_3[PO_4\{WO(O_2)_2\}_4]$ for epoxidation and oxidative cleavage of olefins to carboxylic acids with HP in emulsion or microemulsion systems [34r–u]. Surfactant-type peroxometalates (tungstates or molybdates) form emulsion droplets, because of the lipophilic cation (typically, a quaternary ammonium), assembling at the interface and finally also acting as a phase-transfer agent. For example, $[C_{16}H_{33}N(CH_3)_3]_2W_2O_3(O_2)_4$ gave a 77.8% yield to AA from cyclohexene, 81.3% from 1,2-cyclohexandiol, and 60.3% from cyclohexanone with 30% aqueous hydrogen peroxide, at 90 °C [34v].

Various heterogeneous catalysts have also been investigated for this reaction. Tetrahedral tungstate units can be incorporated at the surface of mesoporous silica, for example, SBA-15, to obtain heterogeneous systems that can effectively oxidize cyclohexene with HP [35a]. A 46% selectivity to colorless, crystalline AA was obtained at 85 °C, with total conversion of cyclohexene, under organic, solvent-free conditions. After 30 h reaction time, the TOF reached 350 $mol_{AA} \, mol_W^{-1} h^{-1}$, which is remarkably higher than that of other heterogeneous catalysts described in the literature, such as W-MCM-41. Interestingly, the latter system yields glutaraldehyde (72%) in the oxidation of cyclopentene with HP [35b].

Titanium framework-substituted aluminophosphate, TAPO-5, gave a 30.3% selectivity to AA at total cyclohexene conversion, in 72 h reaction time at 80 °C [35c]. However, when the filtered TAPO-5 catalyst was reused its activity was significantly diminished. The major by-product was 1,2-cyclohexanediol, which is formed as reaction intermediate. The trans isomer was more slowly transformed into 2-hydroxycyclohexanone than the cis isomer. A detailed investigation of the mechanism indeed showed that the trans-diol formed by ring opening of the cyclohexene epoxide, whereas the cis isomer formed via a free-radical mechanism. Scheme 7.14 shows the mechanism proposed.

The incorporation of Al into a mesostructured silica SBA-15, followed by Ti grafting, led to the formation of a bifunctional catalyst, which afforded an 80% yield to AA in 24 h, at 80 °C, using t-BuOOH as the oxidant [35d, e]. *trans*-1,2-Cyclohexandiol, 2-hydroxycyclohexanone, glutaric and succinic acids were the main by-products of the reaction.

Ti-MMM with a mesopore diameter of 3.4–3.8 nm showed fairly good activity in cyclohexene oxidation with HP in acetonitrile [35f, g]. Cyclohexene oxide and *trans*-cyclohexane-1,2-diol prevailed among the oxidation products. No titanium leaching

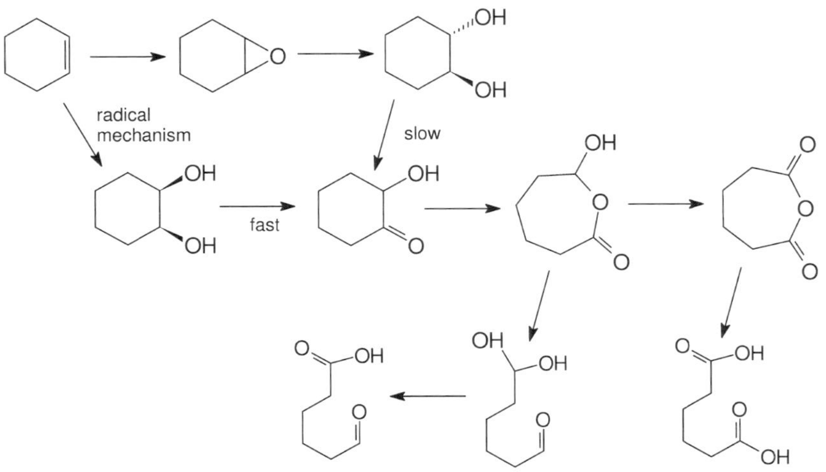

Scheme 7.14 Reaction network in the oxidative cleavage of cyclohexene catalyzed by TAPO-5. Source: elaborated from [35c].

occurred during the oxidation process. However, Ti-MMM, like most known mesoporous Ti, Si catalysts (Ti-MCM-41, Ti-MCM-48, TiO_2–SiO_2 mixed oxides, etc.) suffered from hydrolytic instability, leading to catalyst deactivation and making its recycling problematic [35h–j]. Ti-MMM-2 prepared under weak acidic conditions, in contrast to Ti-MMM prepared under weak alkaline conditions [35k, l], showed quite good stability and recyclability in HP-based oxidation of thioethers, alkenes and alkylphenols under mild reaction conditions. In a solvent-free system, the yield of AA reached 10–15% at a HP/cyclohexene molar ratio of 3.6 (80 °C, 72 h reaction time) [35m]. Stepwise addition of the oxidant to the reaction mixture caused an increase in the yield of the target product by up to 33%. However, the yield of AA decreased significantly in the second run because of both the partial decrease of the mesopore surface area and mesopore volume and the irreversible formation of oligomerized TiO_2-like species on the Ti-MMM-2 surface.

Cyclohexene may even act as a reagent for direct AA synthesis by means of nitric acid oxidation. In this case, a yield of 94.9% has been obtained, in the presence of Cu and V catalysts and an inert solvent (CCl_4) [35n].

The oxidation of cyclohexene with HP allows operation at milder reaction temperatures and reduced pressure. However, the main problem of this process is that the cost for the oxidant alone is already higher than the value of the AA produced. Moreover, cyclohexene is not really an easily available feedstock.

7.4.1.2 Two-Step Oxidation of Cyclohexene to AA Via 1,2-Cyclohexandiol

The syn dihydroxylation of olefins is usually performed in the presence of metal oxide, for example, $KMnO_4$ or OsO_4, and with *t*-BuOOH or HP as terminal oxidants. The anti dihydroxylation can be achieved with peracetic acid and *m*-$ClC_6H_4CO_3H$ in water. However, the atom efficiency of these oxidants is low, and they form equimolar amounts of deoxygenated compounds [36a, b].

Osmium-catalyzed dihydroxylation of olefins involves an Os(VIII)/Os(VI) substrate-selective redox system [36c–g]. In this system, N-methylmorpholine N-oxide (NMMO) can be used for the reoxidation of Os(VI) to Os(VIII), with NMMO being reduced to NMM. In cyclohexene oxidation catalyzed by Os with HP, the yield to *cis*-1,2-cyclohexandiol can be improved remarkably by the use of specific mediators for NMM oxidation to NMMO, for instance by means of catalytic flavin/HP. In this case, a yield to the cis-diol of 91% was obtained, as compared to 50% with the OsO_4/HP system alone [36h]. Mixtures of aqueous HP and acetic acid or formic acid are also effective reagents for the dihydroxylation of olefins, but neutralization of the acid solvent is necessary for the recovery of the product.

Several authors have reported the dihydroxylation of olefins with HP catalyzed by transition metal complexes, for example, H_2WO_4, heteropolyacids or CH_3ReO_3, but selectivity is often not very high, because of the formation of by-products involving C–C bond cleavage. Moreover, these systems require chlorohydrocarbons or other solvents [34i,36i–n]. The use of zeolites, or of other heterogeneous catalysts, allows dihydroxylation in the absence of organic solvents but, again, the selectivity is not higher than 60% because of the formation of epoxides, alcohols, ketones and/or ethers. $(NH_4)_{10}W_{12}O_{41}$/hydrotalcite [36o], Ti-β [36p], Ti-MCM

[36q–r] and Nb-MCM-41 [36s] are examples of catalytic systems that have been investigated.

With Ti-substituted Keggin polyoxometalates, for example, $Na_{5-n}H_nPTiW_{11}O_{40}$, the oxidation of cyclohexene with HP in an acetonitrile solvent yields *trans*-1,2-cyclohexandiol as the main reaction product, via a heterolytic oxygen-transfer mechanism, when $n > 2$ in the compound. If the polyoxometalate contains only one proton, the main products are those of allylic oxidation, namely 2-cyclohexene-1-ol and 2-cyclohexene-1-ol, produced via a homolytic oxidation mechanism [36t].

The use of resin-supported sulfonic acid, an easily recyclable catalyst, makes it possible to conduct the dihydroxylation of cyclohexene with 30% HP without any solvent, at 70 °C, with 98% yield to *trans*-1,2-cyclohexandiol [36u]. The mechanism includes the *in situ* generation of the resin-supported peroxysulfonic acid, which oxidizes the olefin to cyclohexene epoxide; the latter is then quickly hydrolyzed by water to yield the final product.

The second step of this process would be the cleavage of 1,2-cyclohexandiol 37; this reaction occurs with W-based catalysts and HP under homogeneous and heterogeneous conditions [37a, b]. For example, the diol (1:1 mixture of cis and trans) was cleaved to an 88% yield of AA with aqueous HP and with an heteropolyacid tris (cetylpyridinium)-12-tungstphosphate catalyst, at a reflux temperature of *t*-BuOH in 24 h reaction time, under two-phase conditions [37c].

Oxidative cleavage of 1,2-diols to carboxylic acids by HP was achieved using catalytic amounts of tungstate and phosphate ions, under acidic conditions [34c,37v]. The reaction was conducted at 90 °C and pH 2 in aqueous solution, with a slight excess of HP with respect to the stoichiometric amount required. A 94% yield of AA was obtained from *trans*-1,2-cyclohexandiol, and a slightly lower 92% yield from the cis isomer. The reaction proceeded via an initial C—H bond fission of the secondary carbinol to form the related α-ketol, followed by oxidative cleavage of the latter to yield a keto acid intermediate.

With Ti-containing Y zeolites, *trans*-1,2-cyclohexandiol was cleaved with HP to AA, obtained with an 80% selectivity at 50% conversion [37d]; the intermediate product was 2-hydroxycyclohexanone, which was oxidized to lighter diacids. In addition, 1,2-cyclopentanediol was cleaved to glutaric acid with HP and a homogeneous tungstic acid catalyst, with a 92% yield at 80 °C, whereas a 50% yield was obtained under the same conditions with heterogeneous W-MCM-41, which could easily be separated and reused [37e]. An even better yield of glutaric acid (94%) has been obtained under phase-transfer conditions with the $[\pi\text{-}C_5H_5NC_{16}H_{33}]_3\{PO_4[WO_3]_4\}$ catalyst, which dissolved during the reaction and precipitated after the reaction, thus combining the advantages of both homogeneous and heterogeneous catalysts; the catalyst could be easily recovered and reused. With the same system, cyclohexene was oxidized to AA with an 82% yield.

1,2-Cyclohexanediol undergoes aerobic oxidative cleavage (with oxygen) to form dicarboxylic acids when heated together with a ruthenium pyrochlore oxide catalyst $[A_{2+x}Ru_{2-x}O_{7-y}]$ (A = Pb, Bi; $0 < x < 1$; $0 < y < 0.5$), under high oxygen pressure in water at alkaline pH [37f–h]. The Cu(II)/Cu(I) system can also activate dioxygen and perform the oxidation of cycloalkanones to the corresponding oxo acids almost quantitatively [37i] in ethylammonium nitrate (an ionic liquid at room temperature). Less demanding procedures can be carried out with water–acetic acid solvent [37j].

Promising results have also been obtained with vanadium-based heteropolyacids $H_{3+n}[PMo_{12-n}V_nO_{40}]$ [13l,37k, l]. The latter systems or Cu(II) salts were shown to be efficient catalyst precursors for the C–C bond cleavage of α-alkylcycloalkanones [37m, n] and α-hydroxyketones [37o] by oxygen. For instance, 1,2-cyclohexandiol can be oxidized at 75 °C with a $H_5PMo_{10}V_2O_{40}$ catalyst, in an ethanol solvent, to obtain the ethyl ester of AA with a 90% selectivity at 62% conversion [13l,37p]. This class of reaction can also be catalyzed by monomeric V complex analogues of [VO{OCH $(CH_3)_2$}$_3$], such as vanadyl sulfate or VO(acac)$_2$. However, the P/Mo/V heteropolyacids afford better yields. In addition, the redox and acid properties of the catalyst can be better tuned. Mechanistic investigations of these homogeneous systems are consistent with homolytic reactions involving V(IV) and V(V) species [13q]. The reaction also includes O^{2-} insertion from the V oxo complex to the activated reactant; in the redox-type mechanism, molecular oxygen reoxidizes the reduced V complex.

Polyoxometalates are also very effective in α-ketol oxidative cleavage; 2-hydroxycyclohexanone was oxidatively cleaved at 60 °C (and even at room temperature) with a $H_5[PMo_{12-n}V_nO_{40}]$ catalyst in the presence of oxygen, to form AA with a 100% selectivity at 90–97% reactant conversion [37r–t]. A $[VO]^{2+}$/Nafion catalyst has afforded the oxidative cleavage of α-hydroxycyclohexanone with oxygen, without any coreductant, in a methanol solvent, at 65 °C with 1 atm oxygen pressure [37u]. Total conversion of the substrate was achieved in 24 h, with 74% selectivity to dimethyl adipate (after esterification of the reaction mixture with diazomethane). By-products were δ- and γ-valerolactone, methyl 6,6-dimethoxyhexanoate (the compound obtained from the reaction of 6-oxohexanoic acid with diazomethane); some by-products arising from proton-mediated reactions were converted into AA derivatives. Indeed, even Nafion alone, and also phosphoric acid, catalyzed the oxidative cleavage with oxygen to yield dimethyl adipate and 6-oxomethyladipate with an 83% conversion and overall yield of 51% [37u].

7.4.1.3 Three-Step Oxidation of Cyclohexene to AA Via Epoxide

A vast literature exists on the epoxidation of cyclohexene to cyclohexene oxide with either homogeneous or heterogeneous catalysts, using either HP or alkyl hydroperoxides as terminal oxidants 38. Numerous microporous and mesoporous materials, containing transition metal ions isolated in the framework of inorganic matrices or grafted (tethered) onto the surface, have been investigated as catalysts for this reaction [38a–j]. Ti-containing molecular sieves are among the best systems for selective oxidation with hydroperoxides [38k–p], including TS-1 and mesoporous Ti-containing materials (Ti-HMS, Ti-MCM-41, Ti-MCM-48, Ti-MMM-2, etc.) capable of oxidizing large organic substrates. A second class of compounds active and selective in cyclohexene epoxidation are transition metal-substituted polyoxometalates (TM-POMs), because they are thermodynamically stable to oxidation and, furthermore, are hydrolytically stable under appropriate pH conditions [13m,38p–y].

Normally, in cyclohexene epoxidation with HP, Ti-silicates produce products that are typical of both two-electron oxidation mechanisms (cyclohexene epoxide and *trans*-cyclohexane-1,2-diol) and one-electron oxidation mechanisms (2-cyclohexene-1-ol and 2-cyclohexene-1-one) [38p,39a–m]. Indeed, many POMs may be solvolytically unstable to HP and, in fact, they often act as precursors of the true catalyst, the complex $[PO_4\{M(O)(O_2)_2\}_4]_3$, (M = Mo, W), and/or other low nuclearity species [38y,39n–o]. The Venturello complex is well known as a very efficient epoxidizing agent [34b,38x,39p]. For example, $TBA_3\{PO_4[WO(O_2)_2]_4\}$ almost completely converts cyclohexene at 70 °C in 5 h reaction time, with aqueous HP in an acetonitrile solvent, giving an 81% yield of the epoxide, 10% of the diol and less than 5% cyclohexenol and cyclohexenone [38p].

The epoxide can be hydrolyzed to 1,2-cyclohexandiol (Section 7.4.1.2); however, an alternative is its oxidative transformation. Oxidation of epoxides without C—C bond cleavage yields α-functionalized ketones; for instance, the acid-catalyzed oxidation of epoxides by DMSO leads to α-hydroxyketone formation, though with limited synthetic scope, because of the low yield and selectivity [39q]. However, a Bi(0)/O$_2$/DMSO system can efficiently oxidize cyclohexene oxide to the α-diketone, in the presence of an acid additive, for example, triflate anions. In 2-h reaction time, a 74% yield of the diketone can be achieved, at 100 °C and 1 atm oxygen, with the by-product being 1,2-cyclohexanediol [39r–t]. The mechanism involves an initial acid-catalyzed ring opening of the epoxide by DMSO, followed by a redox process involving the *in situ* generated α-ketol intermediate and a Bi(III) species. The reduced state of Bi is then re-oxidized by oxygen.

The third step of this synthetic pathway would include the oxidative ring opening of the diketone to yield AA.

7.4.1.4 An Alternative Oxidant for Cyclohexene: Oxygen
The main product of cyclohexene oxidation with oxygen in the absence of a reduction reagent is 2-cyclohexen-1-one [40a–c]. This may finally be hydrogenated to cyclohexanone, but the route is clearly disadvantageous compared to the oxidation of cyclohexane to cyclohexanone. The reaction follows the classical autoxidation mechanism, with cyclohexene hydroperoxide being the first reaction product. When catalyzed by polyoxoanion-supported organometallic [Ir(I), Ru(II), Re(I)] complexes, in the absence of co-reductants, the reaction predominantly yields 2-cyclohexen-1-yl hydroperoxide, 2-cyclohexen-1-one, 2-cyclohexene-1-ol and cyclohexene oxide, at a low conversion (<25%), but with the additional formation of several by-products (around 70) when the conversion of cyclohexene is increased [40d, e]. It has also been suggested that an optimized, non-metal catalyzed but radical-chain initiated, RCHO/olefin/O$_2$ co-oxidative epoxidation process may be even more selective than a metal-catalyzed process. If the reaction is carried out in the presence of a reducing agent, for example, isobutyraldehyde, the reaction becomes very selective to the epoxide (88–94% selectivity at total or almost total conversion of the substrate) [40f]. An autoxidation mechanism is also invoked for the aerial oxidation of cyclooctene by Fe-substituted polyoxotungstates, in which the epoxide is the main reaction product [40g].

Catalysts based on $[(n\text{-}C_4H_9)_4N]_5Na_3[(1,5\text{-}COD)Ir\cdot P_2W_{15}Nb_3O_{62}]$, $[(n\text{-}C_4H_9)_4N]_5$ $Na_3[(1,5\text{-}COD)Rh\cdot P_2W_{15}Nb_3O_{62}]$ and $[(n\text{-}C_4H_9)_4N]_{4.5}Na_{2.5}[(C_6H_6)Ru\cdot P_2W_{15}Nb_3O_{62}]$

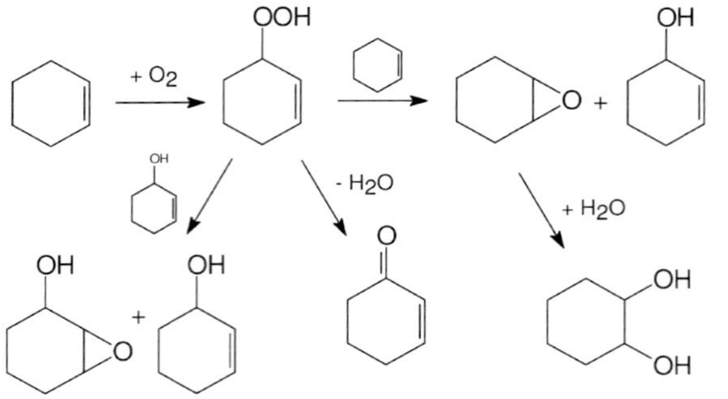

Scheme 7.15 Oxidation of cyclohexene with O_2. Source: elaborated from [40j].

have been shown to catalyze the oxygenation of cyclohexene with molecular oxygen. The polyoxoanion-supported Ir(I) complex showed the highest activity with a TOF of $2.9\,h^{-1}$ at $38\,°C$ in CH_2Cl_2 [40h]. Also, transition metal substituted polyoxomolybdates $(Bu_4N)_4PW_{11}Co(H_2O)O_{39}$ and $(Bu_4N)_4PMo_{11}Ru(H_2O)O_{39}$ were shown to catalyze the epoxidation of alkenes via oxygen transfer from intermediate alkyl hydroperoxides; the latter being formed by catalytic autoxidation with molecular oxygen [40i].

Asahi has proposed the oxidation of cyclohexene with O_2 as a commercially feasible method for AA production [40j]. The reaction, when catalyzed by isopolyoxomolybdates, primarily yields cyclohexene oxide, 1,2-cyclohexandiol and 2-cyclohexene-1-ol. The two former compounds are intermediates in the synthesis of AA. The cyclohexyl hydroperoxide acts as the epoxidizing agent for cyclohexene, finally leading to the generation of the products (Scheme 7.15).

A 90% selectivity to the three product compounds was obtained at 37% cyclohexene conversion, using $(Bu_4N)_2Mo_6O_{19}$ isopolyoxometalate as the catalyst, in a mixed 1,2-dichloroethane–acetonitrile solvent, at $50\,°C$, 1 atm oxygen, in 24 h reaction time. The cyclohexene conversion and the overall selectivity were 30.5% and 63.6%, respectively, over the isopolyoxovanadate $(Bu_4N)_6V_{10}O_{28}$. Conversely, the heteropolyoxometalates $(Bu_4N)_3PMo_{12}O_{40}$ and $(Bu_4N)_3PW_{12}O_{40}$ showed low activities. $(Bu_4N)_4PW_{11}Co(H_2O)O_{39}$ gave the highest conversion of 58%, but the selectivity was only 31%.

The chemistry of the catalytic oxidation of cyclohexene with oxygen may be different from radical-type autoxidation when the Matveev system, made of Pd(II) and P/Mo/V heteropoly-compound [13l,41a–e], is used. In this case, a Wacker-type oxidation of the olefin with oxygen yields cyclohexanone as the primary product. Cyclohexanone can then be oxidized with O_2 to yield AA (Section 7.3.1). Various catalysts belonging to the polyoxometalates class have been investigated: (i) the $PdSO_4/H_3PMo_6W_6O_{40}$ system gave a cyclohexene conversion of 85% after 24 h [41c]; (ii) $Pd(OAc)_2/hydroquinone/Na_nPmo_xV_{12-x}O_{40}$ gave cyclohexanone with a conversion of 58% after 20 h [41d]; (iii) the best results for the chloride ion-free Wacker-type oxidation of cyclohexene have been reported by Kim et al. [41f], who used $Pd(NO_3)_2/CuSO_4/H_3PMo_{12}O_{40}$ as the catalyst in an aqueous solution of acetonitrile at $80\,°C$

and 10 atm of oxygen. They obtained a conversion of 49%, with 97% selectivity for cyclohexanone. More recently, using the same system, after 1 h of reaction at 80 °C and an air pressure of 50 atm, a cyclohexene conversion of 80% was reached, with a selectivity of >99% for cyclohexanone [41b]. Using aqueous HP, the oxidation was more rapid, already giving an 80% conversion after 30 min and a 95% conversion after 60 min without the formation of any by-products.

7.4.2
The Greenest Way Ever: Two-Step Transformation of Glucose to AA

The biocatalytic microbe-based conversion of D-glucose into *cis,cis*-muconic acid and the subsequent hydrogenation of the latter into AA has been proposed by Draths and Frost [42a, b]. This synthesis is emblematic of an environmentally benign process, making use of a renewable raw material for the synthesis of a commodity chemical by means of an intrinsically safe process.

Intermediates in the biosynthetic route are 3-dehydroshikimic acid, protocatechuic acid and catechol (Scheme 7.16). Optimization of microbial *cis,cis*-muconic acid synthesis required the expression of three enzymes not typically found in *Escherichia coli* [42c]. *E. coli* WN1/pWN2.248 was developed that synthesized 36.8 g L^{-1} of *cis,cis*-muconic acid in a 22% (mol/mol) yield from glucose after 48 h of culturing under fed-batch fermentor conditions. Optimization of the carbon flow directed into *cis,cis*-muconic acid biosynthesis and manipulation of enzyme activities were aimed at avoiding accumulation of biosynthetic intermediates.

Muconic acid can then be hydrogenated to AA using supported Rh or Pt catalysts. For instance, hydrogenation of the solution with 10% Pt on carbon at 34 atm of H$_2$ pressure for 2.5 h at room temperature gave a 97% conversion of *cis,cis*-muconic acid into AA [42c].

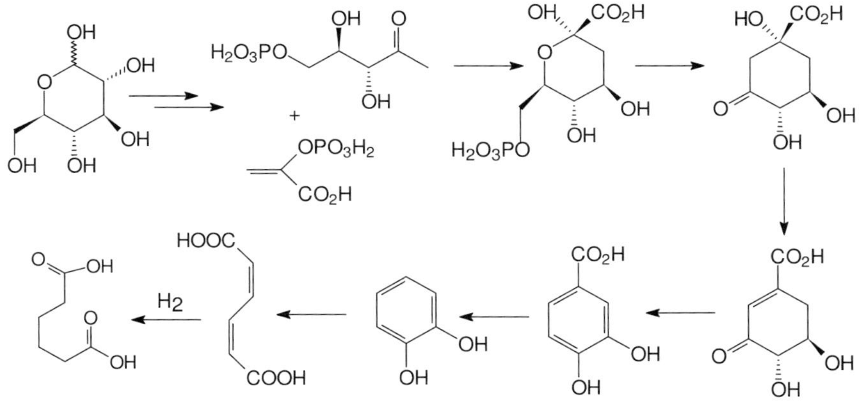

Scheme 7.16 Enzymatic transformation of D-glucose to *cis,cis*-muconic acid and subsequent hydrogenation to AA. Source: elaborated from [42c].

trans,trans-Muconic acid has been hydrogenated to AA using a $Ru_{10}Pt_2$ nanoparticle catalyst supported on mesoporous silica, at 80 °C, 30 bar H_2 pressure, with a 91% conversion and 96% selectivity [42d].

Problems that still need to be solved include scaling-up and process implementation; although this approach can be considered the greenest way ever to be investigated for AA synthesis, these problems may make it economically unsustainable. As pointed out by Thomas [42d], a manufacturing process of this feedstock would be somewhat similar to the microbial synthesis of L-lysine from D-glucose under fermentation conditions; however, the cost of AA, starting from this feedstock, would be considerably higher than that of resin-grade AA currently sold.

7.4.3
The Ultimate Challenge: Direct Oxidation of *n*-Hexane to AA

The oxyfunctionalization of linear alkanes at the terminal position is one of the major challenges of catalysis; in the case of *n*-hexane, oxidation at the two terminal C atoms would lead to AA. However, there is a little amount of *n*-hexane in cracker streams, because it is easily converted to benzene. Therefore, sourcing of this hydrocarbon would be a challenge.

Enzymes with non-heme iron active centers catalyze alkane oxidation with oxygen with a high terminal selectivity, but any attempt to replicate their properties in nonbiological systems has led to catalysts with modest performances. For instance, linear alkanes give around a 20% selectivity to terminal oxidized products with Mn (III) porphyrins [43a] and, in the case of *n*-hexane, oxidation with iodosylbenzene gave a selectivity of 19% to 1-hexanol, the oxidation at the second C atom being, however, the prevailing one. Indeed, the k_{prim}/k_{sec} ratio is greatly affected by the spatial constraints around the active site, which may favor the approach of the less hindered primary C atom.

A constrained environment can also be achieved by using crystalline aluminosilicate or aluminophosphates, which have pores of a specific size [43b]. Redox-type cations can be incorporated in the structure, or accommodated on the porous surface by various techniques, and metalloporphyrin or phthalocyanines complexes can be anchored by exchange, thus developing heterogeneous, recoverable catalysts with restricted access to the active species [24f,43c–g]. However, modest or nil terminal selectivities were also reported for alkane oxidation with HP on zeolites modified by redox-active cations [43h–j]. In the absence of any constrained environment for access to the active site, for example, with silica supported oxovanadium complexes, the oxidation of *n*-hexane with O_2 gave 2-hexanone as the main reaction product, with minor formation of 1-hexanol, at 160 °C and 10 atm oxygen [43k].

With microporous Co(III)- and Mn(III)-aluminophosphates, MnAPO-18 and CoAPO-18, unprecedented terminal selectivities have been reported in *n*-hexane oxidation with oxygen on structures with small eight-ring windows (0.38 nm) [43l–p]. This created a ready access to the Co site by the terminal group of the linear alkane, with the alkane gaining entry into the interior of the porous catalyst with an end-on approach, thus limiting the oxyfunctionalization at the other C atoms. With both

Co^{III}AlPO-18 and CoAlPO-34, 1-hexanol is the major product during the initial stages of the reaction, but this is converted into 1-hexanal and hexanoic acid; 1,6-hexandiol, 1,6-hexanedial and AA appear later. However, the formation of the 1,6-oxidized products was not kinetically related to that of the compounds oxidized at one C atom only. Therefore, attack at the two terminal C atoms occurred concurrently, because the amount and dispersion of Co atoms accommodated on the inner walls of the AlPO framework were such that two Co(III) ions were separated by about 7–8 Å. Clearly, this is affected by the Co/P ratio, which is the main parameter influencing the Co dispersion and hence the distribution of products. With CoAPO-18, a selectivity of 33.6% to AA was obtained at a 9.5% *n*-hexane conversion, at 100 °C after 24 h, with oxygen as the oxidant, the major by-products being hexanoic acid and 2-hexanone. The overall terminal selectivity achieved was 65%.

Despite the relevance of these findings and the implications that they may have [43q, r] these excellent figures were not confirmed using catalysts with an identical composition and structure, namely, MnAPO-5 and MnAPO-18 [43s, t]. It was reported that *n*-hexane oxidation turnover rates (per redox-active Mn center) by oxygen were similar on MnAPO-5 and MnAPO-18, because the reactant may rapidly diffuse and reach the active site, regardless of the pore size in the microporous structure. No regiospecificity was detected for *n*-hexane oxidation to alkanols, aldehydes and ketones (7–8% terminal selectivity), and the relative reactivity of primary and secondary C–H bonds in *n*-hexane was identical in both catalysts and similar to that predicted from relative C–H bond energies in *n*-hexane. The selectivity to terminal acids was very low.

When Mn cations were placed at exchange sites within channels in eight-membered (ZSM-58), ten-membered (ZSM-5 and ZSM-57) and 12-membered ring (MOR) zeolite channels by sublimation of MnI_2, the synthesis rates for hexanols, hexanal/hexanones and acids in *n*-hexane oxidation with oxygen were proportional to the hexyl hydroperoxide concentration on all Mn-zeolite catalysts, with the exception of Mn-ZSM-58, on which products formed via noncatalytic autoxidation because of restricted access to Mn cations present within small channels [43u]. Catalytic decomposition of the hydroperoxide occurred on intrachannel Mn cations. Regioselectivity was influenced by the constrained environment around Mn cations, which increased terminal selectivities above the values predicted from the relative bond energies of methyl and methylene C–H bonds in *n*-hexane. Manganese cations within ten-ring channels gave higher terminal selectivities, for example, 24% on MnZSM-5. However, terminal selectivities decreased with increasing alkane conversion, because non-selective noncatalytic autoxidation prevailed as the hydroperoxide concentration increased concurrently.

7.5
An Overview: Several Possible Green Routes to AA, Some Sustainable, Others Not

Scheme 7.17 summarizes the possible routes for the synthesis of AA. Some of them are "green" but only a few of them can be considered truly sustainable, from

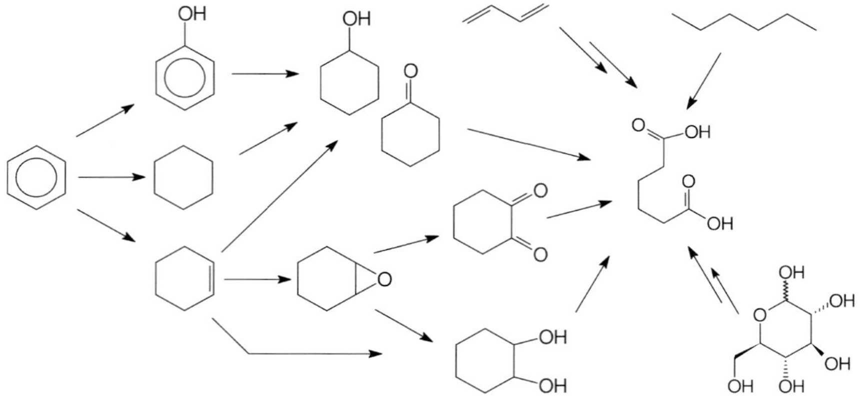

Scheme 7.17 Summary of the various alternative pathways for AA synthesis.

both an environmental and economic point of view, if the cost of reactants and the complexity of the operation are taken into consideration. For instance, the direct oxidative cleavage of cyclohexene into AA with HP is a type of synthesis that abides by the rules of green chemistry, but it is a prohibitive route because of the cost of HP (stoichiometry 4 mole per mole of cyclohexene). The same is true for the combined biocatalytic–catalytic synthesis starting from D-glucose with the intermediate formation of muconic acid.

To date, it seems that the most sustainable approach is the one that combines the use of cheap raw materials, for example, cyclohexane, benzene or phenol, with oxygen as the terminal oxidant. Within this context, a process that does not use acetic acid in the aerial oxidation of the KA Oil into AA or, even better, in the direct oxidation of cyclohexane to AA would represent a significant step forward towards a new and sustainable synthesis. On the other hand, recent examples demonstrate that even the traditional process making use of nitric acid for the oxidation of KA Oil may be turned into an intrinsically green one that is economically sustainable due to the use of the co-produced N_2O in down-stream applications.

Research in this field is still very active, and the various alternative options and strategies currently under investigation increase the likelihood that in the near future a new and sustainable synthetic route will finally be implemented at the commercial level.

References

1 (a) http://www.chemplan.com/ chemplan_demo/sample_reports/ Adipic_acid.pdf; (b) http://www.icis.com/ Articles/2007/04/03/4501573/Chemical-Profile-Adipic-acid.html; (c) http://www.icis.com/Articles/2007/ 07/06/4503569/Adipic-acid.html;

(d) http://www.entrepreneur.com/ tradejournals/article/172133861.html; (e) http://www.the-innovation-group. com/ChemProfiles/Adipic%20Acid.htm.
2 (a) Castellan, A., Bart, J.C.J. and Cavallaro, S., (1991) *Catal. Today*, **9**, 237; (b) Castellan, A., Bart, J.C.J. and Cavallaro, S. (1991)

Catal. Today, **9**, 255; (c) Schuchardt, U., Cardoso, D., Sercheli, R., Pereira, R., da Cruz, R.S., Guerreiro, M.C., Mandelli, D., Spinace, E.V. and Pires, E.L. (2001) *Appl. Catal. A*, **211**, 1; (d) Thomas, J.M. and Raja, R. (2005) *Annu. Rev. Mater. Res.*, **35**, 315; (e) van Asselt, W.J. and van Krevelen, D.W. (1963) *Recl. Trav. Chim.*, **82**, 51 & 429 & 448; (f) van Asselt, W.J. and van Krevelen, D.W. (1963) *Chem. Eng. Sci.*, **18**, 471; (g) Sheldon, R.A. and Kochi, J.K. (1976) *Adv. Catal.*, **25**, 272; (h) Hermans, I., Nguyen, T.L., Jacobs, P.A. and Peeters, J. (2005) *Chem. Phys. Chem.*, **6**, 637; (i) Hermans, I., Jacobs, P.A. and Peeters, J. (2006) *J. Mol. Catal. A Chem.*, **251**, 221; (j) Hermans, I., Jacobs, P.A. and Peeters, J. (2006) *Chem. Eur. J.*, **12**, 4229; (k) Hermans, I., Jacobs, P.A. and Peeters, J. (2007) *Chem. Eur. J.*, **13**, 754; (l) Hermans, I., Peeters, J. and Jacobs, P. (2007) *J. Org. Chem.*, **72**, 3057; (m) Hermans, I., Van Deun, J., Houthoofd, K., Peeters, J. and Jacobs, P. (2007) *J. Catal.*, **251**, 204; (n) Cavani, F. and Teles, J.H. (2009) *ChemSusChem*, **2**, 508.

3 (a) Belkhir, I., Germain, A., Fajula, F. and Fache, E., (1998) *J. Chem. Soc., Faraday Trans.*, **94**, 1761; (b) Tanaka, K. (1974) *Hydrocarb. Process.*, **53**, 114; (c) Suresh, A.K., Sharma, M.M. and Sridhar, T. (2000) *Ind. Eng. Chem. Res.*, **39**, 3958; (d) Bréhéret, A., Lambeaux, C., Ménage, S., Fontecave, M., Dallemer, F., Fache, E., Pierre, J.L.P., Chautemps, P. and Averbursch Pouchot, M.T. (2001) *C.R. Acad. Sci. Ser Iic: Chim.*, **4**, 27; (e) Greene, M.I., Sumner, C. and Gartside, R. (1998) US Patent 5,780,683 (assigned to ABB Lummus Global); (f) Greene, M.I., Sumner, C. and Gartside, R. (1999) US Patent 6,008,415 (assigned to ABB Lummus Global); (g) Mills, P.L. and Chaudhari, R.V. (1999) *Catal. Today*, **48**, 17; (h) Chen, J.R., Yang, H.H. and Wu, C.H. (2004) *Org. Proc. Res. Dev.*, **8**, 252.

4 (a) Wismeijer, A.A., Kieboom, A.P.G. and Van Bekkum, H., (1986) *Recl. Trav. Chim. Pays-Bas*, **105**, 129; (b) Krishna, R. and Sie, S.T. (1994) *Chem. Eng. Sci.*, **49**, 4029;

(c) Bellussi, G. and Perego, C. (2000) *CATTECH*, **4**, 4.

5 (a) Nagahara, H. and Konishi, M., (1987) JP Patent 62-45541; (b) Nagahara, H. and Fukuoka, Y. (1986) JP Patent 61-50930; (c) Mitsui, O. and Fukuoka, Y. (1984) JP Patent 59-184138 and 59-186929; (d) Nagahara, H., Ono, M., Konishi, M. and Fukuoka, Y. (1997) *Appl. Surf. Sci.*, **121**, 448; (e) Nagahara, H. and Konishi, M. (1988) US Patent 4,734,536 (assigned to Asahi Kasei Kogyo)

6 (a) Shul'pin, G.B. (2002) *J. Mol. Catal. A*, **189**, 39; (b) Mizuno, N., Nozaki, C., Hirose, T., Tateishi, M. and Iwamoto, M. (1997) *J. Mol. Catal. A*, **117**, 159; (c) Hayashi, T., Kishida, A. and Mizuno, N. (2000) *Chem. Commun.*, 381; (d) Nozaki, C., Misono, M. and Mizuno, N. (1998) *Chem. Lett.*, 1263; (e) Cramarossa, M.R., Forti, L., Fedotov, M.A., Detusheva, L.G., Likholobov, V.A., Kuznetsova, L.I., Semin, G.L., Cavani, F. and Trifirò, F. (1997) *J. Mol. Catal. A*, **127**, 85; (f) Bonchio, M., Carraro, M., Scorrano, G. and Kortz, U. (2005) *Adv. Synth. Catal.*, **347**, 1909; (g) Kuznetsova, N.I., Kuznetsova, L.I., Kirillova, N.V., Detusheva, L.G., Likholobov, V.A., Khramov, M.I. and Ansel, J.E. (2005) *Kinet. Catal.*, **46**, 204; (h) Shul'pin, G.B., Gradinaru, J. and Kozlov, Y.N. (2003) *Org. Biomol. Chem.*, **1**, 3611; (i) Ohta, T., Tachiyama, T., Yoshizawa, K., Yamabe, T., Uchida, T. and Kitagawa, T. (2000) *Inorg. Chem.*, **39**, 4358; (j) Velusamy, S. and Punniyamurthy, T. (2003) *Tetrahedron Lett.*, **44**, 8955; (k) Schuchardt, U., Pereira, R. and Rufo, M. (1998) *J. Mol. Catal. A*, **135**, 257; (l) Murahashi, S.I., Komiya, N., Hayashi, Y. and Kumano, T. (2001) *Pure Appl. Chem.*, **73**, 311; (m) Okuno, T., Ohba, S. and Nishida, Y. (1997) *Polyhedron*, **16**, 3765; (n) Komiya, N., Naota, T., Oda, Y. and Murahashi, S.I. (1997) *J. Mol. Catal. A*, **117**, 21; (o) Barton, D.H.R., Beviere, S.D., Chavasiri, W., Csuhai, E. and Doller, D. (1992) *Tetrahedron*, **48**, 2895; (p) Kirillov, A.M., Kopylovich, M.N., Kirillova, M.V., Karabach, E.Yu., Haukka, M., Guedes da

Silva, M.F.C. and Pombeiro, A.J.L. (2006) *Adv. Synth. Catal.*, **348**, 159; (q) Kirillov, A.M., Kopylovich, M.N., Kirillova, M.V., Haukka, M., da Silva, M.F.C.G. and Pombeiro, A.J.L. (2005) *Angew. Chem. Int. Ed.*, **44**, 4345; (r) Di Nicola, C., Karabach, Y.Yu., Kirillov, A.M., Monari, M., Pandolfo, L., Pettinari, C. and Pombeiro, A.J.L. (2007) *Inorg. Chem.*, **46**, 221; (s) Komiya, N., Naota, T., Oda, Y. and Murahashi, S.I. (1997) *J. Mol. Catal. A: Chem.*, **117**, 21; (t) Silva, A.C., Lopez Fernandez, T., Carvalho, N.M.F., Herbst, M.H., Bordinhao, J., Horn, A., Jr, Wardell, J.L., Oestreicher, E.G. and Antunes, O.A.C. (2007) *Appl. Catal. A*, **317**, 154; (u) Carvalho, N.M.F., Horn, A. Jr, and Antunes, O.A.C. (2006) *Appl. Catal. A*, **305**, 140; (v) Esmelindro, M.C., Oestreicher, E.G., Marquez-Alvarez, H., Dariva, C., Egues, S.M.S., Fernandes, C., Bortoluzzi, A.J., Drago, V. and Antunes, O.A.C. (2005) *J. Inorg. Biochem.*, **99**, 2054.

7 (a) Trettenhahn, G., Nagl, M., Neuwirth, N., Arion, V.B., Jary, W., Pochlauer, P. and Schmid, W. (2006) *Angew. Chem. Int. Ed.*, **45**, 2794; (b) Nagataki, T., Tachi, Y. and Itoh, S. (2005) *J. Mol. Catal. A*, **225**, 103; (c) Roelfes, G., Lubben, M., Hage, R., Que, L. and Feringa, B.L. (2000) *Chem. Eur. J.*, **6**, 2152; (d) Vincent, J.M., Bearnais-Barbry, S., Pierre, C. and Verlhac, J.B. (1999) *J. Chem. Soc., Dalton Trans.*, 1913; (e) MacFaul, P.A., Arends, I.W.C.E., Ingold, K.U. and Wayner, D.D.M. (1997) *J. Chem. Soc., Perkin Trans. 2*, 135; (f) Barton, D.H.R., Li, T. and MacKinnon, J. (1997) *Chem. Commun.*, 557; (g) Shul'pin, G.B., Nizova, G.V., Kozlov, Y.N., Cuervo, L.G. and Süss-Fink, G. (2004) *Adv. Synth. Catal.*, **346**, 317; (h) Nesterov, D.S., Kokozay, V.N., Dyakonenko, V.V., Shishkin, O.V., Jezierska, J., Ozarowski, A., Kirillov, A.M., Kopylovich, M.N. and Pombeiro, A.J.L. (2006) *Chem. Commun.*, 4605; (i) Shul'pin, G.B. and Lachter, E.R. (2003) *J. Mol. Catal. A*, **197**, 65; (j) Cuervo, L.G., Kozlov, Y.N., Süss-Fink, G. and Shul'pin, G.B. (2004) *J. Mol. Catal. A*, **218**, 171; (k) Kozlov, Y.N.,

Nizova, G.V. and Shul'pin, G.B. (2005) *J. Mol. Catal. A*, **227**, 247; (l) Kozlov, Y.N., Romakh, V.B., Kitaygorodskiy, A., Buglyo, P., Süss-Fink, G. and Shul'pin, G.B. (2007) *J. Phys. Chem. A*, **111**, 7736; (m) Khaliullin, R.Z., Bell, A.T. and Head-Gordon, M. (2005) *J. Phys. Chem. B*, **109**, 17984; (n) Suzuki, Y., Arada, E., Nakamaru, K., Takeda, Y., Sano, M., Hashimoto, K. and Miyake, T. (2007) *J. Mol. Catal. A*, **276**, 1.

8 (a) Thiemens, M.H. and Trogler, W.C. (1991) *Science*, **251**, 932; (b) Weiss, R.F. (1981) *J. Geophys. Res.*, **86**, 7185; (c) (2002) Greenhouse Gases and Global Warming Potential Values, Excerpt from Inventory of U.S. Greenhouse Emissions and Sinks: 1990–2000, April.

9 (a) Kapteijn, F., Rodriguez-Mirasol, J. and Moulijn, J.A. (1996) *Appl. Catal. B*, **9**, 25; (b) Fetzer, T., Buechele, W., Wistuba, H., Witte, C., Buerger, G. and Herrmann, G. (1997) US Patent 5,612,009 (assigned to BASF); (c) Drago, R. and Jurkzyc, K. (1998) US Patent 5,705,136 (assigned to Univ. California); (d) Rajadurai, S. (1996) US Patent 5,562,888, (assigned to Cummins Engine Co); (e) Baier, M., Fetzer, T., Hofstadt, O., Hesse, M., Buerger, G., Harth, K., Schumacher, V., Wistuba, H. and Otto, B. (2004) US Patent 6,723,295 (assigned to BASF); (f) Vernooy, P.D. (2000) WO Patent Appl 00/51715 (assigned to DuPont); (g) Cremona, A., Rubini, C. and Vogna, E. (2004) US Patent 6,683,021 (Assigned to Sud Chemie MT); (h) Byrne, J.W. (1994) WO Patent Appl 94/27709 (Assigned to Engelhard); (i) Li, Y. and Armor, J. (1992) US Patent 5,171,553 (Assigned to Air Prod & Chem); (j) Farris, T., Li, Y., Armor, J. and Braymer, T. (1995) US Patent 5,472,677 (Assigned to Engelhard Co); (k) Shimizu, A., Tanaka, K. and Fujimori, M. (2000) *Chemosphere – Global Change Sci.*, **2**, 425; (l) Fu, C.M., Korchak, V.N. and Hall, V.K. (1981) *J. Catal.*, **68**, 166; (m) Winter, E.R.S. (1974) *J. Catal.*, **34**, 431; (n) Alini, S., Bologna, A., Basile, F., Montanari, T. and Vaccari, A. (2001) Eur Patent 1,262,224 (Assigned to

Radici Chimica SpA); (o) Alini, S., Rinaldi, C., Basile, F. and Vaccari, A. (2003) Eur Patent 1,504,805 (Assigned to Radici Chimica SpA); (p) Centi, G., Perathoner, S., Vazzana, F., Marella, M., Tomaselli, M. and Mantegazza, M. (2000) *Adv. Environ. Res.*, **4**, 325.

10 (a) Alini, S., Frigo, E. and Rinaldi, C. (2002) Eur Patent 1,413,349 (Assigned to Radici Chimica SpA); (b) Alini, S., Frigo, E. and Rinaldi, C. (2003) Eur Patent 1,488,845 (Assigned to Radici Chimica SpA).

11 (a) Kharitonov, A.S., Panov, G.I., Ione, K.G., Romannikov, V.N., Sheveleva, G.A., Vostrikova, L.A. and Sobolev, V.I. (1992) US Patent 5,110,995 (assigned to BIC); (b) Uriarte, A.K., Rodkin, M.A., Gross, M.J., Kharitonov, A.S. and Panov, G.I. (1997) *Stud. Surf. Sci. Catal.*, **110**, 857; (c) Panov, G.I., Sheveleva, G.A., Kharitonov, A.S., Romannikov, V.N. and Vostrikova, L.A. (1992) *Appl. Catal.*, **82**, 31; (d) Kharitonov, A.S., Sheveleva, G.A., Panov, G.I., Sobolev, V.I., Paukshtis, Ye.A. and Romannikov, V.N. (1993) *Appl. Catal. A*, **98**, 33; (e) McGhee, W.D. (1998) US Patent 5,808,167 (assigned to Solutia); (f) Hafele, M., Reitzmann, A., Klemm, E. and Emig, G. (1997) *Stud. Surf. Sci. Catal.*, **110**, 847; (g) Mokrinskii, V.V., Slavinskaya, E.M., Noskov, A.S. and Zolotarskii, I.A. (1998) WO Patent Appl. 9825698 (assigned to Solutia); (h) Parmon, V.N., Panov, G.I., Uriarte, A. and Noskov, A.S. (2005) *Catal. Today*, **100**, 115.

12 (a) (2007) *Chem. Eng.*, **796**, 56; (b) http://berichte.basf.de/basfir/copsfiles/en/2006/20f_bericht/13944_BASF_Form_20-F_2006.pdf?suffix=.pdf; (c) Teles, J., Rößler, B., Pinkos, R., Genger, T. and Preiss, T. (2005) WO Patent Appl 030689 (assigned to BASF); (d) Teles, J., Rößler, B., Pinkos, R., Genger, T. and Preiss, T. (2005) WO Patent Appl 030690 (assigned to BASF).

13 (a) Amend, W.J. (1943) US Patent 2,316,543; (b) Flemming, W. and Speer, W. (1935) US Patent 2,005,183; (c) Kamiya, Y. and Kotake, M. (1973) *Bull. Chem. Soc. Jpn.*,

46, 2780; (d) Kamiya, Y. (1971) *Kogyo Kagaku Zasshi*, **74**, 91; (e) Ogawa, M., Kusunoki, M. and Kitabatake, M. (1967) *Kogyo Kagaku Zasshi*, **70**, 60; (f) Ogawa, M. (1968) *Kogyo Kagaku Zasshi*, **71**, 147; (g) Kamath, S.S. and Chandalia, S.B. (1973) *J. Appl. Chem. Biotechnol.*, **23**, 469; (h) Shen, H. and Weng, H. (1988) *Ind. Eng. Chem. Res.*, **27**, 2246; (i) Rao, D.G. and Raghunathan, T.S. (1984) *J. Chem. Technol. Biotechnol.*, **34**, 381; (j) Constantini, M. and Krumenacker, L. (1983) FR Patent 2541993 (assigned to Rhone Poulenc); (k) Tanaka, K., Matsuoka, Y. and Shimizu, A. (2001) JP 213841; (l) Brégeault, J., Launay, F. and Atlamsani, A. (2001) *C.R. Acad. Sci. Paris, Serie IIc, Chimie*, **4**, 11; (m) Brégeault, J.M. (2003) *Dalton Trans.*, 3289; (n) Srinivas, D., Chavan, S.A. and Ratnasamy, P. (2003) US Patent 6,521,789 (assigned to Council of Scientific and Industrial Research); (o) Shimizu, A., Tanaka, K., Ogawa, H., Matsuoka, Y., Fujimori, M., Nagamori, Y., Hamachi, H. and Kimura, K. (2003) *Bull. Chem. Soc. Jpn.*, **76**, 1993.

14 (a) Atlamsani, A., Brégeault, J.M. and Ziyad, M. (1993) *J. Org. Chem.*, **58**, 5663; (b) Brégeault, J.M., El Ali, B., Mercier, J., Martin, J. and Martin, C. (1988) *C.R. Acad. Sci. Paris*, **307**, 2011; (c) Brégeault, J.M., El Ali, B., Mercier, J., Martin, J. and Martin, C. (1989) *C.R. Acad. Sci. Paris*, **309**, 459; (d) El Ali, B., Brégeault, J.M., Mercier, J., Martin, J., Martin, C. and Convert, O. (1989) *J. Chem. Soc., Chem. Comm.*, 825; (e) Atlamsani, A. and Brégeault, J.M. (1992) *Synthesis*, 79; (f) El Ali, B., Brégeault, J.M., Martin, J. and Martin, C. (1989) *New J. Chem.*, **13**, 173; (g) Seidel, W.C. (1989) US Patent 4,883,910; (h) Brégeault, J.M., Bassam, E.A. and Martin, J. (1991) US Patent 4,983,767 (assigned to Rhone Poulenc Chimie); (i) Neumann, R. and Khenkin, A.M. (2006) *Chem. Commun.*, 2529; (j) Costantini, M. and Krumenacker, L. (2009) FR Patent 83-3649830 (assigned to Rhone Poulenc); (k) Chavan, S.A., Srinivas, D. and Ratnasamy, P. (2002) *J. Catal.*, **212**, 39; (l) Ballarini, N., Cavani, F.,

Casagrandi, L., D'Alessandro, T., Frattini, A., Accorinti, P., Alini, S. and Babini, P.(September 2008) Future feedstocks for fuels and chemicals. Proceed. DGMK-Conference, Berlin, C p. 225; (m) Minisci, F., Fumagalli, C. and Pirola, R.WO Patent 0187815 (assigned to Lonza S.p.A., Minisci, Fumagalli, Pirola).

15 (a) Gallezot, P. (1997) *Catal. Today*, **37**, 405; (b) Besson, M., Gauthard, F., Horvath, B. and Gallezot, P. (2005) *J. Phys. Chem. B*, **109**, 2461; (c) Beziat, J.C., Besson, M. and Gallezot, P. (1996) *Appl. Catal. A*, **135**, L7; (d) Creeze, E., Barendregt, A., Kapteijn, F. and Moulijn, J.A. (2001) *Catal. Today*, **69**, 283; (e) Usui, Y. and Sato, K. (2003) *Green Chem.*, **5**, 373; (f) Ishii, Y., Adachi, A., Imai, R. and Ogawa, M. (1978) *Chem. Lett.*, 611.

16 (a) Tanaka, K. (1975) *Shokubai*, **17**, 197; (b) Tanaka, K. (1974) *CHEMTECH*, 555; (c) Tanaka, K. (1974) *Hydrocarb. Process.*, **53**, 114; (d) Tanaka, K. and Shimizu, A. (2001) JP 1253845 (assigned to Asahi Chem Co)

17 (a) Onopchenko, A. and Schulz, J.G.D. (1977) US Patent 4,032,569 (assigned to Gulf R&D); (b) Schulz, J.G.D. and Onopchenko, A. (1981) US Patent US 4,263,453 (assigned to Gulf R&D).

18 (a) Richardson, D., Xu, C. and Abboud, K. (2001) US Patent 6,258,981 (assigned to University of Florida); (b) Park, C.M. and Goroff, N. (1993) US Patent 5,221,800 (assigned to Amoco Corp); (c) Mall, S. and Kumar, S.S. (2001) US Patent 6,235,932 (assigned to Chemintel India); (d) Kulsrestha, G.N., Saxena, M.P., Gupta, A.K., Goyal, H.B., Prasad, R., Prasada Rao, T.S.R. and Patel, P.D. (1996) US Patent 5,547,905 (assigned to Council of Scientific and Industrial Research); (e) Steinmez, G.R., Lafferty, N.L. and Sumner, C.E. (1988) *J. Mol. Catal.*, **49**, 39; (f) Steinmez, G.R., Lafferty, N.L. and Sumner, C.E. (1990) US Patent 4,902,827 (assigned to Eastman Kodak); (g) Chu, L.A., Fodor, L. and Valdez, D.L. (2004) US Patent Appl. 092767; (h) Kollar, J. (1994) WO Patent 07834 (assigned to Bayer AG).

19 (a) Kollar, J. (1994) US Patent 5,236,561 and 5,321,157 (assigned to Redox Technologies); (b) Kollar, J. (1995) US Patent 5,463,119 (assigned to Redox Technologies).

20 (a) Bonnet, D., Fache, E. and Simonato, J.P. (2004) US Patent Appl. 0242922; (b) Costantini, M. and Fache, E. (2000) US Patent 6,147,256 (assigned to Rhodia Fiber & Resin Int); (c) Costantini, M., Fache, E. and Nivert, D. (1998) US Patent 5,756,837 (assigned to Rhone Poulenc F&I); (d) Bonnet, D., Fache, E. and Simonato, J.P. (2003) WO Patent Appl. 03014055 (assigned to Rhodia Polyamide Int); (e) Amoros, D., Augier, F., Bonnet, D., Broglio, M.I. and Simonato, J.P. (2004) WO Patent Appl. 04041768 (assigned to Rhodia Polyamide Int); (f) Bonnet, D., Ireland, T., Fache, E. and Simonato, J.P. (2006) *Green Chem.*, **8**, 556; (g) Bonnet, D., Fache, E. and Simonato, J.P. (2004) US Patent Appl 0242922 A1 (assigned to Rhodia Polyamide Int); (h) Fache, E. (2006) US Patent 7,041,848 (assigned to Rhodia Polyamide Int); (i) Bonnet, D., Ireland, T. and Simonato, J.P. (2007) US Patent 7,253,312 (assigned to Rhodia Polyamide Int); (j) Bonnet, D., Fache, E. and Simonato, J.P.US Patent Appl. (2004) 0242922; (k) Bonnet, D., Amoros, D., Simonato, J.P., Augier, F. and Broglio, M.I. (2006) US Patent Appl. 0094900; (l) Bonnet, D., Petroff Saint-Arroma, R., Righini, S., Ireland, T. and Simonato, J.P. (2006) WO Patent Appl. 136674; (m) Costantini, M. and Fache, E. (1998) Eur. Patent Appl. 870,751 (assigned to Rhodia Polyamide Int); (n) Fache, E. and Costantini, M. (1998) WO Patent 2000/059858 (assigned to Rhodia Fiber and Resin Intermediates).

21 (a) Dassel, M.W. and Vassiliou, E. (1999) US Patent 5,939,582 (assigned to Twenty-first Century Research Co); (b) Dassel, M.W. and Vassiliou, E. (1996) US Patent 5,502,245 (assigned to Twenty-first Century Research Co); (c) DeCoster, D.C., Vassiliou, E., Dassel, M.W. and Rostami,

A.M. (1999) US Patent 5,929,277 (assigned to Twenty-first Century Research Co); (d) DeCoster, D.C., Vassiliou, E., Dassel, M.W., Rostami, A.M. and Dudgeon, D.J. (1999) US Patent 5,908,589 (assigned to Twenty-first Century Research Co); (e) Dassel, M.W. and Vassiliou, E. (1998) US Patent 5,801,282 (assigned to Twenty-first Century Research Co); (f) Dassel, M.W., Vassiliou, E., DeCoster, D.C., Rostami, A.M. and Aldrich, S.M. (2001) US Patent 6,183,698 (assigned to RPC); (g) Rostami, A.M., Dassel, M.W., Vassiliou, E. and DeCoster, D.C. (1999) US Patent 5,998,572 (assigned to RPC); (h) http://investor.fluor.com/phoenix.zhtml?c=124955&p=irol-newsArticle&ID=59504&highlight; (i) http://www.hydrocarbononline.com/article.mvc/New-Single-Oxidation-Adipic-Acid-Route-Will-S-0001; (j) Hirota, M. and Hagiya, K. (2007) Eur Patent 1,748,042 (assigned to Sumitomo Chemical Company).

22 (a) Kulsrestha, G.N., Shankar, U., Sharma, J.S. and Singh, J.J. (1991) *J. Chem. Technol. Biotechnol.*, **50**, 57; (b) Rao, D.G. and Tirukkoyilur, R.S. (1986) *Ind. Eng. Chem. Prod. Res. Dev.*, **25**, 299; (c) Yuan, Y., Ji, H., Chen, Y., Han, Y., Song, X., She, Y. and Zhing, R. (2004) *Org. Proc. Res. Dev.*, **8**, 418; (d) Chavan, S.A., Srinivas, D. and Ratnasamy, P. (2002) *J. Catal.*, **212**, 39.

23 (a) Lin, S.S. and Weng, H.S. (1993) *Appl. Catal. A*, **105**, 289; (b) Lin, S.S. and Weng, H.S. (1994) *Appl. Catal. A*, **118**, 21; (c) Kraushaar-Czarnetzki, B. and Hoogervorst, W.G.M. (1992) Eur. Patent 519,569; (d) Concepcion, P., Corma, A., Lopez Nieto, J.M. and Perez-Pariente, J. (1996) *Appl. Catal. A*, **143**, 17; (e) Vanoppen, D.L. and Jacobs, P.A. (1999) *Catal. Today*, **49**, 177; (f) Sankar, G., Raja, R. and Thomas, J.M. (1998) *Catal. Lett.*, **55**, 15; (g) Belkhir, I., Germain, A., Fajula, F. and Fache, E. (1998) *J. Chem. Soc., Faraday Trans.*, **94**, 1761; (h) Belkhir, I., Germain, A., Fajula, F. and Fache, E. (1997) *Stud. Suf. Sci. Catal.*, **110**, 577; (i) Selvam, P. and

Mohapatra, S.K. (2005) *J. Catal.*, **233**, 276; (j) Shen, H.C. and Weng, H.S. (1988) *Ind. Eng. Chem. Res.*, **27**, 2254; (k) Mohapatra, S.K. and Selvam, P. (2003) *Topics Catal.*, **22**, 17; (l) Zhou, L., Xu, J., Miao, H., Wang, F. and Li, X. (2005) *Appl. Catal. A*, **292**, 223; (m) Corma, A. and Lopez Nieto, J. (2006) US Patent 7,087,793 (assigned to Sumitomo Chem Co); (n) Kesavan, V., Sivanand, P.S., Chandersekaran, S., Koltypin, Y. and Gedanken, A. (1999) *Angew. Chem. Int. Ed.*, **38**, 3521; (o) Corma, A., Lopez Nieto, J. and Domine, M.E. (2008) US Patent 7,358,401 (assigned to Sumitomo Chem Co).

24 (a) Dugal, M., Sankar, G., Raja, R. and Thomas, J.M. (2000) *Angew. Chem. Int. Ed.*, **39**, 2310; (b) Thomas, J.M. and Raja, R. (2001) *Chem. Commun.*, 675; (c) Raja, R., Thomas, J.M., Xu, M., Harris, K.D.M., Greenhill-Hooper, M. and Quill, K. (2006) *Chem. Commun.*, 448; (d) Thomas, J.M. and Raja, R. (2006) *Catal. Today*, **117**, 22; (e) Raja, R., Sankar, G. and Thomas, J.M. (1999) *J. Am. Chem. Soc.*, **121**, 11926; (f) Modén, B., Oliviero, L., Dakka, J., Santiesteban, J.G. and Iglesia, E. (2004) *J. Phys. Chem. B*, **108**, 5552; (g) Perkas, N., Wang, Y., Koltypin, Y., Gedanken, A. and Chandrasekaran, S. (2001) *Chem. Commun.*, 988; (h) Parton, R.F., Peere, G.J., Neys, P.E., Jacobs, P.A., Claesseus, R. and Baron, G.R. (1996) *J. Mol. Catal. A*, **113**, 445; (i) Selvam, P. and Mohapatra, S.K. (2006) *J. Catal.*, **238**, 88; (j) Mohapatra, S.K., Sahoo, B., Keune, W. and Selvam, P. (2002) *Chem. Commun.*, 1466.

25 (a) Lu, G., Ji, D., Qian, G., Qi, Y., Wang, X. and Suo, J. (2005) *Appl. Catal. A*, **280**, 175; (b) Lu, G., Zhao, R., Qian, G., Qi, Y., Wang, X. and Suo, J. (2004) *Catal. Lett.*, **97**, 115; (c) Xu, L.X., He, C.H., Zhu, M.Q. and Fang, S. (2007) *Catal. Lett.*, **114**, 202; (d) Xu, L.X., He, C.H., Zhu, M.Q., Wu, K.J. and Lai, Y.L. (2007) *Catal. Lett.*, **118**, 248; (e) Xu, Y.J., Landon, P., Enache, D., Carley, A.F., Roberts, M.W. and Hutchings, G.J. (2005) *Catal. Lett.*, **101**, 175; (f) Corma, A. and Lopez Nieto, J. (2007) US Patent, 7,166,751

(assigned to Sumitomo Chem Co); (g) Raja, R. and Ratnasamy, P. (1997) *Catal. Lett.*, **48**, 1; (h) Parton, R.F., Vankelcom, J.F.J., Casselman, M.J.A., Bezouhanova, C.P., Utterhoeven, J.B. and Jacobs, P.A. (1994) *Nature*, **570**, 541.

26 (a) Pires, E.L., Magalhaes, J.C. and Schuchardt, U. (2000) *Appl. Catal. A*, **203**, 231; (b) Pires, E.L., Wallau, M. and Schuchardt, U. (1998) *J. Mol. Catal. A*, **136**, 69; (c) Yao, C.S. and Weng, H.S. (1998) *Ind. Eng. Chem. Res.*, **37**, 2647; (d) Yao, C.S. and Weng, H.S. (1992) *Chem. Eng. Sci.*, **47**, 2745.

27 (a) Selvam, P. and Mohapatra, S.K. (2004) *Microporous Mesoporous Mater.*, **73**, 137; (b) Limtrakul, J. and Sooknoi, T. (2002) *Appl. Catal. A*, **233**, 227; (c) Pires, E.L., Arnold, U. and Schuchardt, U. (2001) *J. Mol. Catal. A*, **169**, 157; (d) da Cruz, R.S., de Souza e Silva, J.M., Arnold, U. and Schuchardt, U. (2001) *J. Mol. Catal. A*, **171**, 251; (e) da Cruz, R.S., de Souza e Silva, J.M., Arnold, U., Sercheli, M.S. and Schuchardt, U. (2002) *J. Braz. Chem. Soc.*, **13**, 170; (f) Carvalho, W.A., Wallau, M. and Schuchardt, U. (1999) *J. Mol. Catal. A*, **144**, 91; (g) Sakthivel, A. and Selvam, P. (2002) *J. Catal.*, **211**, 134; (h) Poladi, R.H.P.R. and Landry, C.C. (2002) *Microporous Mesoporous Mater.*, **52**, 11; (i) Poladi, R.H.P.R. and Landry, C.C. (2002) *Microporous Mesoporous Mater.*, **52**, 11; (j) Huybrechts, D.R.C., Buskens, P.L., Mathys, G.M.K. and Martens, L.R.M. (1998) US Patent 5,739,076 (assigned to Exxon Chem).

28 (a) Mishra, G.P. and Pombeiro, A.J.L. (2005) *J. Mol. Catal. A*, **239**, 96; (b) Pillai, U.R. and Sahle-Demessie, E. (2002) *Chem. Commun.*, 2142; (c) Mohapatra, S.K. and Selvam, P. (2004) *Catal. Lett.*, **93**, 47; (d) Samanta, S., Mal, N.K. and Bhaumik, A. (2005) *J. Mol. Catal. A*, **236**, 7; (e) Fan, W., Fan, B., Song, M., Chen, T., Li, R., Dou, T., Tatsumi, T. and Weckhuysen, B.M. (2006) *Microporous Mesoporous Mater.*, **94**, 348; (f) Mohapatra, S.K., Hussain, F. and Selvam, P. (2003) *Catal. Lett.*, **85**, 217;

(g) Miyake, T., Koike, K., Aoki, I., Muarayama, N. and Sano, M. (2005) *Appl. Catal. A*, **288**, 216.

29 (a) Beckman, E.J. (2003) *Environ. Sci. Technol.*, **37**, 5289;Wu, X.W., Oshima, Y. and Koda, S. (1997) *Chem. Lett.*, 1045; (b) Srinivas, P. and Mukhopadhyay, M. (1994) *Ind. Eng. Chem. Res.*, **33**, 3118; (c) Hou, Z., Han, B., Gao, L., Liu, Z. and Yang, G. (2002) *Green Chem.*, **4**, 426; (d) Ambruster, U., Martin, A., Smejkal, Q. and Kosslick, H. (2004) *Appl. Catal. A*, **265**, 237.

30 (a) Ishii, Y., Sakaguchi, S. and Iwahama, T. (1999) *Yuki Gosei Kagaku Kyokaishi*, **57**, 38; (b) Iwahama, T., Syojyo, K., Sakaguchi, S. and Ishii, Y. (1998) *Org. Process Res. Dev.*, **2**, 255; (c) Ishii, Y. (1997) *J. Mol. Catal. A*, **117**, 123; (d) Ishii, Y., Iwahama, T., Sakaguchi, S., Nakayama, K. and Nishiyama, Y. (1996) *J. Org. Chem.*, **61**, 4520; (e) Hirai, N. (1998) JP 10114702; (f) Nakano, T. and Ishii, Y. (1998) Eur Patent 858,835 (assigned to Daicel Chem Ind); (g) Ishii, Y. and Nakano, T. (1997) JP 9327626; (h) Nakano, T. and Ishii, Y. (1997) WO Patent Appl 97/28897 (assigned to Daicel Chem Ind); (i) Ishii, Y. and Sagagushi, S. (1999) *Catal. Surv. Jpn.*, **3**, 27; (j) Ishii, Y. and Sagakuchi, S. (2006) *Catal. Today*, **117**, 105; (k) Hirai, N. (1997) Eur Patent 825,165 (assigned to Daicel Chem Ind); (l) Ishii, Y. and Nakano, T. (1997) Eur. Patent 824,962 (assigned to Daicel Chem In); (m) Ishii, Y. and Kajikawa, Y. (2006) US Patent 7,015,356 (assigned to Daicel Chem Ind); (n) Kerry Yu, K.M., Hummeida, R., Abutaki, A. and Tsang, S.C. (2006) *Catal. Lett.*, **111**, 51; (o) Ishii, Y. and Nakano, T.Eur Patent 1,258,292 (assigned to Daicel Chem Ind); (p) Ishii, Y. and Nakano, T. (1999) Eur Patent 990,631 (assigned to Daicel Chem Ind); (q) Murata, S. and Tani, N. (2002) US Patent 6,459,002 (assigned to Sumitomo Chem Co and Daicel Chem Ind); (r) 30 Aug 2007) *Jpn. Chem. Week.*, **48** (2430), 1; (s) Ishii, Y., Sakaguchi, S. and Iwahama, T. (2001) *Adv. Synth. Catal.*, **343**, 393; (t) Bassler, P. and Berning, W. (2007) 1st

Int. Symposium on "Selective Oxidation Catalysis: C—H Activation via Molecular Oxygen", Stuttgart, October 31st; (u) Hermans, I., Van Deun, J., Houthoofd, K., Peeters, J. and Jacobs, P. (2007) *J. Catal.*, **251**, 204.

31 (a) Minisci, F., Recupero, F., Pedulli, G.F. and Lucarini, M. (2003) *J. Mol. Catal. A*, **63**, 204; (b) Minisci, F., Punta, C. and Recupero, F. (2006) *J. Mol. Catal. A*, **251**, 129; (c) Ishii, Y. and Sakaguchi, S. (2004) *Modern Oxidation Methods* (ed. J-.E. Backvall), Wiley-VCH Verlag GmbH, Weinheim, p. 119; (d) Sheldon, R.A. and Arends, I.W.C.E. (2004) *Adv. Synth. Catal.*, **346**, 1051; (e) Recupero, F. and Punta, C. (2007) *Chem. Rev.*, **107**, 3800; (f) Sawatari, N., Yokota, T., Sakaguchi, S. and Ishii, Y. (2001) *J. Org. Chem.*, **66**, 7889; (g) Fukuda, O., Sakaguchi, S. and Ishii, Y. (2001) *Tetrahedron Lett.*, **42**, 3479; (h) Baucherel, X., Gonsalvi, L., Arends, I.W.C., Ellwood, S. and Sheldon, R.A. (2004) *Adv. Synth. Catal.*, **346**, 286; (i) Tong, X., Xu, J. and Miao, H. (2005) *Adv. Synth. Catal.*, **347**, 1953; (j) Sheldon, R.A., Arends, I.W.C.E. and Dijksman, A. (2000) *Catal. Today*, **57**, 157.

32 (a) Roper, M. (1991) *Stud. Surf. Sci. Catal.*, **64**, 381; (b) http://www.chemweek.com/ sections/business_finance/europe_ mideast/6951.html; (c) Bruner, H.S. Jr, Lane, S.L. and Murphree, B.E. (1998) US Patent 5,710,325 (assigned to Du Pont de Nemours, DSM N.V.); (d) Denis, P., Patois, C. and Perron, R. (1997) US Patent 5,625,096 (assigned to Rhone-Poulenc Chimie); (e) Drent, E., Eberhard, M.R. and Pringle, P.G. (2003) US Patent 6,639,091 (assigned to Shell Oil Co); (f) Drent, E. and Jager, W.W. (2004) US Patent 6,706,912 (assigned to Shell Oil Co); (g) Drent, E. and Jager, W.W. (2004) US Patent 6,737,542 (assigned to Shell Oil Co); (h) Suykerbuyk, J.C.L.J., Drent, E. and Pringle, P.G. (2000) US Patent 6156934 (assigned to Shell Oil Co); (i) Beller, M., Cornils, B., Frohning, C.D. and Kohlpaintner, C.W. (1995) *J. Mol. Catal. A*,

104, 17; (j) Packett, D.L., Briggs, J.R., Bryant, D.R. and Phillips, A.G. (1999) US Patent 5,892,127 (assigned to Union Carbide Chem & Plastics Technology Co); (k) Dahlhoff, G., Niederer, J.P.M. and Hoelderich, W.F. (2001) *Catal. Rev.*, **43**, 381; (l) Sielken, O.E. and Hovenkamp, H. (1997) US Patent 5,693,851 (assigned to DSM NV and Du Pont de Nemours Co); (m) Lane, S.L. (1998) US Patent 5,840,959 (assigned to Du Pont de Nemours Co and DSM NV).

33 (a) Taniguchi, Y., Hayashida, T., Shibasaki, H., Piao, D., Kitamura, T., Yamaji, T. and Fujiwara, Y. (1999) *Org. Lett.*, **1**, 557; (b) Piao, D., Inoue, K., Shibasaki, H., Taniguchi, Y., Kitamura, T. and Fujiwara, Y. (1999) *J. Organomet. Chem.*, **574**, 116; (c) Jia, C., Kitamura, T. and Fujiwara, Y. (2001) *Acc. Chem. Res.*, **34**, 633; (d) Asadullah, M., Kitamura, T. and Fujiwara, Y. (2000) *Angew. Chem. Int. Ed.*, **39**, 2475; (e) Reis, P.M., Silva, J.A.L., Palavra, A.F., da Silva, J.J.R.F., Kitamura, T., Fujiwara, Y. and Pombeiro, A.J.L. (2003) *Angew. Chem. Int. Ed.*, **42**, 821; (f) Kirillov, A.M., Haukka, M., Kirillova, M.V. and Pombeiro, A.J.L. (2005) *Adv. Synth. Catal.*, **347**, 1435; (g) Reis, P.M., Silva, J.A.L., Palavra, A.F., da Silva, J.J.R.F. and Pombeiro, A.J.L. (2005) *J. Catal.*, **235**, 333; (h) Zerella, M., Mukhopadhyay, S. and Bell, A.T. (2003) *Org. Lett.*, **5**, 3193; (i) Kirillova, M.V., da Silva, J.A.L., Frausto da Silva, J.J.R., Palavra, A.F. and Pombeiro, A.J.L. (2007) *Adv. Synth. Catal.*, **349**, 1765; (j) Kirillova, M.V., Kirillov, A.M., Reis, P.M., Silva, J.A.L., Fraústo da Silva, J.J.R. and Pombeiro, A.J.L. (2007) *J. Catal.*, **248**, 130.

34 (a) Venturello, C., Alneri, E. and Ricci, M. (1983) *J. Org. Chem.*, **48**, 3831; (b) Venturello, C., D'Aloisio, R., Bart, J.C.J. and Ricci, M. (1985) *J. Mol. Catal.*, **32**, 221; (c) Venturello, C. and Ricci, M. (1984) Eur Patent 122,804, (1985) US Patent 4, 532, 079 (assigned to Montedison SpA); (d) Oguchi, T., Ura, T., Ishii, Y. and Ogawa, M. (1989) *Chem. Lett.*, 857; (e) Ishii, Y.,

Yamawaki, K., Ura, T., Yamada, H., Yoshida, T. and Ogawa, M. (1988) *J. Org. Chem.*, **53**, 3587; (f) Aubry, C., Chottard, G., Platzer, N., Brégeault, J.M., Thouvenot, R., Chauveau, F., Huet, C. and Ledon, H. (1991) *Inorg. Chem.*, **30**, 4409; (g) Sakagucchi, S., Nishiyama, Y. and Ishii, Y. (1996) *J. Org. Chem.*, **61**, 5307; (h) Sato, K., Aokil, M. and Noyori, R. (1998) *Science*, **281**, 1646; (i) Noyori, R., Aoki, M. and Sato, K. (2003) *Chem. Commun.*, 1977; (j) Sato, K., Aokil, M. and Noyori, R. (1996) *J. Org. Chem.*, **61**, 8310; (k) Bolm, C., Beckmann, O. and Dabard, O.A.G. (1999) *Angew. Chem. Int. Ed.*, **38**, 907; (l) Oguchi, T., Ura, T., Ishii, Y. and Ogawa, M. (1989) *Chem. Lett.*, 857; (m) Fujitani, T. and Nakazawa, M. (1988) JP Patent 93746; (n) Sato, K. and Usui, Y. (2002) JP Patent 216841; (o) Sato, K. and Usui, Y. (2002) JP Patent 216692; (p) Deng, Y., Ma, Z., Wang, K. and Chen, J. (1999) *Green Chem.*, 275; (q) Jiang, H., Gong, H., Yang, Z., Zhang, X. and Sun, Z. (2002) *React. Kinet. Catal. Lett.*, **75**, 315; (r) Antonelli, E., D'Aloisio, R., Gambaro, M., Fiorani, T. and Venturello, C. (1998) *J. Org. Chem.*, **63**, 7190; (s) Lambert, A., Plucinshi, P. and Kozhevnikov, I.V. (2003) *Chem. Commun.*, 714; (t) Kaur, J. and Kozhevnikov, I.V. (2004) *Catal. Commun.*, **5**, 709; (u) Neumann, R. and Khenkin, A.M. (1994) *J. Org. Chem.*, **59**, 7577; (v) Zhu, W., Li, H., He, X., Zhang, Q., Shu, H. and Yan, Y. (2008) *Catal. Commun.*, **9**, 551.

35 (a) Cheng, C.Y., Lin, K.J., Prasad, M.R., Fu, S.J., Chang, S.Y., Shyu, S.G., Sheu, H.S., Chen, C.H., Chuang, C.H. and Lin, M.T. (2007) *Catal. Comm.*, **8**, 1060; (b) Chen, H., Dai, W.L., Deng, J.F. and Fan, K. (2002) *Catal. Lett.*, **81**, 131; (c) Lee, S.O., Raja, R., Harris, K.D.M., Thomas, J.M., Johnson, B.F.G. and Sankar, G. (2003) *Angew. Chem. Int. Ed.*, **42**, 1520; (d) Lapisardi, G., Chiker, F., Launay, F., Nogier, J.P. and Bonardet, J.L. (2004) *Catal. Commun.*, **5**, 277; (e) Lapisardi, G., Chiker, F., Launay, F., Nogier, J.P. and Bonardet, J.L. (2005) *Microporous Mesoporous Mater.*, **78**, 289;

(f) Trukhan, N.N., Derevyankin, A.Yu., Shmakov, A.N., Paukshtis, E.A., Kholdeeva, O.A. and Romannikov, V.N. (2001) *Microporous Mesoporous Mater.*, **44/45**, 603; (g) Kholdeeva, O.A., Derevyankin, A.Yu., Shmakov, A.N., Trukhan, E.A.N.N., Paukshtis, A., Tuel, V.N. and Romannikov, (2000) *J. Mol. Catal. A*, **158**, 417; (h) Trukhan, N.N., Romannikov, V.N., Paukshtis, E.A., Shmakov, A.N. and Kholdeeva, O.A. (2001) *J. Catal.*, **202**, 110; (i) Kholdeeva, O.A. and Trukhan, N.N. (2006) *Russ. Chem. Rev.*, **75**, 411; (j) Ziolek, M. (2004) *Catal. Today*, **90**, 145; (k) Kholdeeva, O.A., Mel'gunov, M.S., Shmakov, A.N., Trukhan, N.N., Kriventsov, V.V., Zaikovskii, V.I., Malyshev, M.E. and Romannikov, V.N. (2004) *Catal. Today*, **91/92**, 205; (l) Kholdeeva, O.A., Zalomaeva, O.V., Shmakov, A.N., Mel'gunov, M.S. and Sorokin, A.B. (2005) *J. Catal.*, **236**, 62; (m) Timofeeva, M.N., Kholdeeva, O.A., Jhung, S.H. and Chang, J.S. (2008) *Appl. Catal. A*, **345**, 195; (n) Ionescu, E., Matache, S. and Oprescu, I. (1982) RO Patent 83847 (assigned to Centrala Industriala de Fire si Fibre Sintetice).

36 (a) Trost, B.M. (1995) *Angew. Chem. Int. Ed. Engl.*, **34**, 259; (b) Sheldon, R.A. (1997) *Chem. Ind.*, 12; (c) Schröder, M. (1980) *Chem. Rev.*, **80**, 187; (d) Sharpless, K.B. and Akashi, K. (1976) *J. Am. Chem. Soc.*, **98**, 1986; (e) VanRheenen, V., Kelly, R.C. and Cha, D. (1976) *Tetrahedron Lett.*, 1973; (f) Kolb, H.C., VanNieuwenhze, M.S. and Sharpless, K.B. (1994) *Chem. Rev.*, **94**, 2483; (g) Akashi, K., Palermo, R.E. and Sharpless, K.B. (1978) *J. Org. Chem.*, **43**, 2063; (h) Bergstad, K., Jonsson, S.Y. and Bäckvall, J.E. (1999) *J. Am. Chem. Soc.*, **121**, 10424; (i) Hermann, W.A., Fischer, R.W. and Marz, D.W. (1991) *Angew. Chem. Int. Ed.*, **30**, 1638; (j) Schwegler, M., Floor, M. and Van Bekkum, H. (1988) *Tetrahedron Lett.*, **29**, 823; (k) Singh, V. and Deota, P.T. (1988) *Synth. Commun.*, **18**, 617; (l) Oguchi, T., Ura, T., Ishii, Y. and Ogawa, M. (1989) *Chem. Lett.*, 857; (m) Venturello, C.

and Gambaro, M. (1989) *Synthesis*, 295; (n) Herrmann, W.A., Fischer, R.W. and Marz, D.W. (1991) *Angew. Chem. Int. Ed. Engl.*, **30**, 1638; (o) Tatsumi, T., Yamamoto, K., Tajima, H. and Tominaga, H. (1992) *Chem. Lett.*, 815; (p) Corma, A., Camblor, M.A., Esteve, P., Martinez, A. and Perez-Pariente, J. (1994) *J. Catal.*, **145**, 151; (q) Corma, A., Navarro, M.T. and Perez-Pariente, J. (1994) *J. Chem. Soc. Chem. Commun.*, 147; (r) Tatsumi, T., Noyano, K.A. and Igarashi, N. (1998) *Chem. Commun.*, 325; (s) Xin, J., Suo, J., Zhang, X. and Zhang, Z. (2000) *New J. Chem.*, **24**, 569; (t) Kholdeeva, O.A., Trubitsina, T.A., Timofeeva, M.N., Maksimov, G.M., Maksimovskaya, R.I. and Rogov, V.A. (2005) *Appl. Catal. A*, **232**, 173; (u) Usui, Y., Sato, K. and Tanaka, M. (2003) *Angew. Chem. Int. Ed.*, **42**, 5623.

37 (a) Shimizu, M., Orita, H., Suzuki, K., Hayakawa, T., Hamakawa, S. and Takehira, K. (1996) *J. Mol. Catal. A*, **114**, 217; (b) Iwahama, T., Sakaguchi, S., Nishiyama, Y. and Ishii, Y. (1995) *Tetrahedron Lett.*, **36**, 1523; (c) Ishii, Y., Yamawaki, K., Ura, T., Yamada, H., Yoshida, T. and Ogawa, M. (1988) *J. Org. Chem.*, **53**, 3587; (d) Schindler, G.P., Bartl, P. and Hoelderich, W.F. (1998) *Appl. Catal. A*, **166**, 267; (e) Chen, H., Dai, W.L., Gao, R., Cao, Y., Li, H. and Fan, K. (2007) *Appl. Catal. A*, **328**, 226; (f) Felthouse, T.R. (1987) *J. Am. Chem. Soc.*, **109**, 7566; (g) Arts, S.J.H.F., van Rantwijk, F. and Sheldon, R.A. (1996) *J. Carbohydr. Chem.*, **15**, 317; (h) Sheldon, R.A., Arends, I.W.C.E. and Dijksman, A. (2000) *Catal. Today*, **57**, 157; (i) Atlamsani, A. and Brégeault, J.M. (1991) *New J. Chem.*, **15**, 671; (j) Atlamsani, A. and Brégeault, J.M. (1993) *Synthesis*, 79; (k) Atlamsani, A., Brégeault, J.-M. and Ziyad, M. (1993) *J. Org. Chem.*, **58**, 5663; (l) Brégeault, J.-M., El Ali, B. and Martin, J. (1990) Eur Patent, 0,355,075; (m) Brégeault, J.-M., El Ali, B., Mercier, J., Martin, J. and Martin, C. (1988) *C. R. Acad. Sci., Ser. II*, **307**, 2011; (n) El Ali, B., Brégeault, J.-M., Mercier, J., Martin, J., Martin, C. and Convert, O. (1989) *Chem.*

Commun., 825; (o) El Aakel, L., Launay, F., Atlamsani, A. and Brégeault, J.M. (2001) *Chem. Commun.*, 2218; (p) Brégeault, J.M., El Ali, B., Mercier, J., Martin, J. and Martin, C. (1989) *C.R. Acad. Sci. Paris, Ser. II*, **309**, 459; (q) Vennat, M., Herson, P., Brégeault, J.-M. and Shul'pin, G.B. (2003) *Eur. J. Inorg. Chem.*, **5**, 908; (r) Atlamsani, A., Brégeault, J.-M. and Ziyad, M. (1993) *J. Org. Chem.*, **58**, 5663; (s) El Aakel, L., Launay, F., Atlamsani, A. and Brégeault, J.-M. (2001) *Chem. Commun.*, 2218; (t) Brégeault, J.M., Vennat, M., Salles, L., Piquemal, J.Y., Mahha, Y., Briot, E., Bakala, P.C., Atlamsani, A. and Thouvenot, R. (2006) *J. Mol. Catal. A*, **250**, 177; (u) El Aakel, L., Launay, F., Brégeault, J.M. and Atlamsani, A. (2004) *J. Mol. Catal. A*, **212**, 171; (v) Venturello, C. and Ricci, M. (1986) *J. Org. Chem.*, **51**, 1599.

38 (a) Clerici, M.G. (2005) Oxidation, functionalisation: classical, alternative routes, sources. Proceedings of the, DGMK/SCI., Conference, October 12-14, Milan, Italy, p. 165; (b) Centi, G. (ed.) (2001) *Catalysis by Unique Metal Ion Structures in Solid Matrices*, Kluwer, Dordrecht, The Netherlands, (c) Hartmann, M. and Kewvan, L. (1999) *Chem. Rev.*, **99**, 635; (d) Corma, A. (1997) *Chem. Rev.*, **97**, 2373; (e) Saxton, R.J. (1999) *Top. Catal.*, **9**, 43; (f) Arends, I.W.C.E. and Sheldon, R.A. (2001) *Appl. Catal. A*, **212**, 175; (g) De Vos, D.E., Dams, M., Sels, B.F. and Jacobs, P.A. (2002) *Chem. Rev.*, **102**, 3615; (h) Viswanathan, B. and Jacob, B. (2005) *Catal. Rev. Sci. Eng.*, **47**, 1; (i) Taguchi, A. and Schuth, F. (2005) *Microporous Mesoporous Mater.*, **77**, 1; (j) Ratnasamy, P., Srinivas, D. and Knözinger, H. (2004) *Adv. Catal.*, **48**, 1; (k) Notari, B. (1988) *Stud. Surf. Sci. Catal.*, **37**, 413; (l) Clerici, M.G. (1991) *Appl. Catal.*, **68**, 249; (m) Bellussi, G., Carati, A., Clerici, M.G., Maddinelli, G. and Millini, R. (1992) *J. Catal.*, **133**, 220; (n) Clerici, M.G. and Ingallina, P. (1993) *J. Catal.*, **140**, 71; (o) Kholdeeva, O.A. and Trukhan, N.N. (2006) *Russ. Chem. Rev.*, **75**, 411; (p) A: Kholdeeva, O. and

Maksimovskaya, R.I. (2007) *J. Mol. Catal. A,* **262**, 7; (q) Mizuno, N., Yamaguchi, K. and Kamata, K. (2005) *Coord. Chem. Rev.,* **249**, 1944; (r) Neumann, R. (2004) *Transition Metals for Organic Synthesis,* 2nd edn (eds M. Beller and C. Bolm), vol. 2, Wiley-VCH Verlag, Weinheim, p. 415; (s) Neumann, R. (2004) (ed. J-.E. Baeckvall), *Modern Oxidation Methods,* Wiley-VCH Verlag, Weinheim, p. 223; (t) Hill, C.L. (2004) *Angew. Chem.,* **43**, 402; (u) Pope, M.T. (1983) *Heteropoly and Isopoly Oxometalates,* Springer-Verlag, New York, (v) Pope, M.T. and Müller, A. (eds) (1993) *Polyoxometalates: From Platonic Solids to Anti-Retroviral Activity,* Kluwer, Dordrecht, The Netherlands, (w) (1998) *Chem. Rev.,* 98 special issue on POMs; (x) Hill, C.L. and Prosser-McCartha, C.M. (1995) *Coord. Chem. Rev.,* **143**, 407; (y) Neumann, R. (1998) *Prog. Inorg. Chem.,* **47**, 317.

39 (a) Xia, K.H., Chen, X. and Tatsumi, T. (2001) *J. Mol. Catal. A,* **176**, 179; (b) Blasco, T., Corma, A., Navarro, M.T. and Pariente, J.P. (1995) *J. Catal.,* **156**, 65; (c) Dutoit, D.C., Schneider, M., Hutter, R. and Baiker, A. (1996) *J. Catal.,* **161**, 651; (d) Fraile, J.M., Garcia, J.I., Mayoral, J.A. and Vispe, E. (2000) *J. Catal.,* **189**, 40; (e) Wu, P., Tatsumi, T., Komatsu, T. and Yashima, T. (2001) *J. Catal.,* **202**, 245; (f) Ahn, W.-S., Kim, N.-K. and Jeong, S.-Y. (2001) *Catal. Today,* **68**, 83; (g) Kholdeeva, O.A., Derevyankin, A.Yu., Shmakov, A.N., Trukhan, N.N., Paukshtis, E.A., Tuel, A. and Romannikov, V.N. (2000) *J. Mol. Catal. A,* **158**, 417; (h) Trukhan, N.N., Derevyankin, A.Yu., Shmakov, A.N., Paukshtis, E.A., Kholdeeva, O.A. and Romannikov, V.N. (2001) *Microporous Mesoporous Mater.,* **44/45**, 603; (i) Kholdeeva, O.A., Trubitsina, T.A., Timofeeva, M.N., Maksimov, G.M., Maksimovskaya, R.I. and Rogov, V.A. (2005) *J. Mol. Catal. A,* **232**, 173; (j) Swegler, M., Floor, M. and Van Bekkum, H. (1988) *Tetrahedron Lett.,* **29**, 823; (k) Yamase, T., Ishikawa, E., Asai, Y. and Kanai, S. (1996) *J. Mol. Catal. A,* **114**, 237;

(l) Chiker, F., Launay, F., Nogier, J.P. and Bonardet, J.L. (2003) *Green Chem.,* **5**, 318; (m) Dengel, A.C., Griffith, W.P. and Parkin, B.C. (1993) *J. Chem. Soc., Dalton Trans.,* 2683; (n) Salles, L., Aubry, C., Thouvenot, R., Robert, F., Doremieux-Morin, C., Chottard, G., Ledon, H., Jeannin, Y. and Brégeault, J.-M. (1994) *Inorg. Chem.,* **33**, 871; (o) Duncan, D.C., Chambers, R.C., Hecht, E. and Hill, C.L. (1995) *J. Am. Chem. Soc.,* **117**, 681; (p) Venturello, C. and D'Aloisio, R. (1988) *J. Org. Chem.,* **53**, 1553; (q) Santosusso, T.M. and Swern, D. (1975) *J. Org. Chem.,* **40**, 2764; (r) Antoniotti, S. and Duñach, E. (2004) *J. Mol. Catal. A,* **208**, 135; (s) Antoniotti, S. and Duñach, E. (2004) *Eur. J. Org. Chem.,* 3459; (t) Antoniotti, S. and Duñach, E. (2003) *Synthesis,* **18**, 2753.

40 (a) Mizuno, N., Lyon, D.K. and Finke, R.G. (1991) *J. Catal.,* **128**, 84; (b) Neumann, R. and Dahan, M. (1995) *J. Chem. Soc., Chem. Commun.,* 171; (c) Pillai, U.R., Sahle-Demessie, E., Namboodiri, V.V. and Varma, R.S. (2002) *Green Chem.,* **4**, 495; (d) Weiner, H., Trovarelli, A. and Finke, R.G. (2003) *J. Mol. Catal. A,* **191**, 217; (e) Weiner, H., Trovarelli, A. and Finke, R.G. (2003) *J. Mol. Catal. A,* **191**, 253; (f) Mizuno, N., Weiner, H. and Finke, R.G. (1996) *J. Mol. Catal.,* **114**, 15; (g) Bonchio, M., Carraro, M., Farinazzo, A., Sartorel, A., Scorrano, G. and Kortz, U. (2007) *J. Mol. Catal. A,* **262**, 36; (h) Mizuno, N., Lyon, D.K. and Finke, R.G. (1991) *J. Catal.,* **128**, 84; (i) Neumann, R. and Dahan, M. (1995) *J. Chem. Soc., Chem. Commun.,* 171; (j) Liu, Y., Murata, K., Inaba, M., Nakajima, H., Koya, M. and Tomokuni, K. (2004) *Chem. Lett.,* **33**, 200.

41 (a) Kozhevnikov, I.V. and Matveev, K.I. (1983) *Appl. Catal.,* **5**, 135; (b) Melgo, M.S., Lindner, A. and Schuchardt, U. (2004) *Appl. Catal. A,* **273**, 217; (c) Ogawa, H., Fujinami, H., Taya, K. and Teratani, S. (1984) *Bull. Chem. Soc. Jpn.,* **57**, 1908; (d) Yokota, T., Fujibayashi, S., Nishiyama, Y., Sakagushi, S. and Ishii, Y. (1996) *J. Mol. Catal. A,* **114**, 113; (e) Miller, D.G. and

Wayner, D.D.M. (1990) *J. Org. Chem.*, **55**, 2924; (f) Kim, Y., Kim, H., Lee, J., Sim, K., Han, Y. and Paik, H. (1997) *Appl. Catal. A*, **155**, 15.

42 (a) Draths, K.M. and Frost, J.W. (1994) *J. Am. Chem. Soc.*, **116**, 399;(b) Frost, J.W. (1998) *Green Chemistry* (eds P.T. Anastas and T.C. Williamson), Oxford University Press, Oxford, (c) Niu, W., Draths, K.M. and Frost, J.W. (2002) *Biotechnol. Prog.*, **18**, 201; (d) Thomas, J.M., Raja, R., Johnson, B.F.G., O'Connell, T.J., Sankar, G. and Khimyak, T. (2003) *Chem. Commun.*, 1126.

43 (a) Cook, B.R., Reinert, T.J. and Suslick, K.S. (1986) *J. Am. Chem. Soc.*, **108**, 7281; (b) Corma, A. (1995) *Chem. Rev.*, **95**, 559; (c) Moden, B., Da Costa, P., Fonfe, B., Lee, D.K. and Iglesia, E. (2002) *J. Catal.*, **209**, 75; (d) Zhan, B.Z. and Li, X.Y. (1998) *Chem. Commun.*, 349; (e) Knopsgerrits, P.P., Devos, D., Thibaultstarzyk, F. and Jacobs, P.A. (1994) *Nature*, **369**, 543; (f) Balkus, K.J., Eissa, M. and Levado, R. (1995) *J. Am. Chem. Soc.*, **117**, 10753; (g) Herron, N. and Tolman, C.A. (1987) *J. Am. Chem. Soc.*, **109**, 2837; (h) Herron, N. (1989) *New J. Chem.*, **13**, 761; (i) Tatsumi, T., Watanabe, Y., Hirasawa, Y. and Tsuchiya, J. (1998) *Res. Chem. Intermed.*, **24**, 529; (j) Rao, P. and Ramaswamy, A.V. (1992) *J. Chem. Soc., Chem. Commun.*, 1245; (k) Halasz, I., Agarwal, M., Senderov, E. and Marcus, B. (2003) *Catal. Today*, **81**, 227; (l) Mishra, G.S. and Pombeiro, A.J.L. (2006) *Appl. Catal. A*, **304**, 185; (m) Thomas, J.M., Raja, R., Sankar, G. and Bell, R.G. (1999) *Nature*, **398**, 227; (n) Raja, R., Sankar, G. and Thomas, J.M. (2000) *Angew. Chem. Int. Ed.*, **39**, 2313; (o) Thomas, J.M., Raja, R., Sankar, G. and Bell, R.G. (2001) *Acc. Chem. Res.*, **34**, 191; (p) Raja, R. and Thomas, J.M. (2002) *J. Mol. Catal. A*, **181**, 3; (q) Raja, R. and Thomas, J.M. (1998) *Chem. Commun.*, 1841; (r) Labinger, J.A. (1999) *CATTECH*, **3**, 18; (s) Hartmann, M. and Ernst, S. (2001) *Angew. Chem. Int. Ed.*, **39**, 888; (t) Modén, B., Zhan, B.Z., Dakka, J., Santiesteban, J.G. and Iglesia, E. (2007) *J. Phys.Chem. C*, **111**, 1402; (u) Modén, B., Zhan, B.Z., Dakka, J., Santiesteban, J.G. and Iglesia, E. (2006) *J. Catal.*, **239**, 390; (v) Zhan, B.Z., Modén, B., Dakka, J., Santiesteban, J.G. and Iglesia, E. (2007) *J. Catal.*, **245**, 316.

8

Ecofining: New Process for Green Diesel Production from Vegetable Oil

Franco Baldiraghi, Marco Di Stanislao, Giovanni Faraci, Carlo Perego,
Terry Marker, Chris Gosling, Peter Kokayeff, Tom Kalnes, and Rich Marinangeli

8.1
Introduction

Existing technology for producing diesel fuel from vegetable oil has largely centered on the production of FAME biodiesel [1–3]. While FAME has many desirable qualities, such as high cetane, there are other issues associated with its use such as poor stability and high solvency, leading to filter plugging problems. Moreover, there has been little integration of renewable fuels production within petroleum refineries to date, despite the rapidly increasing growth in renewable fuels demand. The segregated production of renewable fuel components increases cost, since the existing infrastructure for distribution and production of petroleum-based fuels is not utilized. In addition, about the 9 vol.% of the product is glycerol, which is a low value product in unrefined form and has a limited market when refined. Methanol is required as a co-feed and feedstocks containing high concentrations of fatty acid can cause operational problems due to saponification reaction with the caustic present as a catalyst [4].

For all of these reasons, UOP LLC and Eni S.p.A. recognized the need for different processing route to convert vegetable oils into a high quality diesel fuel or diesel blend stock that is fully compatible with petroleum derived diesel fuel. The two companies started a collaborative research effort in 2005 to develop such a process based on conventional hydroprocessing technology that is already widely deployed in refineries and utilizes the existing refinery infrastructure and fuels distribution system. The result of this effort is the UOP/Eni Ecofining process. This new technology utilizes widely available vegetable oil feedstocks to produce a high cetane, low gravity, aromatics- and sulfur-free diesel fuel. The cold flow properties of the fuel can be adjusted over a wide range to meet various cloud point specifications in either the neat or blended fuel.

The main improvement of the Ecofining technology compared to the conventional FAME biodiesel is that it allows refiners to obtain a synthetic fuel that has a similar chemical composition and similar chemical-physical properties to petroleum diesel. For this reason the product can all be easily blended with conventional refinery products. Moreover, the integrated production of the green diesel allows the refiner to

Sustainable Industrial Processes. Edited by F. Cavani, G. Centi, S. Perathoner, and F. Trifiró
Copyright © 2009 WILEY-VCH Verlag GmbH & Co. KGaA, Weinheim
ISBN: 978-3-527-31552-9

Table 8.1 Green diesel fuel properties versus mineral ULSD and conventional biodiesel (FAME).

	Ultra low sulfur diesel (ULSD)	Biodiesel (FAME)	Green diesel
Oxygen content (%)	0	11	0
Specific gravity	0.84	0.88	0.78
Sulfur content (ppm)	<10	<1	<1
Heating value (MJ kg^{-1})	43	38	44
Cloud point (°C)	0	−5 to 15	−20 to 20
Distillation range (°C)	200–360	340–370	200–320
Polyaromatics (wt%)	11	0	0
NO$_x$ emission (wt%)	Baseline	+10%	−10%
Cetane number	51	50 ÷ 65	70 ÷ 90
Stability	Baseline	Poor	Baseline

control the quality of the renewable blending products. In addition, all of the Ecofining by-products are already present during normal refinery operation and do not require any special handling. Table 8.1 compares the properties of green diesel fuel with those of mineral ultra low sulfur diesel (ULSD) and of FAME biodiesel.

From Table 8.1 the evident advantages of green diesel advantages over mineral diesel fuels and FAME are:

- Green diesel is a high quality cetane component (CN > 80), which means higher engine efficiency.
- Green diesel is a hydrocarbon mixture, not an oxygenated organic compound, which means it has the same energy content as mineral diesel fuel and higher than FAME.
- Green diesel is a stable blending component without double bonds and oxygenated molecules.
- Green diesel maintains car reliability and reduces distribution costs.
- Green diesel low density is an advantage over FAME because it allows to upgrade low valuable and high density refinery streams, thereby expanding the diesel pool.
- Green diesel has the same boiling range as a mineral one. This prevents vaporization problems in the combustion chamber and it does not impact on the boiling point specification in case of blending with mineral diesel fuel.
- Green diesel is produced by a "refinery" process that permits quality control of biofuel and the use of existing infrastructure and fuel distribution systems.
- Green diesel meets the highest requirements of car manufacturers and can be utilized with all diesel automotives without modification.

8.2
From Vegetable Oil to Green Diesel

In general, the processing of biologically derived feedstocks is complicated by the fact that these materials contain a significant amount of oxygen. The feedstocks of

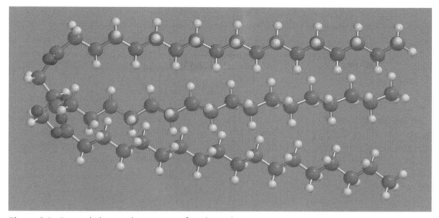

Figure 8.1 General chemical structure of triglycerides.

primary interest in the Ecofining process are primarily vegetable oils such as soybean, palm, jatropha or rapeseed (including canola) oils. Other products such as animal fats and greases can also be used as a feedstock. Vegetable oils mainly consist of triglycerides with typically 1–2% free fatty acid content. Figure 8.1 shows the chemical structure of triglyceride molecule.

Triglycerides and fatty free acids both contain relatively long, linear aliphatic hydrocarbon chains. A fatty acid is a carboxylic acid, often with a long unbranched aliphatic tail (chain), which is either saturated or unsaturated. The aliphatic hydrocarbon always contains an even number of carbon atoms and also corresponds to the carbon number range typically found in the range of the diesel fuels. The triglyceride molecule has a three-carbon "backbone." The general chemical formula is $RCOO-CH_2CH(-OOCR')CH_2-OOCR''$, where R, R', and R'' are long alkyl chains. The three fatty acids $RCOOH$, $R'COOH$ and $R''COOH$ can be all the same ($R = R' = R''$), or all different ($R \neq R' \neq R''$), or only two the same ($R = R' \neq R''$). Chain lengths of the fatty acids in naturally occurring triglycerides can vary, but 16, 18 and 20 carbons are the most common. Most natural fats contain a complex mixture of individual triglycerides; because of this, they melt over a broad range of temperatures.

As an example of the complexity of the vegetable oil composition, the main properties of soybean, rapeseed and palm oil are reported in Table 8.2. In particular, the concentration of the fatty acid type, the chain length and the olefin bonds in each chain are indicated. Moreover, in the last column the same characteristics for crude rapeseed oil are reported. Table 8.2 shows that the vegetable oils can differ from in the amount of oxygen, and the fatty acid distribution as well as in the degree of unsaturation. As far as contaminants are concerned the vegetable oils can contain sulfur, phosphorus, alkali metals and free fatty acids. The amount of these contaminants is strongly affected by the purity grade of vegetable oil.

The quality of vegetable oil, in particular the fatty acid distribution and the degree of unsaturation can affect the properties of FAME biodiesel, but they do not affect the properties of green diesel.

Table 8.2 Main properties and composition of soybean, palm and rapeseed.

Vegetable oil	Soy	Palm	Rapeseed	Rapeseed
Form	Refined	Refined	Refined	Crude
Carbon (%)	77.46	76.93	77.97	78.86
Hydrogen (%)	11.23	11.60	11.42	11.85
Oxygen (%)	11.31	11.47	10.61	9.29
Nitrogen (ppm)	6.8	6.3	3.1	25
Sulfur (ppm)	<3	<3	<3	4
Phosphorus (ppm)	<5	<5	<5	100–600
Alkali metals (ppm)	10–15	10–15	10–15	100–600
Acid number (mg-KOH g^{-1})	0.3	0.5	0.11	2.3
Specific gravity at 15 °C	0.923	0.916	0.921	0.922
Fatty acid distribution				
Palmitic acid (16-0)[a] (%)	12.50	34.30	6.70	6.70
Stearic acid (18-0)[a] (%)	0.65	1.63	1.45	1.45
Oleic acid (18-1) (%)	27.81	43.93	54.76	54.76
Linoleic acid (18-2)[a] (%)	54.19	14.27	27.51	27.51
Linolenic acid (18-3)[a] (%)	4.67	4.51	7.76	7.76
Iodine number	117–143	35–61	94–120	

[a]Numbers in parentheses refer to the chain length followed by the number of double bonds.

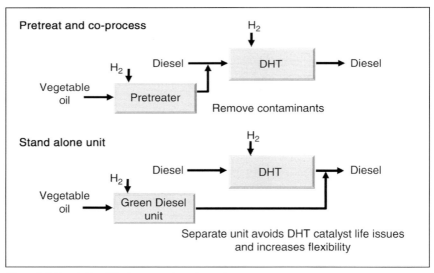

Figure 8.2 Vegetable oil hydroprocessing alternatives.

As stated in the introduction, the main objective of this new technology is to remove all the oxygen from the aliphatic chain. Two options have been investigated to produce green diesel from renewable feedstocks: co-processing in an existing distillate hydroprocessing unit or building a dedicated unit [5]. Figure 8.2 gives a block diagram for the two alternative configurations.

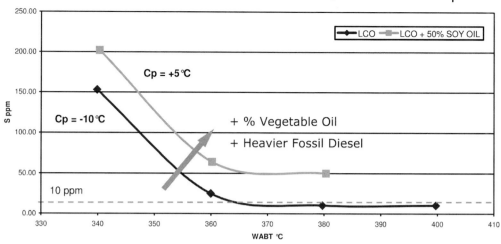

Figure 8.3 Vegetable oil co-feeding lowers HDS activity for LCO: at the same average reactor temperature (WABT) the desulfurization activity is lower.

The co-processing route was initially evaluated as an attractive option, since the existing equipment could be re-used, but after careful evaluation it was recognized that the co-feeding can cause several design problems and some limitations to the final products. These reasons can be summarized as follows:

- catalyst HDS activity reduction due to the high oxygen content of the vegetable oil with the effect of reducing the catalyst life;
- the cold flow property issue will limit the volume of vegetable oil that can be co-processed because in the co-feeding process the vegetable oil is converted into *n*-paraffins having poor cold properties.

This is evident in Figure 8.3, which shows the results of a co-feeding pilot plant test using LCO (light cycle oil) + 50% vegetable oil as feedstock. The extent of vegetable oil effect in the co-feeding process depends on the amount of vegetable oil and on the type of fossil diesel.

Other problems of a co-processing operation are:

- need to add quench to the existing reactors to control temperature rise due to exothermic biovegetable oil hydrotreatment;
- revamp of the recycle H_2 system to account for CO, CO_2 and H_2O production;
- upgrade the metallurgy to handle fatty acids;
- addition of a pre-treatment reactor to handle phosphorus and the alkali metals.

For these reasons, it was evaluated as more cost-effective to implement the process in a dedicated unit optimized for renewable feedstocks to overcome the problems listed above. Therefore, the Ecofining technology is a stand alone process carried out in a two-stage process. In the first stage the triglyceride structure of vegetable oil is cracked, removing the oxygen and saturating the double bonds, while in the second stage the cold flow properties are adjusted according to market requirements.

The most valuable way to target the first specification is to use hydroprocessing technology. This route uses hydrogen to remove the oxygen from the triglyceride molecules. The oxygen is easily removed via three competing reactions: hydrodeoxygenation, decarbonylation and decarboxylation. The three-carbon "backbone" yields propane that can be recovered easily when the process is integrated into a refinery. The oxygen contained in the feed is removed from the fatty acid chain either as CO/CO_2 or water. In addition, all olefinic bonds are saturated, resulting in a product consisting of only n-paraffins. The extent of each reaction depends on the catalyst and process conditions and the global mechanism can be depicted as:

$$RCOO-CH_2CH(-OOCR')CH_2-OOCR'' \xrightarrow{H_2} C_3H_8 + \begin{array}{l} RCH_3+2H_2O \\ R'H+CO+H_2O \\ R''H+CO_2 \end{array}$$

From this scheme it is evident that the hydrodeoxygenation produces a paraffin having the same number of carbon atoms as the fatty acid present in the triglycerides structure, while decarbonylation and decarboxylation produce a paraffin having one carbon atom less than the fatty acid present in the triglycerides.

A bimetallic hydrotreating catalyst, tailored for this kind of feedstock, has been developed and tested under different conditions and the main results are summarized here.

According to the experimental investigation the overall reaction is highly exothermic (e.g., the ΔH of reaction depends on the in-saturation of the fatty acids in the triglycerides and can vary from 97 kcal kg^{-1} for stearic acid and 422 kcal kg^{-1} for linolenic acid) and very fast. A complete conversion of vegetable oil has been observed for a short contact time and moderate temperature (e.g., above 310 °C). Increasing the temperature will affect the decarboxylation reaction while increasing the pressure will affect hydrodeoxygenation. Figure 8.4 shows the ratio of the decarboxylated chain (C_{17}) to the hydrodeoxygenated one (C_{18}) with varying temperature for three different pressures. The hydrogen consumption depends on the type of feed and is proportional to the concentration of the olefin bonds and oxygen content in the triglycerides. Notably, the liquid products are only n-paraffin with a 99% volume yield while the by-products are mainly in the gas phase, made up of CO/CO_2 and H_2O deriving from the oxygen of the fatty acid and propane from the three-carbon "backbone" present in the triglycerides structure.

Despite the high cetane number, the high cloud point of the liquid stream coming out from the hydrotreating reactor has a great impact in limiting the volume that can be blended with mineral diesel. To overcome this restriction these paraffinic streams have to be isomerized in a second stage. For such a purpose an appropriate hydroisomerization catalyst, based on a precious metal loaded on a mild acidic carrier, has been utilized. The reaction occurs at mild operating conditions and the kinetic behavior can be described according to the kinetic model reported for n-C_{16} [6], after proper parameter estimation.

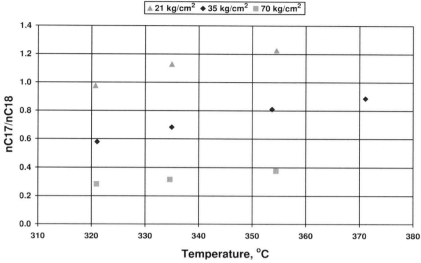

Figure 8.4 Ratio of decarboxylated to hydrodeoxygenated chain versus reactor temperature at three different pressures.

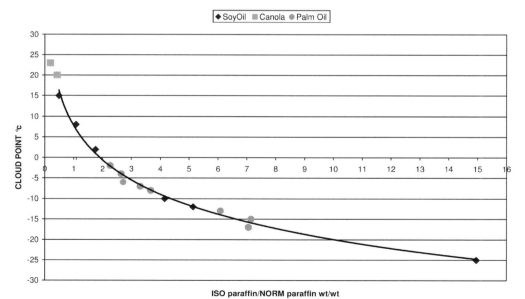

Figure 8.5 Green diesel cloud point as a function of the iso-/n-paraffin ratio.

The scope of this second stage is to control the cold flow properties of the final green diesel. As explained in the open literature [7–9], the diesel yield from the process will depend on the severity required in the isomerization reactor to meet cold flow specifications.

Figure 8.5 reports the correlation between the green diesel cloud point versus the ratio iso-/n-paraffin. Notably, the increase in iso-/n-paraffin ratio is obtained by increasing the severity of the process. The latter results in a different product distribution (diesel, kero and naphtha), with the diesel cut being the larger part (e.g., diesel 88–99 vol.%). Even under more severe conditions the diesel produced has a very high cetane number (>75) and contains no aromatics.

Extensive performance testing has been carried out in pilot plants to determine the optimum process conditions, catalyst stability and product properties. A range of vegetable oils have been processed in the pilot plants, including soybean, rapeseed, palm and jatropha oil. Other potential feedstocks, including tallow and greases derived from animals, have been evaluated.

As far as catalyst stability is concerned, a long pilot plant test has been carried out using soy oil and the results (Figure 8.6) show very good stability and product selectivity during the first 2000 hours of stream.

According to its superior properties, green diesel is a premium blending component. To verify this, the blending of mineral diesel with green diesel and engine tests have been performed. The diesel blends produced fulfill the highest requirements of car manufacturers and can be utilized with all diesel automotives without modification. Reduced tailpipe exhaust emissions were proven, and it was discovered that green diesel can be blended up to 65% to European 10 ppm diesel fuel. Additionally, low-value hydrotreated LCO can be blended with green diesel to be introduced into the typical diesel pool, meeting the European diesel specification.

8.3
UOP/Eni Ecofining Process

According to the above results an integrated process called the Ecofining process has been developed. Figure 8.7 shows a simplified flow diagram.

In the Ecofining process, vegetable oil is combined with hydrogen, brought to reaction temperature, and is then sent to a reactor section where the vegetable oil is converted into the green diesel product. The reactor section can consist of either a deoxygenation reactor or a combination of a deoxygenation and isomerization reactors to achieve better cold flow properties in the green diesel product. The product is separated from the recycle gas in the separator and the liquid product sent to a fractionation section. The design of the fractionation section can vary from a one-column system producing on-specification diesel and unstabilized naphtha to a three-column system producing propane, naphtha and diesel products. It is envisaged that most installations will be a single column and the lighter products will be recovered in other existing refinery process units. The recycle gas is treated in an amine system to remove CO_2. Table 8.3 gives typical product yields.

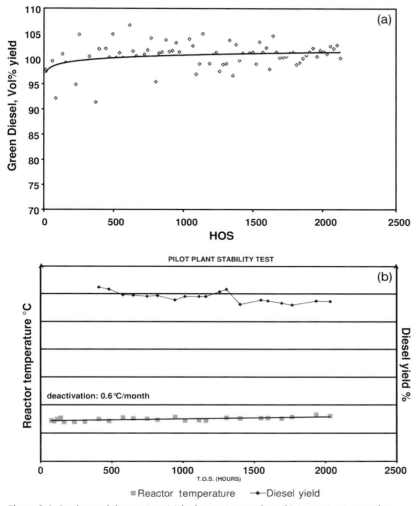

Figure 8.6 Catalyst stability testing: (a) hydrotreating catalyst; (b) isomerization catalyst.

8.4
Life Cycle Assessment

A life cycle analysis (LCA) of different diesel-production routes has been performed [10, 11].

The LCA evaluation program is a method to determine and compare the environmental impacts of alternative products or processes, including impacts of initial
resource extraction to waste disposal. In this study, the scope of the analysis was from
crude-oil extraction through combustion of the refined diesel fuels in a vehicle. It was
assumed that all fuels had the same performance in the vehicle. The primary focus of

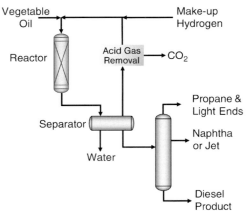

Figure 8.7 Simplified Ecofining process flow-scheme.

Table 8.3 Typical products yield of the Ecofining process

Feed	
Vegetable oil (wt%)	96.5–97
Hydrogen (wt%)	3–3.5
Products	
Diesel (wt%)	75–85
Naphtha (wt%)	1–8
Propane (wt%)	4–5
Water (wt%)	6–8
$CO + CO_2$ (wt%)	3–4
Sulfur (wt%)	<1
Diesel properties	
Cloud point (°C)	0 to −10
Cetane number	>80

the analysis was on fossil energy use and GHG emissions, although other impact categories are included.

Figures 8.8 and 8.9 summarize the results of the LCA. Green diesel compares favorably to biodiesel in the LCA study. Fossil energy consumption (Figure 8.8) over the life cycle is expected to be reduced by between 84–90% for green diesel produced from soybean oil or palm oil, respectively, when H_2 is produced internally from by-products (green diesel-B) rather than from fossil resources (green diesel-A). Thus, green diesel has the potential to displace more petroleum resources per energy content in the fuel compared to biodiesel. Larger reductions in greenhouse gas emissions for green diesel relative to biodiesel have been predicted by this study for soybean feedstocks (Figure 8.9), but lack of verifiable data on palm oil prevented any conclusions being made for this feedstock. Overcoming this omission and inclusion

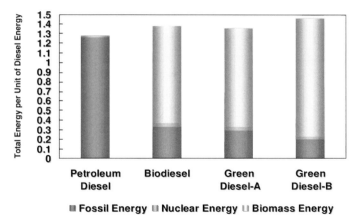

Figure 8.8 LCA analysis comparison of green diesel versus FAME biodiesel and a petroleum diesel: energy consumption per unit of diesel energy.

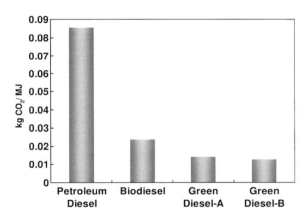

Figure 8.9 LCA analysis comparison of green diesel vs. FAME biodiesel and a petroleum diesel: GHG emissions.

of other environmental impacts will be the subject of future research in green diesel production and use.

8.5
Conclusion

UOP and Eni have developed the Ecofining process, a new, sustainable, route for converting vegetable oil into premium quality diesel fuel. This green diesel product is a superior alternative to FAME, with significantly better diesel product properties and is fully compatible with conventional mineral diesel products. The Ecofining process is fully developed and available for licensing from UOP. Eni is currently evaluating the size and the location of the first commercial unit to be realized within one of its refineries.

References

1 Fangrui, M.A. and Milford, H. (1999) *Bioresour. Technol.*, **70**, 1.

2 Knothe, G. (2005) *Fuel*, **84**, 1059.

3 Bournay, L., Casanave, D., Delfort, B., Hillion, G. and Chodorge, J.A. (2005) *Catal. Today*, **106**, 190.

4 Lotero, E., Liu, Y., Lopez, D.E., Suwannakarn, K., Bruce, D.A. and Goodwin, J.G. Jr (2005) *Ind. Eng. Chem. Res.*, **44**, 5353.

5 Holmgren, J., Gosling, C., Marinangeli, R., Marker, T., Faraci, G. and Perego, C. (September 2007) *Hydrocarb. Process*, **86**, 67.

6 Calemma, V., Peratello, S. and Perego, C. (2000) *Appl. Catal. A-Gen.*, **190**, 207.

7 Alvarez, F., Ribeiro, F.R., Perot, G., Thomazeau, C. and Guisnet, M. (1996) *J. Catal.*, **162**, 179.

8 Girgis, M.J. and Tsao, Y.P. (1996) *Ind. Eng. Chem. Res.*, **35**, 386.

9 Deldari, H. (2005) *Appl. Catal. A-Gen.*, **293**, 1.

10 Kalnes, T., Marker, T. and Shonnard, D.R. (2007) *Int. J. Chem. Reactor Eng.*, **5**, A48, pp. 1–9.

11 Holmgren, J., Gosling, C., Kokayeff, P., Faraci, G. and Perego, C. Green diesel production from vegetable oils. AICHE Spring Conference, April 2007-10-25.

9

A New Process for the Production of Biodiesel by Transesterification of Vegetable Oils with Heterogeneous Catalysis

Edouard Freund

9.1
Introduction

Transport fuels: gasoline (for cars), diesel (for trucks and cars) and kerosene (for aircrafts) are entirely dependent on oil, with no possibility for carbon capture (and sequestration).

Furthermore, if we consider the predicted growth of transport fuel on a world scale, as illustrated in Figure 9.1 and summarized in Table 9.1, we can expect, at least in the 20 years, that the use of all fuels will grow, with kerosene having the fastest growth. In addition, gasoline, the dominant fuel in 2005, will be superseded by middle distillates (kerosene and diesel).

The introduction of biomass-derived components or even substitution fuels may help to alleviate the dependency on oil and to control fossil carbon dioxide emissions from transport, provided some conditions are respected:

- producing the starting biomass material should require limited amounts of fossil fuels (including all steps);
- the process used for producing the final substitute motor fuel should also be energy efficient, to lead to a low overall ratio: fossil energy required/energy content of the final motor fuel.

The introduction of biomass-derived components has been proposed for more than a century:

- ethanol for gasoline, or a derivative of ethanol: ethyl-*tert*-butyl ether (ETBE). The direct introduction of ethanol is at present the preferred solution;
- vegetable oils – not in very common use – and more recently methyl esters derived by transesterification (see discussion below).

Another factor is important for the development of biodiesel: the fast growth of kerosene – kerosene is much more difficult to substitute because of the very rigid specifications of this fuel (necessary for aircraft safety but also imposed by

Sustainable Industrial Processes. Edited by F. Cavani, G. Centi, S. Perathoner, and F. Trifiró
Copyright © 2009 WILEY-VCH Verlag GmbH & Co. KGaA, Weinheim
ISBN: 978-3-527-31552-9

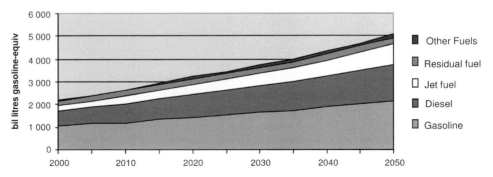

Figure 9.1 An extrapolation of motor fuel use. Source: IFP.

international acceptance). Thus, there will be a competition between diesel and kerosene, leading to constraints on diesel. The introduction of a significant percentage of biomass-derived diesel components will help to fill the gap. Overall, the development of biodiesel is observed not only in Europe – where private car dieselization is increasing – but also in countries like Brazil and the USA, even though gasoline is used mainly for private cars.

Figure 9.1 shows that the worldwide demand for diesel may reach very soon one billion ton per year. A significant substitution of diesel by biodiesel (of the order of 10%, or even more) implies the construction of numerous units (the average capacity of the unit already on stream being about $200\,000\,\mathrm{t\,yr^{-1}}$).

This has important consequences with respect to the conditions employed by the processes used for biodiesel production:

- the yield, as defined by the ratio the number of ton of final biodiesel per ton of raw material (the vegetable oil), should be high, to partly compensate for the low yield per hectare per year of most vegetable oils;
- the by-products should be of the appropriate purity, according to their specific market(s) (which will probably not be the market of motor fuels);
- waste production should be minimal for the process to be environment friendly;

Table 9.1 World fuel use by fuel type (%). Source: IFP.

	2005	2010	2020	2030
Gasoline	46.2	44.8	44.3	43.6
Diesel	31.2	31.6	31.8	31.7
Jet fuel	12.3	13.1	14.3	15.7
Residual fuel	8.0	7.6	6.9	6.5
Other fuels	2.3	2.8	2.6	2.5

- the process should be energy efficient, taking into account all steps: feed purification, catalyst production (and recycling), product and by-product(s) purification, waste treatment;
- the process should be economically competitive, taking into account the rather low differential between the raw material (vegetable oil) and the product (biodiesel).

We can notice, incidentally, that the processes used in the refining and primary petrochemicals industries meet the above conditions for the same reasons (massive production, limited differential between fuel and products). Mostly, in these fields, continuous processes are used (as apposed to batch processes) and, when applicable, heterogeneous catalysis is chosen in preference to homogeneous catalysis.

9.2
Direct Use of Vegetable Oils

The direct use of vegetable oils in diesel would obviously be the simplest way of introducing a substitute for diesel, especially for diesel engines in the range 10–1000 kW [the so-called rapid or semi-rapid diesel engines – for "slow" high power engines such as those used in large ships, other fuels can be used, especially "bunker" (heavy) fuel oil]. Such a direct use is well documented and has been proposed since the origin of the diesel engine (by Rudolf Diesel himself).

However, such a fuel is not well adapted to commercial modern diesel engines, even when used as a low percentage in the normal diesel pool. Several kinds of problems have been observed (and confirmed by recent studies, carried out in Europe, especially in Germany and France):

- cold start is very difficult in all cases but especially with oils having a high cloud point;
- preheating is required to reduce viscosity and facilitate injection;
- combustion is not complete (not even very satisfactory), so that the severe specifications applied to the modern diesel engines (in terms of production of nitrogen oxides, soot particles and organic compounds) cannot be respected;
- the limited thermal stability of partly unsaturated vegetable oils (the general case) will lead to the formation of deposits on the injectors and other parts of the engine and generally to a breakdown.

This is why, contrarily to the case of gasoline for which pure ethanol has long been used as a satisfactory component, vegetable oils have not been in use as a blending component in the diesel pool.

9.3
Methyl Ester Derived from Vegetable Oils

The use of the methyl ester of vegetable oils was proposed as early as 1983 by Guibet *et al.* [1] and one of the first industrial units was started in France near Compiègne based on a design by IFP in 1993.

Table 9.2 Comparison of the properties of diesel, rapeseed oil and its methyl ester.

Characteristics	Diesel	Rapeseed oil	Methyl ester of rapeseed oil
Specific gravity (kg m^{-3})	820–860	920	880–558
Viscosity at 40 °C (mm^2 s^{-1})	2–4.5	30.2	4.5
Octane index	>49	35	50
Flash point (°C)	≥100		170–180
Boiling range (°C)	200–400	Decomposed above 320	~340

Unlike vegetable oils, the methyl esters have properties very near those of diesel oil, as illustrated in Table 9.2 for the case of rapeseed oil (the most common starting material in Europe).

9.4
Homogeneous Process for the Production of Biodiesel

Figure 9.2 gives the general schema of such a process (as developed for instance by IFP). The specifications of the reactants are:

- for the vegetable oil: acid index (expressed in mg KOH/g oil) <1, phosphorus content ≤10 ppm, water <0.1%;
- for methanol: methanol content ≥99.85%, water content <0.1%.

Conventional industrial biodiesel processes are based on homogeneous alkaline catalysts. Sodium hydroxide or sodium methylate are the most often used catalyst in

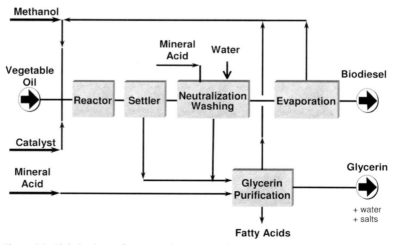

Figure 9.2 Global scheme for a typical continuous homogeneous catalyzed process.

industrial processes. This catalyst is dissolved in the methanol make up of the unit. Catalyst feed can be adjusted to match the required product conversion (i.e., biodiesel final quality). Catalyst feed is generally used in excess to ensure this conversion level. Optimum catalyst concentration level ranges from 0.5 to 1.0 wt%, based on vegetable oil feed. Free fatty acids in the reaction mixture can partly be compensated by the addition of excess catalyst, even if this higher catalyst concentration leads to a higher solubility of methyl ester in the glycerol phase, resulting in an increase in the settling time and an adaptation of the glycerol purification step.

Notably, the fast separation of the glycerol from the reacting mixture also removes most of the catalyst from the reaction mixture, even in the reaction phase, thus increasing the need of high initial catalyst make up.

The removal of homogeneous catalyst from the reactor effluent requires further downstream operations, which are illustrated in Figure 9.2.

The refined oil is mixed with the methanol/alkaline catalyst-mixture, reacting at mild temperature and low pressure and left in a settler for phase separation by gravity. The glycerol phase is decanted and is further refined. The crude ester product is purified by a washing step to remove the last traces of catalyst (Na + K content in the final ester needs to be lower than 5 ppm to meet EN14214). In some cases, a third reactor using acidic water can be used to purify the biodiesel with respect to alkaline and glycerol content. All this settling and neutralizing effluent are sent to the glycerol purification section for further treatment. The FAME remaining methanol–water content is evaporated and recycled into the process.

Most of the catalyst is recovered after the transesterification reaction as sodium glycerate, sodium methylate and sodium soaps in the glycerol phase. An acidic neutralization step with, for example, aqueous hydrochloric acid is required to neutralize these salts: sodium soaps are thus converted into free fatty acids, which can be removed from glycerol by decantation. In this case, glycerol is obtained as an aqueous solution containing sodium chloride. Depending on the process, the final glycerol purity is about 80%.

When sodium hydroxide is used as catalyst, side reactions forming sodium soaps generally occur. This type of reaction is also observed when sodium methylate is employed and traces of water are present. The sodium soaps are soluble in the glycerol phase and must be isolated after neutralization by decantation as fatty acids. The loss of esters converted into fatty acids can be as high as 1% of biodiesel production. The FAME yields obtained can vary from 98.5 up to 99.4 (wt%), depending on the feed quality and the type of catalyst used.

In all cases, for batch homogeneous biodiesel processes, the neutralization step leads to the formation of salts, carried in the glycerol phase, but which need to be disposed of. Glycerol from homogeneous biodiesel processes can to some extent be considered as a waste stream.

Commercial industrial processes can be operated in a either batch or a continuous mode. Batch processes are suitable for small plants, while for larger plants ($>100\,000\,t\,y^{-1}$) continuous process tend to be more economical. In the ESTERFIP batch process (IFP license), transesterification occurs in a single stirred-tank reactor. Continuous transesterification processes include the Ballestra, Connemann CD and

Lurgi PSI processes. These continuous process requires two or three reactors operated in series. After each step of catalysis reaction, glycerol is removed either by gravity or centrifugation [2]. A key point for these processes lies in the final washing of the ester. The role of this final washing step is to remove remaining glycerol, water, methanol and remaining alkaline catalysts. This wash is carried out in some cases by using a counter-current final water wash tower, and in one case using a vacuum flash drying technique. Basic salts coming with the process glycerol and water wash effluent streams are neutralized using strong acidic compound.

For both continuous and batch transesterification processes, around 4 kg of salts is co-produced in the glycerol phase per ton of oil processed. This salt is left to be treated by glycerol end users.

To be acceptable, the ester must meet certain specifications (in Europe: EN14214). The critical specifications are related to the cold properties and stability. These specifications limit the choice of starting vegetable oils, as discussed below. In Europe, rapeseed and oleic sunflower oils are used, in Brazil soybean oil.

These processes are satisfactory in several aspects:

- very active catalyst;
- good thermal efficiency, low energy consumption;
- range of capacity ($100\,000$–$250\,000$ t yr^{-1}) adapted to the capacity of the trituration plants;
- the product can be blended up to 30% wt with diesel, without significant engine modifications.

One limitation of this first generation process is intrinsic (and applicable to the heterogeneous process). It is linked to the specification of the methyl esters, which limits the choice of vegetable oils and the available quantity and price of these vegetable oils:

- to date, only edible vegetable oils have been considered, leading to a competition between diesel and food production;
- among these edible oils, the unsaturation of the fatty chains must be high enough to ensure good cold properties and low enough for acceptable stability. Only pure rapeseed oil and to some extent oleic sunflower oil can be used, as well as palm oil and soybean oil in mixture.

The yield per hectare for these crops does not significantly exceed 1 Toe (ton oil equivalent)/ha/year, with only slight improvements in sight with this limited choice of oils. Furthermore, crop rotation is required to insure pest control and good yield; this leads to much lower yield per hectare of land available each year for such crops.

A specific difficulty associated with processes using homogeneous catalysis is the production of impure glycerol, containing the salt resulting from the neutralization of the soluble catalyst by a mineral acid (either hydrochloric or sulfuric acid), in large amounts as compared to the existing markets of glycerol.

In a typical transesterification process, the final glycerol purity is about 80%. The major impurities of the glycerin produced are water, salts (NaCl, Na_2SO_4, KC1, etc., depending on the base used as catalyst and on the acid used for catalyst neutralization) and organic compounds such as esters and soaps.

The crude glycerol can be refined to obtain the USP grade, used in pharmaceuticals, cosmetics and food applications.

The first step is to get rid of the neutralization salt. The purification scheme is rather complex and costly, especially if distillation is required. Furthermore, the reaction produces wastes (mainly sodium salts polluted by organic compounds) that have to be disposed of.

Massive building of homogeneous biodiesel plants will yield a sharp increase in associated biodiesel wastes, which will need a remedial solution.

Valorization of the by-product glycerol is absolutely essential for the competitiveness of the process. As the classical markets of glycerol are already saturated by the installed biodiesel capacity, new markets have to be developed, such as feeding component for cattle [3], as an intermediate to prepare a diesel component, or as an intermediate for chemical synthesis. For such markets, different and specific level of purities are required, which are not easy to reach economically.

9.5
Improving the Transesterification Route: Esterfip-H

The classical route involves very active soluble catalysts, with the major drawback of a difficult purification of the glycerol produced, as discussed above. A simple way to avoid the problem is to turn to heterogeneous catalysis. Such a process has been developed by IFP and is being commercialized by Axens [4]. The first industrial unit started in 2006, at Sète, in the south of France.

In this new continuous process, the transesterification reaction is promoted by a completely new heterogeneous catalyst. This catalyst consists of a mixed oxide (a zinc aluminate) that promotes the transesterification reaction without catalyst loss [5]. The reaction is performed at higher temperature and pressure than in the homogeneous catalyst process, due the lower activity of the solid catalyst.

Figure 9.3 presents a flow-sheet of this process.

The desired chemical conversion, required to produce biodiesel that meets European specifications, is reached with two successive stages of reaction and glycerol separation to shift the equilibrium of methanolysis. The catalyst section includes two fixed-bed reactors, fed with vegetable oil and methanol at a given ratio. Excess of methanol is removed after each reactor by partial evaporation. Then, esters and glycerol are separated in a settler. Glycerol outputs are gathered and the residual methanol is removed by evaporation. To obtain biodiesel of European specifications, the last traces of methanol and glycerol have to be removed. The purification section of methyl ester output coming from the second decanter consists of a finishing

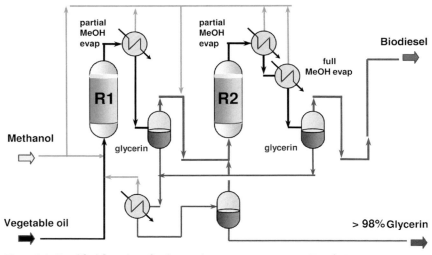

Figure 9.3 Simplified flow-sheet for the new heterogeneous process, Esterfip-H.

methanol vaporization under vacuum followed by a final purification in an adsorber for removing the soluble glycerol.

This new heterogeneous catalyst process offers the main following advantages (Table 9.3):

- a high biodiesel yield can be obtained, since there is no ester loss due to soap formation (FAME yield = 100%);
- the crude glycerol obtained is salt free and has a very good purity (>98%), thus allowing new direct ways of valorization as described above;
- there is neither chemical products consumption nor waste streams.

Table 9.3 Simplified comparison of transesterification processes.

Homogeneous biodiesel process	Heterogeneous biodiesel process
(−) 99.3 Yield	(+) 99.8 Yield
(−) Glycerol quality	(+) Glycerol quality
(−) Complex purification steps	(+) Simplified purification steps
(−) Aqueous and salts side-stream	(+) No side-stream
(−) Chemical added	(+) No chemical added
(−) High catalyst cost	(+) Low catalyst cost
(+) Mild operating conditions	(−) More severe operating conditions
(+) Low MeOH/oil ratio	(−) High MeOH/oil ratio
(+) Proven technology	

9.6
Future Improvements of the Process

Future improvements are planned in three directions:

1. Improve the catalyst.
2. Relax some of the specifications on the feed (purity of the vegetable oil).
3. Develop an economical process for the production of ethyl esters.

9.6.1
Catalyst Improvement

A higher catalyst activity is necessary to lower the reaction temperature, which in turn will allow the operating pressure to be lowered, thus improving the energy requirements of the process. Such desired catalysts are under development at IFP. They can be retrofitted in existing units.

9.6.2
Extension of the Process to other Feeds

The heterogeneous process has potentially more flexibility with respect to the purity of the starting feed. Provided the impurities introduced with the oil have no detrimental effect on the catalyst activity (that can be if necessary protected by a small guard reactor), good purity esters and glycol can still be obtained, due to the more efficient separation section.

9.6.3
Development of a Process for the Production of Ethyl Esters

Replacement of fossil-derived methanol by ethanol is an interesting objective:

- the ethyl ester would be 100% derived from biomass;
- this is one way to incorporate some ethanol in the diesel pool. It would help to fit the motor fuels consumption in Europe, where the gasoline/diesel usage ratio is low and still decreasing, which makes ethanol a non-optimal biofuel in this context.

However, this proves not easy for several reasons. First, the reaction is slower: more catalyst and/or higher temperatures are needed for ethyl esters production to balance the lower ethanol reactivity. Second, ethanol is a better solvent for oil and ethyl esters. With homogeneous catalysis, the higher solvent effect of ethanol induces more severe thermodynamic limitations for oil conversion, since glycerol produced is much more soluble in the ethanolysis reaction mixture than in the case of methanolysis. Conversely, in heterogeneous catalysis the reaction mixture is monophasic, with both methanol and ethanol; in that case, the thermodynamic limitations for oil conversion are roughly the same when using ethanol instead of methanol. Moreover, ethanol acting as a co-solvent makes glycerol extraction from ethyl esters more complex and costly. Finally, the water content of ethanol has to be low, but ethanol dehydration is more complex than methanol, owing to azeotrope formation. In homogeneous catalysis, water is a precursor of soap formation, leading to an increase of catalyst consumption and to a decrease of ester yield. In heterogeneous catalysis, water acts as an inhibitor of the catalyst but does not affect either ester yield or glycerol purity.

Heterogeneous catalysis should be a better way to produce ethyl esters, provided the catalyst improvement looked for in the case of methyl ester is successful. In any case, the process will be less competitive and the product more expensive.

9.7
Conclusion

Although other routes are being developed, some being near commercialization, to produce diesel starting from biomass [6], biodiesel from transesterification of vegetable oils with methanol will remain a very important route. Unlike other routes, the starting materials, vegetable oils, are, chemically, very near the desired product (diesel oil, a mixture of hydrocarbons). Significant improvements are expected on going from a classical "chemical" type of approach, using a batch process and homogeneous soluble catalysts that are consumed (and appear as wastes), to a continuous process with heterogeneous catalyst that can integrate all the progress made in the field of nanotechnology applied to catalysis.

References

1 Stern, R., Guibet, J.C. and Graille, J. (1983) *Rev. Institut Français Pétrole*, **38**, 121–136.

2 Bray, R.G. (2004) *Biodiesel Production*, SRI Consulting.

3 Kijora, C. and Kupsch, R.D. (1996) *Fett/Lipid*, **98** (7–8), 240–245.

4 Bournay, L., Casanave, D., Delfort, B., Hillion, G. and Chodorge, J.A. (2005) *Catal. Today*, **106**, 190–192.

5 Stern, R., Hillion, G. and Rouxel, J.J. (1999) US Patent 5,908,946.

6 Casanave, D., Duplan, J.-L. and Freund, E. (2007) *Pure Appl. Chem.*, **79**, 2071.

10
Highly Sour Gas Processing in a More Sustainable World

François Lallemand and Ari Minkkinen

10.1
Introduction

For decades to come, natural gas will be the energy source of choice to meet ever-greener worldwide environmental standards. Fortunately, gas reserves are growing; but new gas is often found to be of substandard quality in remote and/or stranded areas of the world. When natural wellhead or oil field associated gases are highly loaded with acid gases, the dilemma facing most operators is what to do, how and when to best exploit these poor quality resources. Total is increasingly faced with these choices, together with its operating partners around the world; especially in areas known to have highly sour oil and gas reserves, such as the Caspian sea region.

Fortunately, Total, via past efforts of Elf Aquitaine, has many years of experience in producing gas from slightly sour to very sour gas reserves, notably in the Lacq region of France. This experience has been the ideal proving ground in the development of state-of-the-art acid gas removal technologies. Today the advanced activated MDEA process offers economy and versatility in handling both selective and complete acid gas removal services. The process has a good synergy with modern Claus sulfur recovery processes and remains among the best alternatives even when no sulfur recovery is foreseen.

Nevertheless, there are limitations of even the most advanced amines-only based gas treatment technologies in handling very highly acid gas loaded natural or associated oil field gases – especially for bulk acid gas removal when the acid gases are destined for cycling and/or disposal by re-injection.

Today, cycling and disposal by re-injection offers a promising alternative to avoid sulfur production and reduce CO_2 emissions to the atmosphere simultaneously. To this end, technologies of choice are those that offer maximum simplicity and require least downstream processing intensity and power for re-injection.

Sustainable Industrial Processes. Edited by F. Cavani, G. Centi, S. Perathoner, and F. Trifiró
Copyright © 2009 WILEY-VCH Verlag GmbH & Co. KGaA, Weinheim
ISBN: 978-3-527-31552-9

10.1.1
Background

When Elf Aquitaine, years before becoming Total, decided in the mid-1950s to produce gas from the then discovered Lacq field in south west France several challenges were faced as it was the first highly sour, high pressure and high temperature gas field produced at that time. High performance had to be achieved to remove large quantity of acid gases from raw gases containing over 15% H_2S and up to 10% CO_2. In addition, new steel alloys had to be developed to resist the highly corrosive behavior of the fluids.

The technology first selected in the early days was the so-called SNPA-DEA process using exclusively the reactive chemical solvent diethanolamine (DEA). Much experience was gained while improving the process by nearly doubling the DEA solution concentration. This was done to reduce solvent circulation and operating costs while keeping corrosion under control without inhibitors. The continually improved processes were implemented in several sweetening units between 1957 and 1968, eventually attaining a sales gas production capacity of 25 million standard cubic meters per day (880 MMSCFD).

With the decline in production of the Lacq field and with constantly increasing energy costs, the need for new processes, better suited than DEA to a new competitive gas market environment, became apparent. Fortunately, existing units and equipment made available by the reduction of Lacq gas production conveniently opened the way to conversion, remodeling and testing on the industrial scale new solvent formulations, absorber internals and processing techniques. The new processes then developed were based on the use of *tertiary*-methyldiethanolamine (MDEA) and a flash-procured rich amine regeneration system. This significantly reduced the energy consumption of the amine reboilers.

Since 1977 MDEA has been used by Elf Aquitaine for de-acidification of gases that do not require total CO_2 removal. The possibilities offered by the MDEA process to control the CO_2 slippage from the absorber by proper choice of absorber internals and operating conditions make it suitable for many different applications, allowing tailored amounts of CO_2 in the product gas while making H_2S-richer acid gas for Claus sulfur recovery units. Since 1986, an MDEA process has been used at Lacq to remove H_2S from raw high-pressure sour gas streams to produce H_2S-rich streams used for thio-organic chemical synthesis. The MDEA process for selective H_2S removal has been implemented in more than 20 units worldwide, either operated by the Total group or other operating companies under license. The need to respond also to the requirement of total acid gas removal, to be able to produce product gas with less than 50 ppm CO_2, subsequently led to the development of an activated MDEA process having the removal performance of DEA with significantly reduced energy consumption. Today a series of chemical activators used with MDEA in a split semi-regenerative process scheme offers the most cost-effective answer to complete or controlled acid gas removal from sour to very sour natural gases.

Total has also long been interested in the bulk removal of acid gases; notably, in gas fields operated in the Far East, containing over 40% CO_2, as well as in the Caspian sea region, for gas and oil-associated gases containing over 20% H_2S with CO_2. Recognizing that the world sulfur market today is saturated and CO_2 emissions as a greenhouse gas need to be drastically curtailed, has led to serious consideration of acid gas cycling and/or disposal by re-injection. To this end, technologies other than the classical aqueous amine based ones are expected to have more favorable attributes. Some of these attributes are the delivery of acid gases in a dry state and at higher pressures to reduce recompression power, facilitate re-injection system design and avoid exotic materials.

Finally, Total in collaboration with IFP has embarked upon a comprehensive program with a pooling of resources to develop new gas treatment technologies to better suit the bulk acid gas removal destined for re-injection applications. To this end, pre-treatment techniques and hybrid solvent processes are envisioned as the most cost-effective overall solutions.

10.2
Use of Activated MDEA for Acid Gas Removal

When the MDEA process was developed in the mid-1970s it was principally destined for the sweetening of gases that did not require complete CO_2 removal [1], or required the removal of only a controlled part of the CO_2. Typical applications of the selective MDEA process are:

- H_2S removal from gases already at or close to the allowable CO_2 specification;
- H_2S removal from gases where the CO_2 is not important; as in fuel gas, for example;
- production of an H_2S-rich (i.e., CO_2 poor) acid gas for Claus type sulfur recovery and or thio-chemical synthesis.

The use of MDEA for selective H_2S removal is based on the fact that, unlike DEA, MDEA does not react directly with CO_2, and thus CO_2 absorption kinetics can be controlled by the slow reaction of CO_2 with water. For complete CO_2 removal along with H_2S, the more reactive DEA has traditionally been used but with a serious energy consumption handicap in its regeneration step. The ever tighter constraints imposed on gas processing, stemming mainly from the necessity to reduce operating costs, led to the development of a new MDEA process for complete acid gases removal from sour gases with a substantially lower regeneration energy consumption than conventional DEA.

Two main factors affecting a solvent's performance for CO_2 absorption and its ease of regeneration are solubility and reactivity. While reactivity of CO_2 in MDEA is lower than in DEA, its solubility in MDEA is more strongly influenced by CO_2 partial pressure than its solubility in DEA. This can be shown by the slopes of the equilibrium solubility curves (Figure 10.1).

The basic concept for a new process was then to take advantage of the slope of the equilibrium solubility curves of CO_2 in aqueous MDEA solutions to be able to liberate a maximum amount of the acid gas from the solution by simple physical

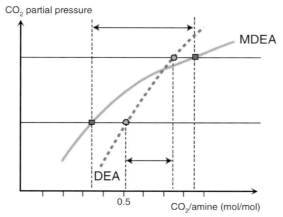

Figure 10.1 Equilibrium solubility of CO_2 in aqueous amines.

pressure let-down flash and thus reduce the thermal regeneration duty substantially. For example, as shown in Figure 10.1, the MDEA solution releases almost twice as much CO_2 as the DEA solution by a pressure letdown from 10 bar to 2 bar. In addition, the equilibrium solubility of H_2S in aqueous MDEA exhibits identical behavior to CO_2, allowing equivalent amount of H_2S liberation by pressure letdown. Unfortunately, unlike with H_2S, MDEA reacts slowly with CO_2. This in particular, which is used to advantage for selective H_2S and controlled CO_2 removal, becomes a handicap for complete acid gas removal – especially when a substantial quantity of CO_2 has to be removed. To overcome this kinetic obstacle, activators were sought among secondary amines having high speeds of reaction with CO_2 to blend into the otherwise desirable MDEA solvent. Figure 10.2 illustrates the reaction mechanism postulated.

With MDEA alone the transformation of CO_2 into bicarbonate is a slow process while the reaction of the carbonic acid with MDEA is instantaneous. With an activator in the MDEA solution the transformation into bicarbonate via a first step formation of

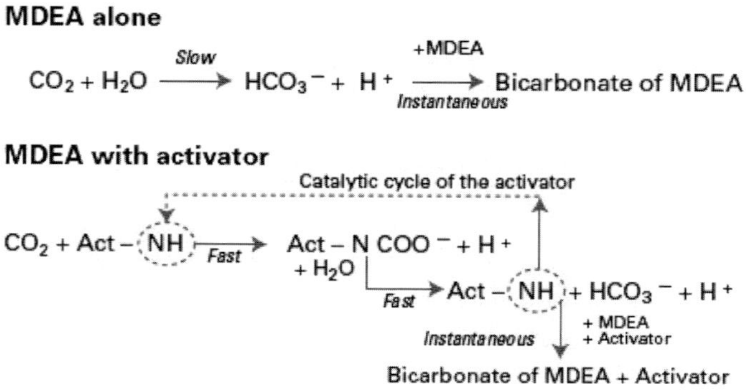

MDEA alone

$$CO_2 + H_2O \xrightarrow{\text{Slow}} HCO_3^- + H^+ \xrightarrow[\text{Instantaneous}]{\text{+MDEA}} \text{Bicarbonate of MDEA}$$

MDEA with activator

Catalytic cycle of the activator

$$CO_2 + Act - NH \xrightarrow{\text{Fast}} Act - N\,COO^- + H^+$$
$$+ H_2O$$

$$\xrightarrow{\text{Fast}} Act - NH + HCO_3^- + H^+$$

$$\xrightarrow[\text{Instantaneous}]{\quad} \begin{array}{l} + \text{MDEA} \\ + \text{Activator} \end{array}$$

Bicarbonate of MDEA + Activator

Figure 10.2 Mechanism of CO_2 activation.

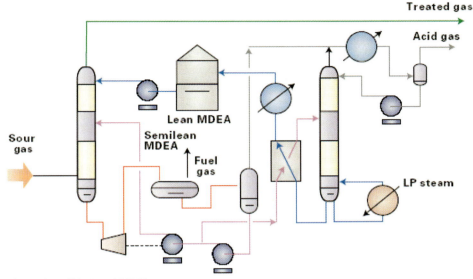

Figure 10.3 Elf Activated-MDEA process.

a carbamate is made faster. Laboratory tests were performed to select amines with carbamate formation rates that could substantially activate the CO_2–MDEA reaction and to determine the concentrations required to optimize solvent efficiency. Several activators were selected from among those best suited to industrial conditions, taking into account commercial availability, cost and impact on the environment [2].

Another process improvement over the conventional DEA process is the introduction of a secondary semi-lean solvent circulation, as shown in the simplified process flow diagram in Figure 10.3. The rich amine solution after letdown through a hydraulic turbine is divested of co-absorbed light hydrocarbons in a first flash drum, as in the conventional process, then further expanded to a low pressure in a second flash drum to partially liberate CO_2 and H_2S. The greater part of the rich amine thus partially regenerated is returned to an intermediate level of the absorber as a semi-lean solvent. This second solvent loop is particularly economic as it reduces the thermal regenerator load and consumes only pumping energy.

When H_2S is present and the treated gas specification calls for the removal of H_2S below 3.0 ppm, or for the removal of CO_2 below the 1–2 vol.% range, thoroughly regenerated, virtually H_2S- and CO_2-free, amine is required. This is accomplished in a conventional thermal regenerator by returning a small flow of the leanest amine to the top section of the absorber.

In 1990 the new solvent and regeneration system were tested on an existing DEA unit at Lacq, which was then converted into the then-called Elf Activated MDEA process [3, 4]. This process was also used in several other locations offshore North Sea such as in Sleipner Vest for CO_2 removal and Elgin Franklin for controlled CO_2 removal. Different activators have been selected and patented to suit each specific case of treatment; total or partial, controlled CO_2 removal with or without H_2S.

10.3
Process Performance Highlights

The performance of the Elf Activated MDEA is closely related to site-specific treatment conditions, notably feed gas composition and treated gas specifications. Below are some practical rules-of-thumb to highlight the general interest of regeneration by flash [5]:

• The greater the H_2S and/or CO_2 partial pressure in the feed gas, the greater is the efficiency of the flash. In the case of the Lacq plant, the acid gas partial pressure of the feed is approximately 15 bar in comparison to a total pressure of 2 bar in the flash drum. This gives a ratio of 7.5 (acid gas partial pressure/flash pressure). In general the advantage of flash starts above a ratio of 3.0.

• The overall energy consumption of the Activated MDEA unit increases as the treated gas specification becomes more stringent, since the amount of lean totally regenerated amine from the thermal regeneration increases. In some CO_2-only cases, as mentioned above, it may even be possible to completely eliminate the thermal regenerator.

As a consequence of the above attributes the Elf Activated MDEA process will be well adapted to the treatment of high pressure and very sour gases where the advantages of flash procured regeneration will be maximized.

10.4
Case Study of the Use of Activated MDEA for Treatment of Very Sour Gas

A noteworthy case study of the use of the Activated MDEA process is given below for the treatment of very H_2S sour gas. It concerns a real plant that was operated by Elf in the 1970s in Alberta Canada. The original design used a conventional 30 wt% DEA solution, which was considered the state-of-the-art at the time the plant was built. Table 10.1 gives the gas composition on a dry basis.

Table 10.1 Sour gas composition: an example.

Component	Volume %
H_2S	34.9
CO_2	7.5
Methane	56.5
Ethane	0.6
Propane	0.1
Butanes	0.1
Pentanes plus	0.3
Pressure [bar, (psig)]	70 (1000)

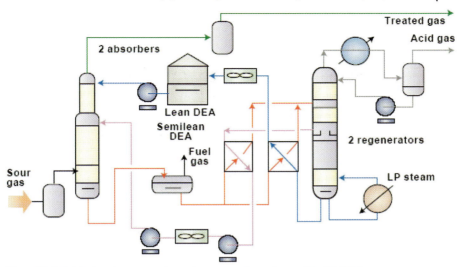

Figure 10.4 Original amine sweetening units process flow-scheme (30 wt% DEA).

The treated gas had to meet typical pipeline gas specifications, that is, less than 4 ppm vol. H_2S and less than 2.0 vol.% CO_2. The acid gas liberated is sent to conventional two-stage Claus units followed by a Sulfreen tail gas treating unit. Figure 10.4 illustrates the original amine sweetening units process flow-scheme. There are two identical trains, each equipped with two high-pressure absorbers and two regenerators.

The raw sour gas capacity of each train is 162 million std.ft^3 per day (MMSCFD) or 4.6 million std.m^3 per day for a sulfur production of over 2000 tons per day. The reboilers duty for each amine train was 135 MW (460 MMBtu h^{-1} or 116 Gcal h^{-1}), which is roughly equivalent to the heat energy produced by the downstream Claus unit as LP steam.

We have revised the original design to reflect current state-of-the-art amine technology, using 48 wt% activated MDEA solution while retaining the same process scheme using a conventional double split flow design with thermal regeneration. This results in a decreased amine solution circulation rate and thus reduced reboiler duty. The reboiler duty is reduced from 135 to 91 MW.

Figure 10.5 shows what would be the design of the amine units using the Elf Activated MDEA process with the addition of partial regeneration by flash. In this case only a third of the total amine solution circulation is sent to thermal regenerator. As a consequence the reboiler duty is reduced to only 46 MW. This uses only one-third of the Claus unit's steam production and only one regeneration column is needed for each train, resulting in additional capital cost savings potential. The quality of the treated gas is identical to the original design and the acid gas produced still has a very good quality, with a hydrocarbon content of less than 1.0 vol.%. The latter specification is important for charging a Claus unit, but is equally important for acid gas cycling and/or re-injection, as will be discussed below. Table 10.2 summarizes the stepwise improvements achieved.

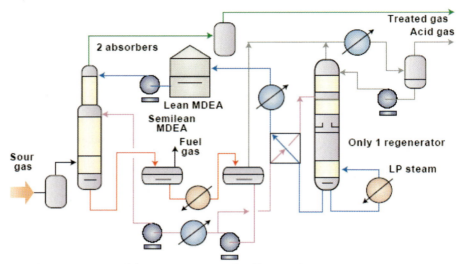

Figure 10.5 Design of the amine units using the Elf Activated MDEA process.

Table 10.2 Stepwise improvements achieved using the Elf Activated MDEA process.

Solvent	Process	Reboil energy consumption [MW (MMbtu h^{-1})]
30 wt% DEA	Original	135 (460)
48 wt% Act. MDEA	Thermal regeneration	91 (310)
48 wt% Act. MDEA	Flash procured regeneration	46 (156)

10.5
Acid Gas Removal for Cycling and/or Disposal

Environmental concern over global warming due to greenhouse gas emissions has given ever rising importance to the re-injection of CO_2 removed from natural gases; either for reutilization to enhance oil recovery (EOR) or just simple disposal to a depleted reservoir to avoid atmospheric venting. Moreover, with the growing acceptance of H_2S re-injection as a feasible alternative to costly sulfur recovery to a diminishing sulfur market, several very sour gas wells can be re-considered as exploitable to produce much needed natural gas. Many of such wells have been blocked-in, waiting either for the gas price increase to justify costly production or a simplification of the overall processing scheme to reduce production costs. Though gas prices today have increased, who knows for how long the present price situation will prevail. In any case, acid gas re-injection to a disposal reservoir will provide the simplification and cost reduction to make exploitation of the sour gas reserves attractive even in a flat gas price scenario. Since H_2S and CO_2 are re-injected underground, back to where they originated from, no CO_2 or sulfur emissions are made to the atmosphere.

Acid gas removal from the desired marketable hydrocarbons is the first step in the sour gas production scheme. Many acid gas removal processes are available to meet current pipeline sales gas-specifications in H_2S and CO_2 content. However, for maximum versatility and economic benefit to an acid gas removal to re-injection or disposal project, the overall process scheme should have quality characteristics, among which are:

- high capacity for acid gas removal with minimum hydrocarbon co-absorption;
- easy regeneration by pressure letdown with minimal thermal input as co-produced heat energy of the Claus unit is no longer available;
- liberation of acid gases at some pressure and preferably cold and dry;
- possibility to adapt regeneration pressures to inter-stage acid gas compression.

Physical solvent processes give some, but not all, of the above qualities. The Selexol process has several industrial applications, most of them for synthesis gas de-acidification and some for natural gas treatment [6–8]. A methanol-based refrigerated solvent process such as the Ifpex-2 process from the Ifpexol technology matrix of IFP is also a good contender [9]. However, physical solvents have a high affinity for hydrocarbons and the separated acid gas stream contains large quantities of valuable hydrocarbon products.

Chemical solvent processes generally have a higher energy requirement than physical solvent processes, but do not absorb hydrocarbons. However, among these processes, the Activated MDEA process from Total discussed in the preceding sections, which removes H_2S completely and CO_2 as required, has a low energy requirement, thanks to the ability of the Activated MDEA to liberate the bulk of the acid gases in a simple flash (flash procured regeneration). With the exception of acid gas dryness and pressure, the Elf Activated MDEA process meets all of the above listed quality characteristics for a re-injection scheme; most remarkably the liberation of acid gases without hydrocarbons. This is important not only to reduce sales gas shrinkage but also to reduce re-compression flowing capacity. The very low hydro-carbon content of the acid gas produced by the Activated MDEA process is another factor impacting positively on the wellhead pressure requirements as the hydro-carbons present reduce the density and water solubility of the pressurized acid gas fluids. The only disadvantage of the Activated MDEA process is the liberation of the acid gases at low pressure.

Hybrid solvent processes use a mixture of a physical solvent with a chemical solvent and combine some of the advantages of physical solvent processes with those of chemical solvent processes. The Sulfinol process has numerous industrial applications in sour gas de-acidification. Its energy requirement is relatively low, but hydrocarbon co-absorption is higher than that of an amine process.

The cost of acid gas removal depends strongly on its concentration and the need for downstream compression. In the case of very sour gas the combination of a bulk removal step ahead of the final sweetening unit can reduce the overall acid gas removal cost. If the acid gas is re-injected the bulk removal process should offer the possibility of pre-extracting and/or recovering acid gas in the liquid phase and at high pressure for pumping to re-injection pressure.

Here we examine the combination of the well-established Elf Activated MDEA process from Total with a new pre-extraction technique under development by IFP.

10.6
Bulk H$_2$S Removal for Disposal

The proposed overall process flow-scheme for the treatment of very sour gases (i.e., with an acid gas content above 20%) with re-injection of the separated acid gases to a disposal reservoir incorporates a special patented H$_2$S pre-extraction step upstream of the Activated MDEA acid gas removal process. In this upstream step [10], called SPREX, a substantial amount of the H$_2$S and some of the CO$_2$ are pre-removed from the wet raw gas as a pumpable liquid stream. This liquid will essentially contain by solubility all the water of saturation that comes with the inlet raw gas. It will also contain some of the incoming hydrocarbons. Figure 10.6 depicts a process flow diagram of this special pre-extraction and Activated MDEA combination process.

The dried gas leaving the SPREX contactor is cooled through a gas/gas heat exchange, gas/condensate heat exchange and then through a refrigerated chiller before being passed to a low temperature separation (LTS) drum operating essentially at line pressure. Since the SPREX liquid removed most of the water from the feed gas by solubility, no further free water condensation is expected in the chilling train. However, in practice some hard-piped methanol injection would be incorporated in case upset trace free water might appear. No methanol recovery is warranted due to the negligible amount of, and only intermittent injection of, methanol, if any. Any

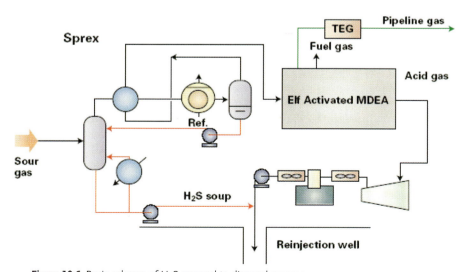

Figure 10.6 Basic scheme of H$_2$S removal to disposal process.

injected methanol is thus considered lost. The refrigerated chiller may be replaced by any other refrigeration process, such as a turbo-expander, if the inlet fluid is available at sufficient pressure.

The separated liquid stream is pumped through a heat exchanger and warmed before being recycled to the SPREX contactor. The H_2S-rich recycle contains dissolved hydrocarbons, which are then stripped from the liquid in contact with ambient temperature feed gas. Owing to the equilibrium recycle, the H_2S partial pressure is substantially increased in the contactor, allowing a good portion of the incoming H_2S to condense and drop out in its bottom section. The H_2S-rich mixed liquid stream or "soup" at line pressure and near ambient temperature, containing also pre-extracted CO_2, some of the hydrocarbons from the inlet gas together with soluble water and any methanol injected, is taken to the suction of the disposal pump shown. Pumping liquid is more energy efficient than compressing vapor to re-injection pressure.

The gas stream exiting the LTS with the remaining acid gas is warmed through gas/gas heat exchange and is then taken to the absorber of the MDEA unit. The MDEA unit is designed using the Elf Activated MDEA process. The selection of an activator or none will depend upon the amount of CO_2 to be allowed to slip into the pipeline gas. In this case, CO_2 removal can be selective with savings on solvent loading and regeneration duty. When the remaining acid gas loads are still significant, the partial flash regeneration may be advantageously used to reduce solvent circulation and reboil duty as described and illustrated in the previous section with Figure 10.3.

The treated gas leaving the MDEA unit will require downstream dehydration since the pipeline gas specifications require a minimum water dew point. This is best achieved with a conventional TEG glycol dehydration unit, as the gas is very lean without condensable hydrocarbons.

10.7
SPREX Performance

Acid gas extraction performance of the SPREX process is dependant on the inlet fluid composition and on the temperature in the LTS. The higher the acid gas content of the inlet gas, or the lower the temperature in the LTS, the higher the pre-extraction performance (Figure 10.7).

Hydrocarbon co-extraction will increase with the acid gas pre-extraction efficiency. For the bulk removal of acid gas from very sour gases, where very high acid gas pre-extraction is achieved, more sophisticated process schemes are being evaluated to maintain hydrocarbon co-extraction at a low and acceptable level. It is envisaged that the hydrocarbon shrinkage can be maintained below 3% in all cases, even when treating gases with a very high H_2S content.

For a LTS temperature of $-30\,°C$, an H_2S pre-extraction of more than 70% may be achieved from a gas with 35% H_2S. The CO_2 pre-extraction performance of the SPREX process is somewhat lower than that of H_2S.

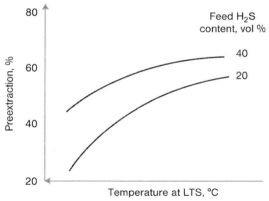

Figure 10.7 SPREX process: H_2S extraction performance.

10.8
Capital Cost and Energy Balance Comparison

The SPREX "soup" is available at line pressure and in the liquid phase, thus considerably reducing the power consumption for re-injecting the acid gas. Since the MDEA process for the remaining acid gas removal is now substantially reduced in its acid gas removal load, its solvent circulation, equipment sizes and investment cost are correspondingly reduced. Table 10.3 summarizes the capital cost and energy savings achieved when using the SPREX process in the Canadian case already described above, assuming that the separated acid gases are re-injected with a wellhead injection pressure of 150 bar (2140 psig). The figures given in Table 10.3 concern the following processing units: de-acidification, acid gas compression and pumping, associated utilities generation, and are exclusive of field costs.

Table 10.3 CAPEX and energy comparisons of the SPREX process.

	Without Sprex	With Sprex
CAPEX (base 100 for conventional treatment)		
Sprex unit	0	23
MDEA + AG compression + utilities generation	100	57
Total processing units	100	80
Energy consumption (MW)		
Power (pumps and compressors drivers)	52	29
Thermal energy (LP steam)	46	34

10.9
Conclusions

Highly sour gas processing is becoming evermore commonplace as the world's appetite for clean burning gas energy is growing and sweet gas reserves are dwindling. Technologies to treat sour natural gas to make it pipeline quality are numerous, but none have shown to be as versatile and economic as those using Activated MDEA.

To be truly competitive, the removal of the acid gas components H_2S and CO_2, be it trim or bulk, complete or partial, requires the optimum choice of an activator together with a carefully crafted know-how in solvent absorption/regeneration process design. The Elf Activated MDEA process developed by Total is probably the most cost-effective solution today to meet the widest range of applications from complete CO_2 removal to bulk H_2S and/or CO_2 removal even for acid gas re-injection projects. The historical R&D efforts of Elf Aquitaine coupled with current resources of Total allows this MDEA process to be credited with the most significant know-how back-up technology base on the market today.

To further meet forthcoming challenges in an ever-greener world, acid gas re-injection for disposal has become an important technology, especially in regard to abatement of greenhouse gas and sulfur emissions. Moreover, today's diminishing sulfur market no longer justifies Claus-type sulfur recovery. To this end, Total and IFP have been collaborating on the development of new and less costly technologies to handle highly H_2S sour gases, especially those intended for bulk acid gas disposal projects.

The upstream acid gas pre-extraction technique from IFP, called SPREX, offers a synergistic combination with the Elf Activated MDEA process in most applications for bulk H_2S rich acid gas disposal projects. The substantial reduction of the acid gas removal and re-compression loads afforded by the SPREX step coupled with energy efficient flash procured Activated MDEA process achieves sour gas purification with bulk acid gas disposal very cost-effectively.

References

1 Blanc, C., Elgue, J. and Lallemand, F. (1981) MDEA process selects H_2S. *Hydrocarb. Process*, **60** (8), 111.

2 Elgue, J., Peytavy, J.L. and Tournier-Lasserve, J. (1991) Recent industrial developments in natural gas sweetening by MDEA. Paper presented at the 18th World Gas Conference, Berlin.

3 Elgue, J. and Lallemand, F. (1996) MDEA based solvents used at the Lacq processing plant. Paper presented at the Gas Processors Association European Chapter Meeting, London, January 18.

4 Elf Activated MDEA (1994) An important improvement in natural gas sweetening processes. 19th World Gas Conference, Milan.

5 Elgue, J., Peytavy, J.L. and Lallemand, F. (1995) The Elf Activated MDEA process: new developments and industrial results. Paper presented at the International Gas Research Conference, Cannes, 1995.

6 Johnson, J.E. and Homme, A.C. (1984) Selexol solvent process reduces lean, high-CO_2 natural gas treating costs. *Energy Prog.*, **4** (4).

7 Shah, V.A. and Huurdeman, T.L. (1990) Synthesis gas treating with physical solvent process using Selexol process technology. *Ammonia Plant Safety, AIChE*, **86**, 279.

8 Epps, R. (1996) Selective absorption of H_2S and removal of CO_2 using Selexol solvent. Paper presented at the January 1996 GPA session, London, England.

9 Minkkinen, A., Rojey, A. Charron, Y. and Lebas, E. (1998) Technological developments in sour gas processing, Les Entretiens IFP, Rueil-Malmaison, France, May 14.

10 Minkkinen, A., Benayoun, D. and Barthel, Y. (1996) U.S. Patent 5,520,249, May 28.

11 Rojey, A., Lebas, E., Larue, J. and Minkkinen, A. (1998) U.S. Patent 5,782,958, July 21.

12 Minkkinen, A. and Jonchere, J.P. (1997) Methanol simplifies gas processing. Paper presented at the 76th Annual Convention of the Gas Processors Association, San Antonio, Texas, March 11.

11
BioETBE: A New Component for Gasoline

Marco Di Girolamo and Domenico Sanfilippo

11.1
Introduction

To face the remarkable growth in worldwide demand for energy in a sustainable way requires strong integration between fossil and renewable fuels. In fact, the uncertainty as to the quantity of crude oil reserves, its availability limited to a few regions affect oil prices and energy supply security and, moreover, combined with need to reduce greenhouse gas emissions, creates pressure to search for renewable or non-CO_2-emitting energy technologies such as wind, photovoltaic, geothermal, nuclear, and so on.

Transportation infrastructures require, or at least privilege, the availability of liquid fuels. In the field of transportation fuel, biofuels are the most feasible alternative to fossil fuels.

Biofuels, in fact, can be produced in a cost effective way, are commercially available, have been successfully tested in various countries, are compatible with the present technologies and, further, are environmentally friendly, with a use in line with sustainable development.

In this scenario biodiesel, bioethanol and ETBE (ethyl *tert*-butyl ether), the ether obtained from bioethanol and isobutene, are the compounds normally utilized as fuels. In particular, ETBE is playing an increasingly important role in the gasoline pool composition owing to its superior properties, which represent a trade-off between the needs of refiners and the severe expectations and regulations of environmental stakeholders.

11.2
High Quality Oxygenated as Gasoline Components

The use of oxygenates compounds as fuel dates back to the introduction of first cars; Henry Ford, in fact, built at the beginning of the twentieth century two different models of car with engines powered with only ethanol (called Quadricycle) or with a

Sustainable Industrial Processes. Edited by F. Cavani, G. Centi, S. Perathoner, and F. Trifiró
Copyright © 2009 WILEY-VCH Verlag GmbH & Co. KGaA, Weinheim
ISBN: 978-3-527-31552-9

flexible fuel: ethanol, gasoline or a combination (called Modet T). Later in the century, gasoline, due to its lower price, became the motor fuel of choice and only in the early 1970s were oxygenates were again considered as fuel and introduced in the gasoline with a double function: as octane booster, to replace alkyl lead compounds, and as volume extenders (produced from sources partially alternative to crude oil).

MTBE (methyl *tert*-butyl ether) appeared immediately as the compound in the position to dominate the oxygenates market due to its superior blending properties, relatively low volatility, complete miscibility with gasoline, low susceptibility to phase separation determining a full fungibility with the existing storage and distribution system. Furthermore, its very simple synthesis is carried out with compounds available but generally not accepted as such in the gasoline pool (methanol and isobutene).

At the end of 1980s, public concern about the high winter CO and summer ozone levels forced the United States Government to regulate the composition and quality of gasoline and diesel fuel with the publication of the Clean Air Act Amendments (CAAA), which, mandating the 2% of oxygen in USA Reformulated Gasoline (RFG), gave to oxygenates the new and important role of "clean air additives" and sealed the success of MTBE as oxygenate of choice.

As a consequence, MTBE, whose industrial production was started in 1973 (Snamprogetti/ANIC plant in Ravenna), became the fastest growing chemical in the 1980s, with a global capacity higher than 20 million-t per year in the 1990s.

However, concern over its wide use in the USA were raised especially in California, where, in 1996, it was detected in some drinking water wells, making the water taste and smell badly. Actually, MTBE accumulates in water, due to its relatively high solubility and very slow biodegradation rate, and as little as 15 ppb of MTBE can be smelt in water with a consequent large, negative impact on the public, even if no negative effects on the human health have been proven.

Several causes have been identified for the presence of MTBE in water but, for sure, the leakage in UST (underground storage tanks) and gasoline pipelines played a major role.

Starting from this problem, there has been a lot of pressure from (and on) the regulators to eliminate the use of MTBE as a blending component in gasoline. California was the first State to ban the use of MTBE (4 million tons consumed) since January 2004 [1]. At present, only about 25 States have approved a legislation for the removal or the restriction of the use of this ether but, in any case, the introduction in 2006 of a new law for the removal of liability protection for refineries using MTBE has practically taken away this ether from the USA market.

As a consequence, the production of ethanol, introduced in gasoline to provide the oxygen loss through displaced MTBE, increased enormously to $13.8 \, Mt \, yr^{-1}$ from a level of about $5.2 \, Mt \, yr^{-1}$ in 2002, exceeding even the Brazilian production (Figure 11.1) [2].

Europe, the second largest market for MTBE, continues to look favorably on ethers as clean gasoline components, with no real safety risk related to the use of MTBE but, notwithstanding this, MTBE consumption is quickly falling due to the competition with biofuels.

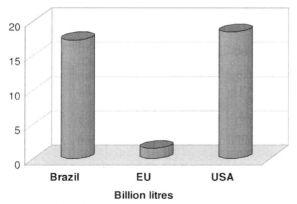

Figure 11.1 2006 production of bioethanol.

Biofuels, obtained from renewable resources, have in fact a central place in the European Union proposals for a Common Energy Policy, with a target of 5.75 vol.% of transportation fuel from biosources by 2010 in EU (Biofuels directive 2003/30/Ec) and a possible target of 10 vol.% in 2020.

The objectives of this program for the promotion of biofuels in Europe are political, economical and environmental and in more detail are aimed at:

- reducing greenhouse gas emissions from the transport sector;
- reducing EU energy import dependence for transport;
- offering new outlets for the agricultural sector (sustainable farming).

The new growing biofuels market in Europe is, however, completely different from that in the USA in terms of both the preferred biocomponent for fuels and the final destination of ethanol [3].

Owing to different trends in the diesel and gasoline market, Europe tends, in fact, to support the production of biodiesel, achieved mainly from vegetable oils, which is nearly four-times greater than from ethanol (Figure 11.2). Moreover, ethanol, which

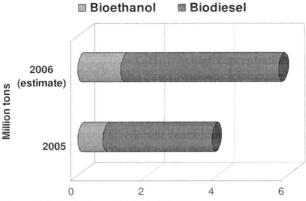

Figure 11.2 Production of biofuels (EU-25).

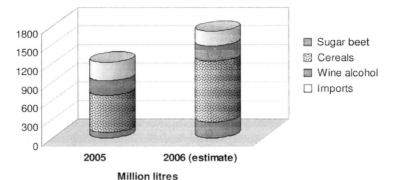

Figure 11.3 Origin of bioethanol in Europe.

can be obtained (Figure 11.3) from any vegetable source containing sugars (like cane or beet) or one that is transformable in sugars (like cereals or cellulose), is used mainly in the form of ETBE; in fact, in 2006 the 75% of ethanol production in Europe was used in the form of ETBE and only the 25% was directly blended in gasoline [2].

As a consequence of this political choice, most MTBE plants in Europe will be (or have been) converted to ETBE production.

11.3
ETBE Technology

11.3.1
ETBE Properties

ETBE is a valuable component for gasoline, with properties similar to, and in some cases better than, MTBE (Table 11.1). ETBE has, in fact, the same blending characteristics as MTBE but it has the additional benefits of a lower solubility in water, a lower volatility (some additional light compounds can be introduced in the gasoline pool without affecting its RVP) and moreover it allows a higher plant production capacity (16% wt).

Table 11.1 Properties of oxygenates.

	MTBE	ETBE	EtOH
Oxygen (wt%)	18.2	15.6	34.7
Boiling point (°C)	55	73	78
Density (kg m^{-3})	740	750	790
Solubility in water (wt%)	4.3	1.2	soluble
RON (Research Octane Number), blending	118	118	125
MON (Motor Octane Number), blending	101	102	101
RVP (Reid Vapor Pressure) (psi)	8	4	18

Bioethanol is also a very attractive compound for increasing gasoline octane quality and it can be added up to 10% to gasoline (E5 is the preferred European solution while E10 is used in USA) without engine modification or also used practically pure, as E85 or E100, in the Flexible Fuels Vehicles.

With respect to ETBE, ethanol has, however, some disadvantages such as volatility and water solubility. Its RVP is more four times that of ETBE and it adversely affects the gasoline volatility; in fact, ethanol addition causes a significant reduction in temperature for the front end evaporation and, as a consequence, the light cheap gasoline components, such as butanes and pentanes, have to be removed to meet the volatility specification limit.

Another important issue of ethanol is its water sensitivity, which affects the water tolerance of gasoline; in the case of water presence in ethanol-added gasoline, there is, in fact, the formation of hydrogen bonds between water and alcohol and the blend separates into two phases, with loss of octane quality. As a consequence, bioethanol requires special handling with different storage and distribution facilities for alcohol and gasoline and the blend has to be carried out just before the final use.

11.3.2
ETBE Synthesis

ETBE is produced by the liquid-phase addition of ethanol (EtOH) to isobutene (IB) in presence of an acid catalyst. Since the reaction is exothermic and limited by the chemical equilibrium, the reactor outlet temperature is maintained as low as allowed by the catalyst activity in order to maximize the isobutene conversion. Operating temperatures range between 40 and 80 °C.

Selectivity to ETBE is usually very high and the main side reactions are isobutene dimerization with formation of 2,4,4-trimethyl-1-pentene and 2,4,4-trimethyl-2-pentene (DIB), ethanol self-condensation to diethyl ether (DEE), water addition to isobutene with formation of *tert*-butyl alcohol (TBA) and the etherification of linear butenes, if present, to produce ethyl *sec*-butyl ether (ESBE).

The overall reaction pattern is the following:

$$EtOH + IB \leftrightarrow ETBE \tag{11.1}$$

$$2IB \rightarrow DIB \tag{11.2}$$

$$2EtOH \rightarrow DEE + H_2O \tag{11.3}$$

$$IB + H_2O \leftrightarrow TBA \tag{11.4}$$

$$EtOH + n\text{-butenes} \leftrightarrow ESBE \tag{11.5}$$

The reaction rate of linear olefins etherification is much lower than isobutene so ESBE presence in the product is limited to a few hundred ppm. Formation of both DIB and DEE is favored at high temperatures and at low molar ratio alcohol/isobutene.

TBA is, instead, the main by-product of ETBE and its formation is faster and approaches the thermodynamic equilibrium. While in the MTBE case the quantity of water is small, in ETBE synthesis there is an increase of TBA production due to the greater amount of water in the feed, deriving from its higher concentration in both fresh (0.2–1 wt%) and recycle ethanol (azeotrope with 6 wt% of water) coming from the alcohol recovery section.

All these by-products do not adversely affect the product quality since TBA, DEE and DIB are also acceptable octane gasoline components.

In addition, the formation of an ethanol/ETBE azeotrope makes difficult it to produce an alcohol-free ether and so a typical commercial ETBE can contain up to 2.5 wt% of alcohols (TBA + EtOH).

Despite the high selectivity of the reaction, the formation of by-products must be thoroughly monitored since severe specifications are usually imposed on the C_4 raffinate (the C_4 stream leaving the plant after the isobutene elimination) used later on as a feedstock to alkylation, metathesis or 1-butene extraction (where oxygenates act as poisons).

The feeds used to produce ETBE (steam cracking, FCC, isobutane dehydrogenation) differ in the levels of poisons and in isobutene and linear butenes contents (Table 11.2).

Table 11.2 Composition (wt%) of typical C_4 streams.

Feed	Isobutene	n-Butenes
Steam cracking	30–50	30–60
FCC	15–20	30–50
Isobutane dehydrogenation	45–55	—

As with MTBE, acidic resins are the catalysts employed in the industrial practice for ETBE synthesis; standard products, made from different manufacturers in the form of spherical beads (e.g., Rohm and Haas, Bayer, Purolite, Dow), have similar characteristics – a macroporous structure, polystyrene-divinylbenzene, functionalized with sulfonic groups (active sites 5.2 eq H^+ per kg).

The life of these industrial catalysts is a function of both operating conditions (space velocity, temperatures, etc.) and level of poisons and/or impurities present in the feed [4].

In any case, the main causes of catalyst deactivation are:

- *Physical degradation*: improper fluid dynamics conditions can cause a loss of the catalyst integrity with sphere breakage, fine powder formation and increase of pressure drop.
- *Thermal degradation*: the temperature inside the reactor has to be carefully controlled to avoid the easy thermal degradation of the resin matrix with breakage of the carbon–sulfur bonds and subsequent release of sulfuric acid. This degradation becomes significant for temperatures above 100 °C.

- *Polymer fouling*: catalyst pore plugging occurs when organic compounds present in the feed (particularly butadiene and isoprene) polymerize inside the hydrocarbon matrix and plug the pores, making inaccessible a number of acidic sites. This type of poisoning is more frequent in the case of low alcohol content or high temperatures inside the reactors.

- The main cause of resin deactivation is, however, the neutralization of acidic sites due to interaction with the contaminants present in the feed. The neutralization may be due to:
  - *Cations*: in this case deactivation takes place by ion exchange with the protons of the functional groups. Main cations in the feed are sodium and calcium (present in the wash water or in the ethanol), iron and chromium (due to solubilized rust), aluminum and silicon (from the zeolite catalyst of FCC units).
  - *Strong nitrogen bases*: this category includes ammonia and low-molecular-weight amines. These compounds have a deactivating action similar to cations.
  - *Weak nitrogen bases*: this type of deactivation takes place due to the action of nitriles (acetonitrile and propionitrile), contained in FCC streams, or of compounds used to extract butadiene from steam cracker cuts like *N,N*-dimethylformamide and of *N*-methylpyrrolidone.

The life of the resins is approximately two years in steam cracker plants and, despite the washing section, 6–12 months in refinery. On the other hand, feeds from isobutane dehydrogenation have a much lower level of contaminants, allowing the catalysts to live up to 4–5 years. Despite the simplicity of this reaction, the kinetics of ETBE are quite complex since the reduction of the alcohol concentration along the catalytic bed leads to a transition in the catalysis mechanism. At the inlet of the bed, there is a high concentration of ethanol and the sulfonic groups are dissociated; the alcohol solvated proton ROH_2^+ becomes the true catalytic species, with a reduction of the catalytic activity. Conversely, as the alcohol is converted, the polarity of the medium gradually decreases and the very active undissociated sulfonic groups, closely interlinked by hydrogen bonds, become the real catalytic species.

Therefore, in the case of etherification reaction, it is the isobutene/alcohol molar ratio that determines the catalytically active species and thus the reaction rate, as shown in Figure 11.4 for MTBE [5].

Moreover, the kinetics are further complicated by the high level of water in the reactor inlet, deriving from its high concentration in both fresh and recycled ethanol, that acts as a temporary poison for the catalyst and strongly reduces the reaction rate [6].

11.3.3
ETBE Reactors

Owing to thermodynamic and kinetic constraints it is very important in the ETBE process to choose the right reactor.

Adiabatic and tubular reactors are the standard solutions employed as front-end reactor (the first reactor where most of the reaction takes place).

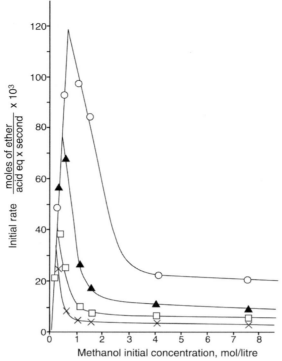

Figure 11.4 Effect of methanol concentration on MTBE initial rate (Isobutene concentration [mol/l]: ○ 6, ▲ 4, ☐ 2, × 1).

The water cooled tubular reactor (WCTR) represents the optimal solution for the etherification because it is the best compromise between kinetics and thermodynamics [7]. The Snamprogetti (now Saipem) WCTR (Figure 11.5) is a bundle-type heat exchanger with the catalyst in the tube side and the tempered cooling water flowing co-current or counter-current in the shell side. The catalyst is self-supporting in the bottom shell of the reactor, in the tubes and above the upper tube sheet.

Up flow operation makes the pressure drop across the reactor almost negligible and avoids catalyst plugging, making catalyst unloading even easier.

WCTR couples all the advantages of a drum reactor in terms of catalyst loading and unloading with the advantages of a tubular reactor with respect to minimum catalyst inventory and optimum temperature control.

As shown in Figure 11.6, where typical profiles for the process temperature and for the isobutene conversion in WCTR are reported, a temperature peak is obtained in the first part of the reactor (usually between 10 and 30% of the bed length), where reactants have the highest concentration, because the rate of heat generation is greater than the rate of heat removal. In contrast, in the final part of the reactor heat exchange prevails and the temperature of the cooling water is approached. As a consequence, in a WCTR there is a higher temperature in the inlet zone, to allow

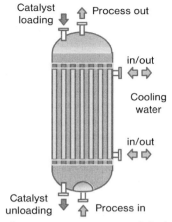

Figure 11.5 Saipem water cooled tubular reactor (WCTR).

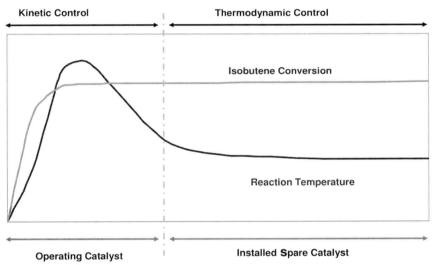

Figure 11.6 WCTR: temperature and isobutene conversion profiles.

high reaction rate, and lower temperature in the outlet zone to approach higher conversion equilibrium; conversely, in an adiabatic fixed bed there is a continuous temperature increase along the reactor axes with the maximum temperature at the end of reactor.

This different temperature profile also has an impact on by-product production; the lower mean temperature achievable in WCTR allows us to maintain at lower extent by-product formation.

Moreover, a WCTR also has a higher resistance to catalyst poisoning since the catalyst used to reach the thermodynamic conversion can be considered as spare catalyst inside the reactor; in fact, less than half of catalyst is really involved at start up conditions and this allows a longer time on stream.

Finally, another important advantage of WCTR is its flexibility since this reactor has three degrees of freedom for continuous optimization of its performances:

1. Process inlet temperature.
2. Tempered water inlet temperature.
3. Tempered water flow rate.

This kind of reactor is highly referenced with more than 20 units built by Saipem and operating world wide.

11.3.4
ETBE Process

All commercial etherification technologies use similar operating conditions but different plant layouts (number and type of reactors) according to the isobutene required conversion [8].

When the feed comes from steam cracking or isobutane dehydrogenation, having a high isobutene concentration (Table 11.2), an isobutene conversion higher than 99% is required to minimize its content in C_4 raffinate leaving the plant. In fact, for steam cracking cuts the residual isobutene can cause undesired by-product formation during the downstream treatments (metathesis or polymerization) of C_4 raffinate; in the case of isobutane dehydrogenation feeds, the residual isobutene forms coke when recycled to the dehydrogenation reactor, with loss of raw material and reduction of the catalyst life cycle. Therefore, to maximize the conversion the plant layout is based on a double reaction stage (Figure 11.7) [9].

In this plant configuration, the C_4 feedstock is mixed with ethanol and passed through the first reaction stage (one or more reactors with intermediate cooling) where the synthesis is carried out under mild temperature conditions. The reactor effluent is then fractionated in a first separation tower to recover the produced ether as a bottom stream and a mixture of unconverted isobutene, C_4 and ethanol (azeotropic concentration) as overhead stream.

This stream is first sent to the second reaction stage (one or more reactors) and then to the second fractionation column.

The bottom of this tower, a mixture of C_4 and ether produced in the second stage, is recycled to the first column while the overhead stream (C_4 and ethanol) is sent to the alcohol extraction tower. In this column, ethanol is removed by counter-current washing with alcohol-free water while C_4 raffinate, isobutene-free, is sent to battery limits. The stream water/ethanol is finally sent to an additional fractionation tower where the ethanol-free water, recovered as bottom stream, is recycled to the washing tower, and the overhead ethanol/water azeotropic stream is recycled to the reaction stages.

When a higher C_4 raffinate purity is required it is possible to perform a more efficient removal of oxygenates (mainly DEE) by including an additional distillation tower or an oxygen removal unit (ORU) based on molecular sieves.

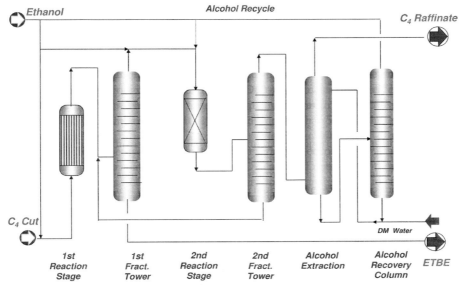

Figure 11.7 Saipem high conversion etherification unit.

As finishing reactors, to complete the isobutene conversion, it is also possible to use, in the second stage, a catalytic distillation tower that combines reaction and fractionation in a single unit operation.

In the case of refinery cuts from FCC units, having a relatively low isobutene concentration (Table 11.2), the plant layout is less sophisticated because it is sufficient to achieve 90–95% of isobutene conversion; in this case the plant configuration is based on a single reaction stage with tubular and adiabatic reactors in series with intermediate cooling.

In any case, independently of plant layout, this reaction is executed with a pressure greater than 8 bar, which is sufficient to keep the C_4 in the liquid phase, and a liquid hourly space velocity (LHSV) of $2–7\,h^{-1}$. The ethanol/isobutene molar ratio used industrially is about the stoichiometric value; clearly, increasing this value it is possible to improve the thermodynamic conversion but, in the industrial operation, this value is restricted by the ETBE specification; in fact, the non-converted alcohol can be recovered either from the bottom of the first fractionation column together with the product or from the top of the second column with the C_4 hydrocarbons (Figure 11.7). For both solutions, the quantity of ethanol in the streams is limited; in the first case by the specifications of the ETBE, and in the second by the C_4/ethanol azeotrope composition (98.5/1.5 wt%).

Industrial experience has shown that this kind of plant, owing to the very high flexibility of the WCTR, can be switched easily from ETBE to MTBE production, and vice versa, without any reduction of feed rate and with few mechanical modifications.

References

1 Higgins, T. (2004) *World Refining,* **14** (5), 4.

2 Vierhout, R. (2007) in Biofuels in Central and Eastern Europe, Prague.

3 Marchionna, M. and Romano, U. (2007) *Chim. Ind.,* **1**, 110.

4 Trotta, R. *et al.* (1994) *Fuel Reformulation,* 40.

5 Ancillotti, F. *et al.* (1978) *J. Mol. Catal.,* **4**, 37.

6 Cunill, F. *et al.* (1993) *Ind. Eng. Chem. Res.,* **32**, 564.

7 Miracca, I. and Tagliabue, L. (1996) *Chem. Eng. Sci.,* **51**, 2349.

8 Sanfilippo, D. *et al.* (2005) *Sustainable Strategies for the Upgrading of Natural Gas: Fundamentals, Challenges and Opportunities* (eds E.G. Derouane *et al.*), Ch. 11, p. 217, Springer.

9 Di Girolamo, M. and Tagliabue, L. (1999) *Catal. Today,* **52**, 307.

12

Olefin/Paraffin Alkylation: Evolution of a "Green" Technology

Anne M. Gaffney and Philip J. Angevine

12.1
Introduction

Today's gasoline has many refining sources: fluid catalytic cracking (FCC) naphtha, straight-run naphtha, coker and visbreaker naphtha, reformate, isomerate and alkylate. High compression engines are pushing the market to increased amounts of high-octane gasoline, and environmental factors are constraining gasoline composition in many other ways. Table 12.1 lists some of the key specifications of gasoline – now and in the near future:

- In addition to SO_x emissions, sulfur causes indirect, inhalable particulates. Since sulfur poisons catalytic converters, lower sulfur dramatically reduces CO, NO_x and unburned hydrocarbon emissions.

- Benzene is a known carcinogen and is carefully regulated. Although other aromatics are less problematic, their acceptable levels are also being lowered. The presence of benzene/aromatics may increase particulate formation and/or the formation of polycyclic aromatic compounds.

- Olefins are photochemically reactive and are a leading cause of smog/ozone formation.

- Although not shown, the evaporative emissions of hydrocarbons (i.e., volatility) are another cause of smog formation and are also regulated via Reid vapor pressure (RVP) specifications.

- The US EPA toxicity equation identifies heavy gasoline as a factor in overall toxicity; therefore, T_{90} – the temperature at which 90 vol.% of the gasoline has been evaporated – must be controlled.

With all of these regulatory restrictions, refiners' gasoline blending options are becoming more limited.

Gasoline's two main blending components are FCC and naphtha reformer gasoline. Both contain large concentrations of aromatic compounds and, in the

Sustainable Industrial Processes. Edited by F. Cavani, G. Centi, S. Perathoner, and F. Trifiró
Copyright © 2009 WILEY-VCH Verlag GmbH & Co. KGaA, Weinheim
ISBN: 978-3-527-31552-9

Table 12.1 Some key gasoline specifications.

	USA Phase II RFG (2004)	California Carb (2005)	EU
RON, min	—	—	91
MON, min	—	—	81
(RON + MON)/2	87/89/91[a]	87/89/91[a]	—
S (ppm, max)	80	30	50 → 10 (2009)
Benzene (vol.% max)	1.0	1.1	1.0
Aromatics (vol.% max)	—	35	35
Olefins (vol.% max)	—	10	18

[a]Set by industry for regular, mid-grade, and premium.

case of FCC gasoline, significant amounts of olefins. Fortunately, alkylate represents the nearly ideal gasoline component. It contains no aromatics, no olefins, essentially no sulfur or nitrogen; it also has a high research octane number (RON) and motor octane number (MON), and low vapor pressure. However, olefin/paraffin alkylation historically has been carried out with mineral acids, such as HF or H_2SO_4, both of which have many negative qualities.

This chapter discusses alkylation and its evolution into a modern refining process. We review the basic chemistry of alkylation, assess the properties and other merits of HF versus H_2SO_4, identify key drivers in the process and discuss the evolution of one particular process – the AlkyClean solid acid catalyst alkylation process.

12.2
Liquid Acid Catalysts

Olefin/isoparaffin alkylation dates back to the 1930s, and the need for high-octane aviation gasoline in World War II was the major impetus for the technology's development. While various catalysts were tried, by the 1940s HF and H_2SO_4 had become the catalysts of choice. The highly active HF enabled refiners to broaden the feedstocks to include $C_3=$ and $C_5=$, thereby boosting the overall alkylate capacity (currently about 30 billion gallons, or 80 million tons, per year on a worldwide basis).

Table 12.2 summarizes the key properties of HF and H_2SO_4. The critical properties for alkylation are acidity and isobutane availability. The catalyst's acidity generally determines the olefin protonation rate. Isobutane availability determines the carbocation formation. H_2SO_4 is a stronger acid than HF. Isobutane is more readily available in HF as it has higher solubility. In addition, isobutane mass transfer from hydrocarbon to acid phase is more expedient in HF versus H_2SO_4.

With HF, water should be completely removed. In the case of H_2SO_4, water and, to a lesser extent, hydrocarbons (acid soluble oils) lower the acid strength. As excess acidity can cause unwanted by-products there is an optimal acid concentration for each set of conditions and feed mix. Since HF has higher isobutane solubility, it provides higher alkylate quality with less isomerization and oligomerization.

Table 12.2 Liquid acid catalyst properties [1–4].

	HF	H$_2$SO$_4$
Molecular weight	20	98
Boiling point (°C)	19.4	290
Freezing point (°C)	−83	10
98% acid	—	3
Specific gravity	0.99	1.84
Viscosity (cP)	0.26 (0 °C)	33 (15 °C)
Surface tension (dyne cm^{-1})	8.1 (27 °C)	55 (20 °C)
Specific heat (Btu lb °F^{-1})	0.83 (−1 °C)	0.33 (20 °C)
Hammett acidity ($-H_0$) at 25 °C	10.0	11.1
98% acid at 25 °C	8.9	9.4
Dielectric constant	84 (0 °C)	114 (20 °C)
Liquid solubility (wt%)		
i-C$_4$H$_{10}$ in 100% acid at 27 °C	2.7	—
i-C$_4$H$_{10}$ in 99.5% acid at 13 °C	—	0.10
HF in i-C$_4$H$_{10}$ at 27 °C	0.44	—
HF in C$_3$H$_8$ at 27 °C	0.90	—

With H$_2$SO$_4$, acid strength needs to be optimized. An acid strength below 90% may suddenly become too low, and polymerization reactions will begin to dominate, leading to a so-called "acid runaway." As such, H$_2$SO$_4$ may oxidize the polymers and form large quantities of SO$_2$. This is a highly undesired situation.

The minimum operating temperature of H$_2$SO$_4$ is constrained by the freeze point, viscosity and other properties, which in turn limits alkylate quality. High surface tension and lower hydrocarbon solubility require very strong mixing within the H$_2$SO$_4$ system to obtain reasonable yields and product quality.

Both acids are spent (diluted) during the process and require regeneration. HF is volatile and can be recovered by distillation inside a refinery alkylation unit. Net consumption of HF is low. While the operating dilution range for H$_2$SO$_4$ is higher than HF, spent H$_2$SO$_4$ is normally regenerated outside the refinery. Dilute acid leaves the refinery and higher concentration acid is brought in. As a practical matter, refiners consider acid use to be the quantity of acid that must be brought in through the gate. By this measure, in H$_2$SO$_4$ alkylation, acid consumption is high (75–150 kg of acid per ton of alkylate produced). HF consumption is low (less than 0.4 kg per ton alkylate). Although HF is more expensive than regenerated H$_2$SO$_4$, the overall acid cost for HF alkylation is normally far lower than H$_2$SO$_4$ alkylation.

Spent sulfuric acid is often recovered offsite in a specially designed regeneration plant by incineration and conversion of the sulfur oxides formed into sulfuric acid. However, large quantities of fresh and spent acid must be transported to the offsite regeneration plant. (The total worldwide consumption of existing H$_2$SO$_4$ alkylation units is about 10–20 billion lb per year or 4–8 million tons per year.) In some cases, refiners have built a captive H$_2$SO$_4$ regeneration unit, which is normally far smaller

Table 12.3 Typical products of liquid acid alkylation [5].

Component	H_2SO_4 alkylate (vol.%)[a]	HF alkylate (vol.%)[a]
Propane	0.05	—
Isobutane	0.04	0.13
n-Butane	0.9	4.9
Isopentane	8.8	5.1
n-Pentane	0.23	0.01
2,2-Dimethylbutane	—	—
2,3-Dimethylbutane	5.4	2.4
2-Methylpentane	1.3	0.9
3-Methylpentane	0.64	0.4
n-Hexane	—	—
2,2-Dimethylpentane	0.25	0.2
2,4-Dimethylpentane	3.6	2.0
2,2,3-Trimethylbutane	0.01	—
3,3-Dimethylpentane	0.01	—
2,3-Dimethylpentane	2.2	1.3
2-Methylhexane	0.22	0.24
3-Methylhexane	0.14	0.12
3-Ethylpentane	0.01	0.01
2,2,4-Trimethylpentane	24.2	38.0
n-Heptane	—	—
2,2,3,3-Tetramethylbutane	—	—
2,2-Dimethylhexane	0.04	—
2,4-Dimethylhexane	2.9	4.2
2,5-Dimethylhexane	4.9	3.6
2,2,3-Trimethylpentane	1.5	1.4
3,3-Dimethylpentane	—	—
2,3,4-Trimethylpentane	13.2	9.6
2,3-Dimethylhexane	3.4	4.9
4-Methylheptane	—	—
2-Methylheptane	0.08	0.09
2,3,3-Trimethylpentane	11.47	8.14
3,4-Dimethylhexane	0.26	0.59
3-Methylheptane	0.23	0.19
2,2,5-Trimethylhexane	7.20	3.20
Other C_9+	6.82	8.38
Total	100.00	100.00

[a]Based on total finished alkylate.

than the size judged economical by commercial acid suppliers. The economics of this approach vary with location and circumstances.

Table 12.3 summarizes the typical products obtained in H_2SO_4 and HF processes. One key observation is that trimethylpentanes/dimethylhexanes (TMP/DMH) formation is far from thermodynamic equilibrium, a desirable factor given the octane spread – TMPs have RON numbers of 100 and higher, DMHs about 50–60. The various products formed are present for all feeds, catalysts and process conditions, only in different proportions.

Table 12.4 Alkylation of isobutane with individual C_3–C_4 olefins (96 wt% H_2SO_4 at 7 °C; well-mixed flow reactor)[a] [6].

Olefin feed	$C_3=$	i-$C_4=$	2-$C_4=$	1-$C_4=$
Feed (vol.%)				
Olefin	15.9	15.8	16.7	14.9
Isoparaffin	77.0[c]	80.5[c]	80.3[c]	78.9[c]
n-Paraffins, C_3–C_5	7.1	3.7	3.0	6.2
Reactor hydrocarbon (vol.%)				
Product[b]	30.2	25.0	29.8	29.6
Isoparaffin (feed type)	62.9[c]	70.3[c]	65.3[c]	64.5[c]
n-Paraffins, C_3–C_5	6.9	4.7	4.9	5.9
Product[b] (vol.%)				
Isobutane	—	—	—	—
Isopentane	3.8	10.0	4.2	4.7
2,3-Dimethylbutane + methylpentane	4.2	5.2	4.6	4.4
2,4-Dimethylpentane	20.8	3.9	2.4	2.6
2,3-Dimethylpentane	50.4	2.6	1.4	1.5
2,2,4-Trimethylpentane	4.4	28.7	30.6	30.5
Dimethylhexane	1.7	9.5	9.0	11.0
2,2,3 + 2,2,4-Trimethylpentane	3.7	23.1	41.6	39.1
2,2,5-Trimethylhexane	0.9	4.9	1.9	1.8
Other C_9s	0.4	1.7	0.5	0.7
C_{10}	5.3	2.5	0.7	0.6
C_{11}	3.7	2.1	0.7	0.7
$C_{12}+$	0.8	5.9	2.6	2.6

[a] 0.22 olefin (liquid-hourly) space velocity.
[b] C_4 and heavier product exclusive of n-paraffins and feed-type isoparaffin.
[c] Isobutane.

Table 12.4 shows the effect of different olefin feeds on the alkylate quality with a sulfuric acid system. All of the butenes give better alkylate yields and product quality than propylene, and 2-butene is the preferred $C_4=$ feed. For an HF unit, 1-$C_4=$ gives a lower RON product because of its higher DMH yield. As such, most HF units have an isomerization unit upstream of the alkylation unit to isomerize the 1-$C_4=$ to 2-$C_4=$.

12.2.1
Reaction Mechanism

The generally accepted alkylation reaction mechanism has four desirable key steps and four undesirable secondary reactions. The four desirable steps are:

1. *Initiation (or olefin protonation)*: In this step, a *t*-butyl cation is formed from isobutene. A *sec*-butyl cation is formed from 1-$C_4=$ or 2-$C_4=$. The *sec*-butyl cation can form a *t*-butyl cation by methyl shift, or it can undergo hydride transfer from isobutane, forming *n*-C_4 and a *t*-butyl cation.

2. *Alkylation (or t-butyl cation/olefin condensation)*: here, the various TMP or DMH carbocations are formed.

3. *Isomerization*: The C_8 carbocations formed in step 2 may isomerize via hydride transfer or methyl shift to form various TMP cations. DMHs are thermodynamically favored; thus, residence time preferably should be short (high temperature reduces the required residence time).

4. *Termination via hydride transfer*: The carbocations react with isobutane to form the various octane products, along with a *t*-butyl cation to continue the reaction sequence.

The following unwanted secondary reactions generally result in reduced yield and quality loss:

1. *Oligomerization*: After the primary reaction forms a C_8+ carbocation, a second olefin reacts to form a higher molecular weight hydrocarbon (e.g., C_{12}) and another *t*-butyl cation. Further reactions can result in even larger products (e.g., C_{16}s, etc.).

2. *Disproportionation*: The typical reaction here is a bimolecular reaction of two equivalent alkylate molecules (e.g., C_8s form a C_7 and C_9 product). Of the major products, 2,3,4-TMP is most reactive for disproportionation.

3. *Cracking*: the larger isoalkyl cations can undergo β-scission to form smaller olefins and isoalkyl cations.

4. *Self-alkylation*: This reaction occurs readily with HF and with most zeolites, albeit to a reduced extent. Using 2-C_4= as a typical reagent, the butene can be protonated to a *sec*-butyl cation, which undergoes hydride transfer from an isobutane molecule to form a low-valued *n*-C_4. Self-alkylation reaction rates increase with molecular weight and alkene branching. With higher alkenes (e.g., C_5+), H_2SO_4 will also become active.

The two major reactants are isobutane and butene. While a wide range of hydrocarbons is formed, the predominant product is the C_8s, and the preferred products are TMPs versus DMHs. The TMPs have excellent RON and MON – both are at and above 100. As such, TMPs represent the highest octane, nonaromatic gasoline component. DMHs have lower octane values, with RONs of 50–60.

A good process and catalyst system will drive alkylation to minimize light and heavy by-products as well as to maximize TMPs versus DMHs. There are several interactive reactions in the proposed mechanism. First, initiation of the alkylation cycle takes place by reaction of olefins, leading to the formation of i-C_4+ species on the acid sites. This adsorbed i-C_4+ further reacts with a C_4= to form an adsorbed i-C_8+ carbocation. The i-C_8+ reacts with i-C4 via hydrogen transfer, and the i-C_8 product is formed along with another i-C_4+ to feed back into the cycle, and the cycle repeats.

As shown in Figure 12.1, there are many more competing reactions. For example, the C_8+ can further react with another olefin (e.g., C_4=) to form a heavier cation, in this case i-$C_{12}+$. This i-$C_{12}+$ can leave the loop via hydrogen transfer, react again

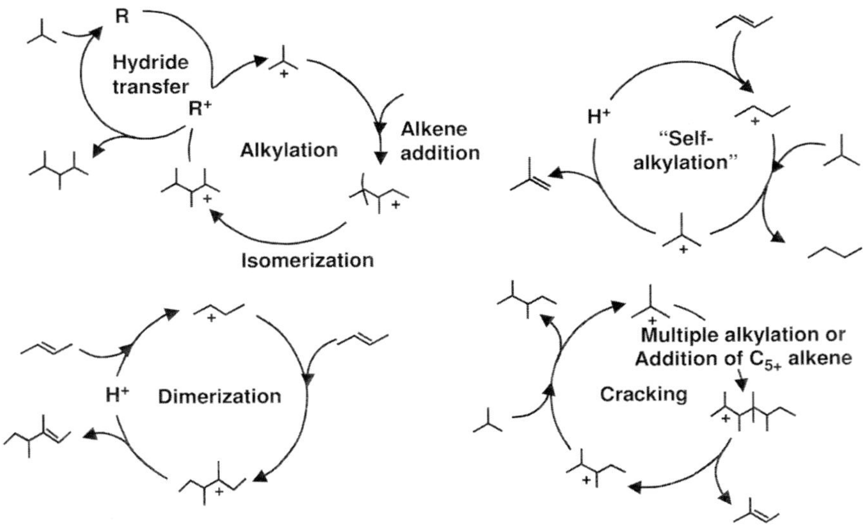

Figure 12.1 Conceptual mechanism for olefin/paraffin alkylation [10].

to form higher molecular weight hydrocarbons or break into smaller products such as an olefin and a sorbed carbocation. The olefin rapidly reacts with the acid sites to form another carbenium ion.

Since hydrogen transfer is essential for the desired product formation, a high isobutane-to-olefin (I/O) ratio is required in the feed. A high hydride transfer rate lowers the formation of non-C_8 (e.g., C_5–C_7 and C_9+) products, impedes the isomerization of kinetically favored TMPs to thermodynamically favored DMHs and limits oligomerization to heavies. A high feed I/O ratio is typically 8–15 for liquid alkylation processes. Owing to internal mixing and relative solubility of olefins and i-C_4, the local I/O ratio can be as high as 1000. The high recycle of isobutane is a key cost factor in the process.

12.2.2
Operating Variables

Several operating variables in both liquid catalyst alkylation processes impact product yields and quality as well as overall costs:

1. *Acid strength and composition*: In H_2SO_4 processes, the optimal acid concentration is about 95–97%. At low levels (e.g., below 90%) catalyst activity is significantly diminished. At high levels (e.g., above 99%) isobutane reacts with SO_3. Acid level is dictated by consumption and fresh catalyst make-up rates. Hence, acid concentration affects kinetics, alkylate yield and quality, and catalyst life. While HF plants are similar, a primary difference is that HF needs to be water-free because any water will rapidly deactivate the HF catalyst and can lead to severe corrosion problems.

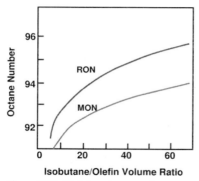

Figure 12.2 Alkylate octane number versus I/O ratio for an HF unit [7].

2. *Isobutane-to-olefin (I/O) ratio*: Since butene–butene reactions can form unwanted dimers and polymers, the I/O ratio is critical for driving yields and alkylate quality. It also dictates acid consumption. Since isobutane is essential for the hydride transfer step, a high I/O ratio drives desired alkylation and hydrogen transfer versus oligomerization reactions. At low I/O, high MW hydrocarbons are formed and catalyst consumption increases. Figure 12.2 [7] shows the alkylate octane number versus I/O ratio for an HF unit. At I/O ratios below 10, the octane and yield (not shown) decrease significantly. Consequently, the standard I/O ratio for HF is 10–15. H_2SO_4 alkylation is more complex. RON will also drop, but H_2SO_4 is less demanding. The external I/O ratio can be 5–8. A high isobutane recycle is required to maintain a very high internal I/O ratio because the olefins adsorb readily into the acid phase and the low olefin concentration is required to mitigate oligomerization.

3. *Reaction temperature*: Figure 12.3 [8] shows a typical octane versus temperature response curve for an HF unit with C_4 feeds. One factor in this effect is the product

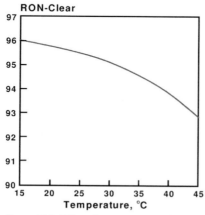

Figure 12.3 Effect of reactor temperature on research octane number (RON) of alkylate produced in an HF alkylation unit [8].

shift in TMP/DMH, which favors DMH at higher temperatures. H_2SO_4 units typically operate at 0 to10 °C, constrained at the lower end by viscosity and hydrocarbon solubility and at the upper end by unwanted oxidation and subsequent acid consumption.

4. *Feedstock effects*:

a) *Olefins*: Table 12.5 [9] shows that the specific olefins used can have a major impact on products and each catalyst system has a different response to each olefin. For sulfuric acid, the preferred reactant is 2-butene, but due to the rapid isomerization of 1-butene to 2-butene, each gives a somewhat similar product. Isobutylene oligomerizes more readily, causing a lower C_8 yield and resulting in higher acid consumption. $C_3=$ (not shown) gives a significantly lower product RON (more C_7 compounds). The HF system is far less impacted by olefin variation. The major effect is the 1-butene/2-butene difference: 1-butene does not isomerize easily to 2-butene and the RON debit is 3–4 numbers for 1-$C_4=$ versus 2-$C_4=$. Consequently, most HF units have an isomerization unit upstream of the alkylation unit for isomerization of 1-$C_4=$ feed and selective hydrogenation of butadiene. $C_3=$ again produces a lower RON product and up to 15% propane by hydrogen transfer reactions, which are typically enhanced by the use of HF. The hydrogen transfer reactions also explain the somewhat higher RON that is observed in the case of $C_3=$ and HF. By hydrogen transfer from i-C_4, the i-C_4+ ions formed will continue to react according to the general mechanism to C_8 compounds.

b) *Feed impurities*: As water is a major issue with HF units because of corrosion, feed pre-drying is critical. Diolefins, such as those found in coker or visbreaker naphtha, lower alkylate quality and increase acid consumption. They can be sharply reduced by a selective hydrogenation pretreatment step. Sulfur compounds (e.g., mercaptans) can increase acid consumption and lower alkylate quality and yields. The mercaptan concentration is often more pronounced for $C_5=$ feeds.

Table 12.5 Isobutane alkylation with different butene isomers (H_2SO_4 catalyst) [9].

Alkylate	Olefinic components		
	i-$C_4=$	1-$C_4=$	2-$C_4=$
Composition (vol.%)			
C_5–C_7	15	6	10
C_8	59	90	77
C_9+	26	4	13
Composition of C_8 product (vol.%)			
2,2,4-TMP	48	41	43
2,2,3-TMP	3	4	4
2,3,4-TMP	15	20	18
2,3,3-TMP	19	25	24
DMHs	15	10	11

5. The effect of space velocity is interrelated with reactor geometry and the effect of olefin concentration. As long as alkylation and hydrogen transfer reaction rates are in balance (effect of T, I/O, etc.), yield and selectivity show little dependence on olefin space velocity (OSV). In general, a very high space velocity will increase acid consumption and formation of acid-soluble oils because of the higher probability of multiple olefins reactions.

12.2.3
Advantages Versus Disadvantages

Both liquid acid processes have major safety issues. H_2SO_4 is very corrosive, but leaks are generally localized because of its lower volatility. Hence, there is reduced catastrophic potential other than that caused by fire from hydrocarbon–air mixtures. HF is more volatile, and even a pinhole leak can cause aerosol formation that can drift for miles. The resulting potential for fatalities from HF inhalation and burns is high, and many communities require extensive mitigation steps to be retrofitted into existing HF units (US$40–50 million). In some areas, new HF units are being banned altogether.

Owing to various compensating costs, the capital associated with HF and H_2SO_4 units are comparable. The operating costs of H_2SO_4 units are higher due to catalyst consumption and the refrigeration costs of the lower temperature process. However, safety mitigation costs are significantly higher for HF. H_2SO_4 has more flexibility within C_4 olefins than HF, but HF is more flexible for handling C_3 and C_5 olefins.

12.3
Zeolite Catalysts

Several types of solid catalysts have been explored for use in alkylation. Several groups have attempted to use "hybrids" (i.e., liquids that have been adsorbed or immobilized on solid supports), such as BF_3/zeolites/aluminas, triflic acid on a carrier, H_2SO_4 on SiO_2, and so on. Since most have shown limited success, we restrict our discussion to the most promising true solid acid materials – zeolite catalysts.

The above-mentioned safety issues summarize the key environmental drivers for solid catalysts. Hybrids have had limited success and ionic liquids are undesirable since they can leach toxic compounds, and so on. Compared with $AlCl_3$ and phosphoric acid catalysts, zeolites are significantly safer and, in many instances, more selective.

The earliest zeolite catalysts used in alkylation date back to the 1960s with rare earth-stabilized X and Y – "REX" and "REY" [11,12]. Historically, the three major catalyst attributes – activity, selectivity and aging stability – could not be achieved. Quite often, stability was the key limitation.

An early, detailed study of the product mix is shown in Figure 12.4 [13], where C_5–C_7 "lights," C_8 alkylate and C_9–C_{12} "heavies" are plotted against time on

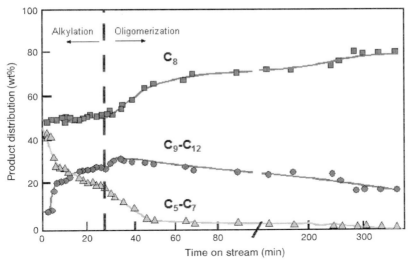

Figure 12.4 Product distribution (carbon number) versus time on stream (2-C$_4$=/isobutane with CeY zeolite) [13].

stream for a cerium-exchanged Y ("CeY") zeolite. For the first 25-minute period, conventional alkylation occurs. Then the hydride transfer activity declines and oligomerization begins to be more important. Unfortunately, the DMH composition increases with time, which is consistent with oligomerization/cracking/re-alkylation (Table 12.6) [13]. As hydride transfer decreases, the product quality shifts downward, consistent with the demise of strong acid sites.

12.3.1
Zeolite Factors Impacting Alkylation Performance

1. Zeolites show strong hydrocarbon sorption, which increases reactant concentration in the zeolite. This effect is more pronounced at low temperature.

2. There is a high concentration of acid sites in the zeolite cage, which drives consecutive bimolecular reactions.

3. The preferential sorption of olefins versus paraffins in zeolites favors oligomerization over alkylation.

4. Diffusional effect in zeolites impact the allowed transitional states of certain TMP isomers (2,3,4- and 2,3,3-TMP have lower steric hindrance), thereby influencing the product distribution. This effect is more pronounced for medium-pore zeolites. Overall product similarity between zeolites and mineral acids suggests that similar or equivalent mechanisms predominate for each.

5. Zeolites have numerous total acid sites, which is more important than the strength of each acid site.

Table 12.6 C_5–C_7 and C_8 product distribution during initial alkylation stage [13].

	Time on stream, min			
	1	5	15	30
C_5–C_7 hydrocarbons (wt%)				
2-Me-butane	51.8	44.3	37.2	35.7
Other C_5 or C_7	0.2	0.3	0.5	1.7
2-Me-pentane	4.0	5.2	5.8	5.2
3-Me-pentane	6.1	6.6	6.7	8.1
2,3-diMe-butane	17.2	17.9	17.4	16.8
2-Me-hexane	1.2	1.8	2.5	1.9
3-Me-hexane	1.4	2.3	3.5	3.5
3-Et-pentane	0.2	0.3	0.4	0.7
2,3-diMe-pentane	6.0	9.1	12.6	13.4
2,4-diMe-pentane	10.7	11.2	12.4	11.1
2,2,3-triMe-butane	1.2	1.0	1.0	1.4
	100.0	100.0	100.0	100.0
C_8 Hydrocarbons (mol.%)				
2-Me-heptane	0.2	0.3	0.3	0.2
3-Me-heptane + 3-ET-hexane	0.4	0.7	0.8	0.7
4-Me-heptane	0.1	0.1	0.2	0.1
2,3-diMe-hexane	4.8	8.4	13.1	10.8
2,4-diMe-hexane	6.2	6.4	7.4	4.5
2,5-diMe-hexane	3.0	3.1	3.8	2.1
3,4-diMe-hexane	6.9	11.3	13.5	22.4
2-Me-, 3-Et-pentane	0.6	1.0	1.3	1.7
2,2,3-triMe-pentane	4.4	3.6	3.2	2.6
2,2,4-triMe-pentane	22.3	18.3	17.7	11.3
2,3,3-triMe-pentane	27.9	24.9	20.7	18.1
2,3,4-triMe-pentane	23.2	21.9	18.0	17.8
Octenes	—	—	—	7.7
	100.0	100.0	100.0	100.0

Zeolites should have the following general properties of a good alkylation catalyst:

- sufficient acidity to form and stabilize carbocation intermediates;
- good hydrogen transfer capability to desorb C_8 carbocations as well as to generate *t*-butyl cations from *sec*-butyl cations and isobutanes;
- sufficiently large pores to enable trimethyl alkanes to exit;
- low concentration of Lewis acid sites, which promote polymerization.

12.3.2
Impact of Reaction Conditions for Zeolites

A high I/O ratio drives hydride transfer versus polymerization. This process variable is a major factor in selectivity and overall operating cost.

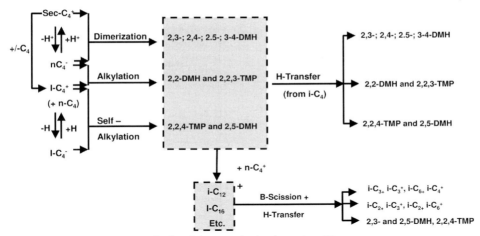

Figure 12.5 Key reaction paths for alkylation and related reactions [9].

Low temperature (e.g., 70 °C and below) yields a higher TMP product slate. This is a kinetic effect, and isomerization reactions are lowered with decreased temperature. The distribution of TMPs is highly dependent on temperature. The preferred high octane, 2,2,4 isomer is abundant at lower temperatures. Figure 12.5 [9] covers many reactions: alkylation, dimerization, self-alkylation, β-scission and H-transfer. Figure 12.6 [14] shows that the C_8 yield goes through a maximum versus T. Other reactions, including oligomerization (at low T) and sequential cracking to C_5–C_7 (at higher T), deplete the C_8 yield.

As mentioned with liquid acid catalysts, selectivity and yield remain at an almost constant high level when the olefin space velocity (OSV) is increased. However, a critical OSV can be identified above which the polymerization reactions rapidly start to increase. This is related to the available number of acid sites and the rate

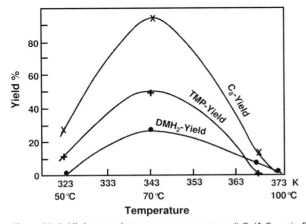

Figure 12.6 Alkylate products versus temperature (i-C_4/1-C_4= via REY catalyst) [14].

at which these sites are regenerated by hydrogen transfer reactions. At a certain OSV, all sites will be covered with hydrocarbon species, and polymerization will become dominant.

12.3.3
Overview of Zeolites in Alkylation

Successful alkylation catalysts have been limited to large-pore zeolites (e.g., zeolite X, zeolite Y, zeolite β, ZSM-20, etc.), probably due to the adsorption/desorption capability of TMPs. Some surprisingly good activity has been reported for medium-pore MCM-22, along with excellent selectivity; however, we surmise that stability was fairly poor. Otherwise, it would have received far more attention.

Although researchers have long sought a solid catalyst substitute for HF and H_2SO_4, they were not successful in finding one that met the common requirements of activity, selectivity and stability. Historically, selectivity was usually thought of as determining yields and product quality only. When researchers eventually understood that the major aging mechanism was a selectivity effect, that is, formation of polyalkylates, they began to assess a way of redefining performance criteria and devising an operating strategy to mitigate the aging issue.

12.4
AlkyClean Alkylation Process: A True Solid Acid Catalyst (SAC) Process

Many attempts have been made to apply a SAC for isobutane alkylation [15–17]; however, in most cases the catalyst and process conditions used resulted in very rapid catalyst deactivation. These attempts also failed because of the absence of viable regeneration procedures. As such, in addition to a suitable catalyst, suitable process and regeneration conditions had to be developed.

Albemarle and ABB Lummus Global began collaboration in 1996 to develop a catalyst/process combination that addresses the aforementioned product selectivity and deactivation/regeneration issues. The following targets were the focal points of our efforts:

- develop a true SAC process, not merely a hybrid process (i.e., not utilizing toxic and/or corrosive volatile compounds adsorbed onto a solid carrier as catalyst);
- ensure no or low environmental hazards compared to current H_2SO_4/HF processes;
- investment and operating costs equal to, but preferably lower than, current processes;
- product quality and yields equal to, but preferably higher than, current processes;
- refinery-compatible catalyst and process technology;
- a robust catalyst with regard to feed impurities;
- no halogens in the catalyst in order to reduce unit corrosion problems and maintenance.

12.4.1
Catalyst Selection and Development

After prescreening of potential candidates in a laboratory alkylation micro-reactor, a "true" SAC was selected as the preferred prototype. By true SAC we mean that the catalytic acid function is intrinsic to the solid itself rather than being a separate species, such as an immobilized liquid deposited on a solid substrate. Starting with this prototype, application research and development work led to novel discoveries, which yielded significant improvements in catalyst performance. The resulting AlkyClean catalyst, which has now been successfully manufactured in commercial-scale trials, is of a type well known and proven in the refining industry. It is based on zeolite Y and the porosity was tailored using a binder [18]. Other zeolites were investigated as well, but they either have an insufficient number of acid sites (e.g., zeolite beta) or a small pore aperture (e.g., ZSM-5).

The AlkyClean catalyst contains no halogens, has acid sites with sufficient strength for alkylation, yields high quality alkylate with minimal side reactions, and exhibits the required activity, stability and regenerability characteristics necessary for a successful process. It is promoted with a low Pt content to assist regeneration and hydrogen transfer.

12.4.2
Process Development Activities

Many bench-scale tests have been carried out to determine favorable process operating and preferred regeneration procedures. To obtain a high I/O at the catalytic sites, the test unit utilized a fixed bed recycle reactor design (Figure 12.7). With this reactor configuration, the feed I/O (i.e., external I/O) could be kept close to that used in commercial liquid acid units (e.g., 8–12). However, the internal I/O (i.e., the I/O at the inlet of the catalyst bed) can be increased to 250 or higher by recycling the isobutene-rich reactor effluent. This higher level of internal I/O, relative to the external I/O, is analogous to the situation for H_2SO_4 and HF units. In those units, the effective (i.e., internal) I/O at the active liquid acid sites is orders-of-magnitude higher than the external I/O. This is due to the effects of intensive mixing and the differences of isobutane and olefins in solubility and reactivity with the acid sites. In the case of H_2SO_4, the effective I/O may be as high as 1000.

Initially, test runs were carried out until the catalyst was deactivated to such an extent that breakthrough of olefins occurred (Figures 12.8 and 12.9) [19]. Unless otherwise stated, pure *cis*-2-butene was used as olefin feed. At the reaction conditions used (reaction temperature 90 °C, I/O = 250), olefin breakthrough occurred after about 8–10 hours. At this point, selectivity, as represented by RON, also started to deteriorate. Alkylate yield was 204 wt% (theoretical yield, based on equimolar reaction of C_4= and i-C_4) or higher until olefin breakthrough.

With reference to Table 12.7 [19], it was found that catalyst activity could be fully recovered repeatedly by vapor phase stripping with hydrogen at 250 °C. In commercial

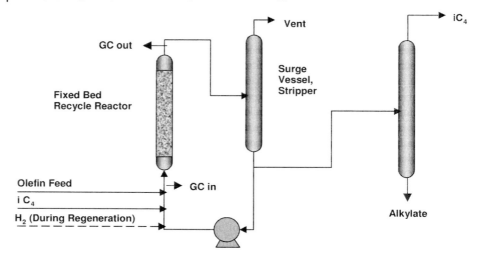

Typical: I/O at reactor inlet 250 or higher
liquid phase 21 bar, 50-90 °C

Figure 12.7 Alkylation bench-scale unit [19].

practice, however, such a regeneration procedure, considering the requirements for draining/heating and cooling/filling, would take too much time compared to the alkylation time period prior to olefin breakthrough. As such, less time-consuming procedures had to be developed. Regeneration conditions that are close to reaction conditions would reduce any heating and cooling periods required at the beginning and end, respectively, of a regeneration cycle.

Prior art/literature claimed that some catalysts could be regenerated in the liquid phase with dissolved hydrogen at alkylation reaction conditions. We investigated this procedure in the case of our catalyst and process combination (cf. Table 12.7, exp. 2–3). After catalyst deactivation (as indicated by olefin breakthrough) and regeneration using this procedure, the regenerated catalyst's life before olefin breakthrough could be restored only to about 40–65% of the initial fresh catalyst period. Thus, a novel regeneration method had to be developed. To this end, various alkylation/regeneration cycles were investigated. A time period with olefin addition was followed by a time period with dissolved hydrogen addition, and so on. It was

Table 12.7 Effect of regeneration procedures [19].

Exp.	Medium	T (°C)	P (bar)	Time (h)	Cat. life (h)
0	Fresh catalyst	—	—	—	10
1	H_2 gas	250	21	1	10
1a	H_2 gas	250	21	1	10
1b	H_2 gas	250	21	1	10
2	i-C_4 liquid with dissolved H_2	90	21	66	6.5
3	i-C_4 liquid with dissolved H_2	115	30	18	4

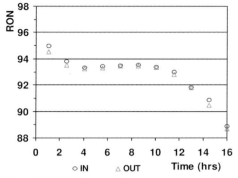

Figure 12.8 RON versus time.

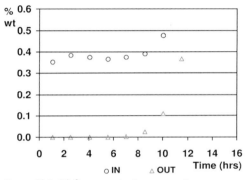

Figure 12.9 Olefin concentration versus time.

observed that this cyclic operation of the catalyst, at alkylation reaction conditions, could be maintained from a week to more than a month, depending on the operating severity. Under this operation, the alkylation step was halted well before any significant olefin breakthrough. A patent covering this new procedure was obtained in 1999 [20].

Figure 12.10 [19] illustrates some of the results under cyclic operation. After every hour of olefin addition, the catalyst was regenerated with dissolved H_2 for one or two hours, depending on the operating I/O. Evidently, good alkylate quality, as represented by RON, can be maintained by utilizing this frequent "mild regeneration" (MR) procedure. Furthermore, high internal I/O and low T are favorable to product quality and catalyst stability.

The catalyst slowly deactivated during the cyclic operation and, after an extended period (two to four weeks), olefin breakthrough occurred during an alkylation cycle. At this point, the catalyst activity was fully restored by treatment with vapor phase H_2 at 250 °C, as described previously. Even after this "high temperature regeneration" (HTR) procedure was carried out 15 times during a pilot run of more than six months with the same catalyst sample, catalyst activity could be fully recovered. A "coke" burn with air was not required.

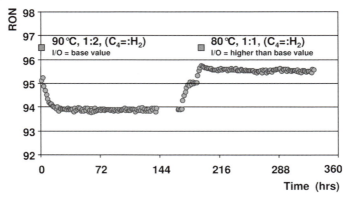

Figure 12.10 Cyclic operation; RON versus time.

12.4.3
Optimization of Process Conditions

After optimizing the catalyst and process conditions, RON numbers as high as 97 to 98 could be obtained for a prolonged time period. Figure 12.11 and Table 12.8 [19] illustrate some of the results.

At sufficiently high I/O, selectivity was optimized by varying reaction temperature. When the temperature was lowered, less C_5–C_7 and more of the desired C_8 compounds were formed. Also, at lower temperature, isomerization of high-octane TMPs to low-octane DMHs was reduced. However, when the temperature was reduced too much at a given olefin space velocity, overall olefin conversion was

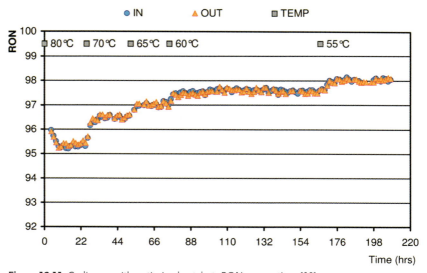

Figure 12.11 Cyclic run with optimized catalyst; RON versus time [19].

Table 12.8 Product distribution and properties versus temperature [19].

T (°C)	C₄= conv. (wt%)	RON	Yield (wt%)	C₅–C₇ (wt%)	C₈ (wt%)	C₉+ (wt%)	TMP/DMH
80	100	95.4	213	22.0	74.2	3.8	6.0
70	100	96.5	208	17.5	78.2	4.2	7.5
65	100	97.0	207	16.0	79.4	4.5	8.2
60	100	97.6	205	14.3	80.7	5.0	9.2
55	100	98.1	204	13.4	81.4	5.1	10.3

reduced and formation of C_9+ compounds sharply increased, leading to more rapid deactivation of the catalyst. Alkylate yield (based on olefin feed) was always somewhat higher than the theoretical 204 wt%. This can be explained by disproportionation of C_8 with i-C_4 via carbenium ions on the catalyst surface, leading to consumption of more than one i-C_4 molecule per C_4 olefin molecule. One drawback is that disproportionation leads to the increased formation of the less desirable C_5–C_7 compounds. At lower temperature, disproportionation activity is lowered, leading to less C_5–C_7 products and therefore lower yields.

Other observations of this test work, with respect to key alkylate product properties, were that neither the Reid vapor pressure (RVP) nor density deviated significantly from values that would be obtained via liquid acid alkylation. Further, acid-soluble oils (ASO), formed as contaminant side products in the case of liquid acid processes, could not be detected among the reaction products in our SAC testing. Compared with the liquid acid technologies, this effect results in both lower feed consumption per unit of alkylate production and eliminates generation of a by-product that can be difficult to dispose of.

12.4.4
Effect of Feedstock Variation

In commercial practice, there will be significant differences in olefin feed composition. Under the AlkyClean process cyclic operation, high RON is obtained over a prolonged time period with various feeds. The use of a refinery-sourced MTBE raffinate gave similar results (alkylate yields and product quality) to a pure *cis*-2-butene feed (Table 12.9). The MTBE raffinate contained about 26 wt% *trans*-2-butene, 15 wt% *cis*-2-butene, 12 wt% 1-butene, 2 wt% isobutene, 40 ppmw (parts per million by weight) of various oxygenates and 3 ppmw of sulfur (balance isobutane and *n*-butane).

In HF alkylation, the processing of 1-butene leads to a significantly lower RON. Therefore, to avoid this octane loss, HF units often employ upstream selective hydrogenation/isomerization to isomerize 1-butene to 2-butene.

The addition of about 25% isobutene on olefins resulted in a loss of less than 0.5 RON. This loss may be attributed to a somewhat higher formation level of C_5–C_7 and C_9+ compounds. For H_2SO_4 alkylation, the same amount of feed isobutene would lead to a loss of about 1 RON.

Table 12.9 Effect of various butene feeds on the AlkyClean process performance at identical process conditions [19].

Feed	*cis*-2-Butene	MTBE raffinate	25/75 i-C$_4$=/*cis*-2-butene
RON	95.5	95.6	95.1
Yield (wt%)	211	210	215
C$_5$–C$_7$ (wt%)	20.2	20.3	25.3
C$_8$ (wt%)	74.5	74.4	67.7
C$_9$+ (wt%)	5.3	5.3	7.0
TMP/DMH	6.1	6.2	6.5

Table 12.10 Estimated impact of feedstock variation [19].

	AlkyClean Process	H$_2$SO$_4$	HF
1-Butene (up to 100 vol.%)	—	—	up to −4.0 RON
Isobutene (25 vol.%)	−0.5 RON	−1.0 RON	−0.5 RON
Propylene (30 vol.%)	−1.0 RON	−1.5 RON	−1.0 RON

Similarly, after blending about 30 vol.% of propylene with *cis*-2-butene, the RON loss was less than 1 number. With H$_2$SO$_4$ alkylation, similar amounts of propylene would lead to a RON about 1.5 lower. Table 12.10 summarizes the estimated impact of feedstock variation on RON relative to a pure *cis*-2-butene feedstock for the AlkyClean process and liquid acid technologies. Based on these results, it can be concluded that our new SAC technology is less sensitive to feedstock variation regarding product quality than either liquid acid technology.

12.4.5
Effect of Impurities

According to the open literature, other solid acid alkylation catalysts are generally susceptible to poisoning/deactivation by water and other common feed impurities (e.g., oxygenates, sulfur compounds, dienes, etc.), thus necessitating (potentially costly) feedstock pretreatment for their removal. In some cases, this requirement is further mandated by the potential corrosion problems associated with the use of halogens in the catalyst system.

In liquid acid units, these impurities increase both liquid acid consumption and side product formation. The feed may also require drying and additional pretreatment, depending on the acid employed and the level of contamination.

In contrast, the AlkyClean solid acid catalyst contains no halogens and it is very robust with regard to water and other potential feed impurities. This was observed even after exposure of the catalyst to high concentrations of oxygenates (250–700 ppmw), sulfur compounds (200–1200 ppmw) and butadiene (400–1800 ppmw). Moreover, after any observed deactivation from these impurities, the catalyst could always be restored to full activity via HTR (i.e., treatment with H$_2$ at 250 °C). Furthermore, in

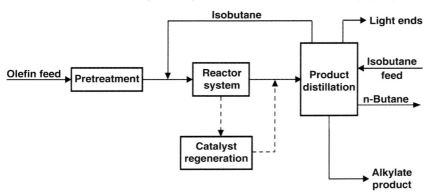

Figure 12.12 Simplified block diagram of the AlkyClean process [19].

commercial practice, after a feed upset requiring HTR, the catalyst does not require additional treatment (e.g., addition of halogens) before being put back on-line. This simplicity reduces turnaround time and eliminates procedures that carry the risk of corrosion problems. In conclusion, because of these catalyst attributes, the AlkyClean process feed pretreatment requirements are significantly less than those for processes based on alternative solid acid catalysts. Required pretreatment levels are projected to be no more than equivalent to that normally associated with sulfuric acid units.

The process flow scheme for the AlkyClean process is similar to that employed for current liquid acid technologies. As illustrated in the block flow diagram in Figure 12.12, the process consists of four main sections: feedstock pretreatment (optional, depending on contaminant level), reactor system, catalyst regeneration and product distillation.

12.4.6
Reactor System/Catalyst Regeneration

AlkyClean reactors operate in the liquid phase in the temperature range 50–90 °C, thereby eliminating the costly refrigeration requirements associated with liquid sulfuric acid technologies. To achieve a high octane alkylate and limit by-product production, H_2SO_4 units typically utilize a total reaction section feed (external) I/O of between 8/1 and 10/1, while HF units run at an I/O of about 12–15/1. In comparison, without any alkylate octane debit, this I/O for the AlkyClean process is in the range 8–10/1, comparable to the H_2SO_4 process, which operates at a significantly lower temperature (Table 12.11). The ability to operate the AlkyClean process at this low I/O is important for two reasons. First, it enables a cost competitive process, as the fractionation requirement associated with isobutane recycle is a major capital investment and operating cost component. Second, it facilitates incorporation of AlkyClean technology in the revamp/de-bottlenecking of an existing liquid acid unit, without major modification to the "back-end" fractionation/recycle facilities.

Key to the AlkyClean technology's superior performance is the coupling of a newly developed catalyst with a novel alkylation reactor system, which minimizes the peak

Table 12.11 Comparison of operating conditions [19].

	AlkyClean process	H$_2$SO$_4$	HF
Operating temperature (°C)	50–90	4–10	32–38
External I/O	8–10/1	8–10/1	12–15/1

olefin concentration in the reaction zone (i.e., maximizes the internal I/O) without requiring extremely high and economically non-viable reactor effluent recycle rates. This is accomplished by utilizing serial reaction stages and a unique (but mechanically uncomplicated) reactor design, which allow for distributed olefin feed injection and the operating conditions essential to both prevent rapid catalyst deactivation and attain high product quality.

In the AlkyClean process, a reactor may remain on-stream for up to 12 hours before olefin breakthrough. In practice, a reactor is regenerated safely before the expected olefin breakthrough time. Multiple reactors enable continuous alkylate production, as individual reactors cycle back and forth between on-line alkylation and mild regeneration, following the inventive procedure established during our process development effort. During mild regeneration, olefin addition is stopped and hydrogen is added to achieve a low reactor concentration of dissolved hydrogen, while maintaining liquid phase alkylation reaction conditions. This enables a seamless switchover between operations and minimizes energy consumption requirements. Over time, however, there is a gradual loss of catalyst activity. To recover this activity, with a frequency depending on the operating severity, a reactor is taken off-stream for high temperature (250 °C) gas-phase regeneration with hydrogen. To allow for high temperature regeneration while maintaining full continuous alkylate production, an additional "swing" reactor is provided.

Figure 12.13 depicts this cyclic operation of the AlkyClean reactor section. Notably, our processing scheme does not require any transfer of catalyst, either between reactor stages or to a separate regeneration vessel. For operability reasons, this was a fundamental design choice made early in the process development effort. Furthermore, the use of a swing reactor provides additional maintenance flexibility and enables the unit to stay on-stream when catalyst replacement (after years of operation) inevitably becomes necessary. At the end of its useful life, the catalyst is returnable to the manufacturer, Albemarle, eliminating any potential catalyst disposal problem for the refiner. The noble metal Pt is reclaimed and the remaining material can be used in the construction industry.

12.4.7
AlkyClean Process Demonstration Unit

Construction of an AlkyClean process demonstration unit at Fortum's facilities in Porvoo, Finland, was completed in 2002. Figure 12.14 shows the process flow schematic of the demo unit, which contains all of the key elements of our proposed commercial design. Three reactors are included – two under cyclic operation (i.e., alternating between alkylation and mild regeneration) allow for continuous production

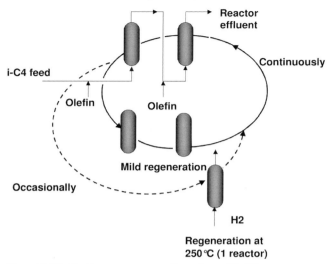

Reactor effluent

Continuously

i-C4 feed

Olefin

Olefin

Mild regeneration

Occasionally

H2

Regeneration at
250 °C (1 reactor)

Figure 12.13 AlkyClean reactor operating scheme [21, 28].

of alkylate and the third allows for swing reactor high temperature regeneration. The process design of the demo reactor section has been set to achieve operating conditions and compositional profiles analogous to those for a commercial design.

Figure 12.15 is a photograph of the installed demo reactor section. The demo reactors are sufficiently large and proportioned to allow for reliable scale-up. As such, each demo reactor represents a "core" of a much larger reactor and provides for the necessary hydrodynamic similarity (e.g., equivalent superficial velocities) to a full commercial-scale reactor system. Equally important, these reactors use AlkyClean catalyst produced in commercial manufacturing trials, not "developmental-scale" catalyst with characteristics that may not be fully duplicable under commercial-scale production conditions.

12.4.8
Demo Unit Operation

The AlkyClean process demonstration unit was built and operated with the following objectives:

- first and foremost, to demonstrate the operability and performance of our new catalyst and process technology;
- to optimize the process through parametric studies;
- to confirm key process, reactor and performance parameters;
- to verify its correspondence to bench scale unit performance;
- to tune key computer models;
- to test alternate olefin feedstocks.

After mechanical completion, the demo unit went through a shakeout and start up period of about one month. During this period, procedures were refined and proven for the *in situ* activation of the catalyst and the reliable start up of the reactor section.

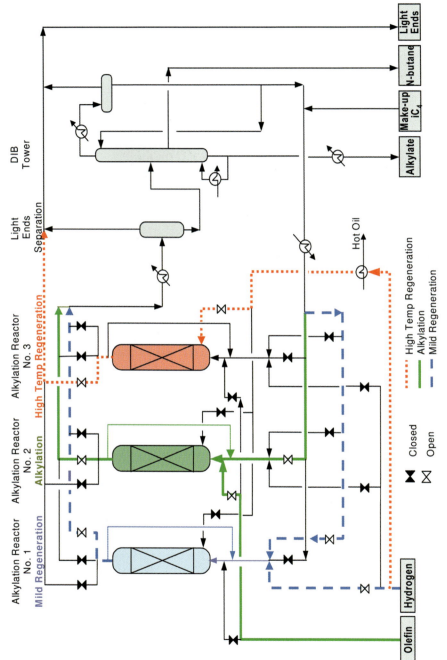

Figure 12.14 Process flow schematic of the demo unit [22].

Figure 12.15 Demo reactor section [22].

Subsequently, over several months, the demo unit reliably operated on a nearly continuous basis (except for holiday shut-down periods) while producing high quality product. The unit utilizes piped slipstreams of actual refinery feeds. To date, we have tested both predominantly C_4 olefin and mixed C_3/C_4 olefin feed streams, and the demo has produced alkylate of comparable quality to that from Fortum's Porvoo Refinery HF alkylation unit.

In addition to proving the operability of the process, key aspects of the technology have also been demonstrated. First and foremost is continuous cyclic operation – alternating reactors between periods of alkylation and mild regeneration – for periods of up to four weeks before taking a reactor off-line for HTR. In doing so, the durability of the AlkyClean catalyst has been demonstrated over hundreds of cycles of mild regeneration and multiple high-temperature regenerations. Regenerated catalyst samples from the demo unit have also been tested in Albemarle's bench-scale unit under benchmark conditions, further confirming our ability to repeatedly regenerate the catalyst and re-establish fresh catalyst activity and performance.

After obtaining performance data over a wide range of conditions, which supported our identified objectives, the demo unit operation was suspended during late 2003. This enabled a more economically efficient and intensive bench-scale unit effort to focus on identified opportunities for catalyst and process optimization that stemmed from insights gained through the analysis of the demo unit's performance. The result was a significantly improved catalyst with an activity advantage that is discussed below. Operational changes were also tested and refined, resulting in the further optimization of process performance.

Figure 12.16 provides a performance comparison of the second-generation catalyst relative to the original (i.e., old) prototype catalyst. This catalyst provided for about a 15–20 °C activity advantage over the old catalyst (i.e., the difference in operating

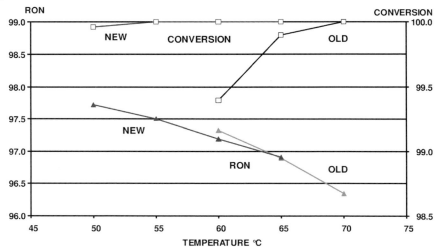

Figure 12.16 Comparison of first- and second-generation AlkyClean catalyst [22].

temperatures for equivalent conversion). In commercial practice, this substantial activity advantage can be used to operate at higher olefin throughputs and/or enable lower operating temperatures that will result in a higher octane alkylate (by up to about 1 RON). Albemarle has made an even more significant improvement in catalyst performance with its new third-generation catalyst.

Based on this successful bench scale program, operation of the demo unit resumed in 2004, following commercial trial manufacture of the newly improved catalyst, demo reactor catalyst replacement, and the completion of required unit modifications to incorporate operational improvements. Several months after its restart, the demo unit continued to run smoothly and continuously under full cyclic operation, with periodic rotational HTR of the reactors.

To date, the benefits of the operational improvements have been verified and the improved activity of the second-generation catalyst has been confirmed, along with stability and full activity recovery over multiple HTRs. Representative performance data from the current demo unit operation processing refinery C_4 olefins is presented in Figures 12.17 to 12.19 – for alkylate RON, RVP and C_5+ yield, respectively. This data, based on automated sampling from the on-line analytical system, show both the high product quality achieved and the stability of the cyclic process over time.

The demo unit successfully operated to fully verify the process's improved performance under targeted commercial design basis conditions. Further, it allowed for parametric optimization and confirmation of bench-scale unit correspondence, and provided necessary support for our correlation and modeling effort. Based on all the positive progress to date, this demo unit campaign was successfully completed after three months.

Finally, after many years of dedicated research carried out to develop a process that fully fulfills the targets set at the start of the quest, commercialization of the technology was started [23–25].

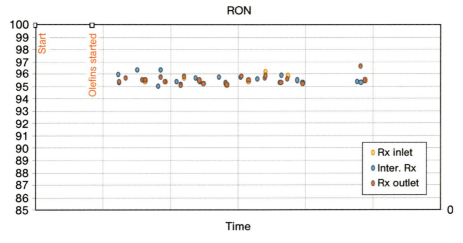

Figure 12.17 Demo unit performance processing refinery C_4 olefins: RON [22].

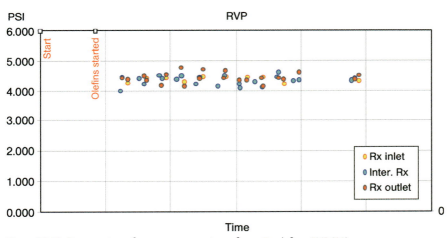

Figure 12.18 Demo unit performance processing refinery C_4 olefins: RVP [22].

Examples of other inventive concepts that were developed during this quest have been mentioned in the patent literature, such as the use of a high-performance zeolite [26], combined use of zeolites imbedded in a mesoporous material [27] and continuous alkylation with intermittent regeneration [28].

12.4.9
Competitiveness versus Liquid Acid Technologies [25]

Table 12.12 provides an overview of the competitiveness of the AlkyClean process versus the established liquid acid technologies. The performance and economics of the

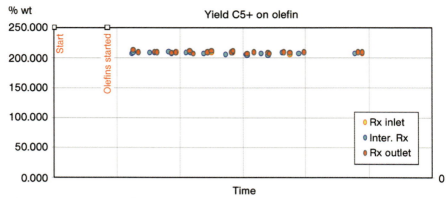

Figure 12.19 Demo unit performance processing refinery C_4 olefins: $C_5 +$ yield based on olefin [22].

AlkyClean process are fully competitive with current liquid acid technologies. High quality product has been produced in an operation that has proven to be reliable and robust. Sensitivity to feedstock variation is low and tolerance to impurities is high. The economic competitiveness of the new SAC process is enhanced by its low mechanical complexity and the use of common (i.e., non-proprietary) refinery process equipment.

Table 12.12 Comparison of AlkyClean process with liquid acid technologies.

Parameter	Modern sulfuric acid	Modern HF acid	AlkyClean process
Base conditions:			
$C_4=$ feedstock			
Product RON	95	95	95
Product MON	92.0–92.5	92.5	92.5
Alkylate yield	Base	Base	Base or better
Total installed cost, ISBL	Base	85% of base	85% of base
Total installed cost, including OSBL (regeneration facilities, and/or safety installations)	Base	70% of base	< 50% of base
Feed treatment	Base	Higher	Base
Product treatment	Yes	Yes	No
ASO yield	Up to 2 wt% on olefins	Less	None
Equipment maintenance	High	High	Very low
Corrosion problems	Yes	Yes	None
Reliability and on stream factor	Average	Average	Expected above average/high
Turnarounds frequency/ duration	Varies/longer	Varies/longer	Match FCC or better/shorter
OPEX	Base	Site specific Typically lower	Base

The low pressure and mild temperatures employed, along with the absence of either a corrosive or erosive environment, allow for the use of carbon-steel construction.

Based on an in-house benchmarking effort, we estimate that the total installed cost (TIC) for an AlkyClean process unit is about 10–15% lower than that for the equivalent H_2SO_4 unit. This comparison excludes off-site costs. When acid regeneration facilities for the H_2SO_4 unit are included, the SAC unit's cost becomes substantially less. We project that this level of investment for the SAC process is about on par with the cost for an HF unit. With respect to total cost of production, the results indicate that the requirements for the new AlkyClean process are comparable to that of the H_2SO_4 process. For both technologies, catalyst consumption cost is a significant component of the variable cost. For the AlkyClean process, this cost was conservatively based on a minimum ultimate catalyst life. Thus, there is considerable upside potential, which would result in the reduction of the AlkyClean process's total production cost. Compared to the AlkyClean and H_2SO_4 processes, production costs for HF units may be judged, on the surface, to still be somewhat lower; however, the increased costs for maintenance, mitigation and monitoring, among others, that the HF technology requires offset any perceived advantage.

12.4.10
Environmental, Cross-Media Effects

Table 12.13 shows the environmental benefits of the AlkyClean process.

Table 12.13 Summary of cross-media effects, waste and safety.

	Sulfuric acid	Hydrofluoric acid	AlkyClean process
Spent catalyst production	High	About 100× lower	About 1000× lower (Pt of spent catalyst can be reclaimed after many years of operation)
Product treatment	Acid removal needed	Acid removal needed	None
With caustic and/or lime	Creates waste water and sludge	Creates waste water and sludge	None
Acid-soluble oil production	Up to about 2 wt% on olefin feed	Less then sulfuric, but still significant	None
Corrosion issues	Yes	Yes	None
Maintenance	High	High	Low
Reliability	Moderate	Moderate	High
Safety	Unit-specific safety precautions as well as transport of H_2SO_4 precautions (accidental acid spills)	Very specific safety precautions required that extend throughout refinery and adjoining neighborhoods (accidental acid spills)	No special precautions other than those for any refinery process unit (inert catalyst)

12.5
Conclusion

The refining and petrochemical industry has seen many acid-catalyzed reactions evolve from liquid acids to solid acids, and each time the benefits were multi-fold. This is one more example of this evolution. As industry adjusts to the psychological hurdle of "first-of-its-kind" technology, the demise of liquid acid alkylation processes will be the cornerstone in making refining a very safe operation.

References

1 Weast, R.C. Astle, M.J. and Beyer, W.H. (eds) (1986) *Handbook of Chemistry and Physics*, 67th edn, CRC Press, Boca Raton, FL.

2 Hyman, H.H., Kirkpatrick, M. and Katz, J.J. (1957) *J. Am. Chem. Soc.*, **79**, 3668.

3 Paul, M.A. and Long, F.A. (1957) *Chem. Rev.*, **57**, 1.

4 Simons, J.H. and Dresdner, R.D. (1944) *J. Am. Chem. Soc.*, **66**, 1070.

5 Simmons, M.C. and Kelly, T.R. (1991) *Gas Chromatography, Second International Symposium*, Academic Press, New York.

6 Cupit, C.R., Gwyn, J.E. and Jernigan, E.C. (1962) *Petro/Chem. Eng.*, **34**, 49.

7 Pfile, M.E. (1987) Alkylation process fundamentals, refining process services. Presented at Symposium on Technology for Gasoline Production and Octane Enhancement, Houston, TX, October 8.

8 Jones, E.K. (1958) *Adv. Catal*, **10**, 185.

9 Corma, A. and Martinez, A. (1993) Chemistry, catalysts, and processes for isoparaffin-olefin alkylation: actual situation and future trends. *Catal. Rev.- Sci. Eng.*, **35** (4), 483–570.

10 Feller, A. and Lercher, J.A. (2004) Chemistry and technology of isobutane/alkene alkylation catalyzed by liquid and solid catalyst. *Adv. Catal*, **48**, 229–295.

11 Garwood, W.E. *et al.* (1966) US 3,251,902.

12 Caesar, P.D. *et al.* (1972) US 3,647,916.

13 Weitkamp, J. (1980) *Proc. Int. Conf. Zeolites*, **5**, 858.

14 Gardos, Gy., Pechy, L., Redey, A. and Sokorai, I. (1980) *Hung. J. Ind. Chem., Veszprem*, **8**, 371.

15 Juguin, B. *et al.* (1988) French Patent 2,631,956.

16 Huss, A. *et al.* (1991) US 4,992,615.

17 Chou, T.S. *et al.* (1990) World Patent 90/00533.

18 van Broekhoven, E.H. and Mas Cabré, F.R. (2005) US 6,855,856.

19 van Broekhoven, E.H. *et al.* (2001) A new solid acid isobutane alkylation technology: AlkyClean. Akzo Nobel Catalysts Eco-Magic Symposium, Noordwijk, The Netherlands, paper G9, June 10–13.

20 van Broekhoven, E.H., Mas Cabré, F.R., Bogaard, P., Klaver, G. and Vonhof, M. (1999) US 5,986,158.

21 D'Amico, V.J. *et al.* (2002) The AlkyClean process: a new solid acid catalyst gasoline alkylation technology. NPRA Spring 2002 Meeting, paper AM-02-19.

22 D'Amico, V.J. *et al.* (2004) AlkyClean solid acid alkylation: will it finally become a reality? Akzo Nobel Catalysts Scope Symposium, Florence, Italy, June 21–23.

23 D'Amico, V. *et al.* (2006) Consider new methods to debottleneck clean alkylate production. Hydrocarbon Processing, February, p. 65.

24 D'Amico, V.J. *et al.* The AlkyClean process – demonstrated new standard for alkylation technology. AIChE Spring National Meeting 2006, Orlando, FL.

25 D'Amico, V. *et al.* (2006) The AlkyClean alkylation process – new technology eliminates liquid acids. NPRA Spring 2006 Meeting, paper AM-06-41.

26 Yeh, C.Y. *et al.* (2005) US 6,844,479.

27 Shan, Z. *et al.* US Appl 20060128555 (filed 2/8/06).

28 van Broekhoven, E.H., Sonnemans, J.W.M. and Zuijdendorp, S. (2007) US 7,176,340.

13
Towards the Direct Oxidation of Benzene to Phenol

Marco Ricci, Daniele Bianchi, and Rossella Bortolo

13.1
Introduction

Phenol is one of the most important intermediates of the chemical industry. Its current global capacity is put at about 8×10^6 t yr^{-1} and is forecast to grow near 4.5% per year through to 2009.

The main consumption of phenol, accounting for nearly 40% of its global demand, occurs in the synthesis of bisphenol A (mainly used to produce polycarbonate for compact discs). This is followed by phenol use in the production of epoxy resins (used, for instance, in protective coatings, in composites for electrical applications and in adhesives), of phenolic resins (which have a broad range of end-uses, including circuit boards) and of caprolactam (to be converted into nylon 6 for fibers and engineering plastics). In addition, phenol is an intermediate in the syntheses of adipic acid, acetyl salicylic acid (aspirin), aniline, hydroquinone, catechol, 2,6-xylenol, alkylphenols, chlorinated phenols and diphenols, polyphenylene oxide engineering plastics and other specialty chemicals. Finally, it is used in plasticizers, in water treating (as a slimicide) and as a disinfectant and anesthetic in medicinal preparations and pharmaceuticals.

Historically, several processes have been developed to an industrial scale to produce phenol, including: (i) sulfonation of benzene and alkali fusion of the benzene sulfonate; (ii) chlorination of benzene and hydrolysis of chlorobenzene; (iii) the cumene process (Section 13.2); (iv) toluene oxidation to benzoic acid and subsequent oxidative decarboxylation of the latter to phenol; and (v) dehydrogenation of cyclohexanol–cyclohexanone mixtures. Today, however, only the cumene process and the toluene oxidation are still run on an industrial scale, all the other processes having been given up due to economic reasons or environmental problems.

Of the two commercially operated technologies, the toluene oxidation route affords not only phenol but also the specialty chemicals benzaldehyde and benzoic acid (Equation 13.1):

Sustainable Industrial Processes. Edited by F. Cavani, G. Centi, S. Perathoner, and F. Trifiró
Copyright © 2009 WILEY-VCH Verlag GmbH & Co. KGaA, Weinheim
ISBN: 978-3-527-31552-9

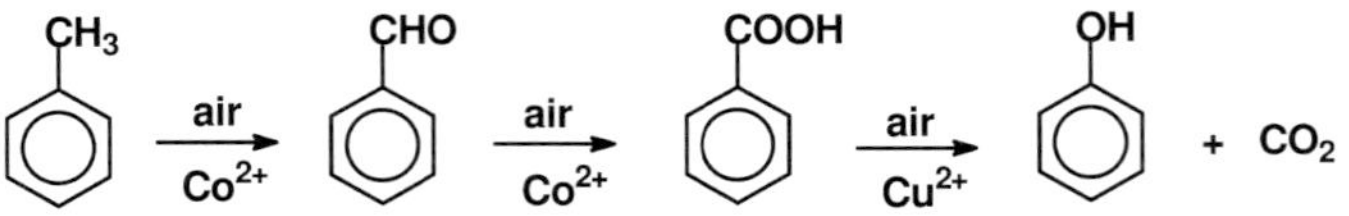

$$
\text{(13.1)}
$$

It is only run, quite successfully, by DSM and in few small plants licensed by DSM: this is because the amount of benzaldehyde and benzoic acid thus produced matches the market demand. Should the technology be adopted by others, the quantity of these specialty derivatives would quickly exceed their demand. Consequently, new plants are almost exclusively based on the cumene process, which, in the following, will be examined into some detail.

13.2
Cumene Process

The cumene process, sometimes referred to as the Hock process, was made possible by the discovery of cumyl hydroperoxide and of its cleavage to phenol and acetone [1]. Shortly after World War II the reaction was developed into an industrial process by the Distillers Co. (BP Chemicals) in the United Kingdom and Hercules in the USA. The first commercial plant was started in Montreal, Canada, in 1952 by M.W. Kellogg.

The process is based upon three different reactions: (i) Friedel–Crafts alkylation of benzene with propene to afford cumene (isopropylbenzene); (ii) cumene oxidation with oxygen to give cumyl hydroperoxide; and (iii) cleavage of cumyl hydroperoxide in acidic medium to afford phenol and acetone (Equation 13.2):

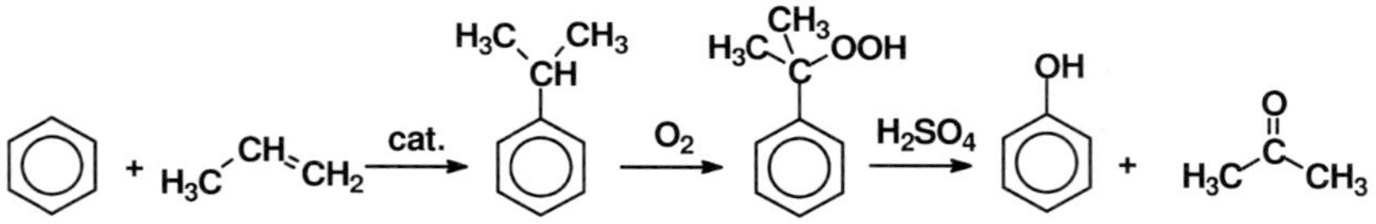

$$
\text{(13.2)}
$$

13.2.1
Alkylation

Cumene is produced by alkylating benzene with propene. The reaction needs a catalyst: in recent years, zeolite-based catalysts have become almost universally used in cumene plants [2], although older plants using aluminum chloride (AlCl$_3$) or phosphoric acid supported on silica are still operating.

Zeolite-based catalysts are definitely less hazardous materials than aluminum chloride or phosphoric acid and, being non-corrosive, they allow a reduction in plant maintenance. Furthermore, they can be regenerated offsite by burning off high molecular weight hydrocarbons deposited on them: the expected catalyst cycle length between regenerations is 36–60 months, and the expected life four or five reaction/ regeneration cycles. At the end of their life, zeolite-based catalysts can be safely disposed of as landfill after hydrocarbon removal, thus eliminating disposal problems associated with AlCl$_3$ or phosphoric acid. Consequently, the use of

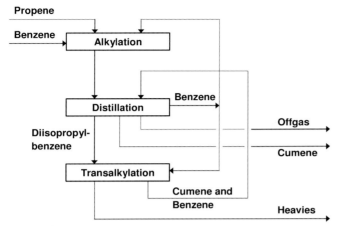

Figure 13.1 Cumene plant: block scheme [3].

zeolite-based catalysts significantly improves the sustainability of the cumene production. However, it requires feedstocks of a relatively high purity and some pretreatment can be necessary.

In all cases, the main by-product of the alkylation reaction is a mixture of diisopropylbenzenes (DIPB) which, with zeolite-based catalysts, accounts for 5–25% of the whole alkylated product. Cumene can be recovered from DIPB by transalkylation with one mole of benzene to form two moles of cumene.

In a typical arrangement (Figure 13.1), alkylation occurs at 130–170 °C in liquid phase on four catalyst beds packed in two reactor shells, arranged in parallel. A mixture of fresh and recycled benzene is charged downflow through the alkylation reactors. Excess benzene is used to minimize polyalkylation and olefin oligomerization. Fresh propene feed (which may contain inert propane in an amount depending on the propene source) is split between the four catalyst beds, and is completely consumed in each bed. The reaction of benzene and propene is exothermic; the temperature rise in each reactor may be controlled by recycling part of the reactor effluent, which absorbs the heat of reaction. Effluent from the alkylation reactors is first sent to a depropanizer column whose bottom stream is sent to a second column, where unreacted benzene is collected overhead and recycled, while the bottom stream is sent to a further column where cumene is recovered overhead. The bottom of the cumene column is sent to a DIPB column, where DIPBs are recovered overhead and sent to the fixed bed transalkylation reactor. In turn, the bottom product from the DIPB column, consisting primarily of heavy aromatics, is typically blended with fuel oil for burning.

Table 13.1 gives a typical material balance of the alkylation section. Notably, however, depending on the specific technology and catalyst offered by each cumene licensor, raw material consumption may be in some cases significantly better. As an example, a total amount of heavies as low as 0.003 t per t of cumene is now claimed by Polimeri Europa.

Table 13.1 Cumene plant: material balance [3].

Material	Metric tonne per metric tonne of cumene
Feed	
Benzene (as 100% purity)	0.652
Propene (as 100% purity)	0.352
Product	
Cumene	1.000
By-product	
Heavies	0.005

13.2.2
Oxidation and Concentration

In the second step of the cumene process, cumene undergoes oxidation with air, possibly oxygen-enriched, to afford cumyl hydroperoxide (CHP). This is a typical radical reaction and the hydroperoxide forms at the expense of the less energetic, tertiary C—H bond. The reaction is carried out in liquid phase, at 85–120 °C and 4–10 bar. Owing to the formation of small amounts of acid by-products (mainly formic acid), unstabilized systems work at pH 3–6. However, these acids promote the decomposition of cumyl hydroperoxide to acetone and phenol: the latter is an excellent inhibitor of radical reactions and its presence is not compatible with the autoxidation. Therefore, it is common practice to stabilize the reaction medium through the addition of an emulsified weakly basic aqueous phase (sodium hydroxide or carbonate, pH 7–8). Initially, the oxidation is quite slow and several catalysts have been described that can speed up this step. However, the reaction is autocatalytic and its rate gradually increases with increasing hydroperoxide concentration. Main by-products are 2-phenyl-2-propanol and acetophenone, both arising from thermal decomposition of cumyl hydroperoxide, and dicumyl peroxide, formed by an equilibrium reaction of cumyl hydroperoxide with 2-phenyl-2-propanol.

The oxidation is mostly carried out in traditional bubble column reactors: series of two to six reactors, up to 20 m high, are common in industry. The reaction is exothermic: ~120 kJ are released per mole of produced hydroperoxide, and must be removed by cooling. The final reaction mixture, containing 20–30% of cumyl hydroperoxide, is then concentrated by distilling off some unreacted cumene to obtain a 65–85% hydroperoxide to be fed to the cleavage step (Figure 13.2).

Significant attention is paid during both the reaction and the concentration to prevent ignition or explosion of the cumene–air mixtures. Furthermore, provisions are needed for a water or a steam quench to the concentrators to prevent hydroperoxide decomposition in case of an emergency.

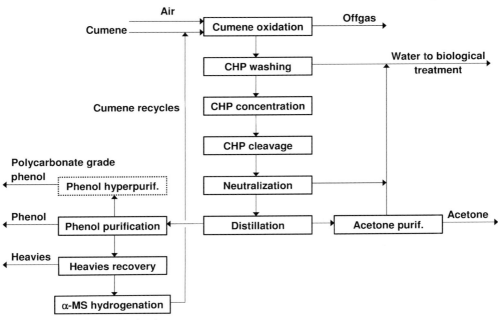

Figure 13.2 Phenol plant: block scheme [3].

13.2.3
Cleavage and Workup

Cumyl hydroperoxide is eventually cleaved in the presence of an acid catalyst, to yield phenol and acetone, together with minor amounts of by-products such as α-methylstyrene, arising from dehydration of 2-phenyl-2-propanol, and dicumyl peroxide. α-Methylstyrene can be recovered to cumene in a hydrogenation stage.

As the catalyst, concentrated (98%) sulfuric acid is almost exclusively used in the industrial practice. The cleavage is run in the presence of 0.2–1% of acid, under reduced pressure, at the boiling temperature of the cumyl hydroperoxide–acetone mixture, which depends on the acetone content. Again, the reaction is strongly exothermic ($\sim$250 kJ mol^{-1}) and the heat is removed by evaporation of acetone from the reaction mixture. At 70–80 °C, cumyl hydroperoxide conversion is virtually quantitative, with a selectivity to phenol up to 94–95%.

Acetone and phenol can be recovered after the neutralization of the acidic mixture from the cleavage reactor with sodium hydroxide or phenolate solution. The neutralized mixture is then subjected to a series of distillations. Acetone is first distilled, then cumene is recovered, together with α-methylstyrene, which is either purified and marketed or hydrogenated back to cumene and recycled to the oxidation. Phenol is finally distilled with a purity up to 99.99%, suitable for the production of polycarbonate grade bisphenol A and other chemicals and polymers.

The material balances of the oxidation/cleavage section and of the overall process are reported in Tables 13.2 and 13.3, respectively. Again, raw material consumption

Table 13.2 Phenol plant: material balance [3].

Material	Metric tonne per metric tonne of phenol
Feed	
Cumene	1.330
Product	
Phenol	1.000
By-products	
Acetone	0.626
Heavies	0.053
Process water	0.493

Table 13.3 Overall material balance.

Material	Metric tonne per metric tonne of phenol
Feed	
Benzene	0.867
Propene	0.468
Product	
Phenol	1.000
By-products	
Acetone	0.626
Heavies	0.060
Process water	0.493

may in some cases be significantly better, depending on the specific phenol technology offered by each phenol licensor.

13.2.4
Cumene Process: Final Considerations

Since 1952, the cumene process has been improved to a fairly impressive extent as its yields to acetone and phenol, based on both benzene and propene, are currently close to the stoichiometric ones. Nor are there problems in building huge plants: 300 000 t yr^{-1} capacity is a common size, while still having a quite reduced environmental impact. Thus, although some precautions are needed to ensure safe operations, the process is fully satisfactory in many aspects.

Some improvements, in terms of reduction of energy consumption, could be attained by increasing the benzene conversion per pass in the alkylation and oxidation stages, thus reducing benzene and cumene recycles. By assuming a reaction molar ratio between benzene and propene as low as 2.5, currently allowed by the most competitive cumene technologies based on the latest generation of zeolite-based catalysts, the benzene conversion per pass in the alkylation stage remains below the 45% while, in the cumene oxidation stage, the conversion per pass is not higher than 35% in order to maintain a safe concentration of the intermediate cumyl hydroperoxide with good selectivity. Consequently, benzene conversion per pass over the benzene alkylation and cumene oxidation stages (cumyl hydroperoxide conversion in the cleavage stage can be considered quantitative) is never higher than about 15%, which means that the cumene process to phenol is inherently affected by relevant recycle volumes.

However, the major problem in cumene process to phenol seems to be the fixed coproduction of about 0.6 t of acetone per t of phenol, while the yearly market growth for phenol applications is nowadays higher than the market growth for acetone, whose sale price is then continuously squeezed by the market. The current market situation trends to a further worsening for the acetone sale price in the future, based mainly on two facts: (i) on one hand, all volatile organic compounds (including acetone) are subjected to a strong, increasing environmental pressure, urging for their substitution in the medium term and (ii) on the other hand, the main current use of phenol is in polycarbonate synthesis, which does not actually require phenol itself but rather bisphenol A, whose production requires just one mole of acetone per two moles of phenol (Equation 13.3), inherently leading then to unbalanced acetone and phenol consumptions:

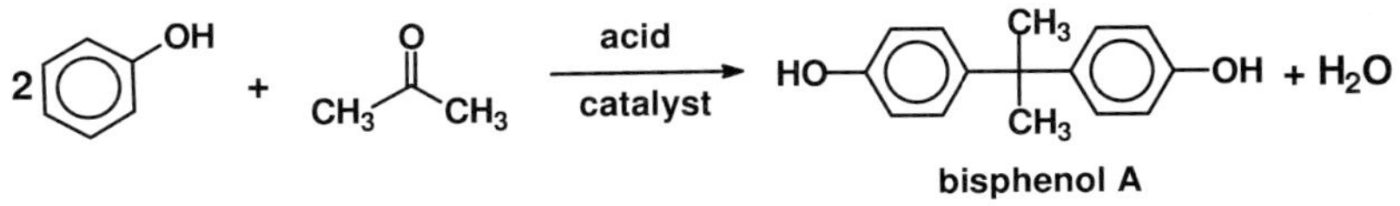

bisphenol A

(13.3)

Any addition of phenol capacities adds a further unbalance on the acetone side.

A first approach to the problem has been the development, by Mitsui Petrochemical, of a recycle scheme for converting the acetone back into propene (via hydrogenation to 2-propanol and subsequent dehydration of the latter) to be added to the feed of the alkylation step (Scheme 13.1).

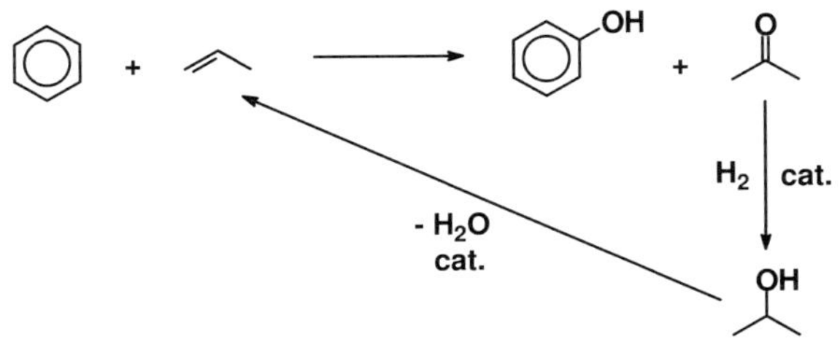

Scheme 13.1 Mitsui Petrochemical acetone recyle process.

Friedel–Crafts alkylation with 2-propanol is also possible [4], without the need of the dehydration step, whose energy requirement accounts for the most part of the overall energy balance for the acetone into propene transformation.

As the recycle of acetone to benzene alkylation to cumene leads to a corresponding saving in propylene, its profitability depends on the acetone to propene sale prices ratio.

Polimeri Europa has developed a technology based on the direct alkylation of benzene with 2-propanol (thus avoiding its dehydration stage) where the acetone recycle versus the acetone sale profitability is claimed at acetone to propylene sale prices ratio equal or lower to 0.6.

Nevertheless, much effort is currently being devoted to decouple phenol and acetone productions and, particularly, to develop effective processes for the direct oxidation of benzene to phenol.

13.3
Solutia Process

The selective insertion of an oxygen atom into a benzene carbon–hydrogen bond to yield phenol is not a classical organic chemistry reaction. The first process for such a reactions was the Solutia process, based on discoveries by Panov and coworkers at the Boreskov Institute of Catalysis in Novosibirsk and then developed in close cooperation with Monsanto. In this process, the oxidant is nitrous oxide, N_2O, while an iron-containing zeolite is used as the catalyst (Equation 13.4):

$$\text{benzene} + N_2O \xrightarrow{\text{Fe-zeolite}} \text{phenol} + N_2 \tag{13.4}$$

The reaction is run in the gas phase at 350 °C and, at 27% of benzene conversion, selectivity for phenol is claimed to be 98% [5]. The main by-products are dihydroxybenzenes (about 1%) and carbon oxides (0.2–0.3%). Selectivity is of paramount importance for this process, since 15 molecules of nitrous oxide are consumed for the total oxidation of just one molecule of benzene (Equation 13.5):

$$C_6H_6 + 15\, N_2O \rightarrow 6\, CO_2 + 3\, H_2O + 15\, N_2 \tag{13.5}$$

The catalyst is an iron-containing ZSM-5 zeolite. Its half-life is three to four days so that, periodically, catalytic activity must be restored by passing air through the deactivated catalyst at high temperature: no performance deterioration has been reported after more than 100 regeneration cycles.

Figure 13.3 shows a plant layout. Recycled benzene along with makeup benzene and nitrous oxide are preheated and continuously fed to a moving bed reactor utilizing the zeolite catalyst. The latter flows vertically down the reactor by gravity, while the reaction gas flows across the annular catalyst beds. The predominant reactions are exothermic: about 250 kJ are released per mole of phenol produced. In addition, significantly more heat can be generated by the deep oxidation of benzene to

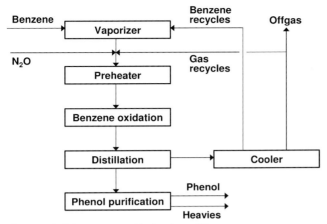

Figure 13.3 Solutia process: block scheme [6].

carbon dioxide. So, the temperature rises as the gas flows radially across the catalyst beds. The reactor effluent is collected and cooled. Offgas (mainly nitrogen) is incinerated or vented to atmosphere, whereas unreacted benzene and phenol are condensed and sent to a crude phenol column. Benzene is fractionated overhead and recycled to the reactor. The crude phenol from the bottom of the column is sent to a phenol purification column, where phenol product is taken overhead and the heavies from the bottom of the column are sent to fuel. At the same time, partially deactivated catalyst is continuously withdrawn from the bottom of the moving bed reactor and transported to the regenerator, where the accumulated coke is burnt off. The regenerated catalyst is then transported again to be fed at the top of the moving bed reactor.

Table 13.4 gives the material balance of the Solutia process.

There is significant debate about the mechanism of this reaction and, in particular, about the nature of the iron sites responsible for the unique reactivity. It is generally

Table 13.4 Solutia process: material balance [6].

Material	Metric tonne per metric tonne of phenol
Feed	
Benzene	0.874
Nitrous oxide	0.551
Product	
Phenol	1.000
By-products	
Dihydroxy benzenes	0.011
CO_2	0.029

accepted that the reaction proceeds over the so-called α-sites – defect sites of the zeolite that are formed in an iron-containing zeolite matrix upon high temperature activation, and where iron atoms migrate. The iron atoms that make up α-sites are in a bivalent state, with a special affinity to nitrous oxide. N_2O decomposition causes the transition Fe^{2+} to Fe^{3+}, producing the so-called atomic α-oxygen species, which can selectively insert oxygen into C−H bonds of alkanes and aromatics [7].

Despite its brilliant results, it seems unlikely that the Solutia process can become a major source of phenol. Nitrous oxide availability is quite limited and its production on-purpose (by the conventional ammonium nitrate decomposition, which enables nitrous oxide of high purity to be produced for medical anesthetic applications, or even by selective oxidation of ammonia) would result too expensive. Therefore, the only reasonable scenario to exploit the Solutia process is its implementation close to adipic acid plants, where nitrous oxide is co-produced by the nitric oxidation of cyclohexanol–cyclohexanone mixtures and where it could be used to produce phenol instead of being disposed of. However, the stoichiometry of the process is such that a relatively small phenol plant would require a world-scale adipic acid plant for its nitrous oxide supply. In fact, a pilot plant has been operated using this technology, but its commercialization has been postponed.

13.4
Direct Oxidation of Benzene to Phenol with Hydrogen Peroxide

A promising alternative to nitrous oxide is provided by hydrogen peroxide, which is finding more and more favor due to the lack of environmental impact and the easy storage and handling. Both homogeneous and heterogeneous catalysts have been developed recently to be used in the direct oxidation of benzene to phenol by hydrogen peroxide. In the first case, soluble iron complexes were used under biphase conditions, while titanium-containing zeolites were selected as heterogeneous catalysts.

13.4.1
Definition of the Problem and First Attempts

The direct oxidation of benzene to phenol is usually affected by a poor selectivity due to the lack of kinetic control. Indeed, phenol is more reactive towards oxidation than benzene itself and consecutive reactions occur, with substantial formation of over-oxidized products like catechol, hydroquinone, benzoquinones and tars. This is the usual output of the oxidation of aromatic hydrocarbons by the classical Fenton system, a mixture of hydrogen peroxide and an iron(II) salt, usually ferrous sulfate, most often used in stoichiometric amounts [8].

Thus, to obtain a selective synthesis of phenol via direct oxidation of benzene, suitable strategies have to be envisaged to slow down the undesired consecutive reactions and to allow phenol to accumulate. The first step in this direction was made by George Olah, who used extremely concentrated (98%) hydrogen peroxide in a superacidic medium ($FSO_3H–SbF_5$, 1:1) at $-78\,^{\circ}C$ and obtained phenol with a 54% yield, based both on the benzene and on the hydrogen peroxide [9]. Owing to the very harsh reaction conditions, this work has only historical value, but the concept that, in the superacidic medium, the phenol was protonated (thus being deactivated towards any further oxidation) was noteworthy.

A few years later, Hubert Mimoun discovered the rather selective oxidation of benzene by a few vanadium(V) peroxo complexes [10]. Using an excess of hydrogen peroxide under phase-transfer conditions transformed this stoichiometric reaction into a true catalytic process, but the turnover numbers remained very low [11].

13.4.2
Homogeneous Catalysis by Iron Complexes: A Biphase Fenton Reagent

Over-oxidation problems are solved efficiently in biological systems by segregating catalysts and products into different environments. So, for instance, the active sites of several oxygenases are buried deeply into hydrophobic pockets where lipophilic substrates are readily oxidized, while the more hydrophilic reaction products, when released into the surrounding aqueous environment, do not have further access to the catalytic site [12].

A simple biphasic system can mimic these important features of biological systems and this observation is the basis for a benzene oxidation with hydrogen peroxide. According to this methodology, the reaction medium is formed by an aqueous phase, containing both the hydrogen peroxide and the oxidation catalyst, and an organic one able to dissolve most of the produced phenol. The use of the aqueous–organic reaction medium dramatically affects the selectivity of the reaction; in particular, a remarkable enhancement is obtained using a biphasic system generated by water and acetonitrile (volume ratio $= 1{:}1$) in the presence of benzene. With this particular medium, the concentration of benzene in the aqueous phase rises from 0.18% (the solubility of benzene in water) to 0.76%, and the produced phenol was extracted for the most part (85%) in the organic phase. As a consequence, the benzene/phenol molar ratio is only 0.25 in the aqueous phase of a water–benzene mixture, whereas it rises to 3.7 in the water–acetonitrile–benzene one. In this way, the biphasic operation minimizes the over-oxidation reactions by reducing the contact between the phenol and the catalyst, which segregates in the aqueous phase.

The catalyst is a water-soluble iron(II) salt, typically iron sulfate, the performances of which are improved by the addition of a suitable, bidentate ligand. N-N ligands (e.g., phenanthroline derivatives) have a detrimental effect on the hydrogen peroxide activation and result in very low conversions. O-O ligands (e.g., catechol derivatives) show a good activity but also promote the hydrogen peroxide decomposition, thus resulting in rather low selectivities. However, N-O ligands, especially pyrazinecarboxylic acid derivatives, give the most interesting results. 2-Methylpyrazine-5-carboxylic acid N-oxide (Figure 13.4) and the corresponding ligand without the methyl group (pyrazine-3-carboxylic acid N-oxide) are the most efficient.

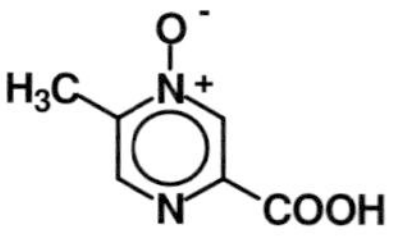

Figure 13.4 2-Methylpyrazine-5-carboxylic acid N-oxide.

Using $FeSO_4$ (1.67×10^{-3} M) in conjunction with equimolar amounts of methyl-pyrazine-5-carboxylic acid N-oxide and trifluoroacetic acid, in a water–acetonitrile–benzene (5:5:1 v/v/v) biphasic system, with benzene–H_2O_2–$FeSO_4 = 620:60:1$, a benzene conversion of 8.6% is achieved (35 °C; 4 h). Hydrogen peroxide conversion is almost complete (95%) and selectivities to phenol are 97% (based on benzene) and 88% (based on H_2O_2) [13]. These values are definitely higher than those described in the literature for the classical Fenton system [14], whereas iron complexes with pyridine-2-carboxylic acid derivatives are reported to be completely ineffective in the oxidation of benzene under the well-known Gif reaction conditions [15].

Regarding the possible mechanism, notably, toluene, ethylbenzene and *tert*-butylbenzene are less reactive than benzene, which is not consistent with the expected order for an electrophilic aromatic substitutions, such as that found with the classic Fenton reagent. There are also other differences with respect to the Fenton chemistry. In particular, under biphase conditions the reaction is definitely more selective: although comparisons are difficult due to the huge amount of data, sometimes inconsistent, on the Fenton system (for which most of the data have been obtained with the iron used in stoichiometric amounts) it seems that selectivities close to those observed under biphase conditions are only attained at a conversion around of 1%. Furthermore, in the biphase system, only a negligible amount (<1%) of biphenyl was detected among secondary products, whereas in the classic Fenton oxidation this compound is formed by radical dimerization of hydroxycyclohexadienyl radicals in typical yields ranging from 8 to 39%.

Is a different mechanism operative under biphase conditions? Although detailed mechanistic information is not currently available, the reaction is likely to start with the oxidation of Fe(II) to Fe(III) (Equation 13.6):

$$L_2 Fe^{II} + H_2O_2 \rightarrow L_2 Fe^{III} OH + OH^{\bullet} \qquad (13.6)$$

Thus, it is also likely that hydroxyl radicals are present in the reaction mixture, where they probably act as oxidizing agents according to the overwhelmingly established Fenton radical mechanism, and differences could simply arise due to the peculiar biphase system: for instance, toluene is less soluble than benzene in the aqueous phase, in which the reaction takes place (98.9 and 142.3 mmol L^{-1}, respectively) and this could help to explain its lower reaction rate. Alternatively, these differences might be the result of a competition with a second mechanism. Following several suggestions [16], the new mechanism could be triggered by the formation of an iron(III) hydroperoxo species, which could undergo heterolytic cleavage to afford an electrophilic, high-valent iron-oxo complex, stabilized by the ligand, able to oxidize aromatic hydrocarbons to phenols (Scheme 13.2).

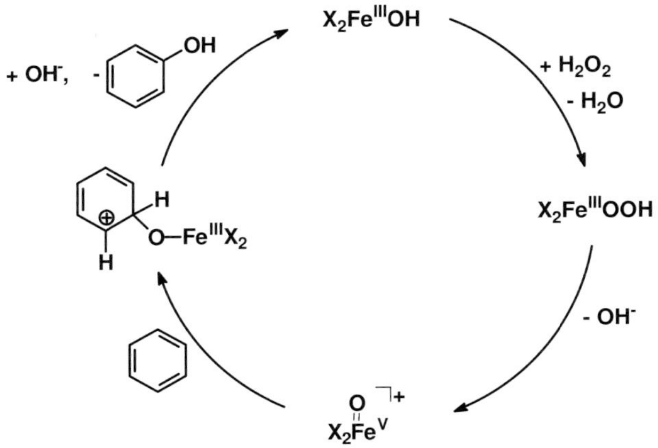

Scheme 13.2 Possible catalytic cycle involving the intervention of high-valent iron species.

13.4.3
Heterogeneous Catalysis by Titanium Silicalite

A process based on the iron catalysis described above would be affected by a low volume productivity, in terms of the amount of produced phenol in a given time per liter of reactor volume. This is a common weakness of homogeneous versus heterogeneous catalysis. However, at the same time the iron-catalyzed reaction was being studied, a second, heterogeneous, catalyst was found that can catalyze the direct oxidation of benzene to phenol with hydrogen peroxide: titanium silicalite (TS-1). It was discovered in 1979 in the laboratories of the ENI group in San Donato Milanese and the relevant patent was filled on December 21, 1979. It is a crystalline, synthetic zeolite in which tetrahedral $[SiO_4]$ and $[TiO_4]$ units are arranged into an orthorhombic MFI structure (ZSM-5 type), with titanium replacing up to 3% of silicon atoms. Owing to its structure, TS-1 shows a three-dimensional system of channels having near-circular section, which constitute the zeolitic micropores of the material: two main sub-systems are present, sinusoidal channels and straight ones, with diameters of 5.6×5.3 and 5.3×5.1 Å, respectively [17].

Soon after its discovery, TS-1 was recognized as a valuable catalyst for many oxidations by hydrogen peroxide, including alkane oxidation, olefin epoxidation, alcohol oxidation, phenol hydroxylation and cyclohexanone ammoximation (Scheme 13.3) [18].

However, the activity of TS-1 in the oxidation of benzene appeared to be very poor. TS-1 does not perform very well in two-phase systems, so that only solvents able to homogenize the hydrophobic substrate and the aqueous hydrogen peroxide can be used. Even in this case, using solvents such as acetone, acetonitrile or *tert*-butanol, the selectivity to phenol rapidly dropped at very low benzene conversion, mainly due to the formation of dioxygenated products and tars: typically, selectivity was already less than 50% at benzene conversion as low as 5%. Even worse results were obtained using methanol, which was oxidized in competition with benzene to give formaldehyde dimethyl acetal.

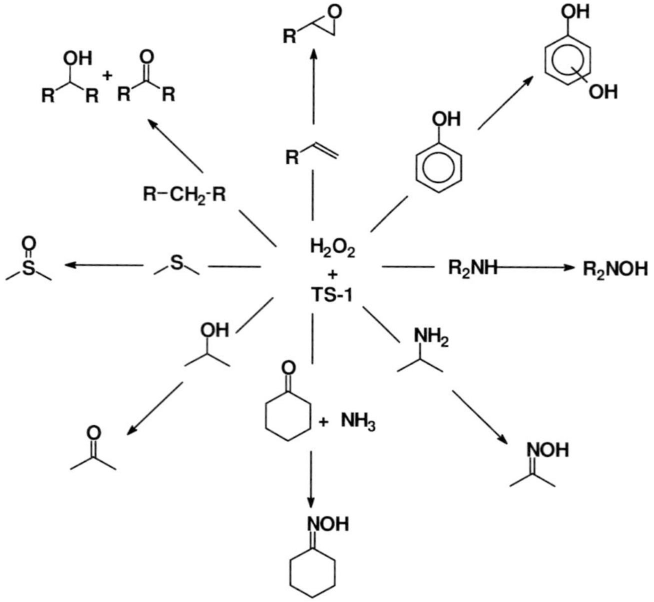

Scheme 13.3 Some oxidation reactions catalyzed by TS-1.

In contrast, the use of sulfolane as solvent allows a conversion of benzene close to 8% while maintaining the selectivity to phenol higher than 80%. Detected by-products are catechol (7%), hydroquinone (4%), 1,4-benzoquinone (1%) and tars (5%) [19].

Even better results are obtained by a post-synthesis treatment of TS-1 with both hydrogen peroxide and ammonium hydrogen fluoride, NH_4HF_2. Upon such a treatment ($H_2O_2/F/Ti = 10:2.5:1$; 60 °C; 4 h), a substantial amount of titanium (up to 75% of the initial value) is removed. Nevertheless, the crystalline structure of the zeolite remains unchanged and the catalytic activity does not decrease. On the contrary, it actually increases since the turnover frequency of residual titanium atoms rises from 31 to 80 h^{-1}. Even more importantly, at 8.6% benzene conversion the selectivities, both on benzene and on hydrogen peroxide, also increase from 83 to 94% and from 67 to 83% respectively, with formation of catechol (4%) and hydroquinone (2%) as the only by-products, without any evidence of further oxidation reactions [19].

The treated catalyst has been named TS-1B. A preliminary characterization of it (with residual titanium 29% of the original amount) shows that it has a peculiar UV/Vis spectrum. In particular, absorption at 48 000–50 000 cm^{-1} (typical of pure TS-1) was strongly reduced and a new titanium species, absorbing at 40 000 cm^{-1}, was generated by the treatment. The formation of amorphous extraframework titanium species (TiO_2), absorbing at 30 000–35 000 cm^{-1}, was not observed (Figure 13.5).

A possible mechanism for the TS-1-catalyzed hydroxylation of phenol with hydrogen peroxide has been proposed [20]: it is an ionic mechanism that, with

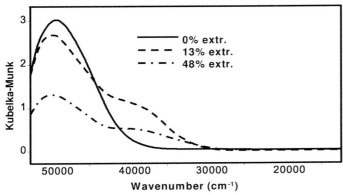

Figure 13.5 UV/Vis DRS (diffuse reflectance spectroscopy) of TS-1 and TS-1B. TS-1 (solid line); TS-1B after extraction of 13% of TiO$_2$ (dashed line); TS-1B after extraction of 48% of TiO$_2$ (dot-dashed line).

minor modifications, can probably hold also for benzene hydroxylation. Alternative radical pathways are also possible.

Why is sulfolane, and only sulfolane, so effective in improving selectivities? Already in 1963, Drago suggested that sulfolane forms a complex with phenol via hydrogen bonding (Figure 13.6) [21].

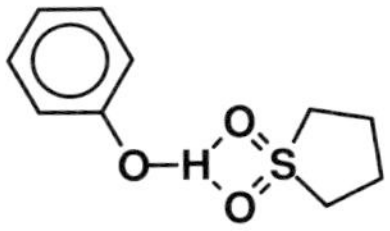

Figure 13.6 Possible structure of the phenol/sulfolane complex.

Such a complex is not stable enough to be isolated but its occurrence was inferred by the IR spectra of the solutions. Drago's suggestion is confirmed by *ab initio* calculations. Therefore, the improved selectivity observed upon carrying out benzene oxidation in sulfolane may be due to the formation of this large species, which can not enter the titanium silicalite pores, thus allowing phenol to remain relatively protected against further oxidation.

This effect was confirmed by calculation of the loading of the free phenol molecule and the complex phenol–sulfolane (expressed as the number of loaded molecules in a crystal elementary unit of TS-1), using the software *Sorption* (Cerius 2), which turned out to be 13.6 and 0.8, respectively. Alternatively, the protective effect exerted by sulfolane can be evaluated by measuring the reaction rate, expressed as the turnover frequency (TOF: moles of reacted substrate/moles of Ti per hour) for the oxidation of benzene and phenol, carried out separately in acetone and sulfolane as co-solvents. In the case of acetone, the phenol oxidation (TOF = 190) was ten times faster than that measured for benzene (TOF = 19); conversely, operating in sulfolane the rate measured for phenol (TOF = 51) was only 1.6 time higher than that measured for benzene (TOF = 31), according to the higher value of the observed selectivity.

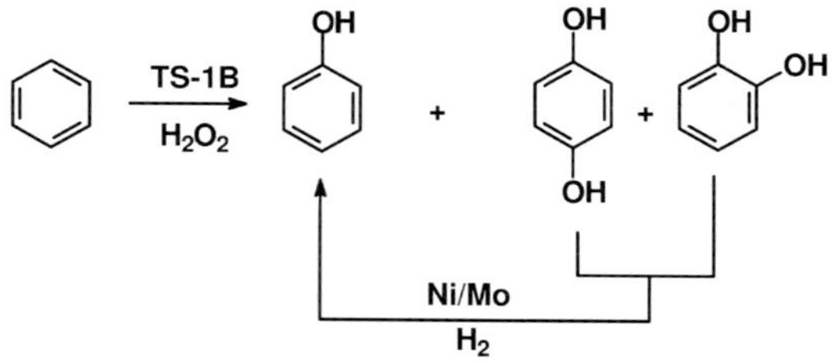

Scheme 13.4 Polimeri Europa TS-1/H_2O_2 process.

Many attempts have been carried out to find catalysts other than TS-1 able to catalyze the oxidation of benzene to phenol with hydrogen peroxide. With mesoporous molecular sieves (Ti-HMS, Ti-MCM-41, Ti-MSA, Ti-ERS8, Ti-AMM, all with pore diameters of 2–5 nm), and with large pores zeolites (like Ti-β or TAPO5, with pore size around 6–7 Å), hydrogen peroxide decomposition was prevalent on benzene conversion, and only traces of phenol were found. Among the zeolites with pore size similar to titanium silicalite, Ti-ZSM-48 (a microporous material with a disordered structure, characterized by a tubular, monodimensional channel system with near circular pores of 5.3–5.6 Å) consumed almost completely the hydrogen peroxide without any phenol production. In contrast, titanium silicalite-2, TS-2, showed a moderate activity in benzene hydroxylation. TS-2 is a synthetic zeolite with a tetragonal MEL structure (ZSM-11 type), with only straight channels with near-circular sections and diameters of 5.4×5.3 Å [17]. The TOF so far obtained in the oxidation of benzene to phenol with hydrogen peroxide catalyzed by TS-2 is only roughly half of that found with TS-1 [22], but it is likely to be improved with a careful optimization of the catalyst (particle size control, post-synthesis treatments, etc.). However, the synthesis of TS-2 is difficult because it is not easy to obtain a pure MEL structure and this handicap can limit its industrial application. Consequently, TS-1 is still the only candidate catalyst for industrial development.

To further increase the overall yield of the process, a second step can be added in which dihydroxylated by-products, hydroquinone and catechol, are treated with hydrogen and partially deoxygenated to phenol, which is recycled back to the process (Scheme 13.4) [23].

The hydrodeoxygenation reaction (HDO) is carried out in the gas phase in a fixed bed reactor (400 °C, 25 bar of hydrogen), using commercial nickel and molybdenum oxides supported on alumina as catalysts. The HDO allows a quantitative transformation of dioxygenated compounds into phenol with a selectivity of 96% [24]. Main by-products are heavy condensed polycyclic aromatic hydrocarbons.

In a typical arrangement, representative of the whole process (Figure 13.7), the oxidation reaction is carried out in a biphasic mixture of benzene, sulfolane and water (30/50/20 w). The hydrogen peroxide is used as an aqueous solution (35% w), with a total molar ratio of H_2O_2/benzene of 0.21.

The oxidation is carried out in fixed bed reactors, operating at 6 atm under adiabatic conditions with an inlet temperature of 95 °C and outlet temperature of 110 °C. The overall oxidation section performances per pass are:

Benzene conversion (%)	15.4
H_2O_2 conversion (%)	100
Selectivity on benzene (%)[a]	84.6
Selectivity on H_2O_2 (%)[b]	61.4

[a] Moles of produced phenol/moles of converted benzene × 100.
[b] Moles of produced phenol/moles of converted H_2O_2 × 100.

Notably, benzene conversion is quite similar to that usually achieved in cumene process over the benzene alkylation and cumene oxidation stages.

The main by-products are catechol (0.12 t per t of phenol), hydroquinone (0.065 t per t of phenol) and phenolic tars (0.02 t per t of phenol).

The reaction mixture, coming from the reaction section, is sent to the separation section for the recovery of benzene, water and phenol, by consecutive distillation.

The by-products (catechol, hydroquinone and tars) are separated from the solvent by salification with NaOH and extraction, thus avoiding the complete distillation of the high boiling sulfolane.

The by-products, obtained as sodium salts in aqueous solution, are recovered by neutralization with H_2SO_4 and extraction with methyl isobutyl ketone (MIBK).

After separation of MIBK by distillation, the by-products, obtained in aqueous solution, are fed to the HDO section for selective hydrogenation. The HDO reaction is carried out in fixed bed reactors, operating under adiabatic conditions at 400 °C inlet temperature and 25 bar of hydrogen pressure. Typically, the hydrogen/dihydroxy benzenes molar ratio is set to 20. The produced phenol is recovered by distillation and recycled to the process cycle, thus avoiding any coproduction of dihydroxybenzenes.

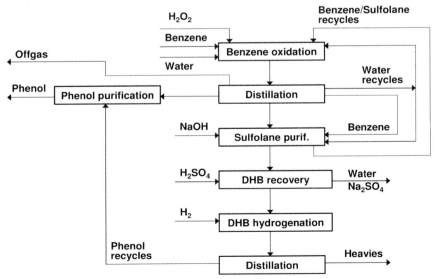

Figure 13.7 Polimeri Europa TS-1/H_2O_2 process: block scheme.

The HDO section performances per pass are:

Catechol conversion:	100%
Hydroquinone conversion:	100%
Selectivity on dihydroxybenzenes:	96%

where the selectivity is given by $100 \times$ (the number of moles of produced phenol/moles of converted dihydroxybenzenes).

The main by-products formed in the HDO reaction are: *ortho*- and *para*-cresols (0.4 kg per t of phenol; arising from toluene impurities of benzene feeding), cyclohexylbenzene (7.2 kg per t of phenol), biphenyl (4.4 kg per t of phenol), dibenzofuran (3.2 kg per t of phenol) and condensed polycyclic aromatic hydrocarbons (25.2 kg per t of phenol).

All the by-products, except cresols, are separated from the phenol produced in the HDO section by distillation. The resulting crude phenol is combined with that produced in the reaction section and fed to the purification section.

Typically, the fraction of phenol, produced in the HDO section corresponds to 13% of the total produced phenol.

The overall process performance, including oxidation and HDO sections, are:

Benzene conversion (%)	100
H_2O_2 conversion (%)	100
Selectivity on benzene (%)[a]	97.7
Selectivity on H_2O_2 (%)[b]	71.0

[a] Moles of produced phenol/moles of converted benzene $\times$ 100.
[b] Moles of produced phenol/moles of converted $H_2O_2 \times$ 100.

A block scheme and the material balance of this Polimeri Europa process are reported in Figure 13.7 and Table 13.5, respectively.

Preliminary economic evaluations suggest that direct oxidation by hydrogen peroxide is not yet competitive with the traditional cumene process as well as with

Table 13.5 Polimeri Europa TS-1/H_2O_2 process: material balance.

Material	Metric tonne per metric tonne of phenol
Feed	
Benzene	0.849
Hydrogen peroxide (100%)	0.509
Product	
Phenol	1.000
By-products	
Na_2SO_4	0.239
Heavies	0.022
Process water	6.830

the acetone recycle technology, but also that it could become convenient if the acetone sale price should further fall close to its fuel value.

13.5
Perspectives

The last developments of hydroxylation of benzene to phenol are directed towards the direct oxidation with molecular oxygen, most often in the presence of a reducing agent such as hydrogen, carbon monoxide or even ammonia [25].

In several cases, the *in situ* formation of hydrogen peroxide is the first step of the process. Thus, phenol can be obtained from benzene, carbon monoxide (5 atm) and oxygen (65 atm) at 70 °C in a benzene–water–methyl isobutyl ketone mixture, with TS-1 and a palladium complex as catalysts [26]. Despite a 91% selectivity to phenol, benzene conversion (3.2%) and productivity are still too low for industrial application. The palladium complex is required to promote hydrogen peroxide formation upon reaction of oxygen, carbon monoxide and water [27].

Carbon monoxide has also been used as the reducing agent by Ishii and coworkers, who reported the oxidation of benzene to phenol using air (15 atm), CO (5 atm) and molybdovanadophosphoric acid as catalysts. The reaction is carried out at 90 °C for 15 h, with a good phenol yield (27%) and a selectivity up to 90% [28]. Daicel Chemical Industries Ltd (Tokyo) recently announced its interest in this process. In this case, however, the authors are inclined to believe that the *in situ* generation of hydrogen peroxide is unlikely: rather, from their experimental results, they conclude that the oxygen is activated on the catalyst and then reacts with the aromatic ring.

The use of oxygen–carbon monoxide or oxygen–hydrogen mixtures presents safety problems due to the occurrence of explosive mixtures: separate supply of the gases into the reaction medium, for example through a permeable membrane, is an approach that may solve this problem, either by producing hydrogen peroxide *in situ* [29] or by using the gas mixture to directly oxidize benzene. For instance, the conversion of benzene into phenol has been reported by supplying hydrogen and oxygen through a palladium membrane [30]. The reaction was carried out in continuous mode at 200 °C: phenol was produced with a selectivity of up to 80% at a benzene conversion around 10–15%.

Despite the attractive perspectives of membrane technology, many basic problems have still to be solved. Beyond the optimization of reaction conditions and catalysts, the chief obstacles to the scaling-up are membrane fragility, deterioration, high cost and manufacture complexity, which restrict, for the moment, this technique to an experimental level.

13.6
Conclusions

It is currently impossible to know if future developments in the direct oxidation of benzene will result in phenol production by a process really competitive with the cumene route. However, the results obtained so far provide a sound basis for future research. In particular, they have definitely shown that it is not unavoidable that the oxidation of phenol proceeds at a rate higher than its formation, thus opening the way to the development of even more selective and, possibly, economic processes.

Acknowledgment

The authors wish to thank Gianni Girotti, from Polimeri Europa S.p.A., for his skilful and valuable revision of the manuscript.

References

1 Hock, H. and Lang, S. (1944) Autoxidation of hydrocarbons. IX. Peroxides of benzene derivatives. *Ber. Dtsch. Chem. Ges. B*, **77**, 257–264.

2 Bellussi, G. and Perego, C. (2000) Industrial catalytic aspects of the synthesis of monomers for nylon production. *Cattech*, **4**, 4–16.

3 Meyers, R.A. (ed.) (2005) *Handbook of Petrochemicals Production Processes*, McGraw-Hill, Ch. 9.1.

4 Girotti, G., Rivetti, F., Ramello, S. and Carnelli, L. (2003) Alkylation of benzene with isopropanol on β-zeolite: influence of physical state and water concentration on catalyst performances. *J. Mol. Catal.*, **204–205**, 571–579.

5 Panov, G.I. (2000) Advances in oxidation catalysis: oxidation of benzene to phenol by nitrous oxide. *Cattech*, **4**, 18–32.

6 Uriarte, A.K., Rodkin, M.A., Gross, M.J., Kharitonov, A.S. and Panov, G.I. (1997) Direct hydroxylation of benzene to phenol by nitrous oxide. *Stud. Surf. Sci. Catal.*, **110**, 857–864.

7 Dubkov, K.A., Ovanesyan, N.S., Shteinman, A.A., Starokon, E.V. and Panov, G.I. (2002) Evolution of iron states and formation of α-sites upon activation of FeZSM-5 zeolites. *J. Catal.*, **207**, 341–352.

8 Symons, M.C.R. and Gutteridge, J.M.C. (1998) *Free Radicals and Iron: Chemistry,* *Biology, and Medicine*, Oxford University Press Oxford.

9 Olah, G.A. and Ohnishi, R. (1978) Oxyfunctionalization of hydrocarbons. 8. Electrophilic hydroxylation of benzene, alkylbenzenes, and halobenzenes with hydrogen peroxide in superacids. *J. Org. Chem.*, **43**, 865–867.

10 Mimoun, H., Saussine, L., Daire, E., Postel, M., Fischer, J. and Weiss, R. (1983) Vanadium(V) peroxo complexes. New versatile biomimetic reagents for epoxidation of olefins and hydroxylation of alkanes and aromatic hydrocarbons. *J. Am. Chem. Soc.*, **105**, 3101–3110.

11 Bonchio, M., Conte, V., Di Furia, F. and Modena, G. (1989) Metal catalysis in oxidation by peroxides. 31. The hydroxylation of benzene by $VO(O_2)(PIC)$ $(H_2O)_2$: mechanistic and synthetic aspects. *J. Org. Chem.*, **54**, 4368–4371.

12 See, for example, Ortiz de Montellano, P.R. (1995) *Cytochrome P450. Structure, Mechanism, and Biochemistry*, 2nd edn, Plenum Press, New York.

13 (a) Vignola, R., Battistel, E., Bianchi, D., Bortolo, R. and Tassinari, R. (2002) Process for the hydroxylation of aromatic hydrocarbons. European Patent EP 861688. to EniChem; (b) Bianchi, D., Bortolo, R., Tassinari, R., Ricci, M. and Vignola, R. (2000) A novel iron-based catalyst for the biphasic oxidation of

benzene to phenol with hydrogen peroxide. *Angew. Chem. Int. Ed.*, **39**, 4321–4323.

14 (a) Walling, C. (1975) Fenton's reagent. V. Hydroxylation and side-chain cleavage of aromatics. *J. Am. Chem. Soc.*, **97**, 363–367; (b) Ito, S., Mitarai, A., Hikino, K., Hirama, M. and Sasaki, K. (1992) Deactivation reaction in the hydroxylation of benzene with Fenton's reagent. *J. Org. Chem.*, **57**, 6937–6941.

15 Barton, D.H.R., Halley, F., Ozbalik, N. and Mehl, W. (1989) Benzylic oxidation by the Gif^{IV} system. *Tetrahedron Lett.*, **30**, 6615–6618.

16 (a) Barton, D.R. and Launay, F. (1998) The selective functionalization of saturated hydrocarbons. Part 44. Measurement of size of reagent by variation of steric demands of competing substrates using Gif chemistry. *Tetrahedron*, **54**, 3379–3390; (b) Chen, K., Costas, M., Kim, J., Tipton, A. K. and Que, L. Jr (2002) Olefin cis-dihydroxylation versus epoxidation by non-heme iron catalysts: Two faces of an Fe^{III}-OOH coin. *J. Am. Chem. Soc.*, **124**, 3026–3035.

17 International Zeolite Association, Atlas of Zeolite Framework Types, http://www.iza-structure.org (last accessed June 3th, 2009).

18 Arends, I.W.C.E., Sheldon, R.A., Wallau, M. and Schuchardt, U. (1997) Oxidative transformations of organic compounds mediated by redox molecular sieves. *Angew. Chem. Int. Ed. Engl.*, **36**, 1145–1163.

19 Balducci, L., Bianchi, D., Bortolo, R., D'Aloisio, R., Ricci, M., Tassinari, R. and Ungarelli, R. (2003) Direct oxidation of benzene to phenol with hydrogen peroxide over a modified titanium silicalite. *Angew. Chem. Int. Ed.*, **115**, 5087–5090.

20 Wilkenhöner, U., Langhendries, G., van Laar, F., Baron, G.V., Gammon, D.W., Jacobs, P.A. and van Steen, E. (2001) Influence of pore and crystal size of crystalline titanosilicates on phenol hydroxylation in different solvents. *J. Catal.*, **203**, 201–212.

21 Drago, R.S., Wayland, B. and Carlson, R.L. (1963) Donor properties of sulfoxides, alkyl sulfites, and sulfones. *J. Am. Chem. Soc.*, **85**, 3125–3128.

22 Unpublished results obtained within the European Project NEOPS (Novel Eco-efficient Oxidation Processes based on H$_2$O$_2$ Synthesis on catalytic membranes).

23 Bianchi, D., Bortolo, R., Buzzoni, R., Cesana, A., Dalloro, L. and D'Aloisio, R. (2004) Integrated process for the preparation of phenol from benzene with recycling of by-products. European Patent EP 1424320 to Polimeri Europa.

24 Dalloro, L., Cesana, A., Buzzoni, R. and Rivetti, F. (2004) Process for the preparation of phenol by means of the hydrodeoxygenation of benzene-diols. European Patent EP 1411038 to Polimeri Europa.

25 Bal, R., Tada, M., Sasaki, T. and Iwasawa, Y. (2006) Direct phenol synthesis by selective oxidation of benzene with molecular oxygen on an interstitial-N/Re cluster/zeolite catalyst. *Angew. Chem. Int. Ed.*, **45**, 448–452.

26 Tassinari, R., Bianchi, D., Ungarelli, R., Battistel, E. and D'Aloisio, R. (1999) Process for the synthesis of phenol from benzene. European Patent EP 894783 to EniChem.

27 Bianchi, D., Bortolo, R., D'Aloisio, R. and Ricci, M. (1999) Biphasic synthesis of hydrogen peroxide from carbon monoxide, water, and oxygen catalyzed by palladium complexes with bidentate nitrogen ligands. *Angew. Chem. Int. Ed.*, **38**, 706–708.

28 Tani, M., Sakamoto, T., Mita, S., Sakaguchi, S. and Ishii, Y. (2005) Hydroxylation of benzene to phenol under air and carbon monoxide catalyzed by molybdovanadophosphoric acid. *Angew. Chem. Int. Ed.*, **44**, 2586–2588.

29 (a) Choudari, V.R., Gaikwad, A.G. and Sansare, S.D. (2001) Nonhazardous direct oxidation of hydrogen to hydrogen

peroxide using a novel membrane catalyst. *Angew. Chem. Int. Ed.*, **40**, 1776–1779; (b) Melada, S., Pinna, F., Strukul, G., Perathoner, S. and Centi, G. (2006) Direct synthesis of H_2O_2 on monometallic and bimetallic catalytic membranes using methanol as reaction medium. *J. Catal.*, **237**, 213–219; (c) Dittmeyer, R. and Svajda, K. (2007) Inherent and reliable selective method for directly synthesising hydrogen peroxide from oxygen and hydrogen with the aid of a catalytically coated wettable porous membrane. World Patent WO 2007028375 to Dechema.

30 Niwa, S., Eswaramoorthy, M., Nair, J., Raj, A., Itoh, N., Shoji, H., Namba, T. and Mizukami, F. (2002) A one-step conversion of benzene to phenol with a palladium membrane. *Science*, **295**, 105–107.

14
Friedel–Crafts Acylation of Aromatic Ethers Using Zeolites

Roland Jacquot and Philippe Marion

14.1
Introduction

The Friedel–Crafts reaction and the related Fries rearrangement of aromatics are the most important methods in organic chemistry for synthesizing aromatic ketones, which are of interest in the synthesis of numerous fine chemicals such as drugs, fragrances, dyes and pesticides.

The conventional method for preparation of these aromatic ketones involves reaction of the aromatic hydrocarbon with a carboxylic acid derivatives using a Lewis acid (AlCl$_3$, FeCl$_3$, BF$_3$, ZnCl$_2$, TiCl$_4$, etc.) or a Brønsted acid (polyphosphoric acid, HF, H$_2$SO$_4$, etc.) as catalyst. The major drawback of the Friedel–Crafts reaction is the need to use the acid catalyst in higher amounts than the stoichiometry relative to the formed ketone. This quantity is required because the ketone (i.e., product of the reaction) forms a stoichiometric, stable complex with the Lewis acid and in this way transformed the catalyst in a reactant. This complex is generally decomposed with water, leading to total destruction of the catalyst and the loss of the Lewis acid. Workup of acylation reaction mixtures could producing metallic salts (Zn, Al hydroxides), which can be recovered, but it is also produces relatively large amounts of hydrochloric acid in the effluent. This hydrochloric acid, which has to be disposed of, could originate both from the catalyst and also from the acyl chloride employed for the acylation. Beside this considerable environmental problem, the corrosion issue due to the presence of hydrochloric acid must be solved. Therefore, a process that could be both environmentally friendly and also competitive is clearly desirable.

This chapter presents an alternative to classical Friedel–Crafts catalysis using heterogeneous catalysts such as zeolites, as we believe them to be the ultimate catalyst for the future production of aromatic ketones. After a short introduction to the literature, we focus on catalyst and process issues for different reactions.

Sustainable Industrial Processes. Edited by F. Cavani, G. Centi, S. Perathoner, and F. Trifiró
Copyright © 2009 WILEY-VCH Verlag GmbH & Co. KGaA, Weinheim
ISBN: 978-3-527-31552-9

14.2
Literature Background

An impressive number of papers and books has been published and numerous patents have been registered on the acylation of aromatic compounds over solid catalysts. Recently Sartori and Maggi [1] have written an excellent review with 267 references on the use of solid catalysts in Friedel–Crafts acylation. In one section of this review, namely acylation of aromatic ethers or thioethers, the authors report work on acylation by solid catalysts such as zeolites, clays, metal oxides, acid-treated metal oxides, heteropolyacids or Nafion. When examining in details these results, it appeared very difficult for us to build upon these experimental results as the reaction conditions differ drastically from one paper to the next. This prompted us to reinvestigate the scope and limitations of the Friedel–Crafts acylation using heterogeneous solids as catalysts, trying as much as we could to rationalize the observed effects.

14.3
Acylation of Anisole by Acetic Anhydride

We focus first on the batch acylation of anisole in the liquid phase. By fixing primary reaction parameters we can focus only on the impact of the nature of the catalyst (Table 14.1).

All tested catalysts are active for this reaction and they all show a very high para selectivity. The zeolites are clearly the preferred catalysts to carry out this acylation under mild conditions. Among them, HY and HBEA exhibit the better activity.

Table 14.1 Acylation of anisole with acetic anhydride over different heterogeneous catalysts.

Entry	Catalyst	Yield (%)[a]	Selectivity (para, %)
1	HZSM5	12	>98
2	HMOR	29	>98
3	HBEA	70	>98
4	HY	69	>98
5	Exchanged clay	14	>98
6	Al clay	16	>98
7	$H_3PW_6MoO_{40}$	21	>98

[a] Yield relative to acetic anhydride (initial molecular ratio anisole:anhydride = 5).

Optimization of the reaction parameters underlines the following trends:

- In general, an increase of the molecular ratio anisole to acetic anhydride is beneficial for the yield calculated over the acetic anhydride but, for the economy of this process, a 2:1 molar ratio mixture of anisole and acetic anhydride is used.
- An increase in temperature is beneficial for the activity but accelerates the irreversible deactivation of the zeolite: 90 °C is an optimal temperature.
- The order of introduction of the reactants is very important: introduction of anisole first gives the better result.

14.3.1
Industrial Processes

Rhodia [2] has operated the first industrial application of zeolite for the acylation of anisole to *para*-methoxyacetophenone, which is a precursor of avobenzone (trade names PARSOL 1789, EUSOLEX 9020, ESCALOL 517, etc.) used in sun creams (Equation 14.1):

$$\tag{14.1}$$

For the acetylation of anisole by acetic anhydride, the zeolite HBEA is the better catalyst, but in an industrial process the separation of this catalyst for recycling is a key issue and it has to be taken into account. Indeed, the filtration rate of HBEA, especially after reaction is too slow and this operation is not industrially and economically viable.

For this reason, the use of powdered zeolite is not possible in a process. That is why Rhodia used as a catalytic phase a zeolite HBEA mixed with alumina in order to shape the solid. On an industrial scale, the catalyst can be used as pellets or extrudates which have the most advantages in terms of efficiency and ease of use.

With extrudates, Rhodia has used a fixed-bed technology and this process is a breakthrough in this field as it enables considerable simplifications of the standard process and thus a reduction of operating cost and a dramatic reduction of effluent volume.

Comparison of the block diagrams of the Rhodia process and the previous classical process in Figures 14.1 and 14.2, respectively, reveals the simplifications resulting from the zeolite process.

Consequently, the new process is not only environmentally friendly but it is also more economic than the previous classical one due to drastic simplifications.

With this catalyst, the fixed bed is the better technology because, in this case, solid–liquid separation, which is a critical step, is avoided.

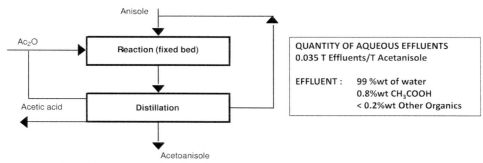

Figure 14.1 Simplified flow chart for the new zeolite recirculating fixed bed process without solvent. Effluent 0.035 tonne per tonne of acetoanisole.

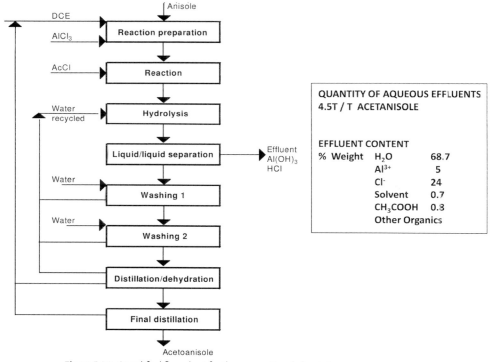

Figure 14.2 Simplified flow chart for the conventional aluminium chloride process using 1,2-dichloroethane as solvent. Effluent 4.500 tonne per tonne of acetoanisole.

A zeolitic catalyst consisting of 40 wt% of binder (alumina) and 60 wt% of a HBEA with a Si:Al ratio of 12.5 was charged into a stainless steel tubular reactor. Anisole is introduced into a stirred reactor and then fed over the fixed bed. The reactor was heated at 90 °C. When 90 °C is reached in the fixed bed, acetic anhydride was added over one hour, keeping the bed under flow. After the end of the addition, the reaction was maintained for 6 hours at 90 °C. When the reaction was over, the medium was cooled to 60 °C and the apparatus was emptied. The fixed bed is ready for another batch. The above operation can be repeated ten times about using the same catalytic bed. After each cycle, the catalytic bed lost a minor part of its activity but an increase in temperature makes it possible to compensate for this loss. When the activity becomes too low even at higher temperature the catalyst can be regenerated under dry air at 500 °C for 12 h. After this treatment the activity of the catalyst is fully recovered.

14.4
Acylation of Veratrole by Acetic Anhydride Over HY Zeolite

Veratrole is the second substrate presented. Table 14.2 shows the reaction and principal results.

Rhodia has manufactured, selectively, acetoveratrole by acetylation of veratrole with acetic anhydride catalyzed by HBEA zeolite in the same process using fixed bed reactor technology. In this case, the best heterogeneous catalyst for acylation of veratrole is the HY zeolite.

Unlike HBEA, it is possible to used powdered HY because even after reaction the filtration of HY CBV 720 sold by Zeolyst is possible and industrially realistic. In this case, the activity of the HY zeolite powder is very high, and as the ratio of veratrole/zeolite in the reaction was very important, the recycling of the catalyst was not a priority. For a new plant dedicated to the acylation of the veratrole we can lower the investment because the recirculating fixed bed is not obligatory. Moreover, HY was cheaper than HBEA.

Table 14.2 Acylation of veratrole with acetic anhydride over different zeolites.

Entry	Catalyst	Yield (%)[a]	Selectivity
1	HZSM5	12	>98
2	HMOR	25	>98
3	HBEA	53	>98
4	HY	95	>98

[a] Yield relative to acetic anhydride (molar ratio veratrole:acetic anhydride = 5).

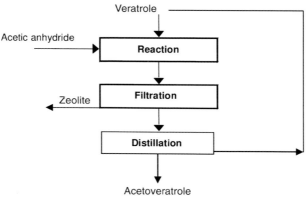

Veratrole

Acetic anhydride

Zeolite

Reaction

Filtration

Distillation

Acetoveratrole

Figure 14.3 Simplified flow chart of acylation of veratrole with acetic anhydride by HY powder.

The veratrole was introduced into a stainless-steel standard reactor with the HY zeolite and the reaction mixture was then heated to 90 °C and acetic anhydride is added over 4 h. The molar ratio veratrole/acetic anhydride was 1.2. When the reaction was over the reactor was cooled to 60 °C and the reaction mixture filtered under 2 bar. With a 5-cm cake, the filtration rate was acceptable. After distillation, the acetoveratrole was isolated in 85% yield. The use of a spray-dried form of HY zeolite, a catalyst consisting of 60 wt% silica and 40 wt% HY zeolite, was another means of increasing the filtration rate. In this last case, the catalyst can easily be recycled.

If we compare process blocks diagrams (Figure 14.3) and amount of effluent of the classical Lewis acid process and the zeolite process for acetoveratrole, we have almost the same figures as with acetoanisole – so the competitive advantage for economics and for sustainable development is obvious. This is also a breakthrough for such technology.

14.5
Deactivation of the Catalysts

Regarding the acylation of anisole by acetic anhydride on HBEA zeolite, various explanations concerning the deactivation of this zeolite have been reported. According to Rohan et al. [3], zeolite deactivation was due to a strong adsorption of the reaction products on the acid sites and to a pore blockage by entrapped di- and tri-acetylated compounds. Therefore, the compounds responsible for the deactivation of the catalysts could be classified in two categories: those that could be recovered by extraction with solvent such as methylene chloride and the non-extractible ones. In the first case the deactivation is reversible while in the second case it is irreversible. Freese et al. [4] have reached the same conclusions. For Derouane et al. [5], the deactivation was essentially due to the preferential adsorption of acetoanisole. These authors have also observed a slight dealumination of the zeolite in their experiments. For veratrole acylation by acetic anhydride, Moreau et al. [6] have noticed that both acetic acid and acetoveratrole led to a similar decrease in activity of the HY catalyst.

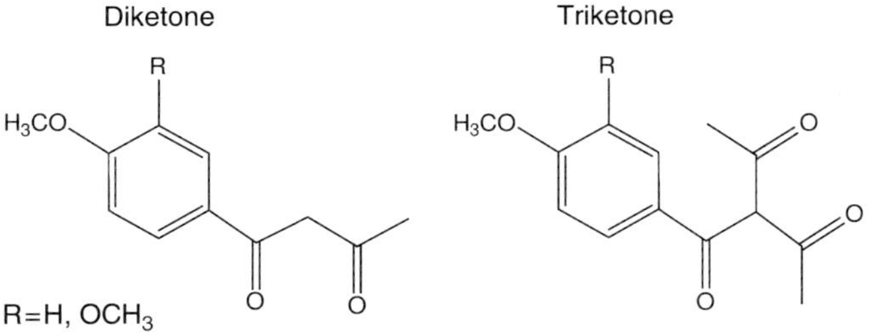

Diketone   Triketone

R=H, OCH$_3$

Scheme 14.1 Structure of compounds obtained after destruction of the zeolites.

Perot *et al.* [7] in collaboration with Rhodia have studied the deactivation of industrial catalysts HBEA and HY during the acylation of veratrole and anisole. After reaction, the spent catalysts were extracted with methylene chloride. This Soxhlet extraction makes possible the elimination of compounds that were not strongly adsorbed on the zeolites. The composition of the residue obtained after evaporation of methylene chloride was practically the same as that of the reaction mixture at the end of the experiment. By this extraction procedure, approximately 80% of the compounds remaining on the catalysts after reaction were recovered. After Soxhlet extraction, the catalyst samples were recovered and dissolved with hydrofluoric acid. The organic compounds released by the catalysts were extracted again by methylene chloride and, after evaporation of solvent, the residues contained di- and triketones as well as cyclization compounds, the structures of which are presented in Scheme 14.1.

Table 14.3 shows the composition of the mixtures obtained after dissolution of the zeolites with HF. For both zeolites, the main product remaining adsorbed after extraction and which could be recovered after their dissolution was a diketone. It could therefore be considered that the diketone adsorbed on the catalysts was to a large extent responsible for their deactivation.

To confirm the inhibition effect of the adsorbed diketone on the acylation of veratrole by acetic anhydride over HBEA and HY zeolites, this product was added to reaction mixtures with both zeolites. Its presence caused a significant decrease in the yield in acylated products. For example, the addition of 3% (by weight with respect to veratrole) of diketone caused a loss in yield of more than 50% in both cases.

For the acylation of anisole by acetic anhydride over HBEA and HY, the reasons for the deactivation accord with those already reported for the reaction with veratrole. Unfortunately, di- or triketones were not extractible and, also, deactivation of the

Table 14.3 Distribution of compounds remaining strongly adsorbed on the zeolites during acylation of veratrole (obtained by dissolution of the zeolite with HF after Soxhlet extraction).

Zeolite	C (wt%) measured	Veratrole (wt%)	Acetoveratrole (wt%)	Diketone (wt%)	Triketone (wt%)	Others (wt%)	Adsorbed diketone (mmol g^{-1})
HBEA	4.7	4	6	78	5	7	0.221
HY	3.2	1	4	87	0	8	0.135

zeolites was irreversible. Only a thermal procedure under dry air at 500 °C can regenerate the spent catalysts.

14.6
Benzoylation of Phenol Ether

Benzoylation under Friedel–Crafts acylation is an important process for the preparation of many industrially valuable chemicals. However, like the acetylation, the use of conventional Lewis acids, for example, $AlCl_3$, $ZnCl_2$, and so on, in the homogeneous Friedel–Crafts acylation of arenes entails problems of materials due to corrosion work up and effluent issues. Thus, for benzoylation, as with acetylation, the use of heterogeneous catalysts seems to be an interesting alternative technique to the homogeneous reaction.

Corma's group [8] has published very interesting work on the activity of zeolites in the benzoylation of anisole with phenylacetyl chloride. Singh et al. [9] have studied the benzoylation of naphthalene to 2-benzoylnaphthalene using zeolite HBEA catalysts. Benzoic anhydride, benzoic acid and benzoyl chloride were employed as benzoylating agents. Benzoic anhydride was also capable of benzoylating naphthalene; however, this benzoylation with benzoic acid did not proceed under the reaction conditions. To compare the reactivity of the different benzoylating agents, we examined the reaction of anisole with zeolites. Table 14.4 presents the results.

According to Singh et al. [9], benzoylation with benzoic acid did not proceed under the reaction conditions but benzoic anhydride and benzoyl chloride gave the benzophenone. Anhydride was better than acid chloride but for reasons of availability, simplification of process and effluent production we focused attention on the use of benzoyl chloride as reactant for the zeolite catalyzed process. Friedel–Crafts

Table 14.4 Screening of benzoylating agent for anisole with HY zeolite (Si:Al = 27).

Entry	Benzoylating agent	Yield[a],[b] of benzophenone (%)
1	Benzoic acid	11
2	Benzoic anhydride	55
3	Benzoyl chloride	47

[a]Yield relative to benzoylating agent.
[b]Reaction conditions: molar ratio anisole:benzoylating agent = 4, 2.5 wt% of powdered catalyst/anisole, 125 °C, 4 h.

Table 14.5 Benzoylation of anisole with benzoyl chloride catalyzed with several zeolites.

Entry	Catalysts	Molar ratio Si:Al	Yield of benzophenone (%)[a]
1	HMOR	10	23
2	HMOR	45	28
3	HBEA	12.5	31
4	HBEA	30	7
5	HY	5.5	13
6	HY	12.5	18
7	HY	27	47

[a]Reaction conditions: molar ratio anisole:benzoyl chloride = 4, 2.5 wt% of powdered zeolite/anisole. Temperature 125 °C, duration 4 h.

acylation with benzoic anhydride involves the loss of one molecule of benzoic acid for every molecule of anhydride consumed by the reaction. With the benzoyl chloride we can remove easily HCl gas.

Table 14.5 shows the results obtained for several zeolites in the reaction of benzoylation of anisole with benzoyl chloride.

The order of activity was HY > HBEA > HMOR for the dealuminated zeolites but HMOR and HBEA have nearly similar activity.

The activity of HY increases with dealumination. This was attributed to the generation of mesoporosity during the dealumination and the corresponding ease of diffusion of the reactants and products. For HBEA, though, the activity decreases with dealumination because we observed a destructuring of the zeolite.

We have developed a batch process, catalyzed by HY zeolite, to produce the pharmaceutical intermediate 4-chloro-4′-ethoxybenzophenone (Table 14.6).

Table 14.6 Benzoylation of phenetole with p-chlorobenzoyl chloride over different zeolites.

Entry	Catalyst	Molar ratio Si:Al	Yield (%)[a]
1	HMOR 20 A	10	23
2	HMOR 90 A	45	28
3	HBEA CP 811	12.5	31
4	HY CBV 760	27.5	47
5	HZSM5 5010	26	11

[a]Yield relative to p-chlorobenzoyl chloride (molar ratio phenetole:p-chlorobenzoyl chloride = 4).

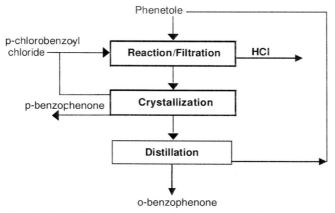

Figure 14.4 Simplified flow chart of the benzoylation of phenetole to o-benzophenone.

All tested zeolites are sold by Zeolyst. For this reaction, HY not only stands out as the best material but, in this case, the reaction was much faster. HBEA, which was as active as the HY in the case of anisole, is now less active in the benzoylation (entries 3 and 4 in Table 14.6). Interestingly, with the HMOR and HZSM5, the benzoylation of phenetole might also proceed at the surface of the catalyst.

For the industrial process (Figure 14.4), Rhodia used HY zeolite powder with molar ratio Si:Al = 27. The phenetole is the reactant and the solvent. During the reaction, HCl was stripped by a nitrogen flow. Under these conditions, we reduced the formation of by-products such as phenol and phenyl chlorobenzoate.

In this batch process, after filtration of the zeolite at 80 °C, we can purify the p-benzophenone by crystallization with a very good yield. The excess phenetole was distilled and recycled.

Under these conditions, the zeolite HY was very stable. During the first cycle (Table 14.7) we have observed a slight dealumination of the HY zeolite, probably by dissolution of extraframework alumina and generation of mesopores, because we did not observed any impact on the integrity of the catalyst. After the first cycle, the kinetics of the reaction increased and then stabilized.

During the reaction, we observed the formation of phenyl p-chlorobenzoate as by-product (Scheme 14.2).

Table 14.7 Recyclability of zeolite HY CBV 760

Recycle	Duration (h)	Conversion (%) benzoyl chloride	Yield (%) o-benzophenone	Yield isolated (%)[a] p-benzophenone
Fresh zeolite	6	>98	8	84
1st	5	>98	8	80
2nd	5	>98	8	83
3rd	5	>98	8	82
4th	5	>98	8	84

[a] Yield relative to benzoyl chloride (molar ratio phenetole:p-chlorobenzoyl chloride = 4).

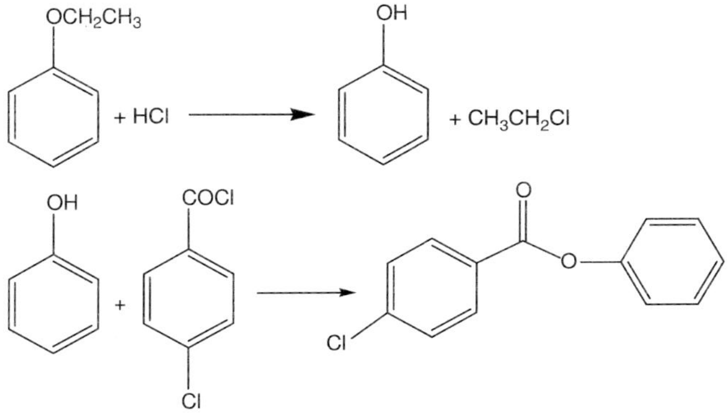

Scheme 14.2 Formation of phenyl *p*-chlorobenzoate during the benzoylation of phenetole.

14.7
Concluding Remarks

The examples shown here are industrial examples of the advantages brought about by catalysis to exploit more sustainable processes. These examples represents a real breakthrough in chemical processes, with a more than 100-fold reduction of effluents compared to the previous processes. The drastic simplification of the process (four less steps) allows a very competitive route. It is a nice example of the role of catalysis as a key for sustainability.

Zeolite catalysts were employed in this chemistry. The different examples show us that different grades of zeolites could be used but it also shows us that catalytic performance is not enough and that process consideration such as catalyst shaping and catalyst separation could drive the choice of catalyst more than the catalyst performance such as conversion and selectivity.

From a process point of view, the operating mode in this chemistry is batch, either with fixed bed technology or slurry. The operating mode teaches us that it could be useful to work with a reactant ratio very far from stoichiometry in order to boost conversion and to impede deactivation by strong adsorption of one partner.

Deactivation is classically observed even if it is not a crucial point here. The reasons for deactivation differ from author to author but we generally find strong adsorption of some product, and plugging of pores by heavies. Extraction of adsorbed products by solvent or regeneration at high temperature (carbon burning) are convenient solutions.

References

1 Sartori, G. and Maggi, R. (2006) *Chem. Rev.*, **106** (3), 1077–1104.
2 Rhodia (1997) Patent PCT, Int. Appl. WO 97 48,665.
3 Rohan, D., Canaf, C., Fromentin, E. and Guinet, M. (1998) *J. Catal.*, **177**, 296.
4 Freese, U., Heinrich, H. and Roessner, F. (1999) *Catal. Today*, **49**, 237.

5 Derouane, E.G., Dillon, C.J., Bethel, D. and Derouane Aɔd Hamid, S.B. (1999) *J. Catal.*, **187**, 209.

6 Moreau, P., Finiels, A. and Meric, P. (2000) *J. Mol. Catal.*, **154**, 185.

7 Perot, G., Guignard, C., Pedron, V., Richard, F., Coustard, J.M., Jacquot, R. and Spagnol, M. (2002) *Appl. Catal. A: Gen.*, **234**, 79–90.

8 Corma, A., Ciment, M.J., Garcia, H. and Primo, J. (1989) *J. Appl. Catal.*, **49**, 109.

9 Bhattacharya, B., Sharma, S. and Singh, A.P. (1997) *Appl. Catal. A Gen.*, **150**, 53.

15
Green Sustainable Chemistry in the Production of Nicotinates

Roderick Chuck

This chapter discusses the fulfillment of the requirements for a green process, together with the importance of the nicotinate vitamin B3. Lonza's contribution to green chemistry is then illustrated by its processes for producing nicotinates. The current niacinamide process running is China employs no fewer than six basic principles of green-sustainable chemistry.

15.1
Requirements for Green Processes

The acceptability of a process should not be governed solely by its cost. An economic process is not necessarily green, especially if waste treatment is ignored or neglected. An industrial process may contravene one or more green principles, and yet still make money, even if complex waste treatment adds to the costs and diminishes economic viability. But the cost of energy today is still low, considering the fact that our present utilizable energy resources are limited. Thus, for a given product, the following guidelines should govern the choice of route:

1. Choice of feedstock (costs are relevant of course, but also total resources, energy, waste, etc. in the manufacture of the given feedstock are important factors).
2. Choice of reaction path (minimize energy requirements by use of selective catalysts).
3. Choice of catalyst (efficiency, separation and recycling of catalyst).
4. Down-stream processing/unit operations (minimizing the number of stages necessary to obtain the product in the state desired by the customer).
5. Minimizing not only the amount of pollutants but also the volume of waste streams (effluent/off-gases and solid waste).
6. Recycling of auxiliary, side-, and intermediate products into the process.
7. Avoidance of hazardous or toxic materials wherever possible.

Sustainable Industrial Processes. Edited by F. Cavani, G. Centi, S. Perathoner, and F. Trifiró
Copyright © 2009 WILEY-VCH Verlag GmbH & Co. KGaA, Weinheim
ISBN: 978-3-527-31552-9

15.2
Significance of Niacin

Co-enzyme I (nicotinamide-adenine dinucleotide NAD) and Co-enzyme II (nicotin-amide-adenine dinucleotide phosphate NADP) are required by all living cells. They enable both the conversion of carbohydrates into energy as well as the metabolism of proteins and fats. Both nicotinamide and nicotinic acid are building blocks for these co-enzymes. The common name for the vitamin is niacin and, strictly speaking, refers only to nicotinic acid.

COOH   CONH$_2$

Nicotinic Acid (Niacin)   Nicotinamide (Niacinamide)

Since the human body produces neither nicotinic acid nor the amide, it is dependent on intake via foodstuffs. Although niacin is found in a bound form naturally in wheat, yeast and pork and beef liver, most niacin today is produced synthetically by chemical oxidation of alkyl pyridines. To demonstrate the economic significance of this, in 1995 worldwide a total of 22 000 metric tonnes of niacin and niacinamide were produced. Today between 35 000 and 40 000 tonnes are produced and the demand for nicotinates is rising. Thus, both economic and ecological factors play a significant role.

15.3
Green Principles in the Manufacture of Niacin

15.3.1
Choice of Feedstock

3-Alkylated pyridines such as 3-picoline, 3(5)-ethylpyridine and 2-methyl-5-ethyl-pyridine (MEP) are the natural choice as starting materials for the nicotinates. The choice of alkyl pyridine is governed by their availability and the process being used.

Pyridine bases such as 3-picoline and MEP are predominantly manufactured by the Chichibabin reaction, where a mixture of aldehydes or ketones is reacted with ammonia. Thus, formaldehyde, acetaldehyde and ammonia react in the gas phase to produce a mixture of pyridine and 3-picoline. By choosing the appropriate aldehyde or ketone, catalyst and phase (liquid or gas phase), the composition of the mixture can be varied at will, depending on the desired end-product. In the gas phase, silica alumina catalysts are often used, while in the liquid phase acid catalysts based on phosphoric or acetic acid are employed. In the 1990s, Reilly patented MFI- and BEA-based zeolite catalyst compositions for ammonia–aldehyde conversions to pyridine, picolines and alkyl pyridines.

Chemically, 3-picoline is the ideal starting-material for nicotinic acid or amide: the methyl group can be selectively and readily oxidized to the carboxyl derivative with few by-products or pollutants. High selectivity coupled with the low molecular weight ratio (1:1.3) compared to the end-products make picoline an attractive industrial starting-material for the production of nicotinic derivatives.

3-Picoline is obtained, typically in a 1:2 ratio along with the main product pyridine, by the gas-phase reaction of acetaldehyde, formaldehyde and ammonia. The lack of selectivity of this reaction to either pyridine or picoline has hitherto meant that the economy of the major product (pyridine) has determined the price and availability of picoline. Consequently, producers of pyridine have been able to control the quantity and prices of picoline on the market. This has led to the search for alternative feedstock and manufacturing processes for picoline.

2-Methyl-5-ethyl-pyridine (MEP) is used as a starting material for the high temperature and pressure liquid-phase oxidation with nitric acid. The reasons for this apparently unlikely choice of starting material are many. MEP can be made in the liquid phase from acetaldehyde and ammonia selectively (around 70%) compared to picoline (20–40%) from the traditional picoline/pyridine process. It is thus considerably cheaper to produce than picoline. The reaction of MEP with nitric acid is also surprisingly selective (>80%). The resulting nitric oxide gases are recycled and reacted with air and water to reconstitute nitric acid. This process has been utilized for nearly 40 years by Lonza for producing niacin.

As mentioned above, the bulk of picoline is produced today by condensation of acetaldehyde, formaldehyde and ammonia in the gas phase, which simultaneously produces large quantities of pyridine. A selective and suitable alternative method starting from these or similar simple molecules has yet to be developed. Given the thermodynamic properties of the molecules and reactions involved it does not seem likely to expect a selective process for 3-picoline in the near future following this strategy, although shape-selective catalysts may hold a key.

15.3.2
Reaction Paths for Producing Niacin

15.3.2.1 Liquid-Phase Oxidation of Nicotine with Permanganate, Chromic Acid, etc.
The classic method of preparing nicotinic acid was by oxidizing nicotine with potassium dichromate (Scheme 15.1). This was discovered over 100 years ago.

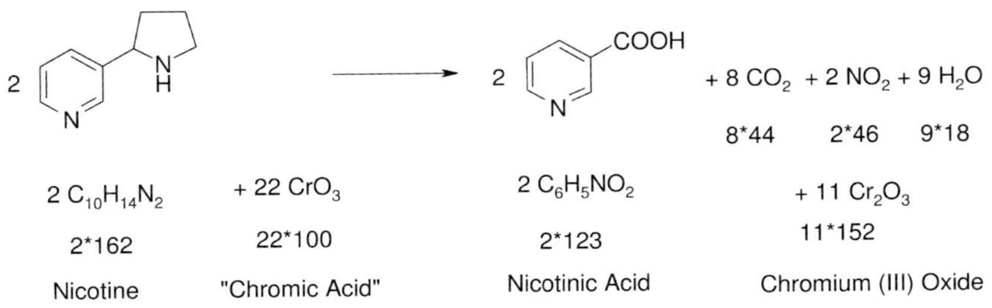

$$2 \quad + \quad \longrightarrow \quad 2 \quad \text{COOH} \quad + 8\ CO_2\ + 2\ NO_2 + 9\ H_2O$$

			8*44	2*46	9*18

$$2\ C_{10}H_{14}N_2 \quad + 22\ CrO_3 \quad\quad\quad 2\ C_6H_5NO_2 \quad\quad\quad + 11\ Cr_2O_3$$

2*162	22*100	2*123	11*152
Nicotine	"Chromic Acid"	Nicotinic Acid	Chromium (III) Oxide

Scheme 15.1 Original chromic acid oxidation of nicotine to nicotinic acid.

Table 15.1 By-products arising from the production of 1 tonne of nicotinic acid from 1.32 tonne of nicotine and 9.02 tonnes of chromic acid.

By-product	Quantity (tonne)
CO_2	1.43
NOx (calculated as NO_2)	0.37
Chromic oxide	6.80

Although no longer relevant today, this reaction serves as an excellent example when considering green technology.

Chromic acid (CrO_3) is carcinogenic and environmentally threatening. Chromic (III) oxide, on the other hand, is extensively used in the tanning industry, and has a higher present value on the market than its precursor. Assuming an ideal chemical reaction (100% yield!), the above reaction gives the figures shown in Table 15.1. Thus almost 9 tonne of by-product are produced for 1 tonne of the desired product.

15.3.2.2 Liquid-Phase Oxidation of 3-Picoline with Permanganate, Chromic Acid or Nitric Acid

The oxidation of picoline with permanganate or chromic acid suffers from the same drawback, albeit in a lesser form, as nicotine (Scheme 15.2).

For 1 tonne of nicotinic acid, 2.8 tonne of inorganic material are produced as waste. With chromium trioxide (neglecting any inorganic acid involved to produce the required chromic acid), 1.24 tonne of Cr_2O_3 are produced per tonne of nicotinic acid. This assumes stoichiometric quantities of oxidant and quantitative yields, both of which in practice are unrealistic. A stoichiometric excess of 50–100% oxidant is usual, and molar yields of 80–90% are generally not exceeded. Thus the inorganic waste for the permanganate process would probably lie around 4 tonne per tonne of niacin produced, and for chromium between 1.7 and 2.0 tonne per tonne. Clearly, ecologically, this situation is untenable. Even though chromium(III) sulfate can be utilized in the leather industry as a tanning agent, several factors argue against this type of process, even if it appears on the surface to be economically attractive:

1. The combining of two economies in one process requires that both end-products (here chromic oxide and nicotinic acid) can be sold. Thus the success of the process is dependent on the demand for both products being sustained. A collapse of one leg will invariably lead to the process as a whole being unsustainable. The dependence of picoline on pyridine illustrates this point.

2. The energy required for the production of chromic acid (or permanganate) is considerable. (Chromite ore is roasted with sodium carbonate at around $1000\,^{\circ}C$ to produce the common starting material for most chromium compounds, namely sodium chromate.)

3. Environmental problems in the leather industry due to chromium pollution can be solved, but are expensive and the alternatives are not without their own problems.

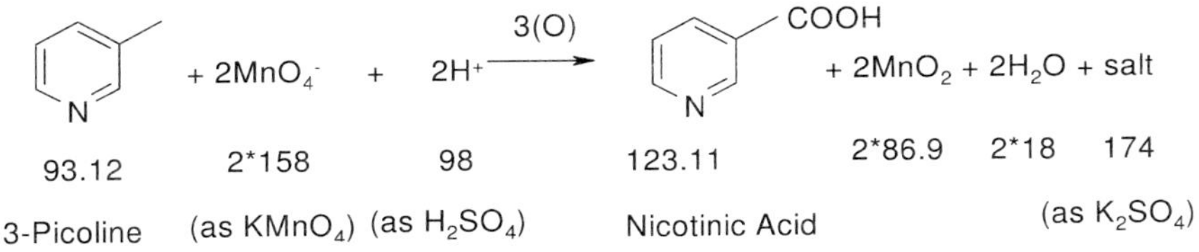
3(O)
COOH
+ $2MnO_4^-$ + $2H^+$
+ $2MnO_2$ + $2H_2O$ + salt
N
N
93.12
2*158
98
123.11
2*86.9
2*18
174
3-Picoline
(as $KMnO_4$)
(as H_2SO_4)
Nicotinic Acid
(as K_2SO_4)

Scheme 15.2 Permanganate oxidation of 3-picoline.

Niacin is used as a feed and food additive. The presence of even small quantities of chromium, however beneficial this may be in practice (chromium is an essential trace metal in the human metabolism), is not likely to be accepted by either today's stringent legislation or by buyers who are geared to high-quality supplies. Removal of last traces of impurity is possible by recrystallization, but this increases the number of unit operations, is therefore expensive and energy consuming, and the problem remains as to what to do with the chromium-containing mother liquors.

15.3.2.3 **Liquid-Phase Oxidation of MEP with Nitric Acid** (Scheme 15.3)

This process has been running for the past 40 years in Lonza's plant in Visp Switzerland and is surprisingly selective. Continuous development and improvement of this process over the years have led to a high-quality product, and to Lonza's ability to maintain their position as the world leader in niacin manufacture.

But, however many improvements and developments have been made to this process, it intrinsically holds some disadvantages, when considered from the "green" stand-point:

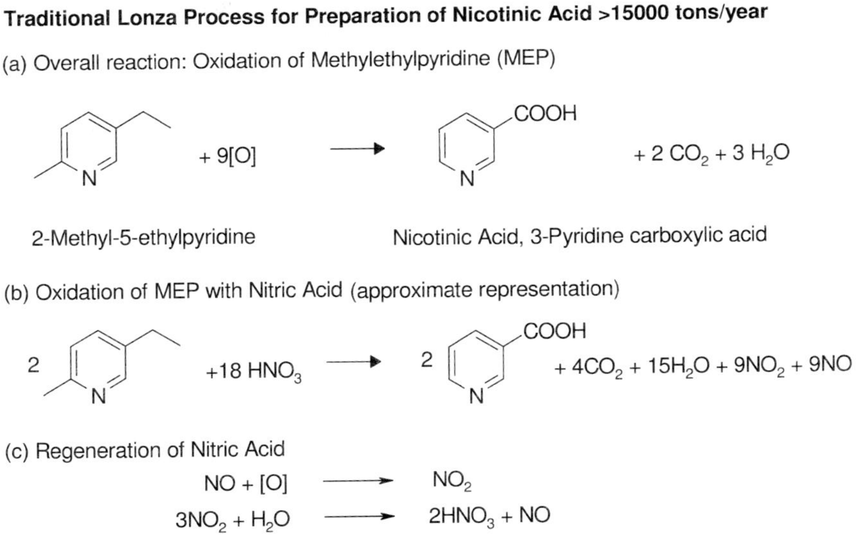
Traditional Lonza Process for Preparation of Nicotinic Acid >15000 tons/year
(a) Overall reaction: Oxidation of Methylethylpyridine (MEP)
COOH
+ 9[O]
+ 2 CO_2 + 3 H_2O
N
N
2-Methyl-5-ethylpyridine
Nicotinic Acid, 3-Pyridine carboxylic acid
(b) Oxidation of MEP with Nitric Acid (approximate representation)
COOH
2
+18 HNO_3
2
+ $4CO_2$ + $15H_2O$ + $9NO_2$ + 9NO
N
N
(c) Regeneration of Nitric Acid
NO + [O] → NO_2
$3NO_2$ + H_2O → $2HNO_3$ + NO

Scheme 15.3 Traditional Lonza process for the preparation of nicotinic acid.

1. *Safety*: using nitric acid at high temperatures and pressures requires a well-conceived and continually executed safety concept, using advanced reaction technology.

2. *Ecology (carbon dioxide and nitric oxides)*: although nitric oxide fumes can be largely regenerated to nitric acid, some nitric oxide (NO) is invariably present in the off-gases, which then have to be catalytically treated to remove the last traces of NOx. Carbon dioxide, however, cannot readily be recycled, and this is vented to the atmosphere. In today's process (including deep oxidation of the starting-material), as can be readily calculated from the reaction equation, over 1 tonne of CO_2 is produced per tonne of niacin.

3. *Downstream processing*: to produce a product quality acceptable to today's standards, extensive processing in the form of recrystallization and decolorizing is necessary. As mentioned above, recrystallization is an energy and labor intensive process.

4. *Starting-material*: although MEP is produced from cheap starting-materials (ammonia and paraldehyde), the process itself produces considerable quantities of by-products and/or waste material, which have to be separated and suitably treated to avoid environmental pollution. Additionally, the carbon efficiency of the MEP conversion into niacin is at best only 0.75 (two carbon atoms are burnt off during oxidation), so that, from the green standpoint, MEP is not the ideal starting material.

15.3.2.4 Direct Oxidation of 3-Picoline to Niacin (Scheme 15.4)

Direct oxidation of 3-picoline with oxygen or air offers the most logical route to nicotinic acid although the reaction technology is not as simple as the ammoxidation. For this reason there is as yet no known industrial plant operating with direct oxidation. However, under controlled conditions and a careful design of catalyst, high selectivity and throughput can be achieved on a semi-commercial scale.

The "green" advantages of the direct oxidation process can be summarized as follows:

1. Use of air as oxidant instead of stoichiometric quantities of chemical oxidizing agents.
2. Use of catalysts to promote reaction.
3. Reaction carried out at atmospheric pressure.
4. Gas-phase reaction means that catalyst does not have to be recovered from solution.
5. Energy from exothermic reaction can be recuperated.
6. Few unit operations necessary to obtain the pure product.
7. The only solvent used is water.
8. Waste is minimized by a highly selective reaction.
9. Conversion is high, leading to an efficient use of equipment, energy and material.
10. Throughput is acceptable for a commercial process.

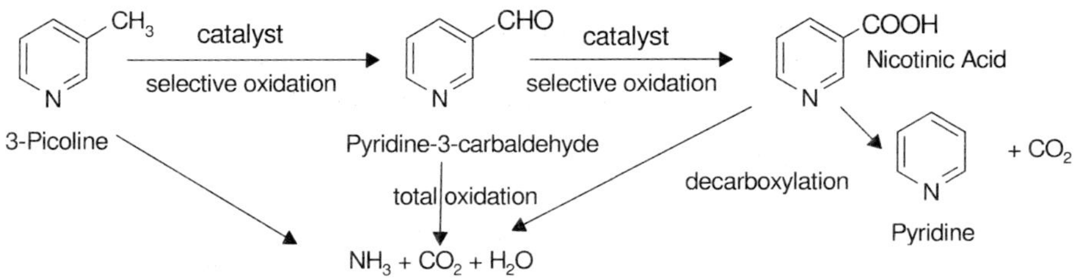

3-Picoline → catalyst selective oxidation → Pyridine-3-carbaldehyde (CHO) → catalyst selective oxidation → COOH Nicotinic Acid

total oxidation → $NH_3 + CO_2 + H_2O$

decarboxylation → Pyridine + CO_2

Scheme 15.4 Reactions in the gas-phase oxidation of picoline to nicotinic acid.

15.3.3
Choice of Catalyst (Efficiency, Separation, Recycling)

The above examples using stoichiometric or excess quantities of chemical oxidants are contrary to the principles of today's green chemistry. The use of carefully designed catalysts enables the reaction to be carried out in the gas phase in a fixed or fluidized bed with air or oxygen as the oxidizing reagent. Gas-phase technology eliminates any catalyst separation or recycling procedures that are necessary in liquid-phase operations. Fixed- and fluidized-bed oxidation catalysts are usually relatively simple in chemical composition, but the development of a robust and selective catalyst is a complex task, often requiring years of patient research and development. Beside the requirement of high selectivity at high throughput, fixed-bed catalysts should possess a long lifetime, since changing the catalyst in a multi-tubular reactor with anything from 5000 to 25 000 tubes is not a trivial undertaking.

15.3.4
Down-Stream Processing/Unit Operations

Goal: to minimize the number of stages necessary to obtain the product in the state desired by the customer.

Separation techniques such as crystallization, distillation and drying generally involve some losses of product and consume considerable amounts of energy, but are usually necessary at some stage, as 100% selectivity is unusual outside biochemical and biological processes. However, the more selective the reaction, the purer the end-product will be and thus require fewer steps or less energy to meet the quality required. The direct oxidation method produces few by-products and the reaction solution already has a high degree of purity.

15.3.5
Minimization of Pollutants and Waste Stream Volume

Since no other materials are involved in the direct oxidation than picoline, water and air or oxygen, any pollutants arising must be by-products of the reaction. It therefore follows that the higher the selectivity the fewer the potential pollutants and the smaller the waste streams. In the direct-oxidation process, the major side-reaction is deep oxidation, so that the major pollutant is carbon dioxide. Green principles dictate here that efforts should continue to be made to further reduce the CO_2 emission by increasing selectivity.

15.3.6
Recycling of Auxiliary, Side and Intermediate Products

In selective gas-phase oxidation processes the conversion often does not reach 100%, in order to reduce total oxidation. This means that some starting material and/or intermediate products are present in the reaction mixture. Following separation, these may be recycled into the reaction to avoid yield losses. Although the above may seem obvious, an economical and efficient recycling procedure is not always simple, depending on the volatility of the components involved and the concentration of any unwanted by-products. In practice, a purge is included in the recycle to maintain an equilibrium concentration of by-products.

15.4
Green Principles in Lonza's Niacinamide Process (5000 mtpa)

Lonza's production of niacinamide in China incorporates aspects of practically all the elements of green chemistry outlined above (Scheme 15.5).

The advantages of 3-picoline as a starting material and the disadvantages of the main manufacturing method have been described above. To avoid the latter, Lonza has developed a viable two-stage alternative starting from 2-methylpentanediamine (MPDA). MPDA is readily obtained by hydrogenating 2-methylglutaronitrile, the major by-product in the adiponitrile process and, as such, a readily available starting-material ($\sim 10^5$ mtpa). In the first stage MPDA is cyclized to methylpiperidine, which is then dehydrogenated to 3-picoline (Scheme 15.6).

Although this route has the same fundamental weakness of being coupled to another product (adiponitrile), the future of nylon 6,6 and the route to its

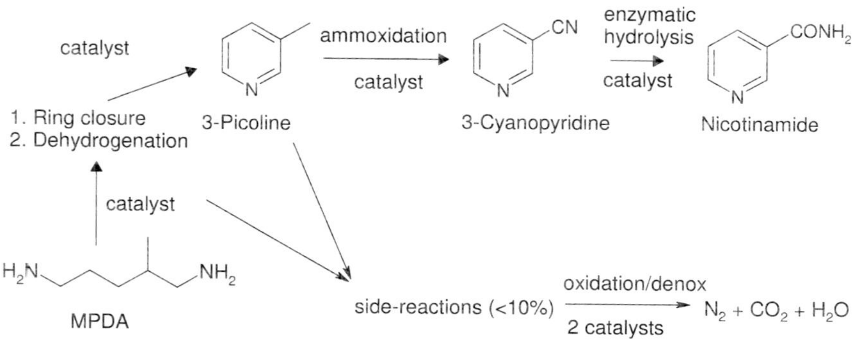

Scheme 15.5 Lonza's nicotinamide process (Guangzhou, China).

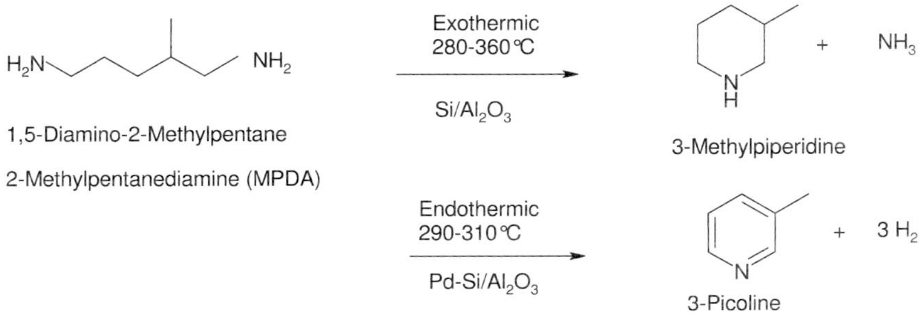

Scheme 15.6 Alternative process for picoline manufacture.

manufacture (hydrogen cyanide addition to butadiene) seem assured for the next 10–20 years or so. Several advantages relevant to green chemistry are apparent in this route:

1. The 3-picoline produced has a very high isomeric purity.
2. Picoline is produced in a two-stage catalytic process that is practically energetically neutral: an endothermic (ring closure) and an exothermic (dehydrogenation) reaction.
3. Ammonia liberated during the ring closure can be utilized in a subsequent process.
4. Utilization of a waste-product (2-methylglutaronitrile can be used as a co-monomer in the production of other polyamides, but the end-product niacin has an intrinsically much higher value).

The advantages of the alternative picoline process are described above. It also illustrates that, in order to reduce the amount of waste produced in reaction, a starting material of the highest quality is desirable, since any by-products in the latter would have to be removed later in the process.

The ammonia liberated in the cyclization step of MPDA is utilized in the ammoxidation step. This is an example of recycling a by-product that would otherwise require disposal.

The ammoxidation step (Scheme 15.7) utilizes a catalyst that selectively converts picoline in the presence of oxygen and ammonia into 3-cyanopyridine. Even at

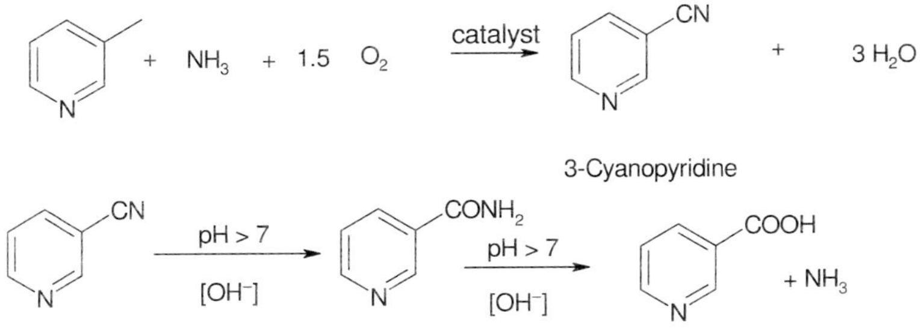

Scheme 15.7 Ammoxidation of 3-picoline and hydrolysis to niacinamide and niacin.

elevated reaction temperatures deep oxidation is slight and the heat generated by the reaction is recuperated and utilized elsewhere in the process.

Cyanopyridine generated by the ammoxidation is hydrolyzed using an enzymatic catalyst with practically quantitative yields. This efficient procedure avoids the consecutive hydrolysis reaction to nicotinic acid (here a by-product!).

Starting from MPDA an overall yield for the process of around 90% is obtained, which means that the carbon efficiency (ratio of carbon in product to reactants) and overall atom efficiency (yield ratio of molecular weight of product to reactant) are 90% and 99%, respectively. The waste and any toxic by-products generated from the process are catalytically treated to give nitrogen, water and low quantities of carbon dioxide (about 200 kg per ton nicotinamide).

The energy-neutral picoline route and the resourceful recuperation and reutilization of energy in the exothermic ammoxidation contribute to low-energy requirements in the process.

Thus the green principles observed in the process can be effectively summarized as follows:

1. Waste is prevented by a highly selective process over four chemical or biochemical steps.
2. Atom and carbon efficiencies are high (90–100%).
3. The reagents used (3-picoline, water, air) are not particularly toxic (picoline $LD_{50} = 420$ mg kg^{-1}). The ammonia necessary for the ammoxidation is predominantly obtained as a by-product in the production of 3-picoline. Any toxic by-products are catalytically converted into benign material.
4. Benign solvents are used (water and toluene for extraction of cyanopyridine). Toluene is practically 100% recycled.
5. Energy requirements are largely covered by the exothermic nature of the main reaction.
6. Catalysis is used in every step to increase efficiency of the reaction and reduce energy requirements. No fewer than six catalysts are employed in the process.
7. Hazardous materials are avoided. Apart from a relatively small quantity of carbon dioxide there is no environmental waste.

16

Introducing Green Metrics Early in Process Development. Comparative Assessment of Alternative Industrial Routes to Elliott's Alcohol, A Key Intermediate in the Production of Resmethrins

Paolo Righi, Goffredo Rosini, and Valerio Borzatta

16.1
Introduction

Early routine quantitative assessment of the "greenness" of processes or even single reactions is very important since it allows the introduction of green metrics at the very beginning of process and reaction design, at the "lab bench" stage, instead of during the late process development. In the end this favors the implementation of processes that are "green" by design, instead of processes turned "green" at a very late stage of process development. Despite the fact that many sophisticated tools are available to assess in great detail the environmental impact of processes, simpler tools to be used by all chemists on a routine basis are far less abundant, and so is the practice of evaluating reactions greenness at the research stage.

Quantitative evaluation of chemical processes in terms of environmental impact and eco-friendliness has gradually become a topic of great interest since the original introduction of the atom economy (AE) by Trost [1], and the E-factor by Sheldon [2]. Since then, other indexes have been proposed for the green metrics of chemical processes, such as effective mass yield (EMY) [3], reaction mass efficiency (RME) [4] and mass intensity (MI) [5], along with unification efforts [6, 7] and comparisons among these indexes [8].

All the above metrics are only based on masses of waste: they do not take into account the nature of the waste. The necessity of metrics that consider not just the amount of the waste but also its environmental impact was first recognized by Sheldon [9]. To put it in his words: "Comparing alternative routes solely on the basis of the amount of waste is to grossly oversimplify." So, he defined the environmental quotient (EQ) as the product of the E-factor (E) and an unfriendliness quotient, Q.

However, despite the simplicity of this definition, the quantitative assessment of Q is not so obvious. Sheldon's original proposal to arbitrarily assign to any substance a Q value relative to that of NaCl, which is set to 1, does not fully solve the problem. This has been followed by a few papers dealing with this problem [10]. However, the methods proposed suffer from a difficult calculation basis and therefore are not

Sustainable Industrial Processes. Edited by F. Cavani, G. Centi, S. Perathoner, and F. Trifiró
Copyright © 2009 WILEY-VCH Verlag GmbH & Co. KGaA, Weinheim
ISBN: 978-3-527-31552-9

amenable for a rapid evaluation and/or screening of the environmental impact of alternative synthetic pathways.

In 2002, Eissen and Metzger proposed EATOS (an environmental assessment tool for organic syntheses), an environmental performance metrics for daily use in synthetic chemistry [11]. This tool allows rapid quantitative assessment of both the E-factor and the potential environmental impact (PEI, Sheldon's Q) of a process. They also provided a PC software application to perform this calculation, which is available from them [12]. With this tool, Sheldon's Q can be quantitatively assessed for both the feedstock and the output (product and wastes) of a multistep synthesis. The assessment is made on the basis of the available substance's ecotoxicological and human toxicological data.

A comparative assessment between two alternative routes to Elliot's alcohol, an industrial intermediate in the preparation of pyrethroids of the resmethrin family, is presented in this chapter. This assessment is made with the aid of the EATOS tool and takes into consideration both the masses (E-factor) as well as environmental impact of the substances employed and released by the processes (Sheldon's Q).

16.2
Elliott's Alcohol

Industrial preparation of resmethrin, one of the first members of the pyrethroid family to be introduced, is achieved via Elliott's alcohol. Available procedures make use of stoichiometric amounts of pyridine, thionyl chloride and fuming nitric acid, and a lot of chlorinated by-products are generated. We present here an alternative route to Elliott's alcohol, based on the Baylis–Hillman reaction in aqueous media. A comparative quantitative assessment of the "greenness" has been performed, using the freeware package EATOS, which takes into account both the mass economy and the environmental impact of the materials involved.

Pyrethroids are synthetic insecticides that possess greater insecticidal potency and enhanced photostability than plant-derived pyrethrins. Thanks to their low mammalian toxicity and biodegradability they have emerged as a replacement for DDT since the late 1960s. The demand for pyrethroids is increasing, also due to the upgraded agriculture technologies of "emerging countries" such as China [13]. Resmethrin (Figure 16.1) was one of the first members of the pyrethroid family to be used. It is commercialized as the mixture of the four stereoisomers, while the most active (1R,3R)-trans stereoisomer is sold under the name bioresmethrin.

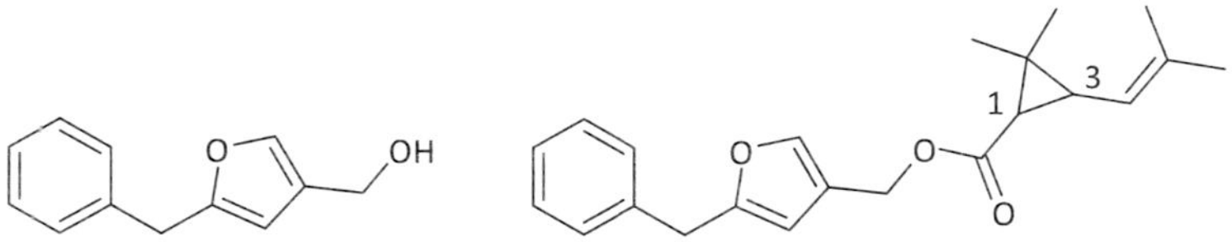

Elliott's alcohol (1)   Resmethrins

Figure 16.1 Elliott's alcohol (1) and resmethrins.

Bioresmethrin is one of the most effective broad spectrum insecticides currently available. It exhibits a high order of insecticidal activity, which is coupled with its excellent toxicological properties [14].

They are both chrysanthemic acid esters of (5-benzylfuran-3-yl)methanol (Elliott alcohol, 1) [15]. Patented methods [16] for the industrial preparation of Elliott's alcohol are demanding or such as to be hardly exploited in industrial-scale plants. For instance, in one of these methods [17] (5-benzyl-3-furyl)methanol is obtained by a sequence of Claisen condensation of benzyl cyanide and a dialkyl succinate, hydrolysis, esterification, protection of the ketone group, formylation, cyclization to 5-benzyl-3-furfuryl ester and reduction to alcohol with lithium aluminium hydride. The process is particularly laborious, it requires anhydrous solvents and uses lithium aluminium hydride, the handling of which requires numerous precautions.

More promising, in terms of "green" efficiency, was the procedure [18] for obtaining 1 depicted in Scheme 16.1. The key step of this procedure is an extremely atom efficient and regiospecific dipolar cycloaddition between an *in situ* generated nitrile oxide and isobutene diacetate (2). The cycloadduct is then converted into 1 by a simple saponification/hydrogenolysis/furan ring formation three-step sequence.

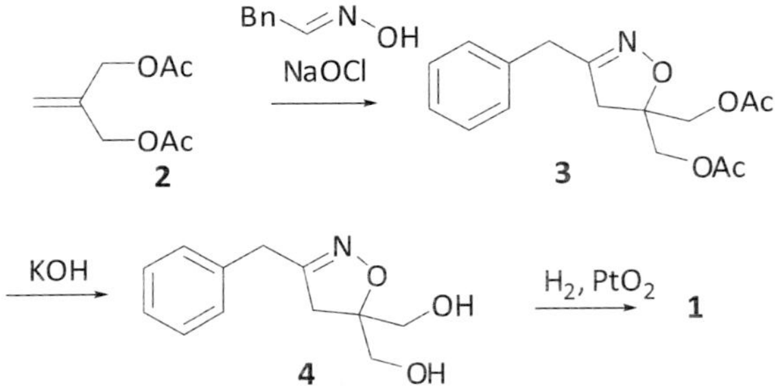

Scheme 16.1 Previous procedure for the preparation of Elliott's alcohol (1).

The weak point of this method is the supply of isobutene diacetate (2), which is expensive and difficult to find in bulk quantities, or can be prepared by a double nucleophilic displacement from the corresponding dichloride (5, Scheme 16.2).

Scheme 16.3 depicts a preparation of dichloride 5 [19] that is claimed to be safer and more convenient.

Of course, this preparation poses a heavy burden on the environment. In the first step, huge amounts of pyridinium salts and SO_2 are generated along with 20% of a tetrachlorinated by-product that has no use but which has to be discarded in the following step. The second step is accompanied by a large NO_x evolution, while the final step releases carbon dioxide and HCl.

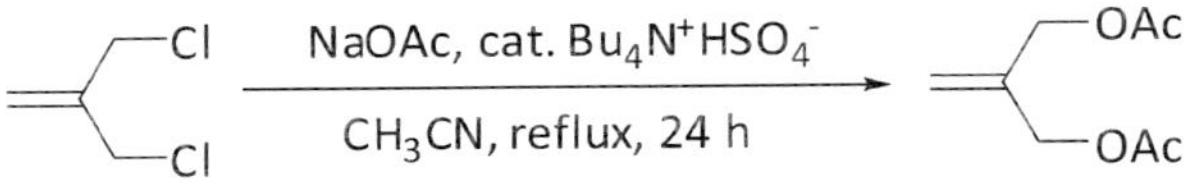

Scheme 16.2 Preparation of diacetate 2.

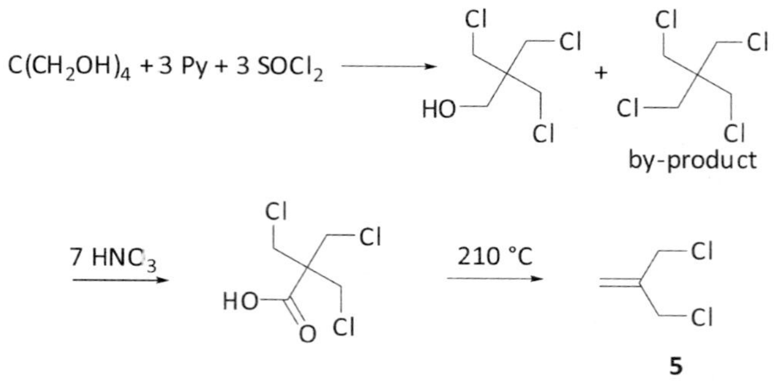

Scheme 16.3 Preparation of isobutene dichloride (5).

16.3
An Alternative Synthesis of Elliott's Alcohol

An alternative procedure [20, 21] for obtaining Elliott's alcohol has been based on the efficient cycloaddition reaction (Scheme 16.1) but avoids the use of isobutene diacetate and all the wastes associated with its preparation (Schemes 16.2 and 16.3).

This new procedure makes use of an alternative building block for the cycloaddition step, namely, ethyl 2-(hydroxymethyl)acrylate (7), which can be easily prepared via a known [22] tandem olefination/Baylis–Hillman sequence (Scheme 16.4) from commercially available and inexpensive triethyl phosphonoacetate (6), paraformaldehyde and potassium carbonate as the base, in water at $40\,^{\circ}C$ for 1 h [23].

Dienophile 7, though structurally similar to diester 2, is electron-poor while 2 is electron-rich. Consequently, the choice of 7 as the new starting material carries with it questions about the viability of the whole process. In particular, (i) Would 7 react at all

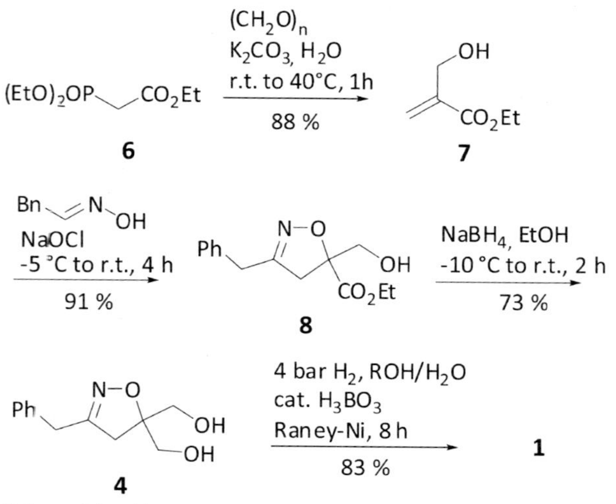

Scheme 16.4 Alternative route to Elliott's alcohol (1).

in the cycloaddition step? (ii) Would this reaction regiospecifically afford the desired isomer? (iii) Can the new cycloadduct be converted into Elliott's alcohol? However, it was found that by treating 7 with phenylacetaldehyde aldoxime and 10% NaOCl, the desired cycloadduct 8 is regiospecifically obtained in high yield (Scheme 16.4). After extractive work-up the product is obtained pure enough to be used in the next step without any further purification.

This cycloaddition reaction was then followed by an effective reduction of the carboxylate moiety to obtain the bishydroxyl derivative 4, the intermediate that is in common with the procedure depicted in Scheme 16.1. This reduction does not need environmentally and safety problematic aluminium hydride reagents, as it is usually necessary for the reduction of an ester moiety. It can be performed, in good yields, with easier-to-handle granular sodium borohydride in lower alcohols. Here we can see another additional "green" bonus of the new starting material chosen: probably the free OH group in compound 8 assists and facilitates the reduction of the ester group, making it possible with the milder sodium borohydride, and avoids the need for aluminium hydride reagents. Finally, 8 is converted into Elliott's alcohol according to the original procedure [18] or by hydrogenation in lower alcohols/water solution in the presence of Raney nickel and orthoboric acid [24].

Summarizing, the alternative route to Elliot's alcohol (1) makes use of 2-(hydroxymethyl)acrylic acid esters, such as 7, as the new starting materials (Scheme 16.4). On the "green" side, the choice of this alternative starting material allows us to keep the atom economic efficient cycloaddition step of the previous procedure (Scheme 16.1), while producing a great reduction of steps, waste amount, and waste toxicity and hazards. In fact, the existing synthetic sequence (Schemes 16.3 and 16.2) is now replaced by one step only: the first step depicted in Scheme 16.4. Thus, the wastes associated with the existing methodology (3 SO_2, 3 pyridinium salts, the tetrachlorinated by-product, 7 NO_x, CO_2, HCl) are now replaced by the wastes associated with the first step of Scheme 16.4 (excess CH_2O and phosphate salts).

However, this kind of "qualitative" greenness assessment, based only on the visual inspection of reaction schemes, may be too limited and may lead to erroneous conclusions.

16.4
Comparative Assessment of the Two Alternative Routes to Elliott's Alcohol

Consequently, a comparative quantitative assessment of the two routes was needed. This "quantitative" greenness assessment compares the two alternative routes: the best existing one (Schemes 16.1 to 16.3) [18] and the new route (Scheme 16.4). Both routes are compared up to the common Elliott's alcohol intermediate 4. Scheme 16.5 depicts the two different routes.

So, which is the greener route? At the early stages of process development or at the discovery stage this evaluation is usually performed qualitatively by visual inspection of the reaction scheme. At first glance, the shorter route B probably has the lower E-factor. But how much lower? And what about the environmental impact? Will the

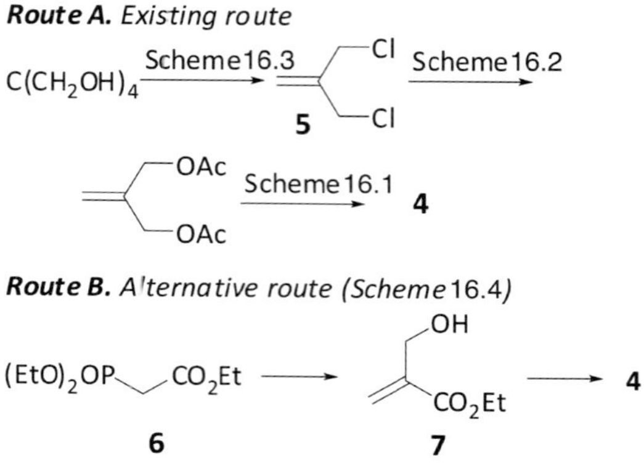

Route A. *Existing route*

C(CH$_2$OH)$_4$ $\xrightarrow{\text{Scheme 16.3}}$ **5** (—Cl, —Cl) $\xrightarrow{\text{Scheme 16.2}}$

(—OAc, —OAc) $\xrightarrow{\text{Scheme 16.1}}$ **4**

Route B. *Alternative route (Scheme 16.4)*

(EtO)$_2$OP $\diagup$ CO$_2$Et $\longrightarrow$ (OH / CO$_2$Et) $\longrightarrow$ **4**

6　　　　　　　**7**

Scheme 16.5 Alternative routes to the common intermediate **4** to Elliott's alcohol.

use of problematic formaldehyde in route B counterbalance the use of problematic pyridine and SOCl$_2$ in route A? Obviously to answer these and other questions a *quantitative* assessment is needed.

16.4.1
Comparison of E-Factors

The E-factor has been calculated for both routes, considering the amounts and yields reported in references [18, 19] for route A and in the Experimental section in reference [21] for route B.

Comparison of the E-factors (Figure 16.2) clearly shows how route B is much less mass intensive than route A. In fact, for each kilogram of compound **4**, route A produces 183 kg of wastes while route B produces only 102 kg, that is, a net 44% reduction.

Inspection of EATOS graphical output (Figure 16.2) also allows us to see where these wastes comes from. Much of the wastes produced in route A (76 out of 183 kg; blue block) are aqueous and come from aqueous reagents and aqueous work-up. Further inspection of this segment with the EATOS tool shows that most of this aqueous waste is produced in the early steps of route A (Scheme 16.3), especially from the use and subsequent removal of mineral acids and their salts. Inspection of the wastes produced in route B shows that most of them come from the auxiliary materials used during isolation steps (41 kg out of 102 kg; brown block) of the process (Scheme 16.4). Overall isolation steps produce most of the wastes in both routes: the figures are 105 kg-waste per kg product (57% of the total waste produced) for route A and 51 kg-waste per kg product (50% of the total waste produced) for route B.

These figures clearly show how a "greenness" evaluation of a process based only on "visual inspection" of synthetic schemes can be misleading: a very large part of the waste comes from isolation steps, which do not appear in synthetic schemes. Moreover, some green metrics, such as atom economy do not consider at all this portion of waste.

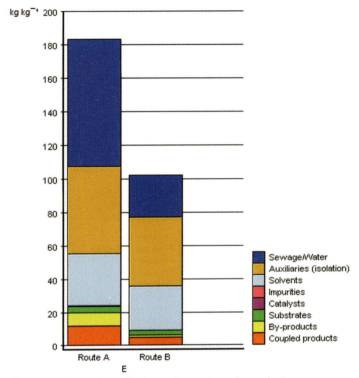

Figure 16.2 Comparison of E-factors (kg waste per kg product).

16.4.2
Comparison of Waste Environmental Impact

To assess the environmental impact of chemical processes, EATOS can take into consideration up to ten different substances' ecotoxicological and human toxicological parameters, and each parameter can be given a different weight. Such substances' parameters are then normalized (each parameter is made to vary from 1 to 10) and then combined to afford an environmental quotient (EI) (much the same as Sheldon's Q). Thus, each different component of the waste can be assigned a quantitative potential environmental impact PEI_{out} (much the same of Sheldon's environmental quotient EQ), defined as the product of its mass (relative to the product unit mass) with its EI.

In this work the parameters taken into account for the assessment of EI_{out} of the waste are: MAK (maximale arbeitsplatz-konzentration) TLV, Hazard Symbol, or LX_{50} for human acute toxicity; any suspected carcinogen, mutagen or teratogen for chronic human toxicity; and WGK wassergefährdungsklassen: "water hazard gas"; http://www.umweltbundesamt.de/wgs-e/index.htm) or LC_{50} to fish for ecotoxicology. These data were obtained from substances' Material Safety Data Sheets, the

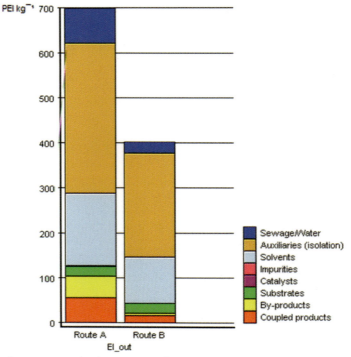

Figure 16.3 Graphical comparison of potential environmental impact of waste.

European Chemicals Bureau (http://ecb.jrc.it/) and the Hazardous Substance DataBank (http://toxnet.nlm.nih.gov/). When insufficient data were available for human chronic toxicity, a prediction of the carcinogenic and mutagenic effects was performed with the OSIRIS tool, available at the Organic Chemistry Portal (http://www.organic-chemistry.org/prog/peo/).

Figure 16.3 gives a quantitative comparison of the potential environmental impact of waste (PEI_{out}).

Figure 16.3 allows us to see immediately that route B also permits a large decrease in waste environmental impact: the figures are 697 and 402 PEI-unit per kg of product for routes A, and B, respectively, that is, route B achieves a net 40% reduction in waste potential environmental impact.

In both routes most of the environmental impact is generated by auxiliary materials (those needed only for product isolation; brown blocks in Figure 16.3). In particular, in route A nearly all (96%) of auxiliary waste is due to the dichloromethane used for product extraction in three different steps of the process. In route B, again most of the auxiliaries waste comes from solvents needed for product isolation.

Further inspection Figure 16.3 and comparison with Figure 16.2 affords some other useful insights. The aqueous waste (blue block) although relevant in mass terms (Figure 16.2) is much less relevant in environmental impact terms: in route A the aqueous waste is more than 40% of the waste total mass (Figure 16.2) while it contributes to only less than 11% of the waste total potential environmental impact (Figure 16.3). Conversely, auxiliary materials of route A account for 25% of the waste mass (Figure 16.2) and for nearly 50% of waste environmental impact. Similar values are found for reaction solvents.

Again, it is important to stress how misleading an environmental assessment could be based only on waste masses: relatively small portions of the waste mass can account for relatively large environmental impacts.

Comparison of the waste produced by the two routes shows that route A suffers from poor reaction conditions more than route B. In fact, in route A the environmental impact of the waste related to reaction – that is, coupled products, by-products and excess substrates – accounts for 124 PEI kg^{-1} (nearly 20% of the total waste). Nearly all of this waste is produced in the early steps of route A (Scheme 16.3). In route B the same figure is 41 PEI kg^{-1} (10% of the total waste of route B). Figure 16.3 shows how route B, compared to route A, produces very little by-products (yellow block) and less coupled products (red block), while it has a similar impact due to excess substrates waste (green block), which for route B is essentially due to excess paraformaldehyde and NaOCl, respectively, used in the first and second steps of Scheme 16.4.

Thus, the environmental analysis performed with the aid of EATOS makes it clear (in quantitative terms) that any effort for the reduction of the environmental impact of both processes should take into consideration, in the first place, an improvement of isolation procedures, that is, using less and/or *less toxic* auxiliaries – a result that would not have been possible at all considering other metrics such as atom economy, or would have been greatly underestimated using metrics based only on masses.

16.4.3
Comparison of Feedstock Environmental Impact

The environmental impact can also be calculated for the feedstock of the process (PEI_{in}), allowing us to assess the hazards and the costs associated with the use of the starting materials of a particular process. This should always be carried out to make sure that a decrease of the environmental impact of waste is not made at the expense of using more hazardous or (environmentally) expensive starting materials.

For the feedstock environmental impact evaluation, EATOS considers risk-phrases and the cost associated with each substance. (For a discussion of the cost as a metrics for starting material environmental impact, see Reference [11].) These data were obtained for all substances, of both routes, from the 2007 Sigma-Aldrich catalogue.

Figure 16.4 reveals that route B also allows a decrease (33%) in the potential environmental impact of materials used for the preparation of Elliott's alcohol intermediate 4. Again, most the hazards and costs come from the use of auxiliaries. For route A a significantly higher proportion of the potential environmental impact of the materials used is due to reaction substrates (green block) and especially to thionyl chloride.

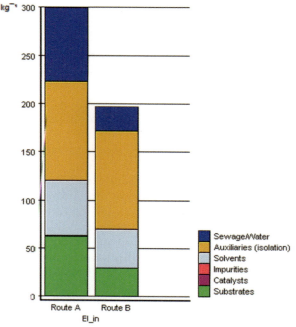

Figure 16.4 Graphical comparison of potential environmental impact of feedstock of synthetic routes depicted in Scheme 16.5.

The results are summarized in Table 16.1. Route B achieves a decrease over route A, in the environmental impact of the waste produced (EI_out: −42%), and this is not done at the expense of more hazardous starting materials. In fact, the environmental impact of starting material is also decreased in route B (EI_in: −34%). From Table 16.1, it is also possible to see the largest relative reduction that route B brought

Table 16.1 Comparison of EATOS environmental indexes of waste (EI_out) and of feedstock (EI_in) for both synthetic routes.

Category	EI_out[a]			EI_in[a]		
	Route A	**Route B**	**Diff[b] (%)**	**Route A**	**Route B**	**Diff[b] (%)**
Aqueous	76.07	25.48	−66	76.07	25.48	−66
Auxiliaries	333.38	230.79	−31	102.31	101.61	−1
Solvents	161.54	103.56	−58	57.39	40.59	−29
Substrates	21.64	21.07	−3	61.83	28.65	−54
By–products	47.90	5.35	−89			
Co–products	54.70	15.23	−72			
Total	696.65	401.63	−42	298.49	196.38	−34

[a] EATOS units.
[b] % Difference between route B and route A.

about was in the environmental impact of the aqueous waste (−66%), in the solvents (−58%), and in the by-products (−89%).

16.5
Driving the "Green" Improvement

EATOS can also be used as a powerful tool to drive the "green" improvement of a chemical synthetic sequence. For example, from Table 16.1 it is clear that, in absolute terms, the greatest contribution to environmental impact of both routes comes from the auxiliaries materials, that is, materials used for isolation and/or purification of the product. Therefore, efforts to further decrease waste environmental impact should be directed, in the first place, towards finding alternative and greener isolation procedures.

In this sense, EATOS can again be very useful since it can be used to rapidly answer a manifold of "what if" questions.

Further insight into the waste environmental impact generated by the auxiliary materials used in route B shows that 225 out of 230 of this impact is due to the solvents (diethyl ether and dichloromethane) used for extractions in the synthesis. For example, what if we could manage to replace those solvents with the *same* amounts of ethyl acetate? Using EATOS, the rapidly obtained answer to this question is that the environmental impact of the waste generated by the auxiliary materials would be reduced dramatically from 230 to only 40 EATOS units. At this point, it is important to stress that the use of any other green metrics based on masses only (atom economy, RME, MI, etc.) leads to the wrong conclusions. In fact, these metrics would record no change in the "greenness" of the process, since toxic solvents are replaced by the *same* amounts of the less toxic ethyl acetate.

16.6
Conclusions

This chapter has presented a quantitative assessment of the environmental impact of two alternative procedures for the preparation of Elliott's alcohol. The assessment was made with the aid of the EATOS tool. This is an easy to use software tool that can be used routinely to assess the "greenness" of reactions or synthetic schemes. Its simplicity allows for a daily routine use, thus favoring the introduction of "green" elements at a very early stage of synthetic process design.

Moreover, the results of this comparison show that the alternative process allows a significant reduction of the mass of waste, the environmental impact of waste and of the hazards and costs associated with the materials used for the process. With the aid of the EATOS tool it was also possible to establish the process zones were most of the environmental impact is produced. From the analysis, it clearly emerged that the use of other green metrics based only on waste mass would have led to different or underestimated results.

References

1 Trost, B.M. (1991) *Science*, **254**, 1471; Trost, B.M. (2002) *Acc. Chem. Res.*, **35**, 695; Trost, B.M. (1995) *Angew. Chem. Int. Ed. Engl.*, **34**, 259.

2 Sheldon, R.A. (1992) *Chem. Ind. (London)*, 903–906; Sheldon, R.A. (2007) *Green Chem.*, **9**, 1273–1283.

3 Hudlicky, T., Frey, D.A., Koroniak, L., Claeboe, C.D. and Brammer, L.E. (1999) *Green Chem.*, **1**, 57–59.

4 Curzons, A.D., Constable, D.J.C., Mortimer, D.N. and Cunningham, V.L. (2001) *Green Chem.*, **3**, 1–6.

5 Constable, D.J.C., Curzons, A.D., Freitas dos Santos, L.M., Green, G.R., Hannah, R.E., Hayler, J.D., Kitteringham, J., McGuire, M.A., Richardson, J.E., Smith, P., Webb, R.L. and Yu, M. (2001) *Green Chem.*, **3**, 7–9.

6 Augé, J. (2008) *Green Chem.*, **10**, 225–231.

7 Andraos, J. (2005) *Org. Process Res. Dev.*, **9**, 149–163; Andraos, J. (2005) *Org. Process Res. Dev.*, **9**, 404–431; Andraos, J. (2006) *Org. Process Res. Dev.*, **10**, 212–240.

8 Constable, D.J.C., Curzons, A.D. and Cunningham, V.L. (2002) *Green Chem.*, **4**, 521–527.

9 Sheldon, R.A. (1994) *Chemtech*, **38**, 38–47.

10 Heinzle, E., Weirich, D., Brogli, F., Hoffmann, V.H., Koller, G., Verduyn, M.A. and Hungerbühler, K. (1998) *Ind. Eng. Chem. Res.*, **37**, 3395–3407; Koller, G., Weirich, D., Brogli, F., Heinzle, E., Hoffmann, V.H., Verduyn, M.A. and Hungerbühler, K. (1998) *Ind. Eng. Chem. Res.*, **37**, 3408–3413; Lepper, P., Keller, D., Herrchen, M., Wahnschaffe, U. and Mangelsdorf, I. (1997) *Chemosphere*, **35**, 2603–2618; Steinbach, A. and Winkenbach, R. (2000) *Chem. Eng.*, **4**, 94–100; Koller, G., Fischer, U. and Hungerbühler, K. (2000) *Ind. Eng. Chem. Res.*, **37**, 960–972.

11 Eissen, M. and Metzger, J.O. (2002) *Chem. Eur. J.*, **8**, 3580–3585.

12 Eissen, M. and Metzger, J.O., EATOS© (Environmental Assessment Tool for Organic Syntheses). http://www.chemie. uni-oldenburg.de/oc/metzger/eatos/ (accessed March 2009).

13 United Nations FADINAP (2002) Agro-chemical Reports 2, 29–35. http://www.fadinap.org/niblist.htm.

14 Ford, M. G. and Pert, D. R. (1974) *Pesticide Science*, **5**, 635–641; Gaines, T.B. and Linder, R.E. (1986) *Fundamental and Applied Toxicology*, **7**, 299–308.

15 (a) Elliott, M., Farnham, A.W., Janes, N.F., Needham, P.H. and Pearson, B.C. (1967) *Nature*, **213**, 493–494; (b) Elliott, M. (1969) *Chem. Ind.*, 776–781; (c) Elliott, M., Janes, N.F. and Pearson, B.C. (1971) *J. Chem. Soc. (C)*, 2551–2554.

16 For alternative procedures for the preparation of 1 see: Naumann, K. (1990) *Synthetic Pyrethroid Insecticides: Chemistry and Patents in Chemistry of Plant Protection*, Vol. 5 (eds W.S. Bowers, W. Ebing, D. Martin and R. Wegler), Springer-Verlag, Berlin, Ch. 2, pp. 112–115.

17 US patent 3,466,304 (1969), assigned to National Research Development Corporation, London, England.

18 EP 0 187 345 (1986), assigned to BASF, Germany; US 4,954,633 (1990), assigned to BASF, Germany.

19 Mondanaro Lynch, K. and Dailey, W.P. (1998) *Org. Synth.*, **75**, 89.

20 ENDURA , PTC WO 02/090341 (2002).

21 Rosini, G., Borzatta, V., Paolucci, C. and Righi, P. (2008) *Green Chem.*, **10**, 1146–1151.

22 Villieras, J. and Rambaud, M. (1988) *Org. Synth.*, **66**, 220.

23 For alternative and efficient preparations of hydroxymethacrylate esters via Baylis–Hillman reactions in aqueous media, see: (a) Dunn, P.J., Hughes, M.L., Searle, P.M. and Wood, A.S. (2003) *Org. Process Res. Dev.*, **7**, 244–253; (b) Mathias, L.J., Kusefoglu, S.H. and Kress, A.O. (1987) *Macromolecules*, **20**, 2326–2328.

24 Curran, D.P. (1982) *J. Am. Chem. Soc.*, **104**, 4024.

17
Basell Spherizone Technology

Maurizio Dorini and Gabriele Mei

17.1
Introduction

The family of polymers, and in particular polyolefins, is important in the modern world because of the very high number of applications in all fields. Among plastic materials, polypropylene is one of the most important, having undergone rapid growth since its discovery in 1954 by the Nobel Prize winner G. Natta of Politecnico di Milano. Today, polypropylene demand is about 40 million ton per year with a market share, among all thermoplastic materials, of about 26%, which is second only to polyethylene (39%).

Competition, in the market of polypropylene licensing technologies, has been a driving force to improve the available processes, with the aim of reducing the investment and variable costs obtained by a simplification of the process, reduction of raw materials and utilities consumptions, improving also the environmental impact with lower gas emission and liquid effluents. The evolution of catalysts and technology has allowed the polypropylene properties to be expanded, to fulfill the market demand and to widen its application.

The predecessor company of BASELL (Montecatini Edison, Himont, Montell, Basell) has always been strongly committed to the research and development of new catalysts for polyolefin, and specifically polypropylene, production and in the continuous improvement of the production processes.

17.2
Technology Evolution

The evolution of the processes for polypropylene production is strictly connected to the improvement of the catalyst system.

By 1950 Ziegler had worked on the growth of alkyl chains by insertion of ethylene into the Al−C bond of trialkyl-aluminium, mainly focused in the field of polyethylene. Natta, from the beginning, attempted propylene polymerization, succeeding in

Sustainable Industrial Processes. Edited by F. Cavani, G. Centi, S. Perathoner, and F. Trifiró
Copyright © 2009 WILEY-VCH Verlag GmbH & Co. KGaA, Weinheim
ISBN: 978-3-527-31552-9

March 1954. Natta's research group fractionated the obtained polymer and found that 40% of it was a hard, high-melting, insoluble fraction.

Natta's discovery soon found industrial application with the first plant, set up by Montecatini in Ferrara, starting production in 1956.

Table 17.1 gives a brief summary of PP catalyst evolution.

In the 1960s, polypropylene processes, operated batchwise, employed first-generation low yield catalysts (<1000 kg-PP per kg catalyst) in mechanically stirred reactors filled with an inert heavy hydrocarbon diluent used to keep the crystalline polymer suspended; propylene was dissolved in the diluent, allowing to operate the reactor at low pressure. Polymer produced with these catalysts had unacceptably high Ti residual metals and contained 10% atactic polypropylene, which required separation. Removal of catalyst residues involved treatment of the polymer with alcohol, multiple organic and/or water washings, polymer/diluent separation via centrifugation or filtration, multistage drying and elaborate diluent/amorphous separation systems. These processes were costly and difficult to operate, and also required extensive water treatment facilities, and catalyst residue disposal systems. The environmental impact was quite high.

As the demand of polypropylene increased, it was necessary to increase the plant capacity and to use continuous processes, this was possible thanks to second-generation catalysts with increased yield (6000–15 000 kg-PP per kg catalyst) and isotacticity, but not yet to an extent that allowed simplification of the production process. Several different processes were developed: slurry, solution, bulk, gas phase.

In the slurry processes the batch reactors were replaced with continuous stirred vessels operated in series, which ran full or under level control. The operating pressure depended on the selected solvent, the most common being hexane, but also heptane, kerosene and butane were used.

Table 17.1 Performance of different catalyst generations.

Generation	Catalyst composition	Yield (kgPP per g-Cat)	I.I. (wt%)	Morphology control	Process requirements
1st	$TiCl_3$/$AlCl_3$ + DEAC	1	90–94	Not possible	Deashing + Atactic removal
2nd	$TiCl_3$ + DEAC	10–15	94–97	Possible	Deashing
3rd	$TiCl_4$/ester/ $MgCl_2$ + AlR_3/ ester	15–30	90–95	Possible	Atactic removal
4th	$TiCl_4$/diester/ $MgCl_2$ + TEA/ silane (HY/HS)	30–60	95–99	Possible	—
5th	$TiCl_4$/diether/ $MgCl_2$ + TEA	70–120	95–99	Possible	—
	$TiCl_4$/ succinate/ $MgCl_2$ + TEA	40–70	90–99	Possible	
6th	Zirconocene + MAO		90–99	Possible	—

Bulk processes (liquid monomer) were operated at higher pressure and have the advantages of a higher reaction rate because of the high monomer concentration, and the absence of diluent.

The solution process was complex and expensive. The product range was somewhat restricted because a special, high-temperature catalyst was required.

The gas-phase process was a simple one but, even with the second-generation catalyst, atactic PP and catalyst residuals were left in the final polymer; consequently, the product quality suffered the presence of atactic polymer and catalyst residuals (stiffness, color, resistance to oxidation).

The first major improvement in the manufacturing process came in the 1970s with the discovery of the milled, active $MgCl_2$ support for PE, the extension to PP with the use of electron donors, and of the combination of internal and external electron donors to promote the iso-index without relinquishing catalyst yield; this brought about third-generation, high yield catalysts (15 000–30 000 kg-PP per kg catalyst), eliminating the need for catalyst residue removal (Ti level below 5 ppm), but the atactic content was still unacceptably high.

The consequences on the manufacturing process were the elimination of catalyst deactivation and removal of a section, with a positive impact on installation and variable costs and also on the environment by removing all the alcohol and water treatment that generated effluents from the plant.

The process still used solvent and the solvent recovery system from the atactic polymer was still in place.

Figure 17.1 shows a block diagram of the notable process simplification on moving from second- (low yield slurry process) to third-generation catalysts (high yield slurry process).

In the 1980s, fourth-generation high yield, high selectivity (HY/HS) catalysts (30 000 kg-PP per kg-catalyst, isotactic index 95–99%) provided a real breakthrough in process simplification, eliminating the need for catalyst and atactic removal.

The polymer flake size and shape is an enlarged copy in size and shape of the catalyst particle. The average diameter of the polymer particles depends on the average diameter of the catalyst and on the extent of polymerization.

The catalyst can have a granular or spherical form, it can be tailored to a very high isotactic index, very high extent, quite broad or narrow molecular weight distribution, and so on. The ratio between the average diameter of the flake and the diameter of the catalyst is called the replication factor.

An additional characteristic of the catalyst was a longer activity that, together with its very high porosity (allowing encapsulation of a large amount of ethylene propylene rubbers inside the particle), opened the way to a large expansion of the PP Impact Copolymer range. Development of the spherical support with controlled morphology was another factor that affected technology simplification, allowing the polymer to be handled through the process using standard control valves, avoiding the handling of a large amount of fines, without fouling or clogging.

The high bulk density and the large average diameter allowed high gas velocities in a fluidized gas-phase reactor with limited entrainment, boosting the possibility of

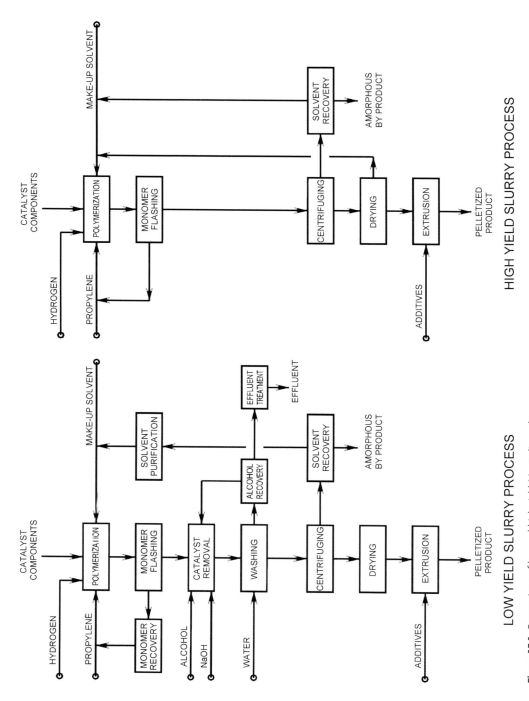

Figure 17.1 Comparison of low and high yield Montedison slurry processes.

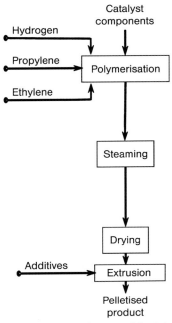

Figure 17.2 Block diagram of the Spheripol process.

heat removal by external gas cooling, and so keeping the fluidized gas reactor small. This further simplified the process and improved product quality.

Refinement of the bulk polymerization reactor and of the gas-phase reactor led to the development of the Spheripol process in 1982 (Figure 17.2).

The Spheripol process was a significant step change in the polypropylene process, allowing a broadening of product capabilities and a considerable saving in terms of investment costs as well as operating costs.

The Spheripol process has been very successful in the PP technology market – since 1982, when the first plant was started up in Brindisi, Italy, over 100 plants have come into operation or are under construction worldwide.

The Spheripol process has been recognized as an environmentally friendly process, being listed in the Polymers BREF, European Commission – IPPC: Integrated Pollution Prevention and Control Best Available Techniques in the production of Polymers dated October 2006.

17.3
Spherizone Technology

Despite the success of the Spheripol process, Basell continue research aimed at improving product characteristics, widening the range of grades and developing the next generation of PP technology, the Spherizone process.

17.3.1
Process Description

The development of the Spherizone technology started in 1995 and was subsequently scaled up from pilot plant to commercial size in 2002, when the new multi-zone circulating reactor (MZCR) was installed at Basell's plant in Brindisi, Italy.

The Spherizone process, using Basell's high yield/high selectivity catalysts, produces spherical polymer particles with an outstanding morphology control directly in the reactor.

It is a modular technology and is typically composed of the following sections:

- Catalyst feeding.

- Polymerization: prepolymerization; polymerization in the MZCR (homopolymer grades, medium–wide–very wide MWD, random copolymers and terpolymers, two composition polymers, homopolymer/random copolymer – twin random copolymers – random/heterophasic copolymers) (Figure 17.3).

- Gas-phase polymerization in a fluidized bed reactor (heterophasic impact and specialty copolymer) as a further option. The gas phase copolymer unit can be added at a later stage without affecting the initial plant configuration or involving significant implementation costs (Figure 17.4).

- Finishing section.

The catalyst is fed continuously to the multi-zone circulating reactor, which is the core of this new technology. This loop reactor consists of two distinct reaction zones, each operating under its own peculiar fluid-dynamic regime.

In the so-called "riser" (Figures 17.3 and 17.4) the polymer particles are entrained upwards by the monomer flow in a fast fluidization regime. Gas superficial velocities are maintained at much higher values than the average particles terminal velocities, so that a highly turbulent flow regime ensues. This generates an optimal heat exchange coefficient between the single particles and the surrounding gas, and ensures that the reaction temperature is kept constant along the reaction bed. Head losses in this area are comparable to those along a fluidized bed of the same solids hold-up, while maintaining a high bed voidage, as typical of fast-fluidized bed.

In the top of the reactor the riser gas is then separated from the solids, which enter the so-called "downcomer." This section operates as a moving packed bed, with the polymer flowing downwards. As it operates close to an adiabatic regime, the reaction heat will increase the temperature of the solid bed as the polymer descends. Therefore, care is taken to recirculate enough polymer to prevent the formation of hot spots and generally excessively high temperatures along the bed, which may jeopardize the flowability of the polymer and the recirculation itself.

The polymer's loop circulation is set up and defined by the pressure balance between the two polymerization zones. As it flows down under gravity, the down-comer polymer bed "pumps" the gas downwards and recovers the head losses developed in the riser, the gas/polymer separator and all other sections of the

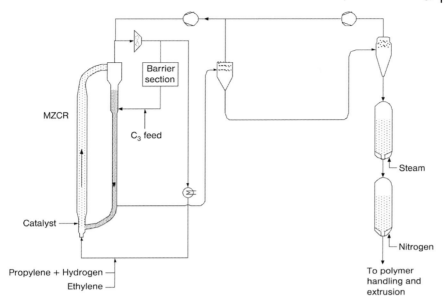

Figure 17.3 Simplified process flow diagram (MZCR only).

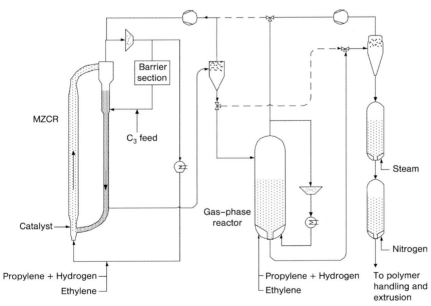

Figure 17.4 Simplified process flow diagram (MZCR plus gas-phase reactor).

reactor. Actually, the moving packed bed develops a pressure profile that increases from top to bottom of the reactor as per Ergun's law, according to the pressure head necessary for recirculation. The overall pressure balance determines the required pressure head, which is at all times is maintained lower than the maximum attainable under conditions of incipient fluidization.

The pressure head, in turn, will determine the differential (slip) velocity between the gas that flows down with the solids and the polymer itself. This is only a function of the reactor's fluodynamics, that is, of the solids flow and the pressure balance to be maintained in the reactor, and is in all cases independent of reaction conditions. The actual fluodynamic conditions of the reactor are such that the gas is in all cases flowing down with the polymer. As the pressure head is determined by the differential in velocities between solids and gas, it is apparent that for the same overall head losses, and therefore the same slip velocity, higher polymer flows in the downcomer will in turn significantly increase the flow of gas entrained with the polymer. This plays an important part in the bimodal operation of the reactor.

At the bottom of the downcomer the polymer particles are recirculated into the riser via a J-valve-like piece of equipment. Furthermore, any suitable valve may also be installed at the bottom of the downcomer as an additional device to further control the polymer flow.

The packed bed fluid-dynamics of the downcomer are essential to bimodal operation: the intergranular gas would normally have the same composition as the riser gas, resulting in a monomodal polymer. However, the two sections of the reactor can be operated at different compositions in hydrogen (used as chain transfer agent) and co-monomer, allowing for the development of a bimodal polymer structure (in terms of MFR and/or comonomer concentration/type) at a macro-molecular level. This can be accomplished by the introduction of a liquid or gas propylene stream on the top of the downer just below the polymer level, so that the riser gas is replaced by one with a different composition (Figure 17.5). Typically, low molecular weights are produced in the riser, while the injection of a hydrogen-poor monomer stream in the liquid or vapor phase allows the production of higher molecular weights polymers in the downcomer.

The polymer recirculation rate has a direct influence on the amount of gas flowing down the downcomer, which results in the need to limit such polymer flow to the minimum allowable by the thermal balance of the downcomer.

For this, the use of a liquid barrier stream is particularly useful, as its latent heat helps remove the heat of reaction developed in the downcomer. The barrier stream is fed and easily distributed in the moving packed bed. The reaction heat and the solid sensible heat vaporize the barrier liquid, which is fed slightly in excess to the theoretical intergranular amount of gases moving downwards, so that a net gas flow upwards is created in the upper point of the downcomer. Thus, the polymer level above the injection point forms a "seal" that prevents contamination between the two reaction zones as long as the packed bed flow regime is maintained.

The reactor can also yield monomodal homo- and random copolymer products by operating the sections under the very same conditions. In this case, the lack of the

High H$_2$ concentration
and/or
High C$_2$ concentration

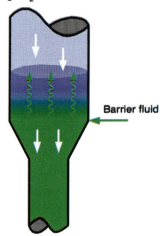

Barrier fluid

Low H$_2$ concentration
and/or
Low C$_2$ concentration

Figure 17.5 Principle of the stripping zone.

cooling effect of the evaporating barrier stream forces the use of higher solids recirculation rates to keep the polymer temperatures all along the downcomer well below the softening point.

The continuous, massive recirculation of the polymer particles between the two zones makes the residence time per pass in each zone one order of magnitude smaller than the overall residence time. In bimodal operation, this allows an intimate mixing of the different polymers being produced, giving a very good homogeneity of the final product. Moreover, the hold-ups and residence times in each leg can be adjusted to vary the split of production between the two areas to suit different requirements.

From the top of the reactor the unreacted monomer is recirculated to the bottom of the riser by a centrifugal compressor. A reactor cooler is positioned in that circuit to remove the heat of reaction and ensures that the desired operating temperature is maintained. The gas flow through the circuit can be varied according to the actual solids recirculation in order to maintain the desired conditions of fast fluidization and solids concentration in the riser.

The pressure in the system is maintained by controlling the fresh monomer flow rates equal to those reacting inside the reactor.

From the delivery of the circulation compressor, a side stream is withdrawn and sent to a monomer distillation section where the desired barrier flow rate is recovered. The top, hydrogen-rich gas is recycled to the riser of the reactor.

The spherical product is continuously withdrawn from the reactor and separated from the unreacted monomer gas in a bag filter separator operating at intermediate pressure. The gas is then compressed and recycled back to the MZCR via a small, single-stage reciprocating compressor.

As an option, the polymer can then be fed to a fluidized bed gas-phase reactor, operated in series with the MZCR, where additional copolymerization can take place to yield high-impact copolymer PP. This gas-phase reactor may be bypassed when homopolymer or random copolymers are produced. In this reactor, the elastomeric phase (ethylene/propylene rubber) is generated within the porous homopolymer matrix that resulted from the first reaction stage. The pores, developed inside the polymer particle in the MZCR upstream, allow the rubber phase to develop without the formation of agglomerates resulting from the sticky nature of the rubber.

Fluidization in the reactor's polymer bed is maintained by adequate recirculation of reacting gas. The reaction heat is removed from the recycle gas by a cooler, while the cooled gas is recycled back to the bottom of the gas-phase reactor for fluidization. This gas-phase reactor maintains a high degree of turbulence and enhances monomer diffusion and reaction rates, and ensures an efficient particle heat removal.

Depending on the adopted configuration, from the intermediate pressure separator or the fluidized bed reactor the product is discharged to a low pressure filter, where the unreacted monomer gas is recovered.

The polymer is then steam-stripped of any residual dissolved monomers in an additional vessel. The same unit also neutralizes the residual active catalyst. The removed residual hydrocarbons are recovered and can be sent back to the reactor system, while the polymer is dried by a closed-loop nitrogen system in a small dryer.

The resulting polymer, free from volatiles, is pneumatically transported by nitrogen to the extrusion unit, where it is mixed with additives and extruded to pellets, typically by a single extrusion line. The product then enters a homogenization silo, before it is stored, bagged or shipped.

The Spherizone technology can be operated using either polymer- or chemical-grade propylene.

Product-wise, the versatility of the Spherizone process is demonstrated by the high-quality product range that includes all standard polypropylene grades, as well as many unique, special products. One key to this versatility is, as mentioned above, the unique design and operation of the MZCR, which, with a very broad range of feasible process conditions, allows for many kinds of polymer structures as well as intimately mixed polymer compositions to be produced.

17.3.2
Process Development and Scale Up

Development of the technology kicked off in late 1995, when a series of basic experiments in a transparent cold model and pioneering work in a pressurized reactor were made at Basell's R&D center in Ferrara, Italy. The resulting prototype reactor, which assumed the final process configuration in 1997, had a relatively small capacity (about 100 kg h^{-1}) and its activity was focused both on process operation and product feasibility. Also, the capability of replicating Basell's existing portfolio of grades with the new technology had to be confirmed.

Less than two years later, a second pilot plant with a capacity four-times as large was built to continue the development of new grades and to address the first scale-up issues. Reactor fluid dynamics such as the operating range of the riser superficial gas velocity, barrier flow rate, solids recirculation between the two legs were demonstrated, verified and found in general agreement with observations gathered from the smaller plant. In addition, several reactor details were examined, such as the riser top exit, the configuration of the reactor bottom and the polymer recirculation valve at the bottom of the downer. Data for the scale-up to industrial size were thus gathered, while product development continued.

While many studies had already been performed or were underway at the time of the scaling of circulating fluidized beds (CFBs) operating at atmospheric pressure and high temperature, no theoretical models had been developed for the MZCR or other reactors operating within similar ranges of particle sizes and gas densities. The experimental observations in the cold model and the pilot plants were also used to understand the basic phenomena and design some reactor details; for instance, they confirmed the existence of a flow pattern resembling that of the core-annulus model developed by Berruti–Kalogerakis [7]. Furthermore, studies on the cold model and the pilot plants allowed exploitation of the solids reflection effect at the top of the riser described by Senior and Brereton [8].

The scale-up to industrial size was based on data gathered from the two pilot plants, which were then elaborated for the design of the commercial unit. The above data were also used for the contemporaneous development of an in-house proprietary reactor simulator for training purposes.

The next scale-up phase was the construction of a 160 ktpa reactor that retrofitted the existing, two-loop reactor Spheripol plant with one MZCR (Spherizone process) in Brindisi, Italy (Figure 17.6). The remainder of the plant remained unchanged. Of course, even at that time, there were no counter indications to building an entirely new plant. However, this choice allowed the industrial demonstration of the new technology with the minimum investment cost.

Conversely, a wider range of superficial gas velocities than that of the pilot plant was chosen to increase the flexibility of the riser–downcomer hold-up ratio. Notably, the industrial-scale plant operates with superficial velocities that are generally higher than those used in the pilot plants, but within the original design range. Further room for maneuver was added by increasing the head of the recirculation compressor relative to that in use at the pilot plants.

Lastly, the riser–downcomer connection was also studied during the scale-up. The thermal stresses due to the temperature increase along the downcomer, and the consequent different elongations of riser and downcomer, were not negligible. To address that issue, the polymer transfer line from the riser into the downcomer was shaped as a 270° bend, to act as an elastic (compensation) element.

The Brindisi Spherizone unit was started up in August 2002 – seven years after research began. The plant has been producing commercially since early 2003 with a high operability. It has proven its competitive operating costs, ease of start-up, shutdown and recovery from operating setbacks as well as short transition times. It

Figure 17.6 Polymerization unit (MZCR), Brindisi.

leverages on proprietary *Avant* Ziegler–Natta catalyst, allowing an outstanding morphology control of the growing polymer particles.

This industrial demonstration of the Spherizone process prompted commercialization of the technology, which was then made available for licensing to third parties. The development of units with capacities of up to $450\,kt\,a^{-1}$, presently under construction, represents a world-scale size for PP plants: Spherizone maintains industry cost leadership with the capacity of the Spheripol technology. Up to the end of 2007, nine Spherizone licenses, for a total capacity near 3 million ton per year, have been granted.

17.3.3
Modular Approach

The design can be made for homopolymer, random copolymer or heterophasic impact copolymer. The modular structure of the process is such that a Spherizone process plant can be upscaled in successive stages, starting with a basic setup for monomodal operations and later adding the barrier recovery section and a fluid bed gas-phase reactor for the production of bimodal and impact copolymer, respectively.

Therefore, the Spherizone process can be designed to meet the particular requirements of individual licensees, yet it is flexible enough to be easily expanded to meet future needs as business develops. New entrants to the polypropylene business might want to build a plant producing only homopolymers and random copolymers as these are the least expensive, are easy to operate and their products account for 75% of all polypropylene sold in the world. Such a basic homopolymer plant can be expanded easily at a later date to produce heterophasic impact and speciality impact copolymers. Impact copolymers are more specialized products, which require additional capital investment and technical support.

17.4
Technology Comparison

As mentioned above, evolution of the production process with continuous simplification has reduced the investment cost of the plant (Figure 17.7) and allowed plants with higher unit capacity.

The first batch process had a capacity of $3 \, kt \, yr^{-1}$, modern processes have a plant capacity of $500 \, kt \, yr^{-1}$ and more.

Simplification of the processes has also improved the specific consumption of raw materials and utilities (Figure 17.8).

Improvement of the processes has been possible only thanks to the development of new catalysts; Figure 17.9 illustrates their performances and level of residual Cl. Thanks to the new catalysts, the properties of the polymer have also improved (Figure 17.9).

Figure 17.10 shows the capability, in terms of product characteristics, of the Spherizone process.

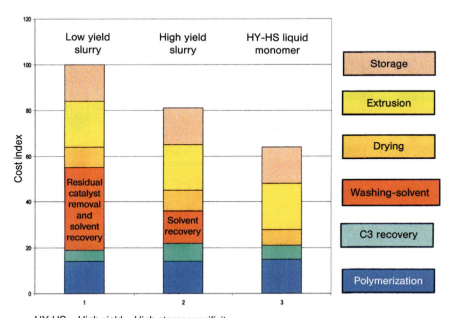

HY-HS = High yield − High stereospecificity

Figure 17.7 Investment cost savings through process simplification.

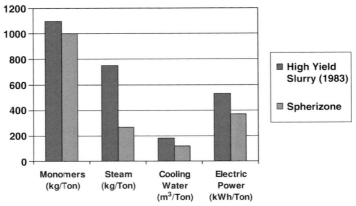

Figure 17.8 Process/catalyst improvement and utilities savings by new technologies.

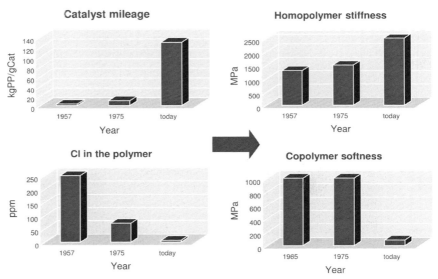

Figure 17.9 Catalyst improvements and product property enhancement.

17.5
Environmental Considerations

Features of the Spherizone process help to reduce both resource consumption and emissions (Figure 17.11). These include use of high yield, highly stereospecific catalysts. recovery and recycle of unreacted monomers, the absence of undesired by-products from the reaction and the low energy consumption. Owing to the MZCR concept, in the polymerization section of the process the overall energy consumption may be reduced by 0–30%, depending on the type of polymer produced.

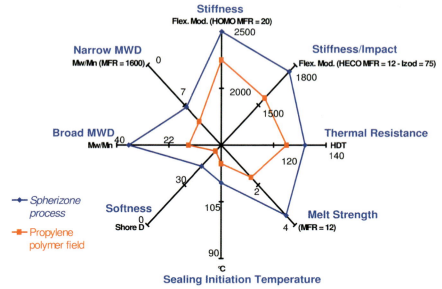

Figure 17.10 Spider diagram of quality parameters.

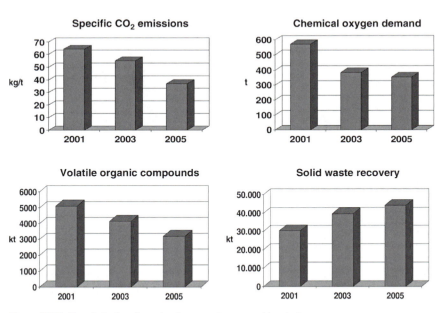

Figure 17.11 Trends in Basell production – environmental key indicators.

References

1 (Oct. 1998) PP- past, present and future: the challenge continues. PP symposium, Ferrara, Italy.

2 Lieberman, R.B., Dorini, M., Mei, G., Rinaldi, R., Penzo, G. and Ten Berge, G. (2006) *Kirk-Othmer Encyclopedia of Chemical Technology*, Vol. 20, John Wiley & Sons, Inc., Hoboken.

3 Pasquini, N. (2005) *Polypropylene Handbook*. Hanser Publisher, New York.

4 Govoni, G., Rinaldi, R., Covezzi, M. and Galli, P., EP-B-0, 782,587 (USP 5,698,642).

5 Govoni, G. and Covezzi, M., EP 1012195.

6 Covezzi, M. and Mei, G. (2001) The multizone circulating reactor technology. *Chem. Eng. Sci.*, **56**, 4059–4067.

7 Berruti, F. and Kalogerakis, N. (1989) Modelling the internal flow structure of circulating fluidized beds. *Can. J. Chem. Eng.*, **67**.

8 Senior, R. and Brereton, C. (1992) Modelling of circulating fluidized bed solids flow and distribution. *Chem. Eng. Sci.*, **47** (2), 281–296.

9 Mei, G., Rinaldi, R., Penzo, G. and Dorini, M. (May 2005) The *Spherizone* process: a new CFB reactor. 8th International Conference on Circulating Fluidized Beds, Hangzhou, China.

10 Schmidt, C.U., Mei, G., Meier, G., Bertolini, S., Busch, M., Wulkow, M., Schwibach, M., Prem, A., Batschkin, T. and Weickert, G. (2003) Training simulator for PP multizone circulating reactor process. Polymer Reaction Engineering V, Quebec, Canada.

11 Mei, G. (September 1996) Modello polimerico multigrain e double grain. Simposio Montell '96, Ferrara, Italy.

Index

Sustainable Industrial Processes. Edited by F. Cavani, G. Centi, S. Perathoner, and F. Trifiró
Copyright © 2009 WILEY-VCH Verlag GmbH & Co. KGaA, Weinheim
ISBN: 978-3-527-31552-9